de Gruyter Textbook

Karl Kraus
Photogrammetry

Karl Kraus

Photogrammetry

Geometry from Images and Laser Scans

Second Edition

Translated by
Ian Harley
Stephen Kyle

Walter de Gruyter
Berlin · New York

Author
o. Univ.-Prof. Dipl.-Ing. Dr. techn. Karl Kraus
formerly
Institute of Photogrammetry and Remote Sensing
Vienna University of Technology
Vienna, Austria

Translators
Prof. Ian Harley
Dr. Stephen Kyle
University College London
London, Great Britain

This second English edition is a translation and revision of the seventh German edition:
Kraus, Karl: Photogrammetrie, Band 1, Geometrische Informationen aus Photographien und Laserscanneraufnahmen. Walter de Gruyter, Berlin · New York, 2004

First English edition:
Kraus, Karl: Photogrammetry, Volume 1, Fundamentals and Standard Processes. Dümmler, Köln, 2000.

∞ Printed on acid-free paper which falls within the guidelines
of the ANSI to ensure permanence and durability.

Bibliographic information published by the Deutsche Nationalbibliothek

The Deutsche Nationalbibliothek lists this publication in the Deutsche Nationalbibliografie; detailed bibliographic data are available in the Internet at http://dnb.d-nb.de.

ISBN 978-3-11-019007-6

Printed in Germany.
Coverdesign: +malsy, kommunikation und gestaltung, Willich.
Printing and binding: Hubert & Co. GmbH & Co. KG, Göttingen.

Foreword to the second English edition

The first edition of Volume 1 of the series of textbooks "Photogrammetry" was published in German in 1982. It filled a large void and the second and third editions were printed soon afterwards, in 1985 and 1990. The fourth edition was published in English in 1992, translated by Peter Stewardson. The following three editions were published in German in the years 1995, 1997, and 2003, making seven editions in all. The English edition was re-printed in 2000.

Volume 1 was additionally translated into several languages, including Serbocroatian, by Prof. Joksics, Technical University Belgrade; Norwegian, by Prof. Oefsti, University of Trondheim; Greek, by Dr. Vozikis and Prof. Georgopoulos, National Technical University of Athens; Japanese, by Prof. Oshima and Mr. Horie, Hosei University; Italian, by Prof. Dequal, Politecnico Torino; French, by Prof. Grussenmeyer and O. Reis, Ecole Nationale Supérieure des Arts et Industries de Strasbourg; Hungarian by Prof. Detreköi, Dr. Mèlykúti, S. Miháli, and P. Winkler, TU Budapest; Ukrainian by S. Kusyk, Lvivska Politechnika; and Turkish, by Prof. Altan, Technical University of Istanbul.

This second English edition is a translation of the seventh, German, edition by Dr. Ian Harley, Professor Emeritus, and Dr. Stephen Kyle, both of University College London. They not only translated the text, but they also made valuable contributions to it; their comments and suggestions led to a clearly improved edition. Compared to the first English edition there are major changes. Analogue and analytical photogrammetry are reduced significantly, most importance is given to digital photogrammetry, and, finally, laser scanning is included. Terrestrial as well as airborne laser scanning have gained great importance in photogrammetry. Photogrammetric methods are, with small adaptations, applicable to data acquired by laser scanning. Therefore, only minor additions to photogrammetry were necessary to cover the chapter on laser scanning. Compared to the previous German edition there are, especially, updates on digital cameras and laser scanners.

The original German version arose out of practical research and teaching at the Vienna University of Technology. Volume 1 first introduces the necessary basics from mathematics and digital image processing. It continues with photogrammetric acquisition technology with special consideration of photo-electrical imaging (CCD cameras). Particular attention is paid to the use of the Global Positioning System (GPS) and Inertial Measurement Units (IMU) for flight missions. The discussion on photogrammetric processing begins with orientation methods including those based on projective geometry. The orientation methods which are discussed for two images are extended to image blocks in the form of photogrammetric triangulation.

In the discussion of stereo-plotting instruments most attention is given to digital softcopy stations. In addition to automatic processing methods, semiautomatic methods, which are widely used in practice, are also explained. This textbook first treats digital orthophoto production, and then includes three-dimensional virtual worlds with photographic texture.

This selection and arrangement of material offers students a straightforward introduction to complex photogrammetry as practised today and as it will be practised in the near future. It also offers practising photogrammetrists the possibility of bringing themselves up-to-date with the modern approach to photogrammetry and saves them at least a part of the tedious study of technical journals which are often difficult to understand. For technically oriented neighbouring disciplines it provides a condensed description of the fundamentals and standard processes of photogrammetry. It lays the basis for that interdisciplinary collaboration which gains ever greater importance in photogrammetry. Related, non-technical disciplines will also find valuable information on a wide range of topics.

For the benefit of its readers, the textbook follows certain principles: didactics are put before scientific detail; lengthy derivations of formulae are put aside; theory is split into small sections alternating with practically-oriented passages; the theoretical basics are made clear by means of examples; and exercises are provided with solutions in order to allow self-checking.

This series of textbooks is a major contribution to photogrammetry. It is very sad that Prof. Kraus, who died unexpectedly in April 2006, cannot see it published. At that time the translation was already in progress. Final editing was performed by Dr. Josef Jansa and Mr. Andreas Roncat from the Vienna Institute of Photogrammetry and Remote Sensing. Thanks are also due to the many people at the Institute of Photogrammetry and Remote Sensing who did major and minor work behind the scenes, such as drawing and editing figures, calculating examples and exercises, making smaller contributions, proofreading, composing the LaTeX text, etc. This book, however, is truly a book by Prof. Kraus.

Karl Kraus was born in 1939 in Germany and became Professor of Photogrammetry in Vienna in 1974. Within these 32 years of teaching, counting all translations and editions, more than twenty textbooks on photogrammetry and remote sensing bearing the name Karl Kraus were published. Many examples and drawings in this textbook were supplied by the students and collaborators of Prof. Kraus in Vienna. With deep gratitude the entire Institute of Photogrammetry and Remote Sensing looks back at the time spent with Karl Kraus and forward to continuing the success story of this textbook.

Norbert Pfeifer
Professor in Photogrammetry
Institute of Photogrammetry and Remote Sensing
Vienna University of Technology

Vienna, Summer 2007

Notes for readers

This textbook provides an introduction to the basics of photogrammetry and laser scanning. References to Volume 2, Chapters B, C, D, and E, refer to

> Kraus, Karl: Photogrammetry, Volume 2, Advanced Methods and Applications, with contributions by J. Jansa and H. Kager. 4th edition, Dümmler, Bonn, 1997, ISBN 3-427-78694-3.

Volume 2 is a completely separate textbook and is currently out of print. It covers advanced topics for readers who require a deeper theoretical knowledge and details of specialized applications.

Contents

Chapter 1

Introduction

1.1 Definitions

Photogrammetry allows one to reconstruct the position, orientation, shape and size of objects from pictures; these pictures may originate as photochemical images (conventional photography) or as photoelectric images (digital photography). Laser scanner images, a third group, have arrived in recent years; laser scanner images have distance information associated with every picture element. The results of a photogrammetric analysis may be:

- numbers—coordinates of separate points in a three-dimensional coordinate system (digital point determination),
- drawings (analogue)—maps and plans with planimetric detail and contour lines together with other graphical representation of objects,
- geometric models (digital)—which are fed in to information systems,
- images (analogue and/or digital)—above all, rectified photographs (orthophotos) and, derived from these, photomaps; but also photomontages and so-called three-dimensional photomodels, which are textured CAD models with textures extracted from photographs.

That branch of photogrammetry which starts with conventional photographs and in which the processing is by means of optical-mechanical instruments is called analogue photogrammetry. That which is based on conventional photographs but which resolves the whole process of analysis by means of computers is called analytical photogrammetry. A third stage of development is digital photogrammetry. In that case the light falling on the focal plane of the taking camera is recorded not by means of a light-sensitive emulsion but by means of electronic detectors. Starting from such digital photographs, the whole process of evaluation is by means of computers—human vision and perception are emulated by the computer. Especially in English, digital photogrammetry is frequently called softcopy photogrammetry as opposed to hardcopy photogrammetry which works with digitized film-based photographs[1]. Photogrammetry has some connection with machine vision, or computer vision, of which pattern recognition is one aspect.

[1]See PE&RS 58, Copy 1, pp. 49–115, 1992.

In many cases interpretation of the content of the image goes hand in hand with the geometrical reconstruction of the photographed object. The outcome of such photointerpretation is the classification of objects within the images according to various different characteristics.

Photogrammetry allows the reconstruction of an object and the analysis of its characteristics without physical contact with it. Acquisition of information about the surface of the Earth in this way is known nowadays as remote sensing. Remote sensing embraces all methods of acquiring information about the Earth's surface by means of measurement and interpretation of electromagnetic radiation[2] either reflected from or emitted by it. While remote sensing includes that part of photogrammetry which concerns itself with the surface of the Earth, if the predominant interest is in geometric characteristics, one speaks of photogrammetry and not of remote sensing.

1.2 Applications

The principal application of photogrammetry lies in the production of topographic maps in the form of both line maps and orthophoto maps. Photogrammetric instruments function as 3D-digitizers; in a photogrammetric analysis a digital topographic model is formed, which can be visualized with the aid of computer graphics. Both the form and the usage of the surface of the Earth are stored in such a digital topographic model. The digital topographic models are input in a topographical information system as the central body of data which, speaking very generally, provides information about both the natural landscape and the cultural landscape (as fashioned by man). A topographic information system is a fundamental subsystem in a comprehensive geoinformation system (GIS). Photogrammetry delivers geodata to a GIS. Nowadays a very large proportion of geodata is recorded by means of photogrammetry and laser scanning.

Close range photogrammetry is used for the following tasks: architectural recording; precision measurement of building sites and other engineering subjects; surveillance of buildings and documentation of damage to buildings; measuring up of artistic and engineering models; deformation measurement; survey of moving processes (for example, robotics); biometric applications (for example, computer controlled surgical operations); reconstruction of traffic accidents and very many others.

If the photographs are taken with specialized cameras, photogrammetric processing is relatively simple. With the help of complex mathematical algorithms and powerful software, however, the geometric processing of amateur photographs has now become possible. This processing technology is becoming more and more widely used, especially now that many people have their photographs available on their computers and, in addition to manipulation of density and colour, are frequently interested in geometric processing.

[2]See DIN 18716/3.

1.3 Some remarks on historical development[3]

Technologies arise and develop historically in response both to need and to the emergence and development of supporting techniques and technologies. With the invention of photography by Fox Talbot in England, by Niépce and Daguerre in France, and by others, the 1830s and 1840s saw the culmination of investigations extending over the centuries into optics and into the photo-responses of numerous chemicals. Also at that time, rapid and cost-effective methods of mapping were of crucial interest to military organizations, to colonial powers and to those seeking to develop large, relatively new nations such as Canada and the USA. While the practical application of new technology typically lags well behind its invention, it was very quickly recognized that cameras furnished a means of recording not only pictorial but also geometrical information, with the result that photogrammetry was born only a few years after cameras became available. Surprisingly, it was not the urgent needs of mapping but the desire accurately to record important buildings which led to the first serious and sustained application of photogrammetry and it was not a surveyor but an architect, the German Meydenbauer[4], who was responsible. In fact it is to Meydenbauer that we should be grateful, or not, for having coined the word "photogrammetry". Between his first and last completed projects, in 1858 and 1909 respectively, on behalf of the Prussian state, Meydenbauer compiled an archive of some 16000 metric images of its most important architectural monuments.

Meydenbauer had, however, been preceded in 1849 by the Frenchman Laussedat, a military officer, and it is he who is universally regarded as the first photogrammetrist despite the fact that he was initially using not a camera but a camera lucida, working on an image of a façade of the Hôtel des Invalides in Paris. The work of both of these scientists had been foreshadowed by others. In 1839 the French physicist Arago had written that photography could serve "to measure the highest and inaccessible buildings and to replace the fieldwork of a topographer". Earlier than this, in 1759, Lambert, a German mathematician, had published a treatise on how to reconstruct three-dimensional objects from perspective drawings.

The effective production of maps using photogrammetry, which was to become a technological triumph of the 20th century, was not possible at that time, nor for decades afterwards; that triumph had to wait for several critical developments: the invention of stereoscopic measurement, the introduction of the aeroplane and progress in the development of specialized analogue computers. For reasons which will become clear to readers of this book, buildings provided ideal subjects for the photogrammetric techniques of the time; topographic features most certainly did not. Without stereoscopy, measurement could be made only of very clearly defined points such as are to be found on buildings. Using cameras with known orientations and known positions, the three-dimensional coordinates of points defining a building being measured photogrammet-

[3] Permission from the publishers to use some of the historical material from "Luhmann, T., Robson, S., Kyle, S., Harley, I.: Close Range Photogrammetry. Whittles Publishing, 2006" is gratefully acknowledged. That material and this present section were both contributed by one of the translators, Ian Harley.

[4] A limited bibliography, with particular reference to historical development, is given at the end of this chapter.

rically were deduced using numerical computation. The basic computational methods of photogrammetry were established long ago.

By virtue of their regular and distinct features, architectural subjects lend themselves to this technique which, despite the fact that numerical computation was employed, is often referred to as "plane table photogrammetry". When using terrestrial pictures in mapping, by contrast, there was a major difficulty in identifying the same point on different photographs, especially when they were taken from widely separated camera stations; and a wide separation is desirable for accuracy. It is for these reasons that so much more architectural than topographic photogrammetry was performed during the 19th century. Nonetheless, a certain amount of topographic mapping by photogrammetry took place during the last three decades of that century; for example mapping in the Alps by Paganini in 1884 and the mapping of vast areas of the Rockies in Canada by Deville, especially between 1888 and 1896. Jordan mapped the Dachel Oasis in 1873.

In considering the history of photogrammetry the work of Scheimpflug in Austria should not be overlooked. In 1898 he first demonstrated double projection, which foreshadowed purely optical stereoplotters. In particular his name will always be associated with developments in rectification.

The development of stereoscopic measurement around the turn of the century was a momentous breakthrough in the history of photogrammetry. The stereoscope had already been invented between 1830 and 1832 and Stolze had discovered the principle of the floating measuring mark in Germany in 1893. Two other scientists, Pulfrich in Germany and Fourcade in South Africa, working independently and almost simultaneously[5], developed instruments for the practical application of Stolze's discovery. Their stereocomparators permitted stereoscopic identification of, and the setting of measuring marks on, identical points in two pictures. The survey work proceeded point by point using numerical intersection in three dimensions. Although the landscape could be seen stereoscopically in three dimensions, contours still had to be plotted by interpolation between spot heights.

Efforts were therefore directed towards developing a means of continuous measurement and plotting of features, in particular of contours—the "automatic" plotting machine, in which numerical computation was replaced by analogue computation for resection, relative and absolute orientation and, above all, for intersection of rays. Digital computation was too slow to allow the unbroken plotting of detail, in particular of contours, which stereoscopic measurement seemed to offer so tantalisingly. Only analogue computation was fast enough to provide continuous feedback to the operator. In several countries during the latter part of the 19th century, much effort and imagination was directed towards the invention of stereoplotting instruments, necessary for the accurate and continuous plotting of topography. In Germany Hauck proposed such an apparatus. In Canada Deville developed what was described by E. H. Thompson as "the first automatic plotting instrument in the history of photogrammetry". Deville's instrument had several defects, but its design inspired several subsequent workers to overcome these,

[5]Pulfrich's lecture in Hamburg announcing his invention was given on 23rdSeptember 1901, while Fourcade delivered his paper in Cape Town nine days later on 2ndOctober 1901.

including both Pulfrich, one of the greatest contributors to photogrammetric instrumentation, and Santoni in Italy, perhaps the most prolific of photogrammetric inventors.

Photogrammetry was about to enter the era of analogue computation, a very foreign idea to surveyors with their long tradition of numerical computation. Although many surveyors regarded analogue computation as an aberration, it became a remarkably successful one for a large part of the 20th century.

In Germany, conceivably the most active country in the early days of photogrammetry, Pulfrich's methods were very successfully used in mapping; this inspired von Orel in Vienna to design an instrument for the "automatic" plotting of contours, leading ultimately to the Orel-Zeiss Stereoautograph which came into productive use in 1909. In England, F. V. Thompson was slightly before von Orel in the design and use of the Vivian Thompson Stereoplotter; he went on to design the Vivian Thompson Stereoplanigraph, described in January 1908, about which E. H. Thompson was to write that it was "the first design for a completely automatic and thoroughly rigorous photogrammetric plotting instrument". The von Orel and the Thompson instruments were both used successfully in practical mapping, Vivian Thompson's having been used by the Survey of India which bought two of the instruments.

The advantages of photography from an aerial platform, rather than from a ground point, are obvious, both for reconnaissance and for survey; in 1858 Nadar, a Paris photographer, took the first such picture, from a hot-air balloon 1200 feet above that city, and in the following year he was ordered by Napoleon to obtain reconnaissance photographs in preparation for the Battle of Solferino. It is reputed that balloon photography was used during the following decade in the American Civil War. The rapid development of aviation which began shortly before the first World War had a decisive influence on the course of photogrammetry. Not only is the Earth, photographed vertically from above, an almost ideal subject for the photogrammetric method, but also aircraft made almost all parts of the Earth accessible at high speed. In the first half, and more, of the 20th century these favourable circumstances allowed impressive development in photogrammetry, although the tremendous economic benefit in air survey was not fully felt until the middle of that century. On the other hand, while stereoscopy opened the way for the application of photogrammetry to the most complex surfaces such as might be found in close range work, not only is the geometry in such cases often far from ideal photogrammetrically but also there was no corresponding economic advantage to promote its application.

In the period before the first World War all the major powers followed similar paths in the development of photogrammetry. After the war, although there was considerable opposition from surveyors to the use of photographs and analogue instruments for mapping, the development of stereoscopic measuring instruments forged ahead remarkably in very many countries; while the continental European countries broadly speaking put most of their effort into instrumental methods, Germany and the Austro-Hungarian Empire having a clear lead in this field, the English-speaking countries focused on graphical techniques. It is probably true that until about the 1930s the instrumental techniques could not compete in cost or efficiency with the British and American methods.

Zeiss, in the period following WWI, was well ahead in the design and manufacture of photogrammetric instruments, benefiting from the work of leading figures such as Pulfrich, von Orel, Bauersfeld, Sander and von Gruber. In Italy, around 1920, Santoni produced a prototype, the first of many mechanical projection instruments designed throughout his lifetime, while the Nistri brothers developed an optical projection plotter, shortly afterwards founding the instrument firm OMI. Poivilliers in France began the design and construction of analogue photogrammetric plotters in the early 1920s. In Switzerland the scene was dominated by Wild whose company began to produce instrumentation for terrestrial photogrammetry at about the same time; Wild Heerbrugg very rapidly developed into a major player, not only in photogrammetric instrumentation, including aerial cameras, but also in the wider survey world. As early as 1933 Wild stereometric cameras were being manufactured and were in use by Swiss police for the mapping of accident sites, using the Wild A4 Stereoautograph, a plotter especially designed for this purpose. Despite the ultra-conservative establishment in the British survey world at that time, E. H. Thompson was able to design and build a stereoplotter in the late 1930s influenced by the ideas of Fourcade. While the one such instrument in existence was destroyed by aerial bombing, the Thompson-Watts plotter was later based on this prototype in the 1950s.

Meanwhile, non-topographic use was sporadic for the reasons that there were few suitable cameras and that analogue plotters imposed severe restrictions on principal distance, on image format and on disposition and tilts of cameras.

The 1950s saw the beginnings of the period of analytical photogrammetry. The expanding use of digital, electronic computers in that decade engendered widespread interest in the purely analytical or numerical approach to photogrammetry as against the prevailing analogue methods. While analogue computation is inflexible, in regard to both input parameters and output results, and its accuracy is limited by physical properties, a numerical method allows virtually unlimited accuracy of computation and its flexibility is bounded only by the mathematical model on which it is based. Above all, it permits over-determination which may improve precision, lead to the detection of gross errors and provide valuable statistical information about the measurements and the results. The first analytical applications were to photogrammetric triangulation, a technique which permits a significant reduction in the amount of ground control required when mapping from a strip or a block of aerial photographs; because of the very high cost of field survey for control, such techniques had long been investigated. In the 1930s, the slotted template method of triangulation in plan was developed in the USA, based on theoretical work by Adams, Finsterwalder and Hotine. Up until the 1960s vast areas were mapped in the USA and Australia using this technique in plan and one of the many versions of the simple optical-projection Multiplex plotters both for triangulation in height and for plotting of detail. At the same time, precise analogue instruments such as the Zeiss C8 and the Wild A7 were being widely used for analogue triangulation in three dimensions.

Analytical photogrammetric triangulation is a method, using numerical data, of point determination involving the simultaneous three-dimensional orientation of all the photographs and taking all inter-relations into account. Work on this line of development had appeared before WWII, long before the development of electronic computers. Analytical triangulation demanded instruments to measure photo coordinates. The first stereocomparator designed specifically for use with aerial photographs was the Cambridge Stereocomparator designed in 1937 by E. H. Thompson. Electronic recording of data for input to computers became possible and by the mid-1950s there were five automatic recording stereocomparators on the market and monocomparators designed for use with aerial photographs also appeared.

Seminal papers by Schmid and Brown in the late 1950s laid the foundations for theoretically rigorous photogrammetric triangulation. A number of block adjustment programs for air survey were developed and became commercially available, such as those by Ackermann.

Subsequently, stereoplotters were equipped with devices to record model coordinates for input to electronic computers. Arising from the pioneering ideas of Helava, computers were incorporated in stereoplotters themselves, resulting in analytical stereoplotters with fully numerical reconstruction of the photogrammetric models. Bendix/OMI developed the first analytical plotter, the AP/C, in 1964; during the following two decades analytical stereoplotters were produced by the major instrument companies and others.

Photogrammetry has progressed as supporting sciences and technologies have supplied the means such as better glass, photographic film emulsions, plastic film material, aeroplanes, lens design and manufacture, mechanical design of cameras, flight navigation systems. Progress in space technology (both for imaging, in particular after the launch of SPOT-1 in 1986, and for positioning both on the ground and in-flight by GNSS) and the continuing explosion in electronic information processing have profound implications for photogrammetry.

The introduction of digital cameras into a photogrammetric system allows automation, nowhere more completely than in industrial photogrammetry, but also in mapping. Advanced computer technology enables the processing of digital images, particularly for automatic recognition and measurement of image features, including pattern correlation for determining object surfaces. Procedures in which both the image and its photogrammetric processing are digital are often referred to as digital photogrammetry.

Interactive digital stereo systems (e.g. Leica/Helava DSP, Zeiss PHODIS) have existed since around 1988 (Kern DSP 1) and have increasingly replaced analytical plotters. To some extent, photogrammetry has been de-skilled and made available directly to a wide range of users. Space imagery is commonplace, as exemplified by Google Earth. Photogrammetric measurement may be made by miscellaneous users with little or no knowledge of the subject—police, architects, model builders for example.

Although development continues apace, photogrammetry is a mature technology with a history of remarkable success. At the start of the 20th century topographic mapping of high quality existed, in general, only in parts of Europe, North America and India. Although adequate mapping is acknowledged as necessary for development but is still lacking in large parts of the world, such deficiencies arise for political and economic reasons, not for technical reasons. Photogrammetry has revolutionized cartography.

Further reading. Adams, L.P.: Fourcade: The centenary of a stereoscopic method of photographic surveying. Ph.Rec. 17(99), pp. 225–242, 2001 • Albertz, J.: A Look Back—Albrecht Meydenbauer. PE&RS 73, pp. 504–506, 2007 • Atkinson, K.B.: Vivian Thompson (1880–1917): not only an officer in the Royal Engineers. Ph.Rec. 10(55), pp. 5–38, 1980 • Atkinson, K.B.: Fourcade: The Centenary—Response to Professor H.-K. Meier. Correspondence, Ph.Rec. 17(99), pp. 555–556, 2002 • Babington-Smith, C.: The Story of Photo Intelligence in World War II. ASPRS, Falls Church, Virginia, 1985, reprint • Blachut, T.J. and Burkhardt, R.: Historical development of photogrammetric methods and instruments. ISPRS and ASPRS, Falls Church, Virginia, 1989 • Brown, D.C.: A solution to the general problem of multiple station analytical stereotriangulation. RCA Data Reduction Technical Report No. 43, Aberdeen, 1958 • Brown, D.C.: The bundle adjustment—progress and prospectives. IAPR 21(3), ISP Congress, Helsinki, pp. 1–33, 1976 • Deville, E.: Photographic Surveying. Government Printing Bureau, Ottawa, 1895. 232 pages • Deville, E.: On the use of the Wheatstone Stereoscope in Photographic Surveying. Transactions of the Royal Society of Canada, Ottawa, 8, pp. 63–69, 1902 • Fourcade, H.G.: On a stereoscopic method of photographic surveying. Transactions of the South African Philosophical Society 14(1), pp. 28–35, 1901. Also published in: Nature 66(1701), pp. 139–141, 1902 • Fourcade, H.G.: On instruments and methods for stereoscopic surveying. Transactions of the Royal Society of South Africa 14, pp. 1–50, 1903 • Fraser, C.S., Brown, D.C.: Industrial photogrammetry—new developments and recent applications. Ph.Rec. 12(68), pp. 197–216, 1986 • von Gruber, O., (ed.), McCaw, G.T., Cazalet, F.A., (trans.): Photogrammetry, Collected Lectures and Essays. Chapman & Hall, London, 1932 • Gruen, A.: Adaptive least squares correlation—a powerful image matching technique. South African Journal of Photogrammetry, Remote Sensing and Cartography 14(3), pp. 175–187, 1985 • Harley, I.A.: Some notes on stereocomparators. Ph.Rec. IV(21), pp. 194–209, 1963 • Helava, U.V.: New principle for analytical plotters. Phia 14, pp. 89–96, 1957 • Kelsh, H.T.: The slotted-template method for controlling maps made from aerial photographs. Miscellaneous Publications no. 404, U.S. Department of Agriculture, Washington, D.C., 1940 • Landen, D.: History of photogrammetry in the United States. Photogrammetric Engineering 18, pp. 854–898, 1952 • Laussedat, A.: Mémoire sur l'emploi de la photographie dans le levé des plans. Comptes Rendus 50, pp. 1127–1134, 1860 • Laussedat, A.: Recherches sur les instruments, les méthodes et le dessin topographiques. Gauthier-Villars, Paris, 1898 (vol. 1), 1901 (vol. 2 part 1), 1903 (vol. 2 part 2) • Luhmann, T., Robson, S., Kyle, S. and Harley, I: Close Range Photogrammetry. Whittles Publishing, 2006. 510 pages • Mason, K.: The Thompson stereo-plotter and its use. Survey of India Departmental Paper No. 5, Survey of India, Dehra Dun, India, 1913 • Meier, H.-K.: Fourcade: The Centenary—Paper by L.P. Adams. Correspondence, Ph.Rec. 17(99), pp. 554–555, 2002 • Meydenbauer, A.: Handbuch der Messbildkunst. Knapp, Halle, 1912. 245 pages • Poivilliers, G.: Address delivered at the opening of the Historical Exhibition, Ninth International Congress of Photogrammetry, London, 1960. IAPR XIII(1), 1961 • Pulfrich, C.: Über neuere Anwendungen der Stereoskopie und über einen hierfür bestimmten Stereo-Komparator. Zeitschrift für Instrumentenkunde 22(3), pp. 65–81, 1902 • Sander, W.: The development of photogrammetry in the light of invention, with special reference to plotting from two photographs. In: von

Gruber, O., (ed.), McCaw, G.T., Cazalet, F.A., (trans.): Photogrammetry, Collected Lectures and Essays. Chapman & Hall, London, pp. 148–246, 1932 • Santoni, E.: Instruments and photogrammetric devices for survey of close-up subject with particular stress to that of car bodies and their components photographed at various distances. Inedito. 1966. Reprinted in Selected Works, Ermenegildo Santoni, Scritti Scelti 1925–1968. Società Italiana di Fotogrammetria e Topographia, Firenze, 1971 • Schmid, H.: An analytical treatment of the problem of triangulation by stereophotogrammetry. Phia XIII(2/3), 1956–57 • Schmid, H.: Eine allgemeine analytische Lösung für die Aufgabe der Photogrammetrie. BuL. 1958(4), pp. 103–113, and 1959(1), pp. 1–12 • Thompson, E.H.: Photogrammetry. The Royal Engineers Journal 76(4), pp. 432–444, 1962. Reprinted as Photogrammetry, in: Photogrammetry and surveying, a selection of papers by E.H. Thompson, 1910–1976. Photogrammetric Society, London, 1977 • Thompson, E.H.: The Deville Memorial Lecture. Can.Surv. 19(3), pp. 262–272, 1965. Reprinted in: Photogrammetry and surveying, a selection of papers by E.H. Thompson, 1910–1976. Photogrammetric Society, London, 1977 • Thompson, E.H.: The Vivian Thompson Stereo-Planigraph. Ph.Rec. 8(43), pp. 81–86, 1974 • Thompson, V.F.: Stereo-photo-surveying. The Geographical Journal 31, pp. 534–561, 1908 • Wheatstone, C.: Contribution to the physiology of vision—Part the first. On some remarkable, and hitherto unobserved, phenomena of binocular vision. Philosophical Transactions of the Royal Society of London for the year MDCCCXXXVIII, Part II, pp. 371–394, 1838

Chapter 2

Preparatory remarks on mathematics and digital image processing

Section 2.1 concerns itself entirely with introductory mathematics. Section 2.2 includes notes on digital image processing in preparation for procedures of digital photogrammetric processing.

2.1 Preparatory mathematical remarks

The various techniques for photogrammetric processing assume knowledge of basic mathematics. While important mathematical matters may have been treated in lectures and textbooks, some mathematical themes of importance for photogrammetry are compiled in what follows.

2.1.1 Rotation in a plane, similarity and affine transformations

Given, a point $P(x, y)$ in a plane coordinate system (see Figure 2.1-1) which has been rotated through an angle α in an counterclockwise direction relative to a fixed coordinate system, we wish to find the coordinates (X, Y) of the point P with respect to the fixed coordinate system.

$$\begin{aligned} X &= x\cos\alpha - y\sin\alpha \\ Y &= x\sin\alpha + y\cos\alpha \end{aligned} \tag{2.1-1}$$

If we introduce the cosines of the angles between the coordinate axes and use matrix notation we have:

$$\begin{pmatrix} X \\ Y \end{pmatrix} = \begin{pmatrix} \cos(\angle xX) & \cos(\angle yX) \\ \cos(\angle xY) & \cos(\angle yY) \end{pmatrix} \begin{pmatrix} x \\ y \end{pmatrix} \tag{2.1-2}$$

Representing matrices and vectors by means of bold symbols this becomes:

$$\mathbf{X} = \mathbf{R}\mathbf{x}, \qquad \mathbf{R} = \begin{pmatrix} r_{11} & r_{12} \\ r_{21} & r_{22} \end{pmatrix} \tag{2.1-3}$$

$\mathbf{R}$ is called a rotation matrix. It is square but not symmetric. The elements of $\mathbf{R}$ are the cosines of the angles between the coordinate axes.

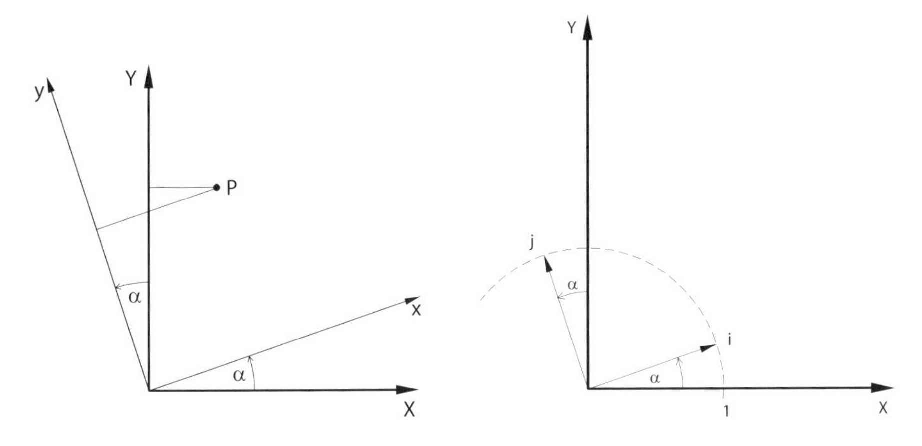

Figure 2.1-1: Plane rotation

Figure 2.1-2: Introduction of the unit vectors

Properties of the rotation matrix R

The question arises whether the four elements r_{ik} may be freely chosen or whether they must satisfy certain conditions. To answer this question we introduce the unit vectors **i** and **j** (Figure 2.1-2) along the coordinate axes x and y. We express their components in the XY system as:

$$\mathbf{i} = \begin{pmatrix} \cos\alpha \\ \sin\alpha \end{pmatrix}, \qquad \mathbf{j} = \begin{pmatrix} -\sin\alpha \\ \cos\alpha \end{pmatrix} \tag{2.1-4}$$

A comparison between Equations (2.1-1) and (2.1-4) shows that the elements r_{ik} of the rotation matrix are none other than the components of the unit vectors **i** and **j**.

$$\mathbf{R} = (\mathbf{i},\ \mathbf{j}) \tag{2.1-5}$$

The two mutually orthogonal unit vectors must, however, satisfy the orthogonality conditions in Equation (2.1-6)[1]. These conditions are formulated in Equation (2.1-6) as inner, or scalar, products of the two vectors in which transposition is signified by the superscript $^\top$.

$$\begin{aligned} \mathbf{i}^\top\mathbf{i} &= \cos^2\alpha + \sin^2\alpha = 1 = r_{11}^2 + r_{21}^2 \\ \mathbf{j}^\top\mathbf{j} &= \sin^2\alpha + \cos^2\alpha = 1 = r_{12}^2 + r_{22}^2 \\ \mathbf{i}^\top\mathbf{j} &= -\cos\alpha\sin\alpha + \sin\alpha\cos\alpha = 0 = r_{11}r_{12} + r_{21}r_{22} \end{aligned} \tag{2.1-6}$$

A matrix which satisfies the orthogonality conditions is known as an orthogonal matrix[2]. If the four elements of the rotation matrix must satisfy the three orthogonality

[1] More strictly, one orthogonal condition and two normalizing conditions.

[2] In amplification of the preceding footnote, when $\det\mathbf{R} = 1$ the matrix is said to be a proper orthogonal matrix and when $\det\mathbf{R} = -1$ the matrix is said to be improper. In the latter case the matrix represents a rotation and a reflection.

conditions, only one parameter may be freely chosen; in general this parameter is the rotation angle α.

Numerical Example.

$$\begin{pmatrix} X \\ Y \end{pmatrix} = \begin{pmatrix} 0.36 & 0.69 \\ 0.19 & 0.27 \end{pmatrix} \begin{pmatrix} x \\ y \end{pmatrix}$$

In the above case, the orthogonality conditions are not fulfilled; the transformation does not represent a rotation. We deal with this transformation at the end of this section.

$$\begin{pmatrix} X \\ Y \end{pmatrix} = \begin{pmatrix} 0.6234 & -0.7819 \\ 0.7819 & 0.6234 \end{pmatrix} \begin{pmatrix} x \\ y \end{pmatrix}$$

Here the orthogonality conditions are fulfilled; this means that under this transformation a field of points will be rotated.

Exercise 2.1-1. Consider a rectangle in an xy system, the vertices being transformed into the XY system using the matrices of both numerical examples. Using the results, consider the characteristics of both transformations.

Exercise 2.1-2. Think about the characteristics of the transformation when just one of the three orthogonality conditions of Equation (2.1-6) is not fulfilled.

Exercise 2.1-3. Consider a matrix which brings about both a rotation and a mirror reflection. (Answer: $r_{11} = \cos\alpha$; $r_{12} = \sin\alpha$; $r_{21} = \sin\alpha$; $r_{22} = -\cos\alpha$).

Inverting the rotation matrix R

By definition, multiplication of the inverted matrix $\mathbf{R}^{-1}$ by the matrix $\mathbf{R}$ gives the unit matrix $\mathbf{I}$:

$$\mathbf{R}^{-1}\mathbf{R} = \mathbf{I}$$

On the other hand multiplication of the transposed matrix $\mathbf{R}^\top$ with the matrix $\mathbf{R}$ also gives the unit matrix (Equations (2.1-5) and (2.1-6)):

$$\begin{pmatrix} \mathbf{i}^\top \\ \mathbf{j}^\top \end{pmatrix} (\mathbf{i},\ \mathbf{j}) = \begin{pmatrix} \mathbf{i}^\top\mathbf{i} & \mathbf{i}^\top\mathbf{j} \\ \mathbf{j}^\top\mathbf{i} & \mathbf{j}^\top\mathbf{j} \end{pmatrix} = \begin{pmatrix} 1 & 0 \\ 0 & 1 \end{pmatrix}$$

As a consequence, we see that the following important result holds for the rotation matrix:

$$\mathbf{R}^{-1} = \mathbf{R}^\top \tag{2.1-7}$$

Reverse transformation

If one wishes to transform points from the fixed XY system into the xy system, one obtains the desired rotation matrix as follows:
From Equation (2.1-3):

$$\mathbf{X} = \mathbf{R}\mathbf{x}$$

Premultiplication by $\mathbf{R}^\top$ gives:

$$\mathbf{R}^\top\mathbf{X} = \mathbf{R}^\top\mathbf{R}\mathbf{x} = \mathbf{I}\mathbf{x} = \mathbf{x}$$

Rewriting this result:

$$\mathbf{x} = \mathbf{R}^\top\mathbf{X} = \begin{pmatrix} r_{11} & r_{21} \\ r_{12} & r_{22} \end{pmatrix} \begin{pmatrix} X \\ Y \end{pmatrix}$$

Exercise 2.1-4. How would Equations (2.1-1), (2.1-2), (2.1-4) and (2.1-5) appear if the rotation of the xy system had been made in a clockwise sense with respect to the XY system?

A transformation with a non-orthogonal matrix (see the first numerical example) is known as an affine transformation. It has the following characteristics:

- orthogonal straight lines defined, for example, by three points in the xy system are no longer orthogonal after the transformation.
- parallel straight lines defined, for example, by four points in the xy system, remain parallel after the transformation.
- line-segments between two points in the xy system exhibit a different length after the transformation.
- on the other hand, the ratio of the lengths of two parallel line segments is invariant under the transformation.

The affine transformation is of the form:

$$\begin{pmatrix} X \\ Y \end{pmatrix} = \begin{pmatrix} a_{10} \\ a_{20} \end{pmatrix} + \begin{pmatrix} a_{11} & a_{12} \\ a_{21} & a_{22} \end{pmatrix} \begin{pmatrix} x \\ y \end{pmatrix}; \qquad \mathbf{X} = \mathbf{a}_0 + \mathbf{A}\mathbf{x} \tag{2.1-8}$$

in which

- a_{10} and a_{20} are two translations (or, more exactly, the XY coordinates of the origin of the xy system) and
- a_{11}, a_{12}, a_{21} and a_{22} are four elements which do not satisfy the orthogonality conditions of Equation (2.1-6) and which consequently allow not only different scales in the two coordinate directions but also independent rotations of the two coordinate axes.

In order to determine the six parameters a_{ik} one requires at least three common points in both coordinate systems.

Numerical Example. Given three points with their coordinates in both systems, we wish to find the six parameters a_{ik} of the affine transformation.

Pt.No.	x	y	X	Y
23	0.3035	0.5951	3322	1168
24	0.1926	0.6028	3403	2061
50	0.3038	0.4035	1777	1197

Using the linear Equations (2.1-8) we obtain the following system of equations

$$\begin{pmatrix} 1 & & 0.3035 & 0.5951 & & \\ & 1 & & & 0.3035 & 0.5951 \\ 1 & & 0.1926 & 0.6028 & & \\ & 1 & & & 0.1926 & 0.6028 \\ 1 & & 0.3038 & 0.4035 & & \\ & 1 & & & 0.3038 & 0.4035 \end{pmatrix} \begin{pmatrix} a_{10} \\ a_{20} \\ a_{11} \\ a_{12} \\ a_{21} \\ a_{22} \end{pmatrix} = \begin{pmatrix} 3322 \\ 1168 \\ 3403 \\ 2061 \\ 1777 \\ 1197 \end{pmatrix}$$

of which the solution is

$$\begin{pmatrix} X \\ Y \end{pmatrix} = \begin{pmatrix} -1425 \\ 3713 \end{pmatrix} + \begin{pmatrix} -171.5 & 8063.4 \\ -8063.7 & 164.0 \end{pmatrix} \begin{pmatrix} x \\ y \end{pmatrix}$$

One obtains a similarity transformation by replacing the nonorthogonal matrix $\mathbf{A}$ of Equation (2.1-8) with an (orthogonal) rotation matrix $\mathbf{R}$ and introducing a unit scale factor, m. The similarity transformation is of the form:

$$\begin{pmatrix} X \\ Y \end{pmatrix} = \begin{pmatrix} a_{10} \\ a_{20} \end{pmatrix} + m \begin{pmatrix} r_{11} & r_{12} \\ r_{21} & r_{22} \end{pmatrix} \begin{pmatrix} x \\ y \end{pmatrix}; \qquad \mathbf{X} = \mathbf{a}_0 + m\mathbf{R}\mathbf{x} \tag{2.1-9}$$

In order to determine the four parameters of the plane similarity transformation (two translations a_{10} and a_{20}, a scale factor m and, for example, a rotation angle α of the rotation matrix $\mathbf{R}$) one requires at least two common points in each coordinate system. The solution of this problem is discussed in Section 5.2.1.

Note: a square in the xy system remains a square after the transformation; it is simply shifted, rotated and changed in scale. Against that, after an affine transformation it becomes a parallelogram.

2.1.2 Rotation, affine and similarity transformations in three-dimensional space

Based on Equation (2.1-2), the rotation in space of a point P with coordinates (x, y, z) in a fixed coordinate system, XYZ, may be formulated as follows, using the cosines of the angles between the coordinate axes:

$$\begin{pmatrix} X \\ Y \\ Z \end{pmatrix} = \begin{pmatrix} \cos(\angle xX) & \cos(\angle yX) & \cos(\angle zX) \\ \cos(\angle xY) & \cos(\angle yY) & \cos(\angle zY) \\ \cos(\angle xZ) & \cos(\angle yZ) & \cos(\angle zZ) \end{pmatrix} \begin{pmatrix} x \\ y \\ z \end{pmatrix} \tag{2.1-10}$$

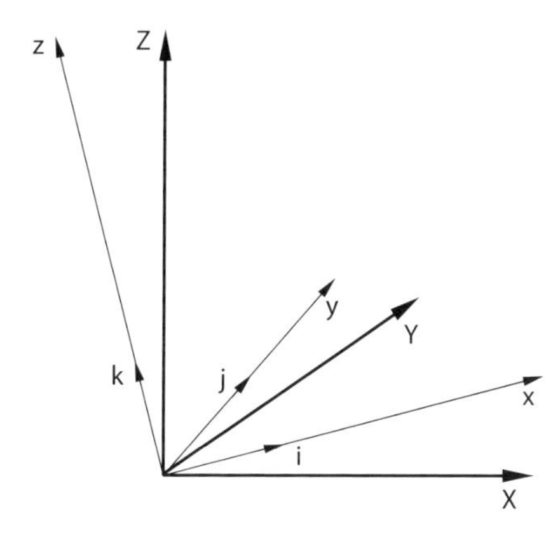

Figure 2.1-3: Rotation in three-dimensional space

$$\mathbf{X} = \mathbf{R}\mathbf{x}; \qquad \mathbf{R} = \begin{pmatrix} r_{11} & r_{12} & r_{13} \\ r_{21} & r_{22} & r_{23} \\ r_{31} & r_{32} & r_{33} \end{pmatrix} \tag{2.1-11}$$

In a similar manner to that of Equation (2.1-5), the matrix $\mathbf{R}$ can be formed from the three unit vectors shown in Figure 2.1-3; $\mathbf{R} = (\mathbf{i}, \mathbf{j}, \mathbf{k})$.[3] It is simple to write out the following six orthogonality[4] relationships among the nine elements r_{ik} for the three-dimensional case.

$$\begin{aligned} \mathbf{i}^\top\mathbf{i} &= \mathbf{j}^\top\mathbf{j} = \mathbf{k}^\top\mathbf{k} = 1 \\ \mathbf{i}^\top\mathbf{j} &= \mathbf{j}^\top\mathbf{k} = \mathbf{k}^\top\mathbf{i} = 0 \end{aligned} \tag{2.1-12}$$

That is, a rotation in three dimensions is prescribed by three independent parameters. In photogrammetry we frequently use three rotation angles ω, φ and κ about the three coordinate axes. In this case, a hierarchy of axes is to be observed, as can be clearly demonstrated with gimbal (or Cardan) axes (Figure 2.1-4):

[3]The three unit vectors with their components r_{ik} are related, through their vector products, as follows:

$$\mathbf{i} = \begin{pmatrix} r_{11} \\ r_{21} \\ r_{31} \end{pmatrix} = \mathbf{j} \times \mathbf{k} = \begin{pmatrix} +\begin{vmatrix} r_{22} & r_{23} \\ r_{32} & r_{33} \end{vmatrix} \\ \\ -\begin{vmatrix} r_{12} & r_{13} \\ r_{32} & r_{33} \end{vmatrix} \\ \\ +\begin{vmatrix} r_{12} & r_{13} \\ r_{22} & r_{23} \end{vmatrix} \end{pmatrix} = \begin{pmatrix} r_{22}r_{33} - r_{32}r_{23} \\ r_{32}r_{13} - r_{12}r_{33} \\ r_{12}r_{23} - r_{22}r_{13} \end{pmatrix}; \qquad \begin{aligned} \mathbf{j} &= \mathbf{k} \times \mathbf{i} \\ \mathbf{k} &= \mathbf{i} \times \mathbf{j} \end{aligned}$$

[4]More accurately, three orthogonality conditions and three normalizing conditions (see also the footnote to Equation (2.1-6) on page 11).

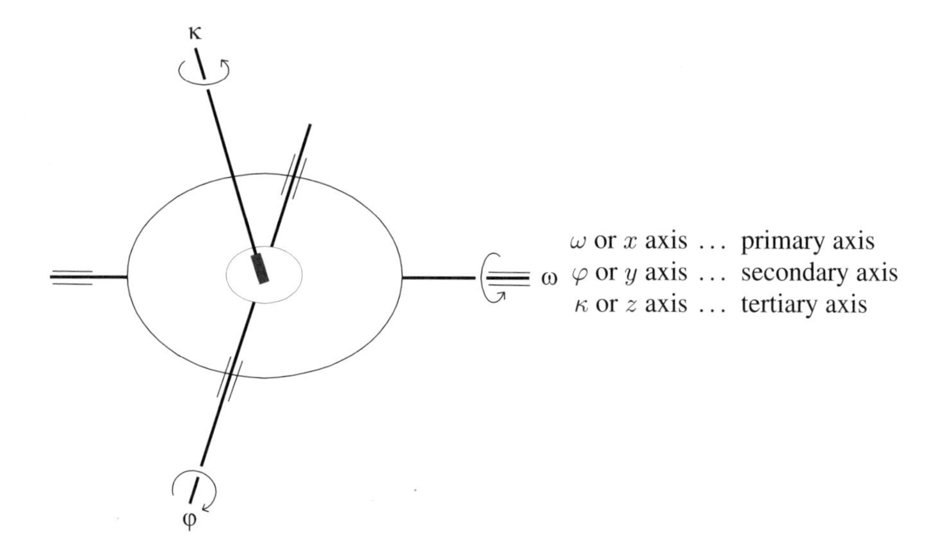

Figure 2.1-4: Rotations about the axes in gimbals

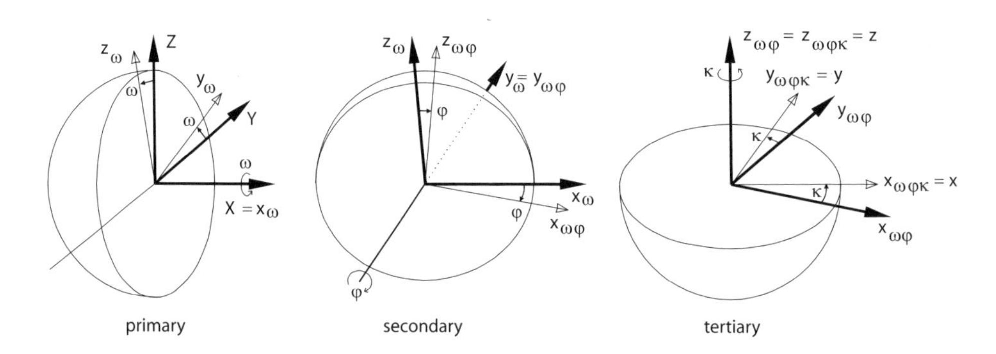

Figure 2.1-5: Hierarchy of the three rotations about the coordinate axes

If one performs an ω rotation, the attitudes in space of the other two axes are changed accordingly. If, however, one rotates in φ, only the κ axis, and not the ω axis, is affected. Rotation about the κ axis changes the attitude of neither of the other two axes. An arbitrary rotation of the xyz system as illustrated in Figure 2.1-3 can therefore be effected by means of three rotations ω, φ and κ. In each case the rotation is to be seen as counterclockwise when viewed along the axis towards the origin.

The transformation into the XYZ system of a point P, given in the xyz coordinate system, may therefore be defined in terms of the three rotation angles ω, φ and κ. In this case the matrix $\mathbf{R}$ of Equation (2.1-11) has the form[5] (see Appendix 2.1-1):

[5]In the following equation, the functions cos and sin are abbreviated by c and s, respectively.

$$\mathbf{R}_{\omega\varphi\kappa} = \begin{pmatrix} \mathrm{c}_\varphi \mathrm{c}_\kappa & -\mathrm{c}_\varphi \mathrm{s}_\kappa & \mathrm{s}_\varphi \\ \mathrm{c}_\omega \mathrm{s}_\kappa + \mathrm{s}_\omega \mathrm{s}_\varphi \mathrm{c}_\kappa & \mathrm{c}_\omega \mathrm{c}_\kappa - \mathrm{s}_\omega \mathrm{s}_\varphi \mathrm{s}_\kappa & -\mathrm{s}_\omega \mathrm{c}_\varphi \\ \mathrm{s}_\omega \mathrm{s}_\kappa - \mathrm{c}_\omega \mathrm{s}_\varphi \mathrm{c}_\kappa & \mathrm{s}_\omega \mathrm{c}_\kappa + \mathrm{c}_\omega \mathrm{s}_\varphi \mathrm{s}_\kappa & \mathrm{c}_\omega \mathrm{c}_\varphi \end{pmatrix} \tag{2.1-13}$$

Exercise 2.1-5. Show, using trigonometrical relationships, that the nine elements of the rotation matrix (2.1-13) fulfill the orthogonality conditions (2.1-12).

If the sequence of the rotations is defined in a different order, the elements of the matrix (2.1-13) are also changed (see Appendix 2.1-1 and especially Section B 3.4, Volume 2). As in Equation (2.1-7) the inverse of the rotation matrix, $\mathbf{R}^{-1}$, is the transposed matrix $\mathbf{R}^\top$, by virtue of the orthogonality conditions (2.1-12). To summarize, three different interpretations have been given for the elements of the three-dimensional rotation matrix $\mathbf{R}$:

- cosines of the angles between the axes of the two coordinate systems
- components of the unit vectors of the rotated coordinate axes with respect to the fixed system
- trigonometric functions of rotation angles about the three axes of a gimbal system

Two successive rotations
First rotation: $\mathbf{X}_1 = \mathbf{R}_1\mathbf{x}$
Second rotation: $\mathbf{X}_2 = \mathbf{R}_2\mathbf{X}_1$
Complete rotation:

$$\mathbf{X}_2 = \mathbf{R}_2\mathbf{R}_1\mathbf{x} = \mathbf{R}\mathbf{x} \tag{2.1-14}$$

By multiplication, the rotation matrices, $\mathbf{R}_1$ and $\mathbf{R}_2$, are combined as a single rotation matrix $\mathbf{R}$:

$$\mathbf{R} = \mathbf{R}_2\mathbf{R}_1 \tag{2.1-15}$$

Since matrix multiplication is not commutative, the order of multiplication of the matrices must be strictly observed. In the transposed matrix $\mathbf{R}^\top$ the sequence must be reversed:

$$\mathbf{R}^\top = (\mathbf{R}_2\mathbf{R}_1)^\top = \mathbf{R}_1^\top\mathbf{R}_2^\top \tag{2.1-16}$$

Example (of Equation (2.1-13)).
Given:

$$\begin{aligned} \omega &= -1.3948\,\text{gon} &&= -1^\circ 15' 19'' \\ \varphi &= 0.1041\,\text{gon} &&= 5' 37'' \\ \kappa &= -0.8479\,\text{gon} &&= -45' 47'' \end{aligned}$$

Required: elements of the rotation matrix $\mathbf{R}$

$$\mathbf{R} = \begin{pmatrix} 0.999910 & 0.013319 & 0.001635 \\ -0.013351 & 0.999671 & 0.021907 \\ -0.001343 & -0.021927 & 0.999759 \end{pmatrix}$$

Checks (see also Appendix 2.1-1):

$$\begin{aligned}
\sin\varphi &= r_{13} &\Rightarrow \varphi &= 0.1041\,\text{gon} &= 5'37''\\
\tan\kappa &= -r_{12}/r_{11} &\Rightarrow \kappa &= -0.8479\,\text{gon} &= -45'47''\\
\tan\omega &= -r_{23}/r_{33} &\Rightarrow \omega &= -1.3948\,\text{gon} &= -1^\circ 15'19''
\end{aligned}$$

The nine elements r_{ik} must satisfy the orthogonality conditions[6]; one checks that $\mathbf{R}^\top\mathbf{R} = \mathbf{I}$:

$$\begin{matrix} & \begin{pmatrix} 0.999910 & 0.013319 & 0.001635\\ -0.013351 & 0.999671 & 0.021907\\ -0.001343 & -0.021927 & 0.999759 \end{pmatrix}\\
\begin{pmatrix} 0.999910 & -0.013351 & -0.001343\\ 0.013319 & 0.999671 & -0.021927\\ 0.001635 & 0.021907 & 0.999759 \end{pmatrix} & \begin{pmatrix} 1.000000 & 0.000001 & 0.000000\\ 0.000001 & 1.000000 & 0.000000\\ 0.000000 & 0.000000 & 1.000001 \end{pmatrix}\end{matrix}$$

Example (of Equation (2.1-14)). We are given a point P in an xyz system which is rotated by ω_1, φ_1, κ_1 relative to an $X_1Y_1Z_1$ coordinate system. The $X_1Y_1Z_1$ system is then rotated by ω_2, φ_2, κ_2 relative to an $X_2Y_2Z_2$ system.

We wish to find the final coordinates X_2, Y_2, Z_2 of the point P and the angles ω, φ, κ by which the xyz system is rotated relative to an $X_2Y_2Z_2$ system.

Given coordinates of P:

$$\mathbf{x} = \begin{pmatrix} -43.461\\ -83.699\\ 152.670 \end{pmatrix}$$

First rotation:

$$\begin{aligned}
\omega_1 &= -1.3948\,\text{gon} &= -1^\circ 15'19''\\
\varphi_1 &= +0.1041\,\text{gon} &= 5'37''\\
\kappa_1 &= -0.8479\,\text{gon} &= -45'47''
\end{aligned}$$

$$\mathbf{R_1} = \begin{pmatrix} 0.999910 & 0.013319 & 0.001635\\ -0.013351 & 0.999671 & 0.021907\\ -0.001343 & -0.021927 & 0.999759 \end{pmatrix}$$

Second rotation:

$$\begin{aligned}
\omega_2 &= -0.1726\,\text{gon} &= -9'19''\\
\varphi_2 &= -1.0853\,\text{gon} &= -58'36''\\
\kappa_2 &= -101.3223\,\text{gon} &= -91^\circ 11'24''
\end{aligned}$$

$$\mathbf{R_2} = \begin{pmatrix} -0.020770 & 0.999639 & -0.017047\\ -0.999782 & -0.020727 & 0.002710\\ 0.002355 & 0.017100 & 0.999851 \end{pmatrix}$$

[6] For didactical reasons, the notation of a matrix multiplication $\mathbf{AB} = \mathbf{C}$ is often written in "Falk's scheme": $\begin{matrix} & \mathbf{B}\\ \mathbf{A} & \mathbf{C}\end{matrix}$

First solution (two stage)
First rotation of P (to $X_1Y_1Z_1$)

$$\begin{pmatrix} 0.999910 & 0.013319 & 0.001635 \\ -0.013351 & 0.999671 & 0.021907 \\ -0.001343 & -0.021927 & 0.999759 \end{pmatrix} \begin{pmatrix} -43.461 \\ -83.699 \\ 152.670 \end{pmatrix} = \begin{pmatrix} -44.3223 \\ -79.7467 \\ 154.5268 \end{pmatrix} = \mathbf{X}_1 = \mathbf{R}_1\mathbf{x}$$

Second rotation of P (to $X_2Y_2Z_2$)

$$\begin{pmatrix} -0.020770 & 0.999639 & -0.017047 \\ -0.999782 & -0.020727 & 0.002710 \\ 0.002355 & 0.017100 & 0.999851 \end{pmatrix} \begin{pmatrix} -44.3223 \\ -79.7467 \\ 154.5268 \end{pmatrix} = \begin{pmatrix} -81.432 \\ 46.384 \\ 153.036 \end{pmatrix} = \mathbf{X_2} = \mathbf{R_2X_1}$$

Second solution (one stage)
Rotation matrix $\mathbf{R} = \mathbf{R}_2\mathbf{R}_1$ for the combined rotations:

$$\begin{pmatrix} -0.020770 & 0.999639 & -0.017047 \\ -0.999782 & -0.020727 & 0.002710 \\ 0.002355 & 0.017100 & 0.999851 \end{pmatrix} \begin{pmatrix} 0.999910 & 0.013319 & 0.001635 \\ -0.013351 & 0.999671 & 0.021907 \\ -0.001343 & -0.021927 & 0.999759 \end{pmatrix} = \begin{pmatrix} -0.034091 & 0.999407 & 0.004822 \\ -0.999419 & -0.034096 & 0.000621 \\ 0.000784 & -0.004798 & 0.999988 \end{pmatrix}$$

Transformation of P (to $X_2Y_2Z_2$):

$$\begin{pmatrix} -0.034091 & 0.999407 & 0.004822 \\ -0.999419 & -0.034096 & 0.000621 \\ 0.000784 & -0.004798 & 0.999988 \end{pmatrix} \begin{pmatrix} -43.461 \\ -83.699 \\ 152.670 \end{pmatrix} = \begin{pmatrix} -81.432 \\ 46.384 \\ 153.036 \end{pmatrix} = \mathbf{X}_2 = \mathbf{R}\mathbf{x}$$

From the definition of the elements of the rotation matrix $\mathbf{R}$ (see Equation (2.1-13)) calculate the angles ω, φ and κ through which the point P has been rotated with respect to the $X_2Y_2Z_2$ system.

$$\begin{aligned} \omega &= -0.0396\,\text{gon} = -2'08'' \\ \varphi &= 0.3070\,\text{gon} = 16'35'' \\ \kappa &= -102.1708\,\text{gon} = -91°57'13'' \end{aligned}$$

Note: $\omega_1 + \omega_2$ is not equal to ω; similarly for φ and κ.

Exercise 2.1-6. Transform the rotated point $P\ (X_2, Y_2, Z_2)$ back into the xyz system, in two stages and in one stage.

In three-dimensional space also, a transformation with a non-orthogonal matrix is called an affine transformation. The characteristics of the plane affine transformation apply also to a three-dimensional affine transformation (see Section 2.1.1). The three-dimensional affine transformation has the form:

$$\begin{pmatrix} X \\ Y \\ Z \end{pmatrix} = \begin{pmatrix} a_{10} \\ a_{20} \\ a_{30} \end{pmatrix} + \begin{pmatrix} a_{11} & a_{12} & a_{13} \\ a_{21} & a_{22} & a_{23} \\ a_{31} & a_{32} & a_{33} \end{pmatrix} \begin{pmatrix} x \\ y \\ z \end{pmatrix}; \quad \mathbf{X} = \mathbf{a}_0 + \mathbf{A}\mathbf{x} \qquad (2.1\text{-}17)$$

in which:

- a_{10}, a_{20} and a_{30} are three translations (alternatively, the XYZ coordinates of the origin of the xyz system).
- $a_{11}, a_{12}, \ldots, a_{33}$ are the nine elements of $\mathbf{R}$; they do not satisfy the orthogonality conditions (2.1-12) and consequently they admit not only different scales in the three coordinate directions but also six independent angles of rotation of the three coordinate axes (Note: a coordinate axis is defined by two angles).

In order to determine the 12 parameters a_{ik}, one requires at least four corresponding points in each coordinate system (by reason of this number of parameters one speaks sometimes of a 12 parameter transformation. Section 4.4.3 contains a numerical example).

The three-dimensional similarity transformation follows from this if one substitutes an (orthogonal) rotation matrix $\mathbf{R}$ for the non-orthogonal matrix $\mathbf{A}$ and introduces a uniform scale factor m. The three-dimensional similarity transformation has the following form:

$$\begin{pmatrix} X \\ Y \\ Z \end{pmatrix} = \begin{pmatrix} a_{10} \\ a_{20} \\ a_{30} \end{pmatrix} + m \begin{pmatrix} r_{11} & r_{12} & r_{13} \\ r_{21} & r_{22} & r_{23} \\ r_{31} & r_{32} & r_{33} \end{pmatrix} \begin{pmatrix} x \\ y \\ z \end{pmatrix}; \quad \mathbf{X} = \mathbf{a}_0 + m\mathbf{R}\mathbf{x} \qquad (2.1\text{-}18)$$

For the determination of the seven parameters of a three-dimensional similarity transformation (three translations a_{10}, a_{20} and a_{30}, one scale factor m and three rotation angles ω, φ and κ defining $\mathbf{R}$ as in Equation (2.1-13)), at least seven suitable equations are required. These equations (for example, two in X, two in Y and three in Z) may be obtained from three corresponding points in the two systems. The solution of this non-linear problem is dealt with in Section 4.1.1.

Note: a cube remains a cube after a three-dimensional similarity transformation; it is simply translated, rotated and changed in scale. After an affine transformation it becomes a parallelepiped.

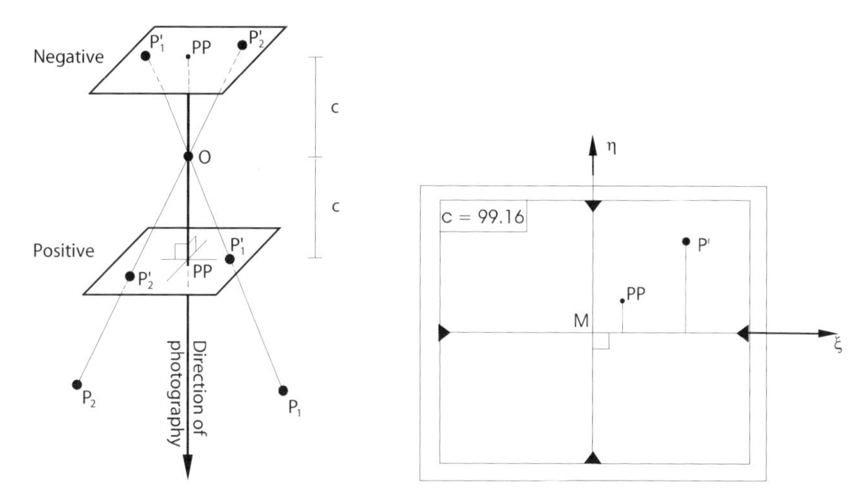

Figure 2.1-6: Positive and negative positions

Figure 2.1-7: Metric image

2.1.3 Central projection in three-dimensional space

To be able to reconstruct the position and shape of objects we must know the geometry of the image forming system. Many of the cameras used in photogrammetry, sometimes known as metric cameras, produce photographs which can be considered, with adequate accuracy, as central projections of the three-dimensional objects in view. (In Sections 2.1.3 to 2.1.7 we generally assume that we are dealing with analogue pictures.) Figures 2.1-6 and 2.1-7 show some definitions.

O ... centre of perspective of a three-dimensional bundle of rays (also, the camera location)
PP ... principal point with coordinates ξ_0,η_0
c ... principal distance (sometimes referred to as the camera constant)
M ... fiducial centre (as a coarse approximation, the point of intersection of the straight lines joining the fiducial marks)

The relationship between the coordinates ξ and η of an image point P' and the coordinates X,Y,Z of an object point P is illustrated in Figure 2.1-8 and is mathematically formulated in Equation (2.1-19)[7] (for the derivation of these collinearity equations, as they are usually called, see Appendix 2.1-2).

[7]Notation used in this book:

ξ,η ... two-dimensional image coordinates
x,y,z ... coordinates in a local three-dimensional coordinate system (frequently model coordinates)
X,Y,Z ... coordinates in a control coordinate system (sometimes called a global system; frequently the national coordinate system)

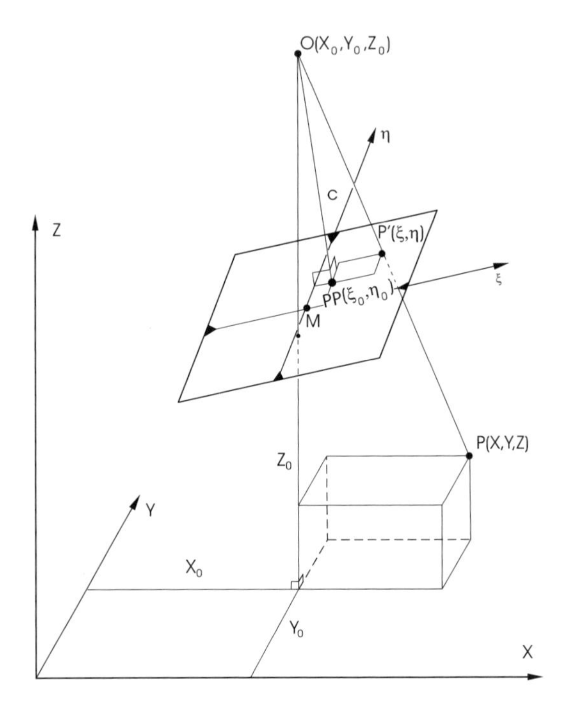

Figure 2.1-8: Relationship between image and object coordinates

$$\xi = \xi_0 - c\frac{r_{11}(X - X_0) + r_{21}(Y - Y_0) + r_{31}(Z - Z_0)}{r_{13}(X - X_0) + r_{23}(Y - Y_0) + r_{33}(Z - Z_0)}$$

$$\eta = \eta_0 - c\frac{r_{12}(X - X_0) + r_{22}(Y - Y_0) + r_{32}(Z - Z_0)}{r_{13}(X - X_0) + r_{23}(Y - Y_0) + r_{33}(Z - Z_0)} \tag{2.1-19}$$

The parameters r_{ik} appearing in Equations (2.1-19) are the elements of the rotation matrix $\mathbf{R}$ which in this case describes the three-dimensional attitude, or orientation, of the image with respect to the XYZ object coordinate system. If so desired, the elements r_{ik} can be expressed in accordance with Equation (2.1-13) in terms of the three angles ω, φ and κ, which are, respectively, rotations about the X axis, the Y_ω axis and the $Z_{\omega\varphi}$ axis, as defined in Figure 2.1-5.

Solving the Equations (2.1-19) for the object coordinates X and Y gives:

$$X = X_0 + (Z - Z_0)\frac{r_{11}(\xi - \xi_0) + r_{12}(\eta - \eta_0) - r_{13}c}{r_{31}(\xi - \xi_0) + r_{32}(\eta - \eta_0) - r_{33}c}$$

$$Y = Y_0 + (Z - Z_0)\frac{r_{21}(\xi - \xi_0) + r_{22}(\eta - \eta_0) - r_{23}c}{r_{31}(\xi - \xi_0) + r_{32}(\eta - \eta_0) - r_{33}c} \tag{2.1-20}$$

Equations (2.1-19) mean that to each object point there is one image point. Equations (2.1-20) draw our attention to the fact that, because the Z coordinates are on the right hand side, to each image point there are infinitely many possible object points. From a single metric image alone it is not possible to reconstruct a three-dimensional object. To do so one also needs either a second metric image of the same object taken from a different place or additional information about the Z coordinate (for example the information that all object points lie on a horizontal plane of known height).

The transformations formulated in Equations (2.1-19) and (2.1-20) assume a knowledge of the following independent values:

$$\begin{array}{ll} \xi_0, \eta_0 & \ldots \text{ image coordinates of the principal point } PP \\ c & \ldots \text{ principal distance} \end{array} \qquad (2.1\text{-}21)$$

The above three parameters are known as the elements of interior orientation[8]. They fix the centre of projection of the three-dimensional bundle of rays with respect to the image plane.

The following six parameters are the elements of exterior orientation. They define the position and attitude of the three-dimensional bundle of rays with respect to the object coordinate system.

$$\begin{array}{l} X_0, Y_0, Z_0 \ \ldots \text{ object coordinates of the camera station} \\ \text{3 parameters defining the rotations of the image (for example, } \omega, \varphi, \kappa) \end{array} \qquad (2.1\text{-}22)$$

To specify the central projection of an image a total of nine parameters is required, which may be determined in various ways. The values of the three constants of interior orientation are specific to the camera and are normally determined, at least in the first instance, by the manufacturer in the laboratory. He tries to ensure that, as closely as possible, the fiducial centre coincides with the principal point ($\xi_0 = \eta_0 = 0$). In terrestrial photogrammetry the six elements of exterior orientation can be established directly. On the other hand, the elements of exterior orientation of an individual image from a photographic flight are not known with sufficient accuracy—unless GPS (Global Position System) and an IMU (Inertial Measurement Unit), both very expensive, are installed. An alternative, indirect method must be used, involving control points; these are points for which both image coordinates and object coordinates are known. If the interior orientation is known one requires three control points, for each control point yields two Equations (2.1-19) from which the exterior orientation may be computed.

[8] In normal English, the orientation of an object implies direction or angular attitude. Photogrammetric usage, deriving from German, applies the word to groups of camera parameters. Exterior orientation parameters incorporate this angular meaning but extend it to include position. Interior orientation parameters, which include a distance, two coordinates and a number of polynomial coefficients, involve no angular values; the use of the terminology here underlies the connection between two very important, basic groups of parameters.

Example (with Equations (2.1-19)).
Given:

Interior orientation:

$$\begin{aligned} c &= 152.67\,\text{mm} \\ \xi_0 &= 0.00\,\text{mm} \\ \eta_0 &= 0.00\,\text{mm} \end{aligned}$$

Projection centre O:

$$\mathbf{X}_0\,[\text{m}] = \begin{pmatrix} 362530.603 \\ 61215.834 \\ 2005.742 \end{pmatrix}$$

Rotation matrix:

$$\mathbf{R} = \begin{pmatrix} -0.034091 & 0.999407 & 0.004822 \\ -0.999419 & -0.034096 & 0.000621 \\ 0.000784 & -0.004798 & 0.999988 \end{pmatrix}$$

Object coordinates of two points:

$$P_1\,[\text{m}] : \begin{pmatrix} 363552.124 \\ 61488.048 \\ 588.079 \end{pmatrix} \qquad P_2\,[\text{m}] : \begin{pmatrix} 362571.087 \\ 61198.320 \\ 596.670 \end{pmatrix}$$

To find: image coordinates of both points

Solution with Equations (2.1-19):

$$P_1'\,[\text{mm}] : \begin{pmatrix} \xi = -33.288 \\ \eta = 110.074 \end{pmatrix} \qquad P_2'\,[\text{mm}] : \begin{pmatrix} \xi = 1.628 \\ \eta = 5.182 \end{pmatrix}$$

2.1.4 Central projection and projective transformation of a plane

Without restricting the generality of the statements, one can consider all the object points to lie in a plane ($Z = 0$ in Figure 2.1-8)[9]. Equations (2.1-20) then read as follows:

$$\begin{aligned} X &= \frac{\bar{a}_1\xi + \bar{a}_2\eta + \bar{a}_3}{\bar{c}_1\xi + \bar{c}_2\eta + \bar{c}_3} \\ Y &= \frac{\bar{b}_1\xi + \bar{b}_2\eta + \bar{b}_3}{\bar{c}_1\xi + \bar{c}_2\eta + \bar{c}_3} \end{aligned} \qquad (2.1\text{-}23)$$

[9] One finds the derivation for a sloping plane, which leads to the same result, in Mikhail, E., Bethel, J., McGlone, C.: Modern Photogrammetry. John Wiley & Sons, 2001.

The coefficients $\bar{a}_i$, $\bar{b}_i$ and $\bar{c}_i$ are related to the parameters of Equations (2.1-20) as follows:

$$\begin{aligned}\bar{a}_1 &= X_0 r_{31} - Z_0 r_{11}\\ \bar{a}_2 &= X_0 r_{32} - Z_0 r_{12}\\ &\vdots\end{aligned}$$

Dividing the numerators and denominators of Equations (2.1-23) by $\bar{c}_3$, we obtain the following expressions for the relationship between image ξ,η and object coordinates X, Y:

$$\begin{aligned}X &= \frac{a_1\xi + a_2\eta + a_3}{c_1\xi + c_2\eta + 1}\\ Y &= \frac{b_1\xi + b_2\eta + b_3}{c_1\xi + c_2\eta + 1}\end{aligned} \tag{2.1-24}$$

From these equations it follows that:

- a single picture suffices for the reconstruction of a plane object.
- eight independent parameters define the central projection of a plane object.

The reduction in the number of independent parameters from nine to eight may, at first sight, seem surprising. It arises from the fact that, in the case of a plane object, relationships exist among the original nine elements. In the special case when the photograph and the plane object are parallel, it is easy to see (Figure 2.1-9) that Z_0 and c are no longer independent of each other; we need only to know the ratio Z_0/c.

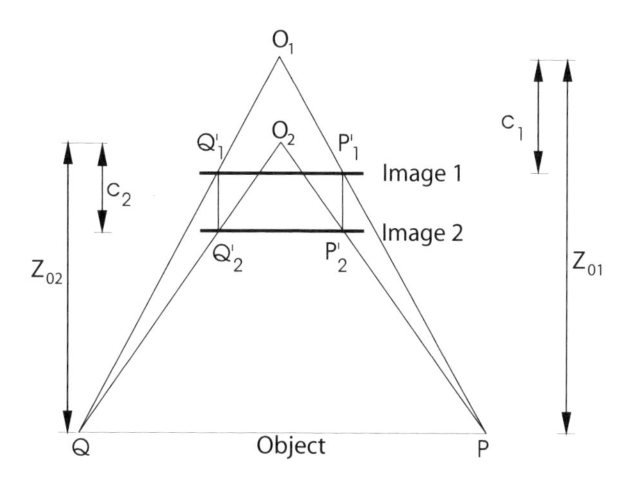

Figure 2.1-9: Two geometrically identical metric images with different values of Z_0 and c, but with the same value of the ratio Z_0/c

We now turn to the question of how the eight parameters can be found in the general case and how other details may be determined from the photographs. Supposing that we have four control points[10] (image and object coordinates known), then the eight coefficients can first be determined from Equations (2.1-24). Then the object coordinates X_i and Y_i of each new point P_i can be determined from its image coordinates ξ_i and η_i.

Numerical Example. We are given both the image coordinates and the object coordinates of four control points A, B, C, D and the image coordinates of P, one of many new points for which the object coordinates are to be found.

	Image coordinates		Object coordinates	
	ξ [mm]	η	X [m]	Y
A	−33.288	110.074	1488.05	3552.12
B	32.183	101.785	2229.38	3507.46
C	−45.762	−74.337	1376.40	1899.76
D	28.472	−96.643	2086.48	1600.12
P	1.628	5.182	?	?

After multiplication by the denominator from Equations (2.1-24) we obtain eight simultaneous linear equations of the form:

$$\begin{aligned} \xi a_1 + \eta a_2 + a_3 \qquad\qquad\qquad\qquad - X\xi c_1 - X\eta c_2 &= X \\ \xi b_1 + \eta b_2 + b_3 - Y\xi c_1 - Y\eta c_2 &= Y \end{aligned}$$

$$\begin{pmatrix} -0.033288 & 0.110074 & 1 & 0 & 0 & 0 & 49.534 & -163.796 \\ 0 & 0 & 0 & -0.033288 & 0.110074 & 1 & 118.243 & -390.996 \\ 0.032183 & 0.101785 & 1 & 0 & 0 & 0 & -71.748 & -226.917 \\ 0 & 0 & 0 & 0.032183 & 0.101785 & 1 & -112.881 & -357.007 \\ -0.045762 & -0.074337 & 1 & 0 & 0 & 0 & 62.987 & 102.317 \\ 0 & 0 & 0 & -0.045762 & -0.074337 & 1 & 86.937 & 141.222 \\ 0.028472 & -0.096643 & 1 & 0 & 0 & 0 & -59.406 & 201.644 \\ 0 & 0 & 0 & 0.028472 & -0.096643 & 1 & -45.559 & 154.640 \end{pmatrix} \begin{pmatrix} a_1 \\ a_2 \\ a_3 \\ b_1 \\ b_2 \\ b_3 \\ c_1 \\ c_2 \end{pmatrix} = \begin{pmatrix} 1488.05 \\ 3552.12 \\ 2229.38 \\ 3507.46 \\ 1376.40 \\ 1899.76 \\ 2086.48 \\ 1600.12 \end{pmatrix}$$

The solution of these equations is:

$$\begin{aligned} a_1 &= 8021.065 & b_1 &= -4066.292 & c_1 &= -1.330 \\ a_2 &= -1084.217 & b_2 &= 7360.815 & c_2 &= -0.728 \\ a_3 &= 1821.069 & b_3 &= 2479.221 \end{aligned}$$

From Equations (2.1-24) the object coordinates of P are:

$$\begin{aligned} X &= \frac{8021.065 \times 0.001628 - 1084.217 \times 0.005182 + 1821.069}{-1.330 \times 0.001623 - 0.728 \times 0.005182 + 1} \\ &= 1839.43\,\mathrm{m} \\ Y &= \frac{-4066.292 \times 0.001628 + 7360.815 \times 0.005182 + 2479.221}{-1.330 \times 0.001628 - 0.728 \times 0.005182 + 1} \\ &= 2525.74\,\mathrm{m} \end{aligned}$$

[10] If the interior orientation is known, three control points suffice (see Section 4.2.1).

In the special case where the image and object planes are parallel (that is to say, $\omega = \varphi = 0$), the three-dimensional rotation matrix takes the form:

$$\mathbf{R} = \begin{pmatrix} \cos\kappa & -\sin\kappa & 0 \\ \sin\kappa & \cos\kappa & 0 \\ 0 & 0 & 1 \end{pmatrix}$$

Inserting $\mathbf{R}$ in Equation (2.1-20) gives:

$$X = X_0 + \frac{Z_0}{c}(\cos\kappa\,(\xi - \xi_0) - \sin\kappa(\eta - \eta_0))$$

$$Y = Y_0 + \frac{Z_0}{c}(\sin\kappa\,(\xi - \xi_0) + \cos\kappa(\eta - \eta_0))$$

Writing this in matrix notation and introducing the quantity $m_B = Z_0/c$ we arrive at:

$$\begin{pmatrix} X \\ Y \end{pmatrix} = \begin{pmatrix} X_0 \\ Y_0 \end{pmatrix} + m_B \begin{pmatrix} \cos\kappa & -\sin\kappa \\ \sin\kappa & \cos\kappa \end{pmatrix} \begin{pmatrix} \xi - \xi_0 \\ \eta - \eta_0 \end{pmatrix} \tag{2.1-25}$$

Taking Section 2.1.1 into account, one reaches the conclusion that in this special case, the photographic image is geometrically equivalent to a map; it is only an object plane reduced in scale (as well as rotated and translated). Equation (2.1-25) is a plane similarity transformation (2.1-9). The image scale, or the map scale[11], is $1 : m_B$ where:

$$m_B = \frac{Z_0}{c} \tag{2.1-26}$$

Equation (2.1-26) can also be derived geometrically (Figure 2.1-10):

If we assume not only that the image plane is parallel to the XY plane of the object coordinate system but also that

$$\kappa = X_0 = Y_0 = \xi_0 = \eta_0 = 0$$

then

$$\frac{s}{S} = \frac{c}{Z_0} = \text{const.} = \frac{1}{m_B}$$

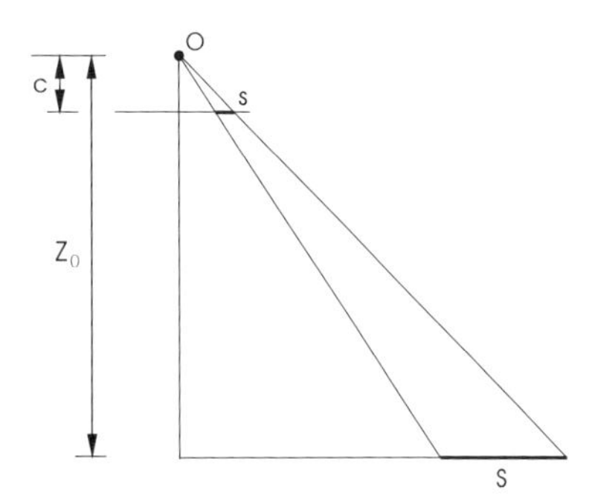

Figure 2.1-10: Image as map

Example. Camera axis perpendicular to the façade ($c = 157.65\,\text{mm}$, $Z_0 = 50.4\,\text{m}$). See Figure 2.1-11.

[11] A scale is defined by the ratio $1 : m_B$. We call m_B the scale number.

Figure 2.1-11: Image scale = map scale

Exercise 2.1-7. What would be the maximum error to be found in this "photographic map" (an orthophoto) if an element of the façade projected 50 cm from the plane defined as $Z_0 = 50.4$ m? (For the solution see Equation (7.2-1) or Exercise 7.2-1.)

Comment: With the expression (2.1-24), which portrays the relationship between XY coordinates in the object plane and $\xi\eta$ coordinates in the image plane, we depart from the idea of a central perspective bundle of rays. In place of the central perspective bundle of rays with its nine elements of interior and exterior orientation, projective geometry provides the mathematical relationship between a plane object and its image. (Employing projective geometry one can write the mathematical relationship between an object (a plane or a straight line) and its image as created by central perspective without making use of the position and orientation of the image in relation to the object.)

Projective geometry is an alternative to central perspective with the following characteristics:

- the mathematical relationship is linear as opposed to the non-linear bundle Equation (2.1-19).
- as a result, no initial values are necessary when, for example, the transformation parameters have to be computed by means of control points.
- the elements of interior orientation are not necessary; this means that projective geometry is very suitable for non-metric images, such as amateur pictures.
- even if the elements of interior orientation are known, as with metric images, one cannot make use of them, except indirectly.

- the affine transformation, the characteristics of which are enumerated in Section 2.1.1, is included as a special case in the projective transformation Equations (2.1-24). (If one puts $c_1 = c_2 = 0$, one obtains the affine transformation with its six parameters. Against that, central perspective (collinearity) equations (2.1-19) or (2.1-20) cannot be applied to images which have been produced by means of an affine process.)

Exercise 2.1-8. Given four control points in the $\xi\eta$ image system $(1(0,0),\ 2(1,0),\ 3(0,1),\ 4(1,1))$ and in the XY object system $(1(0,0),\ 2(2,0),\ 3(0,4),\ 4(2,4))$, as well as the $\xi\eta$ coordinates of a new point $5(0.5, 0.5)$, find the eight parameters of the projective transformation using Equations (2.1-24) and also the object coordinates of the new point. (Solution: $a_1 = 2$, $a_2 = 0$, $a_3 = 0$, $b_1 = 0$, $b_2 = 4$, $b_3 = 0$, $c_1 = 0$, $c_2 = 0$; XY coordinates of the new point, 5, are $(1, 2)$. Comment: The projective transformation (2.1-24) even deals with this typical affine deformation.)

2.1.5 Central projection and projective transformation of the straight line

Without loss of generality, the X axis ($Y = 0$) and the ξ axis ($\eta = 0$) can be adopted as the corresponding straight lines in the object plane and the image plane respectively. From the first of the two Equations (2.1-24) we then obtain the following equation:

$$X = \frac{a_1\xi + a_3}{c_1\xi + 1} \qquad (2.1\text{-}27)$$

The three coefficients a_1, a_3 and c_1 describe the central projection of a straight line. They can be determined from three control points; every other point on the line in the image can subsequently be transformed into the object line.

Equation (2.1-27) evolves from the projective transformation (2.1-24) in which the elements of interior and exterior orientation are unknown. In projective geometry one may prefer to use the cross-ratio, rather than Equation (2.1-27); the cross-ratio for four collinear points is invariant under projective transformation and, it goes without saying, under central projection (see the second solution of the exercise below for a definition of the cross-ratio).

Example (of the reconstruction of a straight line in the object space). We are given a photograph of a street with lane markings in a straight line $APBQC$ (Figure 2.1-12). The lane marking AB is 4.50 m in length and the separation BC is 5.00 m. Points P and Q on the skid marks are to be reconstructed.

1st solution (using Equation (2.1-27))
Image and object coordinates of the three control points A, B and C:

	ξ [mm]	X [m]
A	0.00	0.00
B	38.40	4.50
C	68.30	9.50

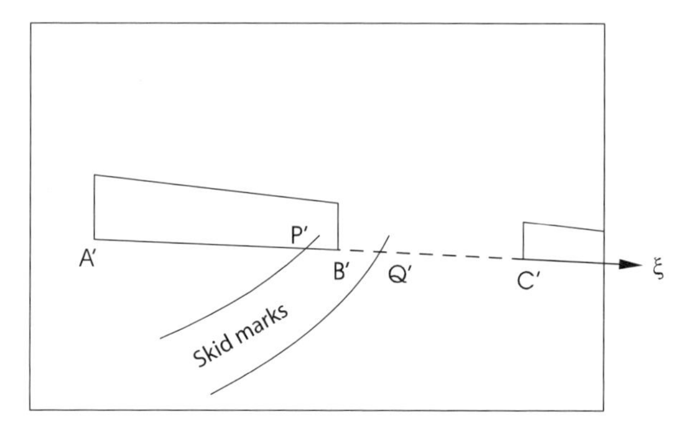

Figure 2.1-12: Straight lines in the image

$$\left.\begin{aligned}
&\text{(2.1-27) for } A\text{:} & 0 &= \frac{a_1 0 + a_3}{c_1 0 + 1} \\
&\text{(2.1-27) for } B\text{:} & 4.50 &= \frac{a_1 38.40 + a_3}{c_1 38.40 + 1} \\
&\text{(2.1-27) for } C\text{:} & 9.50 &= \frac{a_1 68.30 + a_3}{c_1 68.30 + 1}
\end{aligned}\right\} \begin{aligned} a_3 &= 0 \\ a_1 &= 0.09747 \\ c_1 &= -0.00438 \end{aligned}$$

Substitution of the image coordinates ξ_P and ξ_Q in Equation (2.1-27) gives X_P and X_Q:

	ξ [mm]	X [m]
P	33.50	3.83
Q	44.90	5.45

2nd solution (using the cross-ratio)

$$\frac{AB}{PB} : \frac{AC}{PC} = \frac{A'B'}{P'B'} : \frac{A'C'}{P'C'}$$

$$\frac{4.50}{PB} : \frac{9.50}{5.00 + PB} = \frac{38.40}{4.90} : \frac{68.30}{34.80} \quad \Rightarrow \quad PB = 0.67$$

$$\frac{AB}{QB} : \frac{AC}{QC} = \frac{A'B'}{Q'B'} : \frac{A'C'}{Q'C'}$$

$$\frac{4.50}{QB} : \frac{9.50}{5.00 - QB} = \frac{38.40}{6.50} : \frac{68.30}{23.40} \quad \Rightarrow \quad QB = 0.95$$

Exercise 2.1-9. Consider the solution if, instead of point C, the vanishing point F, with coordinate $\xi_F = 221.0\,\text{mm}$ is given; that is, $X_F = \infty$. (Hint: X_F must be infinite in Equation (2.1-27), as is achieved when $221.0c_1 + 1 = 0$.) (Solution: $X_P = 3.82\,\text{m}$, $X_Q = 5.46\,\text{m}$.)

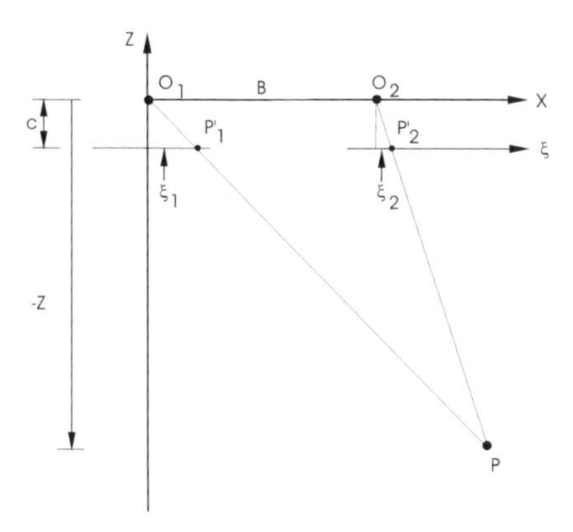

Figure 2.1-13: The "normal case"

Exercise 2.1-10. Show that for four points the four expressions (2.1-27) can be inserted in the cross-ratio formulation. (Hint: Choose the origin of coordinates for both X and ξ in the point A: $a_3 = 0$.)

2.1.6 Processing a stereopair in the "normal case"

Photogrammetry is used above all for the reconstruction of three-dimensional objects from metric images. In Section 2.1.3 we reached the conclusion that two photographs of the same object are necessary for this. The analysis proves to be especially simple if both camera axes are normal to the base (the base is the straight line between the two perspective centres) and parallel to each other (Figure 2.1-13). This is known as the "normal case". This condition is very difficult to achieve in the case of aerial photographs although one seeks to approximate it as closely as possible.

$$\begin{aligned}
X_{01} &= Y_{01} = Y_{02} = Z_{01} = Z_{02} = 0 \\
X_{02} &= B \\
\xi_{01} &= \eta_{01} = \xi_{02} = \eta_{02} = 0 \\
\omega_1 &= \omega_2 = \varphi_1 = \varphi_2 = \kappa_1 = \kappa_2 = 0
\end{aligned}$$

In the "normal case" the three-dimensional rotation matrix $\mathbf{R}$ (2.1-13) becomes the unit matrix for both pictures:

$$\mathbf{R} = \begin{pmatrix} 1 & 0 & 0 \\ 0 & 1 & 0 \\ 0 & 0 & 1 \end{pmatrix}$$

In the "normal case" the relationship between image and object coordinates as formulated in Equations (2.1-20) simplifies as follows:

Image 1

$$X = Z\frac{\xi_1}{-c} \tag{2.1-28}$$

$$Y = Z\frac{\eta_1}{-c} \tag{2.1-29}$$

Image 2

$$X = B + Z\frac{\xi_2}{-c} \tag{2.1-30}$$

$$Y = Z\frac{\eta_2}{-c} \tag{2.1-31}$$

Equations (2.1-29) and (2.1-31) imply that:

$$\eta_1 = \eta_2 \quad \Rightarrow \quad \eta_1 - \eta_2 = p_\eta = 0 \qquad \text{(no } \eta\text{-parallaxes)}$$

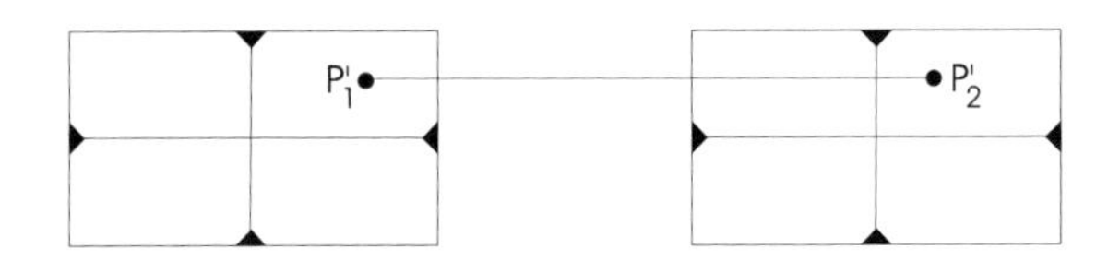

Figure 2.1-14: Two metric images without η-parallaxes

The final formulae for the calculation of the object coordinates X, Y, Z from the image coordinates ξ and η follow from Equations (2.1-28) to (2.1-31). We begin with the Equations (2.1-28) and (2.1-30); that is, $-Z\frac{\xi_1}{c} = B - Z\frac{\xi_2}{c}$:

$$\begin{aligned} -Z &= \frac{cB}{\xi_1 - \xi_2} = \frac{cB}{p_\xi} \\ Y &= -Z\frac{\eta_1}{c} = -Z\frac{\eta_2}{c} \qquad \text{(check)} \\ X &= -Z\frac{\xi_1}{c} \end{aligned} \tag{2.1-32}$$

The difference $\xi_1 - \xi_2 = p_\xi$ (ξ-parallax) can be measured directly in some photogrammetric instruments (Section 6.4.1); in others the original image coordinates ξ_1 and ξ_2 are measured and the difference p_ξ computed. Not only can the Formulae (2.1-32) be derived from Equations (2.1-20) but also, more simply, using ratios directly obvious from Figure 2.1-15.

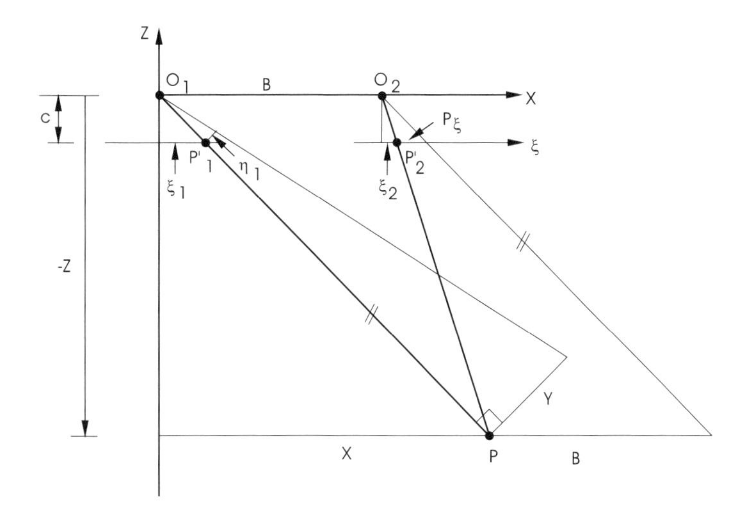

Figure 2.1-15: Geometrical derivation of the relationships (2.1-32)

Exercise 2.1-11. A statue is photographed with a stereometric camera (Section 3.8.2) conforming exactly to the "normal case" with base $B = 1.20$ m and principal distance $c = 64.20$ mm. The values in the table below are measured in the image for points P and Q. Calculate the length of the line-segment $\overline{PQ}$ in object space. (Answer: 2.655 m.)

	ξ_1 [mm]	η_1 [mm]	p_ξ [mm]
P	−3.624	34.202	18.321
Q	29.876	14.809	16.983

2.1.7 Error theory for the "normal case"

Formulae (2.1-32) describe how we arrive at the object coordinates X, Y, and Z from the quantities which can be measured in the picture, ξ_1, η_1 and ξ_2 or, in other cases, p_ξ (the image coordinate η_2 is usually also measured and provides a check). In this section we enquire into the accuracy of these indirectly acquired object coordinates. In so doing we assume that the values for both the principal distance c and the base B are without error.

From the first relationship in the Equations (2.1-32) one obtains the following expression for the mean-square error σ_Z:

$$\sigma_Z = \frac{cB}{p_\xi^2}\sigma_{p_\xi} = \frac{Z}{c}\frac{Z}{B}\sigma_{p_\xi} \tag{2.1-33}$$

The ratio B/Z is known as the base/distance ratio (or in the case of aerial photographs the base/height ratio). The ratio Z/c is called, from Equation (2.1-26), the photo-scale number m_B although only in the special case of parallelism between object and image planes can one speak with justification of a scale number for the whole picture (Section 2.1.4). The mean-square errors σ_x and σ_y of the X and Y coordinates are then derived using the rules of error propagation applied to the corresponding Equations (2.1-32).

$$\begin{aligned} \sigma_Z &= & & m_B \frac{Z}{B} \sigma_{p_\xi} = \frac{Z^2}{cB} \sigma_{p_\xi} \\ \sigma_Y &= \sqrt{\left(\frac{\eta_1}{c}\sigma_Z\right)^2 + \left(\frac{Z}{c}\sigma_\eta\right)^2} &=& \sqrt{\left(\frac{\eta_1}{c} m_B \frac{Z}{B} \sigma_{p_\xi}\right)^2 + (m_B \sigma_\eta)^2} \\ \sigma_X &= & & \sqrt{\left(\frac{\xi_1}{c} m_B \frac{Z}{B} \sigma_{p_\xi}\right)^2 + (m_B \sigma_\xi)^2} \end{aligned} \qquad (2.1\text{-}34)$$

Numerical Example.

Given[12]: Image coordinates $\xi_1 = \eta_1 = 50\,\text{mm} \pm 7\,\mu\text{m}$
Accuracy of measured parallaxes $\sigma_{p_\xi} = \pm 5\,\mu\text{m}$
Principal distance $c = 150\,\text{mm}$

To be found: the root mean square errors of the object coordinates as functions of the photo scale number and the base/distance ratio B/Z. (Take note of the units of measurement shown in the last column.)

	$B/Z = 1:1$		$B/Z = 1:3$		$B/Z = 1:10$		$B/Z = 1:20$		
m_B	$\sigma_{X,Y}$	σ_Z	$\sigma_{X,Y}$	σ_Z	$\sigma_{X,Y}$	σ_Z	$\sigma_{X,Y}$	σ_Z	Units
50000	0.36	0.25	0.43	0.75	0.90	2.50	1.70	5.00	m
10000	0.72	0.50	0.86	1.50	1.81	5.00	3.41	10.00	dm
1000	0.72	0.50	0.86	1.50	1.81	5.00	3.41	10.00	cm
100	0.72	0.50	0.86	1.50	1.81	5.00	3.41	10.00	mm
25	0.18	0.13	0.22	0.38	0.45	1.25	0.85	2.50	mm

Table 2.1-1: Accuracy of photogrammetry as a function of photo scale $1 : m_B$ and of base-distance ratio B/Z

Using this table and the Formulae (2.1-34) we can make the following generalized statements concerning photogrammetric accuracy:

- assuming a constant base/distance ratio, the root mean square errors in all three coordinates are directly proportional to the photo scale number. Any desired accuracy can, therefore, be achieved by means of an appropriate choice of photo scale.

[12] Measurement accuracy will be more closely considered in Sections 4.6 and 6.1.1.

- for a constant photo scale, the root mean square error in the Z coordinate is inversely proportional to the base/distance ratio. The root mean square error in the XY coordinates, however, increases only slowly as the base/distance ratio decreases. If the base ratio is somewhat less than $1:1$, all three object coordinates will be equally accurate.
- with a constant base, the root mean square error in the Z coordinate increases as the square of the distance, Z, from the camera.

For very rough estimation of accuracy:

- for the moment, ignore the first term in the expressions for σ_X and σ_Y from (2.1-34) and
- replace the accuracies σ_ξ, σ_η and σ_{p_ξ} with a generalized accuracy figure for image measurement σ_B.

As a result we derive the following easily remembered rules of thumb:

$$\begin{aligned} \sigma_Z &= m_B \frac{Z}{B} \sigma_B \\ \sigma_X = \sigma_Y &= m_B \sigma_B \end{aligned} \qquad (2.1\text{-}35)$$

Exercise 2.1-12. Using a metric camera and the “normal” disposition of photographs a land-slide is to be monitored from a slope on the opposite side of a valley. An accuracy of $\sigma_X = \sigma_Y = \sigma_Z = \pm 10\,\text{cm}$ is required. What is the greatest distance from which this land-slide can be monitored (measurement accuracy $\sigma_B = \pm 6\,\mu\text{m}$)? What base length should be chosen? Hint: One should use the Formulae (2.1-35). (Answer: $Z = B = 1666\,\text{m}$.) Supplementary exercise: How much can the base be reduced for areas at half the distance? (Answer: $B = 416\,\text{m}$, $B/Z = 1:2$; in this case the root mean square error in the X and Y coordinates will be smaller than that demanded: $\sigma_X = \sigma_Y = \pm 5\,\text{cm}$.)

Exercise 2.1-13. What is the root mean square error in the length of the line-segment $\overline{PQ}$ of Exercise 2.1-11? The values in the above table should be used for σ_ξ, σ_η and σ_{p_ξ}. (Answer: $2.655\,\text{m} \pm 1.0\,\text{mm}$.)

2.2 Preliminary remarks on the digital processing of images

The mathematical methods described in Section 2.1 are also, of course, necessary in digital photogrammetry. In addition one requires some procedures from digital image processing. In this section we cover some aspects of digital image processing which are of interest in photogrammetry.

2.2.1 The digital image

An analogue photograph appears in a light-sensitive coating on a (tough) supporting layer. (In analogue or analytical photogrammetry the supporting layer is usually film.) A digital photograph is recorded by electronic means.[13]

In analogue pictures geometrical shapes such as points, straight lines and so on are interpreted visually and are defined in an abstract way, as is customary in analytical geometry. The left of Figure 2.2-1 shows a straight line and six points as represented in analytical geometry or in an illustration. In Sections 2.1.3 – 2.1.6 everything is dealt with on the basis of such geometry.

In digital images there are no abstract points and lines. Picture elements, usually square, take the place of image points. Instead of a straight line, which in analytical geometry runs from a point, we find a stepwise string of adjacent picture elements.

The usual word for picture element is pixel.

The right hand side of Figure 2.2-1 shows a digital image corresponding to the six points $A-F$ and the straight line G. A digital image can be formed such that all pixels which lie on points and lines receive a value which stands out against the value of the background pixels. The two images in Figure 2.2-1 approach each other more closely the smaller the size of the picture element becomes; that is, the better the geometrical resolution of the digital image. The high degree of abstraction of analytical geometry is, however, never achieved in pixel geometry.[14]

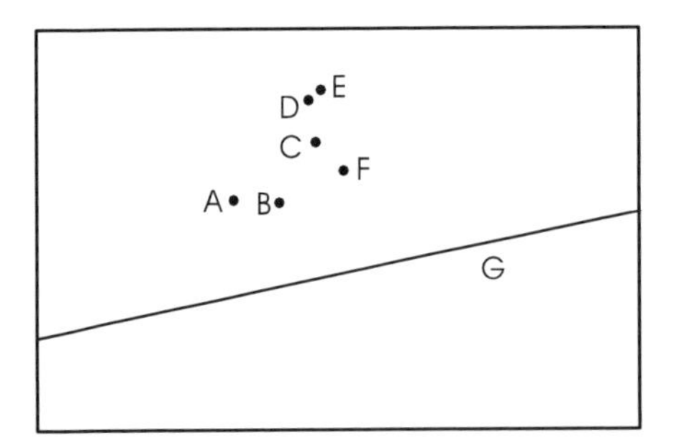

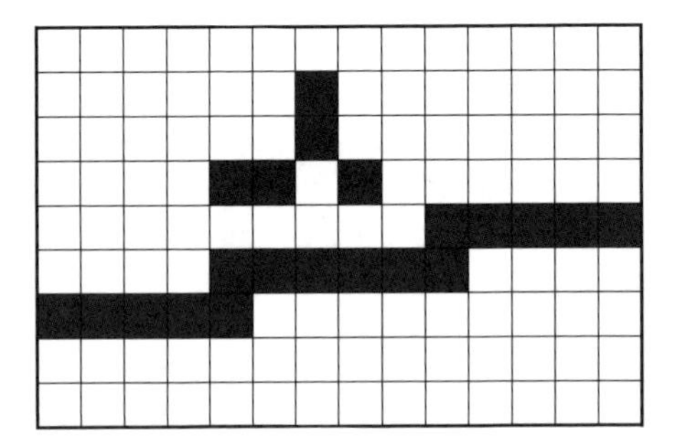

Figure 2.2-1: Analytical geometry (left) and pixel geometry (right)

A few special features of pixel geometry should be touched on with the help of Figure 2.2-1:

a) the two separate points D and E merge together in the digital image.

[13]For that reason digital photogrammetry is also known in English as "soft copy photogrammetry" as opposed to "hard copy photogrammetry" which uses photography recorded on film (PE&RS 58(1), pp. 49–115, 1992).

[14]Limits to the numerical representation of numbers in a computer, e.g. the image coordinates ξ and η, lead to similar (negative) effects in analytical photogrammetry as does the finite pixel size in digital photogrammetry.

b) it is strictly necessary that the pixels belonging to one geometrical shape be specially labeled, for example with what is called a chain code; starting from the position of the initial pixel this defines the next pixel by means of a pointer (right, left, above, below, ...). The criterion "all neighbouring pixels" is in fact inadequate. In pixel geometry, for example, two neighbouring points of the straight line G are joined together just as much as the two separate points A and B (Figure 2.2-1).

c) the formation of geometrical objects with the help only of neighbour relationships, in this context better described as connectivity relationships, leads to different results, depending on the criterion used. 4-neighbourhoods (4N) and 8-neighbourhoods (8N) are common in digital image processing (Figure 2.2-2).

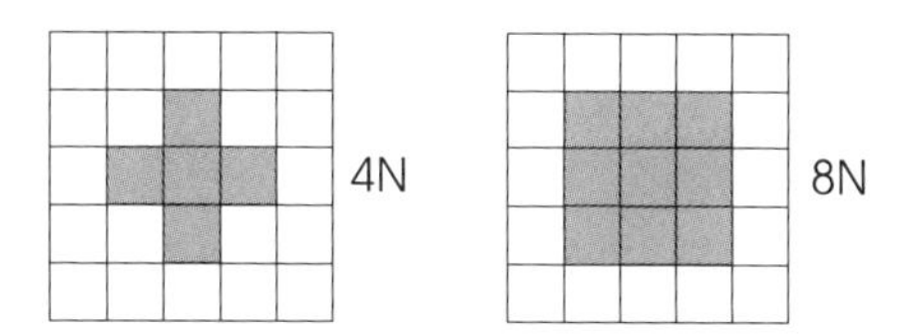

Figure 2.2-2: 4N and 8N relationships

Using a 4N the points B and C remain as separate points; using an 8N all the points $A-F$ are combined. Incidentally, if the pixels representing the points $A-F$ were to be shifted one pixel to the right, then using an 8N connectivity all the relevant pixels of Figure 2.2-1 would be assigned to a single geometrical object.

d) In analytical geometry, separations are usually determined by means of Euclidean distance defined in the direction of the normal to the straight line. In pixel geometry, on the other hand, distance measure is determined using criteria which are indicated in Figure 2.2-2. In Figure 2.2-2 the Euclidean distance of point B from the straight line G is somewhat less than the corresponding distance of point F; that is to say, point F is further removed from the straight line than point B. Application of an 8N, however, results in the pixel representing the point F being closer to the string of pixels representing the straight line G than the pixel representing point B[15].

The picture elements arranged in a raster or a matrix carry information. Their value range depends on the recording equipment and on the computer used. Very commonly the values range between 0 and 255 which clearly exceeds the ability of the human eye which can discriminate between about 50 different shades. Information with 256 different states can be represented by 8 bits (2^8 combinations of bits). A group of eight bits is combined as a byte in most computers. In very sensitive recording equipment the

[15] In this connection one also speaks of a Euclidean metric and of raster data metrics (see, for example, Bill, R.: Grundlagen der Geo-Informationssysteme. Band 2, Wichmann, 1999, and the literature cited therein for further reading).

image data may even be registered with 16 bits, that is with 2 bytes ($2^{16} = 65536$ combinations).

In information technology, besides binary representation, one also comes across octal and hexadecimal forms. The table below sets out pixel values in the different notations opposite each other:

Decimal	Octal	Hexadecimal	Binary bit pattern
0	000	00	00000000
1	001	01	00000001
2	002	02	00000010
.	.	.	.
.	.	.	.
127	177	7F	01111111
128	200	80	10000000
.	.	.	.
.	.	.	.
254	376	FE	11111110
255	377	FF	11111111

Table 2.2-1: Different representations of pixel values

In normal black and white images the pixel values are known as grey values (usually black is coded as 0 and white as 255). In binary images there are only two grey values; the value zero can represent the extraneous background and the value one the significant image information. The right hand side of Figure 2.2-1 shows a binary image. Colour images have three spectral layers which are registered on three image matrices of equal size. These images are usually represented by 24 bits (eight bits for each colour layer with Red, Green and Blue as the primary colours for a so-called RGB reproduction).

Exercise 2.2-1. In Figure 2.2-1 place a line below and parallel to the existing line such that in the corresponding digital picture it does not touch the string of pixels representing the straight line G. Avoidance of contact should be established once in a 4N pattern and once in an 8N pattern.

2.2.2 A digital metric picture

A digital metric image is a digital photograph which meets the requirements of photogrammetry. The basic prerequisites are that the image should have been formed by central projection and that the perspective centre should be fixed in relation to the image. Such a metric image was defined in Section 2.1.3 in which the analogue image was discussed. We now turn to the digital metric image.

The digital metric image is formed in the image plane of a digital (metric) camera. The digital metric image consists of a (two-dimensional) matrix $\mathbf{G}$ with picture elements

g_{ij} (Figure 2.2-3). The row index i runs from 1 in steps of 1 to I, the corresponding column index j from 1 to J. The picture element dimensions are $\Delta\xi \times \Delta\eta$.

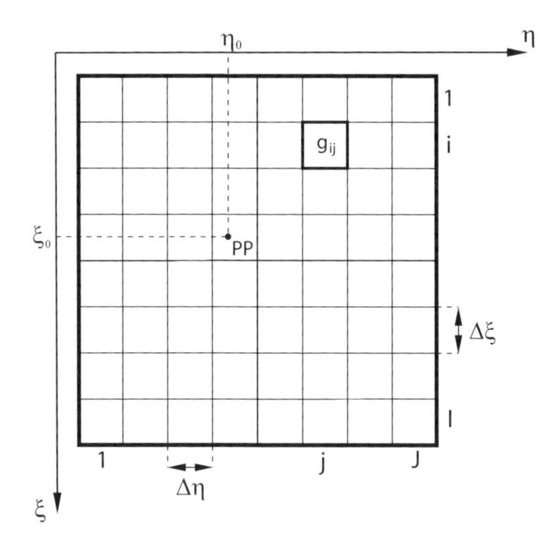

Figure 2.2-3: Digital metric image

In a metric digital image a relationship is required between pixel position and $\xi\eta$ image coordinate values. In Figure 2.2-3 we have introduced an image coordinate system lying half a pixel width outside the image matrix and rotated 100 gon ($\hat{=}$ 90°) clockwise from our accustomed image coordinate system. In a digital metric camera no special fiducial marks are necessary to define the image coordinate system as are required in a metric film camera.

Multiplication of the index i of the matrix by $\Delta\xi$ gives the image coordinate ξ of the mid-point of the pixel g_{ij}; correspondingly the index j multiplied by $\Delta\eta$ gives the image coordinate η. The conventional measurement of image coordinates in analogue photographs is replaced in digital photographs by location of the respective pixel. Location of the pixel by a human operator is completed with a mouse click. If possible, however, pixel identification and measurement should be automatic—even into the subpixel range; that is to say, with accuracy better than the pixel dimensions.

For photogrammetric processing of digital photographs one requires the interior orientation exactly as with analogue photographs. In Figure 2.2-3 the position of the principal point *PP* is given in the $\xi\eta$ image coordinate system. In the case of appropriately small pixels it is adequate just to know the pixel in which the principal point lies. One can continue with this idea by regarding the indices i and j directly as the image coordinates ξ and η. In this case, assuming square pixels, the principal distance c is introduced in units of $\Delta\xi (= \Delta\eta)$. It is easy to see that the collinearity equations (2.1-20) and the Equations (2.1-32) relating to the "normal case" retain their validity despite these unfamiliar units of measurement in both image and camera space.

2.2.3 Digital processing in the "normal case" and digital projective rectification

After finding homologous points in two digital metric images which conform to the "normal case", and using the relationships (2.1-32), the object coordinates X, Y and Z of individual points can be found from the image coordinates, the base (which is part of the exterior orientation) and the interior orientation. (Note: the image coordinates in (2.1-32) are referred to the principal point *PP* as origin.) It should be remembered that, in the processing of a "normal" image pair as outlined, the same mathematical relationships are called upon in the case of digital metric images as in the analogue case. In the processing of digital image pairs the identification of homologous points in the two photographs invites complete automation. Not until Section 6.8.3 do we approach this subject.

In this section we turn to the evaluation of an individual digital photograph of a plane object. We have already learnt, in Section 2.1.4, not only the mathematical relationship between the image plane and the object plane but also the analytical procedure. Four control points are necessary for the determination of eight independent parameters; after the solution of the relevant equations the object coordinates X and Y can be calculated for new points from their measured image coordinates ξ and η (see the numerical example in Section 2.1.4).

With the support of the same mathematics we can also, as we shall shortly see, transfer the whole digital photograph into the object plane. Since projective geometry comes into use (see the comment at the end of Section 2.1.4), in this case we speak of projective rectification. The result is a digital orthophoto, therefore a geometrically correct (undistorted) digital photograph. The conversion of the distorted digital image (the original picture) into the digital orthophotograph, which is in the (two-dimensional) object coordinate system, is done with the help of Equations (2.1-24). This conversion, typical of operations with digital images, will be clarified in what follows.

Because of their arrangement within the image matrix, we know the $\xi\eta$ coordinates of the mid-points of all the pixels of the original distorted image. In Figure 2.2-4 these mid-points are labeled with small black-filled circles. We find the corresponding positions in the object coordinate system using Equations (2.1-24); in Figure 2.2-4 this is signified by the abbreviated notation $(X, Y) = f(\xi, \eta)$. Naturally, on account of the distortion of the original image, the XY positions found for all the original pixels do not fall in an orthogonal raster. The required mid-points of the pixels in the resultant image, the digital orthophoto, with their indices i and j, are shown in Figure 2.2-4 with small open circles. We have a set (a sample) of arbitrarily arranged points which are to be brought into a new pattern, an orthogonal raster; in digital image processing this is known as resampling.

Nearest neighbourhood assignment of grey values

One of the possible techniques for assigning the grey value within such a resampling is the nearest neighbour method. In this technique the grey value of the original pixel is given to that position in the new image matrix which lies nearest, as measured by coordinate differences, to the transformed mid-point of the pixel. In the example of Figure 2.2-4 the grey value $g_{2,4}$ of the original image will be brought to position $(i, j) = (2, 3)$, that of $g_{3,4}$ to position $(i, j) = (3, 2)$, and so on.

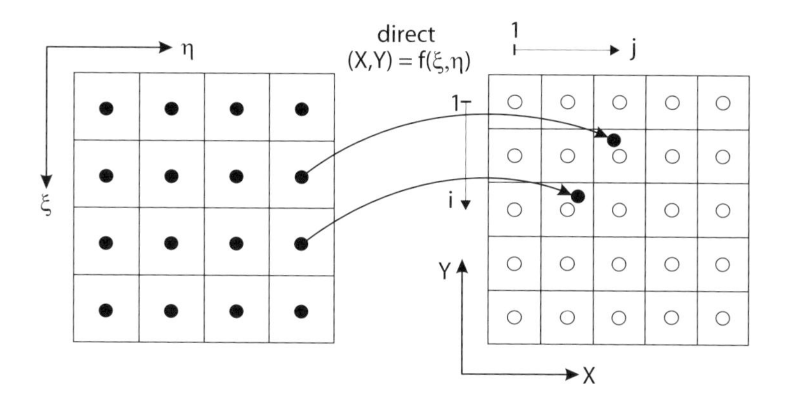

Figure 2.2-4: Direct rearrangement

The rectification outlined in Figure 2.2-4 is called a direct rectification or, in a general form, as a direct rearrangement. It has the disadvantage among other things that it can lead to gaps in the resulting digital image. The much more widely used technique is that of indirect rectification or rearrangement. It is summarized in Figure 2.2-5.

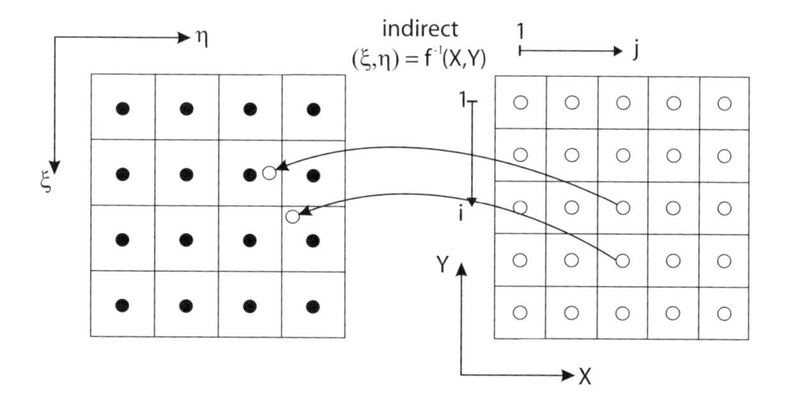

Figure 2.2-5: Indirect rearrangement

It starts with an originally empty image matrix with indices i and j in the object coordinate system. In general the number of pixels in this image matrix will be somewhat larger than that in the original image. Subsequently the mid-points of these pixels are transformed into the original image using the inverse transformation, which we go into more closely at the end of this section. The grey value at that position in the original image can be found using, for example, the nearest neighbour method and can be copied into the new image, the digital orthophotograph. For example, in position $i = 3$, $j = 3$ of the resulting image of Figure 2.2-5, the grey value computed at the inverse point shown within the cell $\xi = 2$, $\eta = 3$ will be inserted; likewise in position $i = 4$, $j = 3$, the grey value computed in $\xi = 3$, $\eta = 4$ will be inserted, and so on.

Bilinear interpolation of grey values

A disadvantage of the nearest neighbour method is that in the most unfavourable case the grey values assigned may be displaced from their correct positions by as much as half a pixel. These displacements may lead to a shift, in the worst case, of up to one pixel of the transformed image; in general it leads to a sawtooth effect and to variable widths of bands of pixels representing lines.

An interesting alternative to nearest neighbour is bilinear interpolation, in which case the desired grey value is found, as a function of $\xi\eta$ coordinates, from the four neighbouring values g_1, g_2, g_3 and g_4 using bilinear interpolation (Figure 2.2-6).

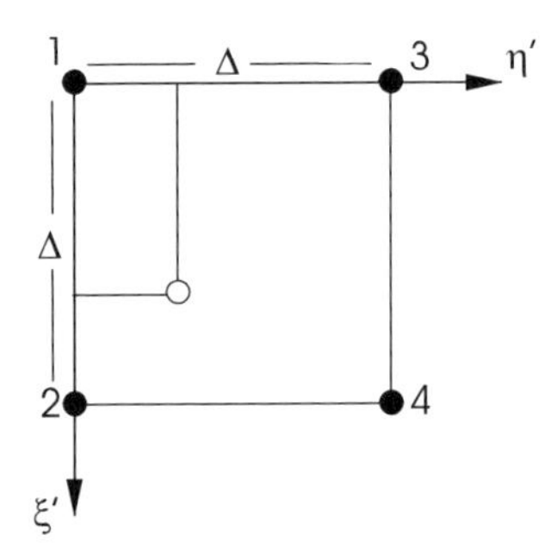

Figure 2.2-6: Bilinear grey value interpolation

The equation for bilinear interpolation is:

$$g = g(\xi, \eta) = a_0 + a_1\xi + a_2\eta + a_3\xi\eta \tag{2.2-1}$$

If one were to consider the grey value as height above the $\xi\eta$ plane, then Equation (2.2-1) would describe a surface of which the height would be linear with respect to ξ for a fixed η and vice versa. Such a surface is known as a hyperbolic paraboloid. We introduce local coordinates $\xi'\eta'$ referred to the upper left-hand corner of the square comprised of the mid-points of the four neighbouring pixels of the original image. The determination of the parameters a_0, a_1, a_2 and a_3 from the four grey values g_1, g_2, g_3 and g_4 then becomes especially simple. If the pixel size is Δ (Figure 2.2-6), then:

$$\begin{pmatrix} g_1 \\ g_2 \\ g_3 \\ g_4 \end{pmatrix} = \begin{pmatrix} 1 & 0 & 0 & 0 \\ 1 & \Delta & 0 & 0 \\ 1 & 0 & \Delta & 0 \\ 1 & \Delta & \Delta & \Delta^2 \end{pmatrix} \begin{pmatrix} a_0 \\ a_1 \\ a_2 \\ a_3 \end{pmatrix} \tag{2.2-2}$$

The simple structure of the coefficient matrix makes it easy to write a general solution:

$$\begin{pmatrix} a_0 \\ a_1 \\ a_2 \\ a_3 \end{pmatrix} = \begin{pmatrix} 1 & 0 & 0 & 0 \\ -1/\Delta & 1/\Delta & 0 & 0 \\ -1/\Delta & 0 & 1/\Delta & 0 \\ 1/\Delta^2 & -1/\Delta^2 & -1/\Delta^2 & 1/\Delta^2 \end{pmatrix} \begin{pmatrix} g_1 \\ g_2 \\ g_3 \\ g_4 \end{pmatrix} \qquad (2.2\text{-}3)$$

Exercise 2.2-2. Check that the general solution given in (2.2-3) is correct. (The product of the matrices of (2.2-2) and (2.2-3) should be the unit matrix.)

Substituting expressions for a_0, a_1, a_2 and a_3 from Equation (2.2-3) in Equation (2.2-1) leads to the desired interpolation formula:

$$g = g(\xi', \eta') = \left(1 - \frac{\xi'}{\Delta} - \frac{\eta'}{\Delta} + \frac{\xi'\eta'}{\Delta^2}\right) g_1 + \left(\frac{\xi'}{\Delta} - \frac{\xi'\eta'}{\Delta^2}\right) g_2 + \\ + \left(\frac{\eta'}{\Delta} - \frac{\xi'\eta'}{\Delta^2}\right) g_3 + \frac{\xi'\eta'}{\Delta^2} g_4 \qquad (2.2\text{-}4)$$

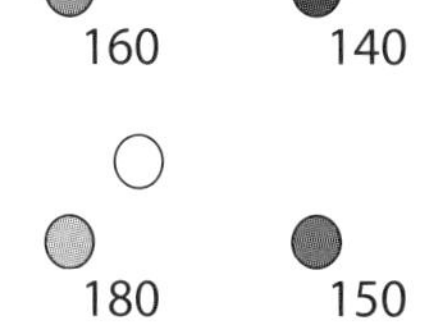

Numerical Example. We choose $\Delta = 1$; ξ' and η' vary between 0 and 1. With respect to an origin at the upper left-hand corner, the coordinates of the point whose grey value we wish to interpolate are $\xi' = 0.75$ and $\eta' = 0.25$. It follows from Equation (2.2-4) that: $g(0.75, 0.25) = 0.1875 \times 160 + 0.5625 \times 180 + 0.0625 \times 140 + 0.1875 \times 150 = 168$

Exercise 2.2-3. Check the result found in this numerical example by means of linear interpolation along lines parallel to the sides of the square and passing through the interpolated point.

Bilinear interpolation involves more computation than nearest neighbour but has the advantage that no shifts appear. The contrast in the original image is, however, somewhat reduced. If this attenuation in contrast is to be avoided one should move to higher order interpolation including, for example, 16 pixels[16].

Inverse transformation equations for indirect projective rectification

The inverse transformation equations $(\xi, \eta) = f^{-1}(X, Y)$, which are necessary for indirect transformation (Figure 2.2-5), are still to be formulated. In the projective transformation, from a mathematical view point, neither image plane nor object plane takes precedence over the other. That is to say, in the Equations (2.1-24) ξ and X as well as η and Y can be interchanged. The eight parameters for the inverse transformation can

[16]See, for example: Luhmann, T., Robson, S., Kyle, S., Harley, I.: Close Range Photogrammetry. Whittles Publishing, 2006.

therefore be found in the same way from eight linear equations, assuming four control points.

It is also possible, however, to formulate the inverse transformation equations for indirect rectification in terms of the eight parameters of the direct projective transformation, $(X, Y) = f(\xi, \eta)$, Equations (2.1-24). After much re-arrangement of Equations (2.1-24) one obtains:

$$\begin{aligned}\xi &= \frac{(b_2 - c_2 b_3)X + (a_3 c_2 - a_2)Y + (a_2 b_3 - a_3 b_2)}{(b_1 c_2 - b_2 c_1)X + (a_2 c_1 - a_1 c_2)Y + (a_1 b_2 - a_2 b_1)} \\ \eta &= \frac{(b_3 c_1 - b_1)X + (a_1 - a_3 c_1)Y + (a_3 b_1 - a_1 b_3)}{(b_1 c_2 - b_2 c_1)X + (a_2 c_1 - a_1 c_2)Y + (a_1 b_2 - a_2 b_1)}\end{aligned} \qquad (2.2\text{-}5)$$

Exercise 2.2-4. Starting with Equations (2.1-24) derive Equations (2.2-5). (A very elegant solution is possible using homogeneous coordinates. (Appendix 2.2-1).

Numerical Example. Let us take up once more the numerical example of Section 2.1.4. Equations (2.2-5) give the following inverse transformation equations, in which both denominator and numerator have been divided by $(a_1 b_2 - a_2 b_1)$:

$$\begin{aligned}\xi &= \frac{0.00016776X - 0.00000441Y - 0.2945585}{0.00023331X + 0.00013323Y - 1} \\ \eta &= \frac{0.00001409X + 0.00019114Y - 0.4995346}{0.00023331X + 0.00013323Y - 1}\end{aligned}$$

If one exchanges the $\xi\eta$ coordinates and the XY coordinates in Equations (2.1-24) and solves the corresponding linear equations as shown in Section 2.1.4, one arrives at identical transformation parameters. With object coordinates $X = 1839.43$ m and $Y = 2525.74$ m for the point P, it emerges that the image coordinates are $\xi = 1.628$ mm and $\eta = 5.182$ mm as already known from Section 2.1.4.

A practical example of digital projective rectification

Figure 2.2-7 is an oblique picture of a façade, which can be regarded as a plane. The photograph was taken with a Kodak DCS 460c digital camera (Section E 3.5, Volume 2). Coordinates were known for four control points in the object plane; their image coordinates were found after identification of the corresponding points in the original image matrix. The transformation parameters for a projective rectification were subsequently evaluated and the whole of the picture was digitally rectified, the result being shown in Figure 2.2-8. The size of the pixels at object scale is 2 cm × 2 cm.

Instead of the projective transformation of an arbitrary quadrangle, it seems at first glance that affine transformation of two triangles might offer an alternative. The two triangles to be transformed in this way are reproduced in Figure 2.2-7. Figure 2.2-9 illustrates the error occurring in the case of the affine transformation. This error can be estimated with the help of Equation (2.1-27) or of the cross-ratio. It is that much larger, the larger the triangle chosen and the larger the tilt of the picture (see Section 7.3.2d).

Figure 2.2-7: Oblique digital photograph

Figure 2.2-8: Rectangular section of the façade projectively rectified

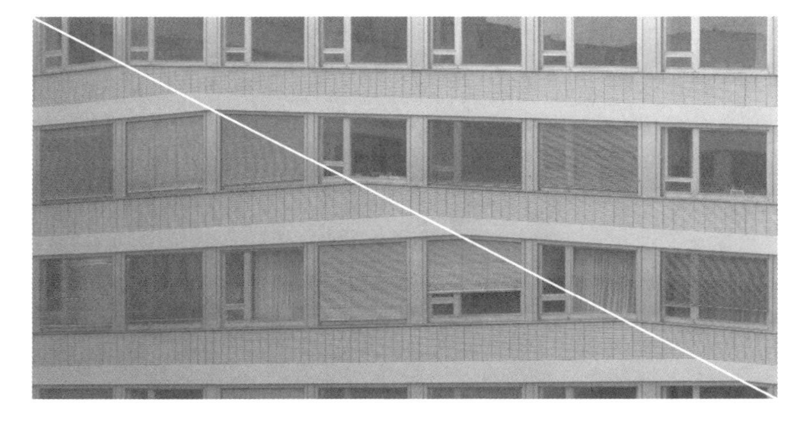

Figure 2.2-9: Affine transformation of two triangles

Exercise 2.2-5. In the above example estimate the error of the affine transformation along the diagonal running from top left to bottom right. Hint: Choose a point in Figures 2.2-7 and 2.2-8 somewhere in the middle of the diagonal. One then has three points on a straight line with known image and object coordinate ξ and X, respectively; using the method of Section 2.1.5 one can state the projective transformation relationship for points on this diagonal. One then transforms the midpoint of this diagonal into the object plane and compares its X coordinate with that of the midpoint in the object plane. The difference can be verified as the error arising as a result of moving from a projective to an affine transformation (Figure 2.2-8 to Figure 2.2-9).

Chapter 3

Photogrammetric recording systems and their application

Metric images are produced by a metric camera which lies at the centre of Section 3.1. If the photons arriving at the image plane of a metric camera are recorded by a chemical sensor, then an analogue metric image is produced. This analogue technology is discussed in Section 3.2. If the photons in the image plane are recorded by an electronic sensor, then a digital metric image is obtained. Digital image recording is handled in Section 3.3. Section 3.4 describes a hybrid technique. It starts with an analogue film in the camera; a subsequent digitization of the film negative or corresponding photograph also results in a digital metric image.

3.1 The basics of metric cameras

A metric photograph has been defined so far as an exact central projection (Sections 2.1.3 and 2.2.2) in which the perspective centre is at a distance c from the principal point of the photograph. The parameters of this simplified mathematical-geometric model, namely the principal distance c and the image coordinates ξ_0 and η_0 of the principal point (*PP*) of the photograph, are defined as the elements of interior orientation. This idealized model does not correspond exactly to reality, however. The inevitable errors of the lens, the camera and the photograph itself must be considered if the highest accuracy is to be achieved.

3.1.1 The interior orientation of a metric camera

The geometric theory of optical systems postulates for a combination of lens elements two principal planes H, H' (the object-space and image-space principal planes) in which the one reproduces the other at a scale of $1 : 1$ perpendicular to the axis. For an optical system consisting of air-glass-air the two optical principal points, i.e. the intersections of the principal planes with the optical axis OA, coincide with the two nodal points N and N'. These are defined in such a way that the central rays, to be discussed in more detail with the help of Figure 3.1-3, pass through the system without deviation and form the same angles τ to the optical axis at N and N' (Figure 3.1-1, $\tau' = \tau$).

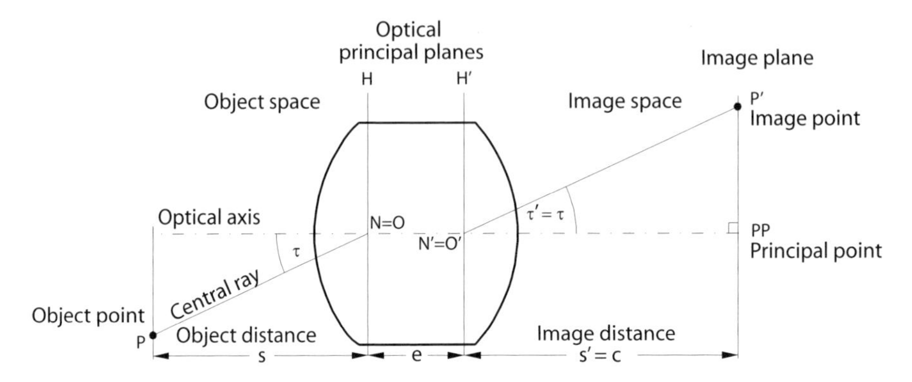

Figure 3.1-1: Idealized geometric image formation of an optical system. N, N' = nodal points = perspective centres O, O'

In this idealized case, N is the object-space perspective centre O, N' is the corresponding image-space perspective centre O' and the principal distance c is equal to the image distance s'. The image distance s' and the object distance s are always measured from the principal planes. In practical amateur photography, however, the distance to be set on the focusing mechanism of the camera is $D = (s + e + s')$, i.e the distance of the object from the image plane.

The optics of photogrammetric cameras are thick, usually asymmetric objectives. The individual lenses are made from different types of glass so as to ensure that imaging errors are corrected to the greatest possible extent. The aperture stop AS is usually not in the centre of the objective (Figure 3.1-2). We must therefore pose the question: where is the physical perspective centre?

All the rays from an object point that pass through the objective must pass through the aperture. The apparent image of the aperture stop, as seen from the object, therefore

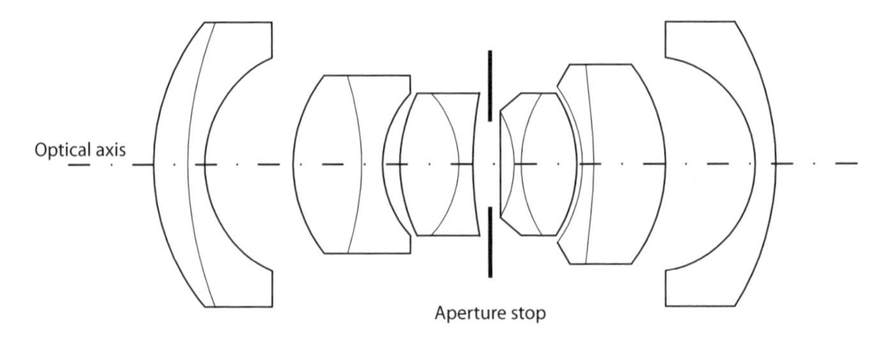

Figure 3.1-2: Cross-section of a typical photogrammetric objective (Wild 21 NAg II, $f/4$)

limits the effective bundle of rays forming the image point; this apparent image is called the entrance pupil (EP). Its centre is the object-space perspective centre O. The analogous exit pupil (EP') lies in the image space of the objective.

Exercise 3.1-1. How can the distance $\overline{VO}$ of the entrance pupil from the vertex V of the objective be determined by theodolite?

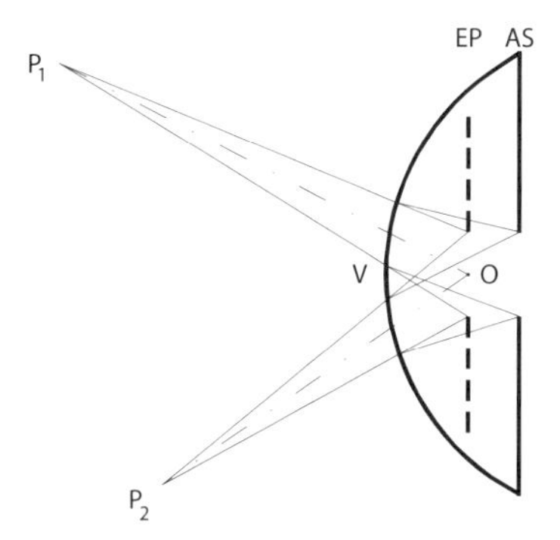

Figure 3.1-3: The definition of the centre of the entrance pupil EP as the object-space perspective centre O. Dot/dashed lines show central rays. V = vertex

The real photogrammetric objective and the idealized model of it described above (Figure 3.1-1) differ significantly:

a) the optical axis should contain the centres of all spherical lens surfaces. After the cementing and assembly of all individual lenses and the mounting of the objective in the camera relative to the mechanical focal-plane frame, small errors will inevitably have accumulated. The reference axis of photogrammetry is therefore not the optical axis OA, but a calibrated, (i.e. standardized) principal ray PR_A which in object space is perpendicular to the image plane and passes through the centre of the entrance pupil (Figure 3.1-4). Its physical extension intersects the image plane in the principal point of autocollimation PP_A (see explanation and definition below).

b) the angles τ are defined at the centre of the entrance pupil and not at the nodal points. Since the entrance pupil usually does not lie in the principal plane H, it follows that τ' is not equal to τ.

c) the mechanically realized principal distance s'_m defined by the focal-plane frame of the camera differs slightly from the optical principal distance s' which provides the sharpest image.

d) the image plane is not rigorously perpendicular to the optical axis.

In essence, the result of these small errors is that the angles τ' in image space are not equal to the angles τ in object space. We therefore define a mathematical perspective

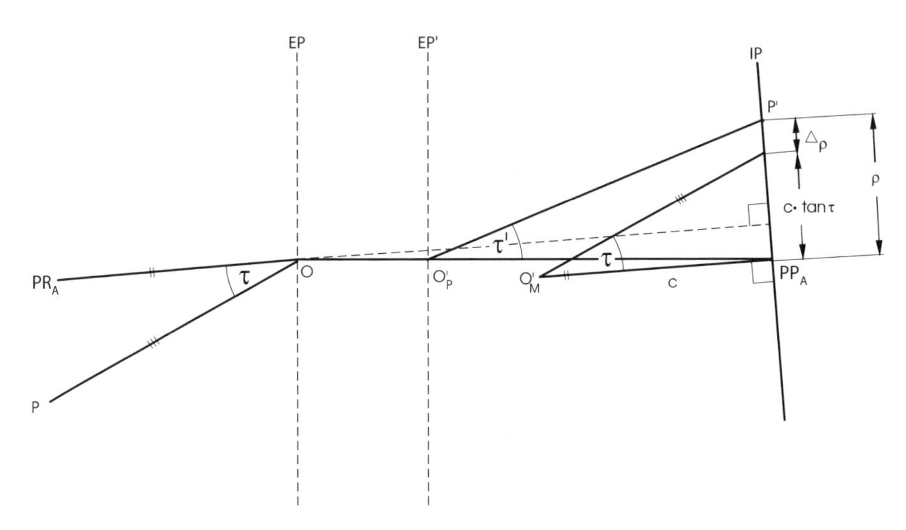

Figure 3.1-4: The definition of the image-space perspective centre O'_M. PR_A = (autocollimation) principal ray, PP_A = principal point of autocollimation in the image plane IP, EP = entrance pupil, EP' = exit pupil, O'_P = physical perspective centre, c = principal distance, ρ = image height = 1, $\Delta\rho$ = (radial) optical distortion

centre O'_M which lies at a perpendicular distance c, the principal distance, from the principal point of autocollimation PP_A and which reproduces the angles τ as closely as possible. Residual errors lead to optical distortions $\Delta\rho$.

The elements of interior orientation, so far defined as ξ_0, η_0 and c, must therefore be extended to include the radial optical distortion $\Delta\rho$:

Equation of interior orientation

$$\rho = c \tan\tau + \Delta\rho \tag{3.1-1}$$

Photogrammetric cameras are mostly calibrated in a laboratory with the help of an optical goniometer (Figure 3.1-5). Firstly, before mounting the camera in the instrument, the observing telescope T_1 is set in its zero position as defined by autocollimation with the telescope T_2. The camera is then mounted with the centre of the entrance pupil EP, i.e. its object-space perspective centre O, in the axis of rotation and rotated about EP until the mirror image of the cross-hairs of the telescope T_2 is superimposed on the cross-hairs (autocollimation). A flat glass plate with a small reflecting surface is placed on the image plane (focal-plane frame) of the camera for this purpose. On the side facing object space this plate also carries precise graduations. In the previously defined zero position of telescope T_1, the operator now observes the principal point of autocollimation PP_A which will be considered as the origin of the ρ-scale. The operator then points T_1 to various graduations on the ρ-scale along each semi-diagonal of the square image frame and observes the corresponding angles τ. The radial optical distortions can then be computed from the differences $\Delta\rho = \rho - c_0 \tan\tau$, where c_0 is the best known value of the principal distance (Figure 3.1-6).

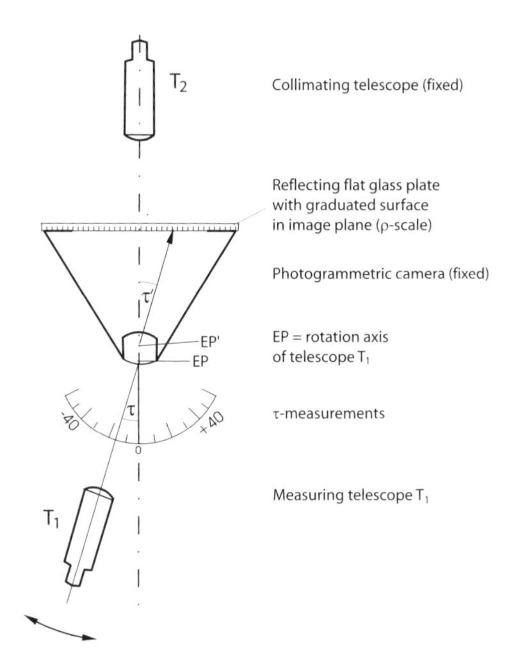

Figure 3.1-5: Schematic diagram of a photo-goniometer

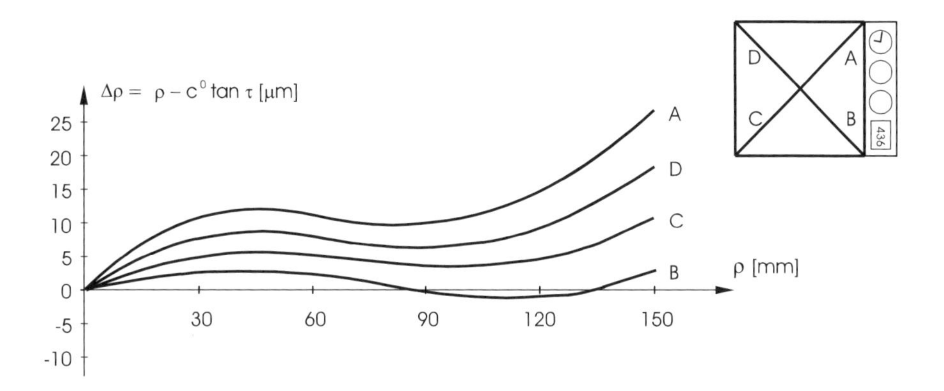

Figure 3.1-6: Radial distortion of the four semi-diagonals $A - D$, referred to PP_A

The results are usually asymmetric, the curves do not coincide. Amongst other reasons, this asymmetry is caused by errors of centering of the individual lens elements. The asymmetry can be greatly reduced by choosing another reference point, slightly different from the principal point of autocollimation PP_A. The reference point resulting from best symmetry is known as the principal point of best symmetry PP_S.

The juxtaposition of two reference points, the principal point of autocollimation PP_A and principal point of best symmetry PP_S, has the following practical consequences:

a) the distortions must be recalculated and given with respect to the PP_S (Figure 3.1-7).

b) when correcting for distortion (Section 3.1.3), the PP_S is used as reference point.

c) the mathematical perspective centre lies a distance c in front of the PP_A; ξ_0 and η_0 in the central projection equations (2.1-19) and (2.1-20) are the coordinates of the PP_A.

d) the rotations incorporated in the matrix elements r_{ik} of the central projection equations (2.1-19) and (2.1-20) are with respect to the principal axis of autocollimation. (Its component in object space, determined by autocollimation (Figure 3.1-5), is perpendicular to the image plane, as required by the central projection (Figure 3.1-4).)

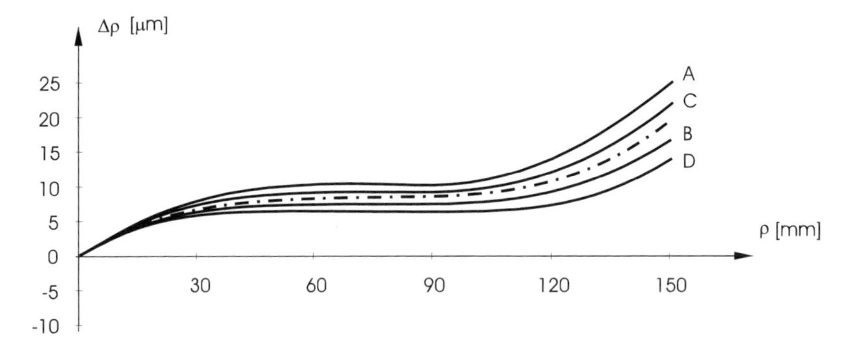

Figure 3.1-7: Radial distortion of the four semi-diagonals of Figure 3.1-6, referred to PP_S, and the resulting mean curve

The change of principal distance Δc is finally computed in such a way as to bring the mean curve as close as possible to the ρ-axis (balanced radial distortion, Figure 3.1-8).

Exercise 3.1-2. How does a change of principal distance affect distortion? (Solution: linearly dependent on ρ.)

The camera manufacturers are obliged to deal with radially symmetric distortion (e.g. Figure 3.1-8). For the most demanding accuracies the radial distortion is separately given for the 4 semi-diagonals, which effectively means details are given about radially

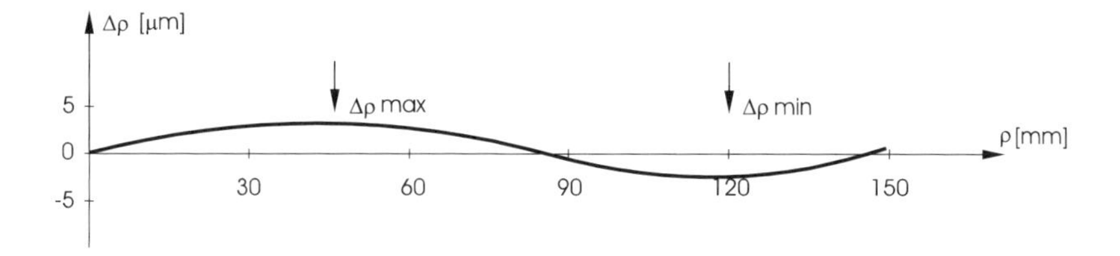

Figure 3.1-8: Mean radial distortion with $|\Delta\rho_{\max}| = |\Delta\rho_{\min}|$

asymmetric distortion. In addition to radial distortion there also exists tangential distortion. This originates principally from the centering errors of individual lenses in the objective. It is always an asymmetric distortion and is generally an order of magnitude less than radial distortion.

Modern photogrammetric objective lenses have a radial distortion within $\pm 5\,\mu$m, and in film-based aerial metric cameras, radial distortion is, in fact, smaller than $\pm 3\,\mu$m[1]. Photogrammetric images taken with old objective lenses have radial distortion up to $30\,\mu$m. Objectives which have not been specially developed for metric cameras can display radial distortions of up to $100\,\mu$m (Figure 3.1-9).

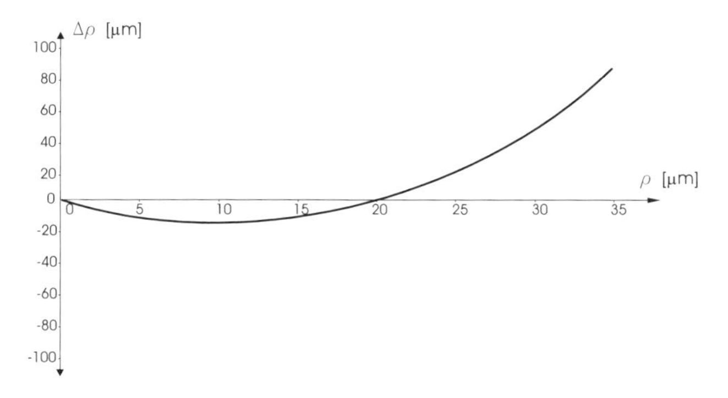

Figure 3.1-9: Distortion curve for Sonnar 4/150 objective from Rollei (focal length= 150 mm, smallest aperture (Section 3.1.4)= 4)

From the point of view of analytical and digital photogrammetry, larger optical distortions, if accurately known, are not a significant disadvantage. However, changes in distortion values are certainly a problem since, in general, such changes cannot be determined. These changes occur mainly in unstable cameras due to changes of focus, vibrations and impacts, etc.

In a metric camera there must exist a coordinate system in which, amongst other parameters, the principal point is given. This photo or image coordinate system is realized differently in analogue and digital cameras.

[1]E.g. Light, D.: PE&RS 58, pp. 185–188, 1992.

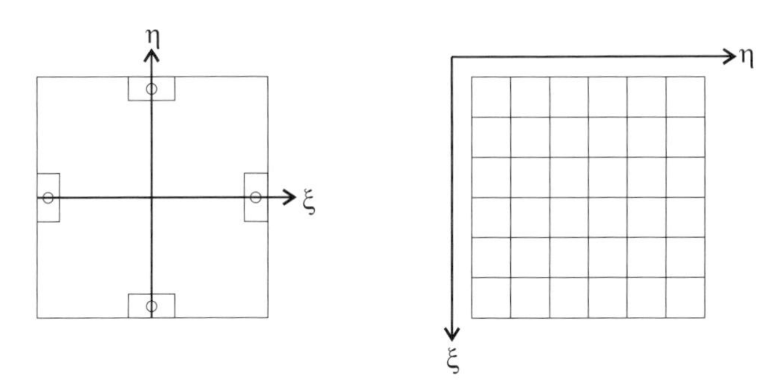

Figure 3.1-10: Image coordinate system in an analogue metric camera (film camera, left) and a digital metric camera (digital camera, right)

In analogue metric cameras which work with film, fiducial marks are located in the image plane and are imaged on every photograph (Figure 3.1-10, left). With the help of these fiducial marks, measurements taken in the image, and normally located in an arbitrary comparator coordinate system, can be related to the camera used to take them (Section 3.2.1). In digital metric cameras, the mathematical relationship between image matrix and camera is never lost: the image points are measured within the image matrix whose coordinate system is defined in the camera (Figures 2.2-3 and 3.1-10, right).

The interior orientation of a metric camera is specified in a calibration certificate by the following data. Note that there are differences between film cameras (F) and digital cameras (D).

- F/D: date of calibration
- F/D: principal distance, c
- F: calibrated coordinates of the fiducial marks stated in a coordinate system which, in principle, can have an arbitrary origin and rotation but in practice corresponds very closely to the ideal configuration (Figure 3.1-10, left)
- F: the principal point of autocollimation PP_A, and the principal point of best symmetry PP_S, are given in this (arbitrary) coordinate system. In most metric cameras, both these principal points and the fiducial centre FC (intersection of connecting lines between fiducial marks, see Figure 3.1-10, left), lie within a circle of radius $< 0.02\,\text{mm}$.
- D: coordinates of principal point of autocollimation PP_A and principal point of best symmetry PP_S are given in the $\xi\eta$ image coordinate system (Figure 3.1-10, right).
- F/D: radial distortion, for example in the form of a piecewise linear function (Figure 3.1-11). In higher quality cameras, the radial distortion is given on 4 semi-diagonals.

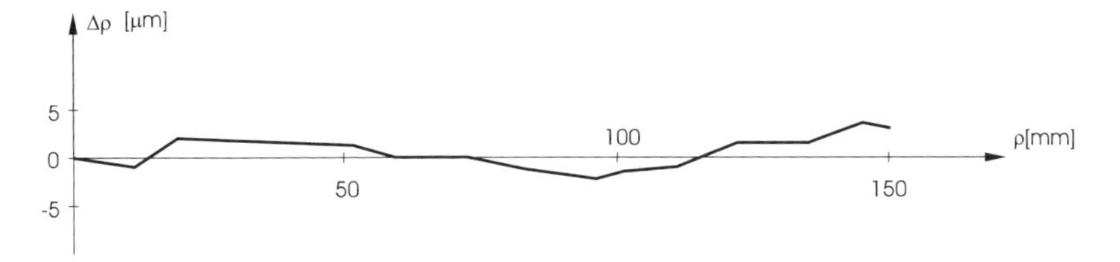

Figure 3.1-11: Radial distortion of a 21 NAgII objective lens manufactured by Wild (now Leica Geosystems)

- F/D: information on image quality (see Section 3.1.5)
- D: exact pixel sizes $\Delta\xi$ and $\Delta\eta$ (Figure 2.2-3)
- D: for a 3-line camera (see Section 3.3.1) at least the ξ coordinates of the three linear arrays
- F/D: type of light used in the laboratory calibration, since the wavelength influences the imaging properties of the optics and therefore also the distortion. (From Figure 3.1-3 it can be seen that the aperture defines the perspective centre O; the physical perspective centre plays a decisive role in the calibration, Figure 3.1-4.)
- F/D: the range used in the calibration, since focusing influences the imaging geometry of the optics and therefore also the distortion

The last three mentioned items of information are of some importance to close-range photogrammetry. Standard photogrammetric analysis methods do not require this information. A short selection of further reading related to the specialized methods in close-range photogrammetry: Fraser, C., Shortis, M.: PE&RS 58, pp. 851–855, 1992. Fryer, J., Brown, D.: PE&RS 52, pp. 51–58, 1986. Luhmann, T., Robson, S., Kyle, S., Harley, I.: Close Range Photogrammetry. Whittles Publishing, 2006.

3.1.2 Calibration of metric cameras

In the previous section we learnt about laboratory calibration with a goniometer. This is used for very high quality cameras in the expectation that the elements of interior orientation remain unchanged over a long period of time.

Calibration can also be carried out by photographing a test field. Such test fields have a relatively large number of control points with known XYZ object coordinates. The $\xi\eta$ photo coordinates of the control points are first measured. With this information, and by means of the equations of central projection (2.1-19), both the elements of exterior orientation and the elements of interior orientation can be calculated in a single analysis. (For details, see Section 5.3.4). The system of equations (2.1-9) can also be extended by polynomial coefficients in order to determine also the unknown optical distortions (Section 3.1.3). When using test field calibration it is assumed that, between making the test field exposures and recording the region of interest or object to be measured, the elements of interior orientation remain unchanged.

In many photogrammetric tasks, a calibration of the camera during the execution of the project is expected. The procedure in this case may be called self-calibration or on-the-job calibration. Mathematically, self-calibration proceeds in the same way as a test field calibration. Self-calibration merges together the elements of calibration and objection reconstruction. The self-calibration method makes use not only of control points with known XYZ coordinates but also of the unknown target points of interest which appear in multiple images. Self-calibration at close-range additionally makes use of conditions of orthogonality and planarity imposed by artifacts and also, or alternatively, an array of plumb lines.

Volume 2 in this series of textbooks has a full Chapter E devoted to calibration. Concepts and practical guidelines for close-range calibration can be found in: Luhmann, T., Robson, S., Kyle, S., Harley, I.: Close Range Photogrammetry. Whittles Publishing, 2006.

Exercise 3.1-3. Without determining distortion, how many control points are required for a test field calibration using a single image? (Solution: 5.) What happens if all control points lie in a plane perpendicular to the camera axis? (Solution: singularity.)

3.1.3 Correction of distortion

A correction for the objective lens distortion, as determined by laboratory calibration with a goniometer, by test field calibration or by self-calibration, must be applied to the image coordinates of the measured target points. The simplest case in which the image coordinates are corrected for radially symmetric distortion in the form of a piecewise linear function is first discussed (Figure 3.1-11).

For each individual image point P', whose image coordinates ξ and η have been determined, the calculation is as follows. It is assumed that the image coordinate system has provisionally been located with its origin at the principal point of best symmetry, PP_S (Figure 3.1-12):

- calculation of radial offset $\rho = \sqrt{\xi^2 + \eta^2}$
- extraction of radial distortion value $\Delta\rho$ at position ρ on the distortion curve of Figure 3.1-11
- reduction of radial distortion correction $\Delta\rho$ into components $\Delta\xi$ and $\Delta\eta$ for image coordinates ξ and η (these relationships can be seen in Figure 3.1-12):

$$\Delta\xi = \frac{\xi}{\rho}\Delta\rho \qquad \Delta\eta = \frac{\eta}{\rho}\Delta\rho \tag{3.1-2}$$

When correcting distortion in a digital metric image, the finite size of the pixels must be taken into account. The correction values $\Delta\xi$ and $\Delta\eta$ (3.1-2) do not, in general, cause

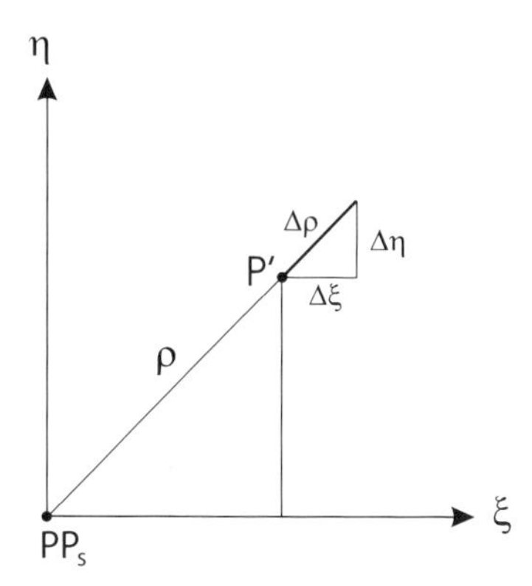

Figure 3.1-12: Correction of radial distortion

an "integer shift" of the original pixel; a resampling of the original image matrix on the orthogonal grid of a new matrix would be required, as explained in Section 2.2.3. Since some potentially significant loss of information is associated wtih every resampling, this should not be done solely for distortion correction. Instead, all image-related corrections (e.g. also refraction, Section 4.5.1) should be accumulated and only at the end of the process should the unavoidable resampling take place (Section 4.5.4).

The test field calibration or self-calibration generates a radial distortion polynomial. The following polynomial has proved itself in practice[2]:

$$\Delta\rho = g_{13}\rho(\rho^2 - \rho_0^2) + g_{14}\rho(\rho^4 - \rho_0^4) \qquad (3.1\text{-}3)$$

g_{13}, g_{14} ... (known) polynomial coefficients (indices have been deliberately chosen to emphasize the link with Equations B (5.2-9) and B (5.2-10) in Volume 2).

ρ ... Radial offset of image point from the principal point of best symmetry PP_S

ρ_0 ... (known) radial offset at which $\Delta\rho$ is zero. In the distortion curve of Figure 3.1-8, $\rho_0 = 85\,\text{mm}$, in the distortion curve of Figure 3.1-9, $\rho_0 = 20\,\text{mm}$.

Correction of radial distortion defined by a polynomial curve is done in the same way as distortion defined by a piecewise linear function. In the former case, $\Delta\rho$ is determined from Equation (3.1-3) at position ρ, in the latter case from the piecewise linear function in Figure 3.1-11. The process then continues with Equation (3.1-2).

[2]E.g. Fryer, J.: In Atkinson (ed.): Close Range Photogrammetry and Machine Vision. Whittles Publishing, pp. 156–179, 1996.

Individual polynomials are applied in cases of radially asymmetric distortion components or a possible tangential distortion (e.g. Equations B (5.2-11) and B (5.2-12) in Volume 2)[3].

Exercise 3.1-4. A point with image coordinates $\xi = -8.342$ mm and $\eta = 23.593$ mm, referred to the principal point of best symmetry PP_S, is to be corrected for distortion as defined in Figure 3.1-11. (A graphical estimation of $\Delta\rho$ from Figure 3.1-11 is sufficient.) (Solution: $\xi = -8.343$ mm, $\eta = 23.595$ mm.)

3.1.4 Depth of field and circle of confusion

This section is devoted to the question of the extent to which an objective lens can sharply image an object which is extended in depth, i.e. in the direction of imaging. The familiar basic lens equation of optics for sharp imagery is:

$$\frac{1}{s} + \frac{1}{s'} = \frac{1}{f} \tag{3.1-4}$$

f ... (constant) focal length of the optical system
s ... (variable) object distance
s' ... image distance, which only corresponds to principal distance c in idealized case (Figure 3.1-1). In reality, it agrees only approximately with the principal distance which is purely geometrically defined (Figure 3.1-4)

In the alternative Newtonian form, with $s = (x+f)$ and $s' = (x'+f)$, the lens equation is:

$$xx' = f^2 \tag{3.1-5}$$

Thus, for every object distance s there is a defined optical image distance s', at which, apart from the effects of lens errors and diffraction effects, all rays emanating from an object point at distance s and passing through the aperture of diameter d meet to form a theoretical image point (Figure 3.1-13). This focused optical image plane IP therefore contains the sharpest possible image of the corresponding focused object plane OP.

Objects in front of and behind the preset object distance will therefore be imaged with some degree of unsharpness, since the rays meet in a point behind or in front of IP. In IP there exists therefore a "circle of confusion" of diameter u. We seek now those object distances s_n (near limit of depth of field) and s_f (far limit) which produce the same diameter u of a circle of confusion, so that we can say that all objects between s_f and s_n, the depth of field, will be imaged in IP with a circle of confusion less than or equal to u. The dimension u is the measure of the sharpness of the image (circle of confusion), and will not be exceeded in the range of the depth of field. We try to achieve a diameter of the circle of confusion of less than 20 μm in a film camera. The pixel size of CCD sensors may vary between 10 and 2 μm. Therefore, the diameter of the circle

[3] An example of distortion correction using piecewise linear functions along the 4 semi-diagonals can be found in: Kraus, K., Stark, E.: BuL 41, pp. 50–56, 1973.

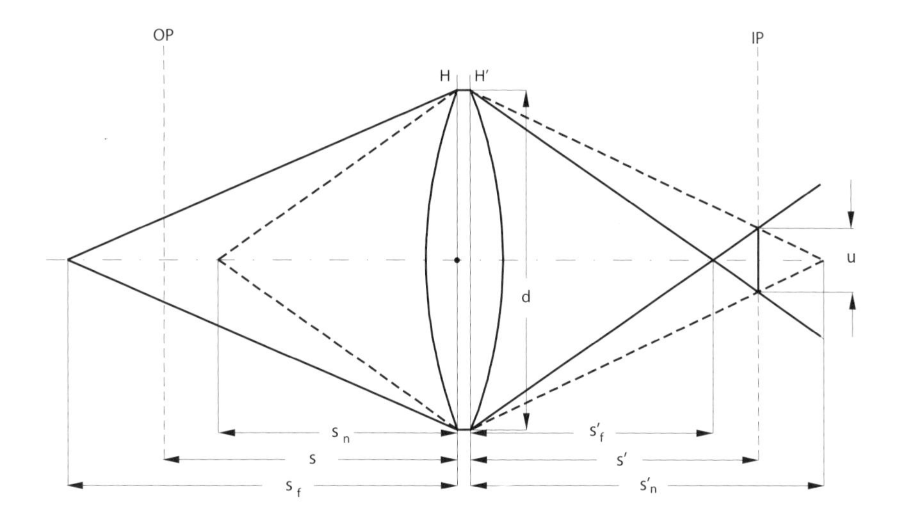

Figure 3.1-13: Derivation of depth of field and circle of confusion

of confusion should vary between 2 and 10 pixel, respectively. From Equation (3.1-4) we have:

$$s' = \frac{sf}{s-f} \qquad s'_n = \frac{s_n f}{s_n - f} \qquad s'_f = \frac{s_f f}{s_f - f} \tag{3.1-6}$$

We see from Figure 3.1-13 that:

$$\frac{d}{s'_n} = \frac{u}{s'_n - s'} \qquad \frac{d}{s'_f} = \frac{u}{s' - s'_f} \tag{3.1-7}$$

Introduce Equation (3.1-6) in (3.1-7) and rewrite to obtain:

$$s_n = \frac{dsf}{df + u(s-f)} \qquad s_f = \frac{dsf}{df - u(s-f)} \tag{3.1-8}$$

We normally multiply numerator and denominator by f/d, so that the f-number or aperture stop $k = f/d$ (= focal length / aperture diameter) appears in one position, only, with a clear effect:

$$s_n = \frac{sf^2}{f^2 + \frac{f}{d}u(s-f)} \qquad s_f = \frac{sf^2}{f^2 - \frac{f}{d}u(s-f)} \tag{3.1-9}$$

If s_n and s_f are given for a particular standpoint, we can find s from Equation (3.1-9), as follows:

$$s = \frac{2 s_n s_f}{s_n + s_f} \tag{3.1-10}$$

So we have the object distance to be set for which u is as small as possible.

To derive the (finite) distance s for which $s_f = \infty$, so that the depth of field extends from s_n to infinity, we must set the denominator of Equation (3.1-9 b) to zero:

$$s_\infty = f\left(\frac{d}{u}+1\right) \approx f\frac{d}{u} = \frac{f^2}{\left(\frac{f}{d}\right)u} \tag{3.1-11}$$

An important practical question is the diameter of the circle of confusion u_i when the object distance is s_i and the preset focusing distance is s. We define $u_i < 0$ for $s_i > s$ and solve Equation (3.1-9) for u_i:

$$u_i = \frac{\left(\frac{s}{s_i}-1\right)f^2}{(s-f)\frac{f}{d}} \tag{3.1-12}$$

Numerical Example.

a) Given: $s_n = 16.8\,\text{m}$, $s_f = 36.8\,\text{m}$, $f = 100\,\text{mm}$, $f/d = 8$. Required: s, u_n, u_f. Equation (3.1-10) gives $s = 23.2\,\text{m}$. Equation (3.1-12) gives $u_n = 0.020\,\text{mm}$ and u_f must equal u_n.

b) Given: A camera of focal length 64 mm with a fixed focusing distance of 25 m (Wild P32, Section 3.8.3). It is to be used to photograph objects at distances from $s_1 = 8\,\text{m}$ to $s_2 = 12.5\,\text{m}$.

Required: u_1, u_2 and u_∞ for the aperture stops $f/8$, 11, 16, 22.

With the assumption that $c \approx s' \approx f$, which can be used for larger object distances as indicated by Equation (3.1-4), Equation (3.1-12) leads to the results in Table 3.1-1.

f/d	$s_1 = 8\,\text{m}$	$s_2 = 12.5\,\text{m}$	$s_3 = \infty$
8	44	21	-21
11	32	15	-15
16	22	11	-11
22	16	8	-8

Table 3.1-1: Diameters of circle of confusion for the Wild P32 camera for $s = 8$, 12.5 and ∞ (see Section 3.8.3). Tabulated values of u in μm

Exercise 3.1-5. What is the significance of the negative sign in the last column of Table 3.1-1? (Solution: The object distance, in this case $s_3 = \infty$, is larger than the object distance covered by the focusing mechanism, in this case $s = 25\,\text{m}$.)

Finally, we need a relation which will yield the setting for that object distance, known as the hyperfocal distance, which will give a depth of field from s_n to ∞. In Equation (3.1-10) divide denominator and numerator by s_f; then, as s_f tends to infinity:

$$s_\infty = 2s_n \qquad s_n = \frac{s_\infty}{2} \tag{3.1-13}$$

The corresponding circle of confusion u_∞ is derived from Equation (3.1-11):

$$u_\infty = \frac{f^2}{\left(\frac{f}{d}\right) s_\infty} \tag{3.1-14}$$

Exercise 3.1-6. A metric camera has a focal length of 64 mm and a fixed focus of 10 m. Compute the depth of field (s_n, s_f) for a maximum permissible circle of confusion of 0.05 mm diameter for the aperture stops $f/4$, 5.6, 8, 11, 16, 22. (Solution: For example, for $k = 8$, $s_n = 5.1$ m, $s_f = 336$ m.)

Exercise 3.1-7. An aerial camera with $c = 305$ mm is required to give equally sharp photographs for the photo scales of 1 : 1500 and 1 : 50000. Compute the best focusing distance s and the maximum circle of confusion to be expected for this range for the aperture stops $f/4$, 5.6, 8, 11. (Solution: $s = 900$ m, $u = 12\,\mu$m, e.g. for $f/8$.)

The necessity to consider depth of field very closely is an important constraint in close-range photogrammetry which does not occur in "long range photogrammetry" where fixed focus cameras are used (Section 3.7.1 ff.). The following possibilities exist in close-range photogrammetry to accommodate the optical requirement for sharp imaging with the photogrammetric requirement for interior orientation which, to achieve optimal accuracy, does not, in principle, change:

a) fixed-focus cameras: s and c are fixed, the camera is used only for objects within its depth of field (Example: Wild and Zeiss stereocameras, Section 3.8.2).

b) calibrated focusing rings: An additive constant to the principal distance is specified for each ring. The rings fit so precisely that the other elements of interior orientation are not significantly changed (Example: Wild P31, Figure 3.1-14, Section 3.8.3).

c) cameras with variable principal distance: Certain principal distances are calibrated and the corresponding settings indicated by index marks on the focusing knob of the camera (Example: Zeiss UMK, Figure 3.1-15, Section 3.8.3 and the widely used semi-metric cameras, Section 3.8.4).

In practice, a circle of confusion of $50\,\mu$m is often adopted. Depending on the pixel size, the corresponding dimension in pixel units can be derived. More stringent demands are often impossible to meet in close-range photogrammetry.

Exercise 3.1-8. The Wild P31 precision terrestrial camera (see Table 3.8-2) has, with the focusing ring for $s = 25$ m, a principal distance of 99.13 mm. What is the focal length of the lens? (Solution: 98.74 mm.) There is another focusing ring which is 3.05 mm thicker. At what object distance will this ring give the sharpest images? (Solution: 2.93 m.) What is then the depth of field ($s_f - s_n$) at $f/8$ and with $u = 0.05$ mm? (Solution: 0.69 mm.) What is the constant to be added to 99.13 mm to give the principal distance for sharp images of objects at a distance of 10 m? (Solution: +0.59 mm.)

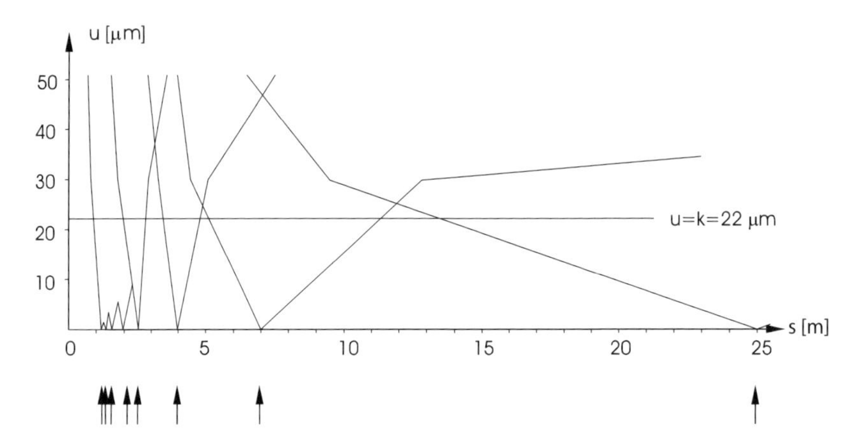

Figure 3.1-14: Circle of confusion of the Wild P31 precision metric camera (c = 100 mm) for the various focusing rings at $s = \uparrow$ and for an aperture stop $f/22$. (The line $u = k$ is explained in Section 3.1.5.1 on diffraction blurring.)

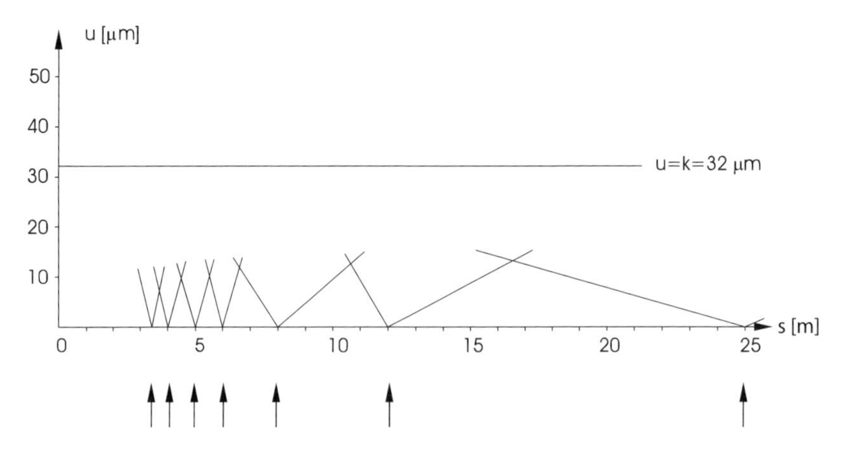

Figure 3.1-15: Circle of confusion of the focusable Zeiss UMK 10/1318 universal metric camera (c = 100 mm) for the calibrated distances $s = \uparrow$ and for an aperture stop $f/32$. (The line $u = k$ is explained in Section 3.1.5.1 on diffraction blurring.)

3.1.5 Resolving power and contrast transfer

The wave nature of light and the consequent diffraction at a circular aperture make it impossible to generate an ideal, i.e. dimensionless, image point. Instead, the wave energy forming a "point" image is distributed in a central diffraction disc containing about 84% of the total energy, together with interference rings around the disc containing $7\%, 3\%, \ldots$ (see Figure 3.1-16). In general, the energy in the interference rings is insufficient to have an effect on the emulsion or sensor located in the image plane of the camera; the sensor therefore records only the central diffraction disc. Also, at the edge of the central disc the energy is too low to produce an image. Experience shows that of the theoretical total diameter u_{th} of the central diffraction disc[4]

$$u_{th}\,[\mu\text{m}] = 2.44k\lambda \quad \text{where} \quad \begin{aligned} k &= f/d \text{ aperture stop} \\ \lambda &= \text{ wavelength in } [\mu\text{m}] \end{aligned} \tag{3.1-15}$$

only a diameter of about

$$u = 0.75u_{th} \tag{3.1-16}$$

is visible on a photographic emulsion used as sensor (see Figure 3.1-16).

3.1.5.1 Diffraction blurring

A very practical rule of thumb can be derived for the diffraction blurring u: if we set Equation (3.1-15) in (3.1-16), we have for the average wavelength of visible light $\lambda = 0.55\,\mu$m

$$u\,[\mu\text{m}] \approx k = \frac{f}{d} \tag{3.1-17}$$

Diffraction theory shows, therefore, that no photographically imaged point can be smaller than the aperture stop number, expressed in $[\mu\text{m}]$.

Numerical Example (using a photographic emulsion as sensor).

$$\begin{array}{lll} k = 8 & \text{Equation (3.1-15): } u_{th} = 2.44 \times 8 \times 0.55 & = 11\,\mu\text{m} \\ & \text{Equation (3.1-16): } u\text{(photograph)} = 0.75 \times 11 & = 8\,\mu\text{m} \\ k = 22 & \text{Equation (3.1-16): } u\text{(photograph)} & = 22\,\mu\text{m} \end{array}$$

The minimum distance $\delta_{\min}$ between two diffraction discs which can just be distinguished as two points depends on the diffraction blurring u and the minimum energy difference which can be distinguished in the image from the corresponding sensor (Figure 3.1-16).

$$\delta_{\min} \approx \frac{u}{2} \tag{3.1-18}$$

This minimum distance due to diffraction cannot quite be achieved in practice.

[4]Serway/Beichner: Physics for Scientists and Engineers. Saunders College Publishing, 5th ed., p. 1222, 2000.

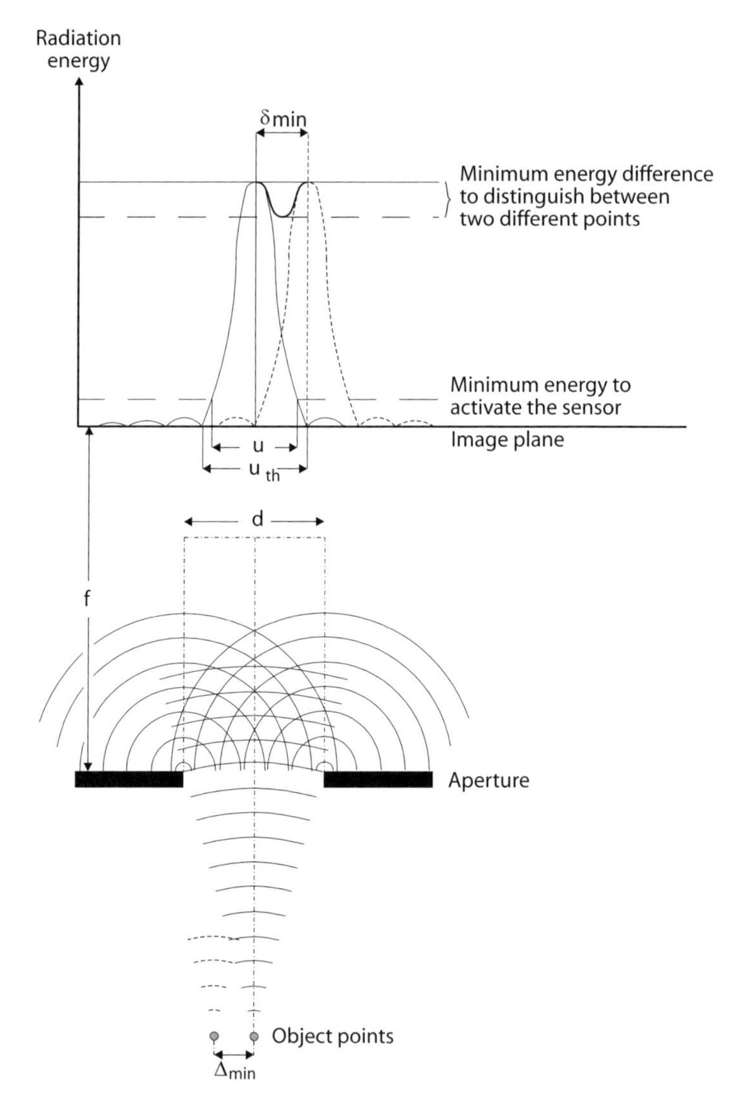

Figure 3.1-16: Distribution of light energy for two object points caused by diffraction at the aperture

3.1.5.2 Optical resolving power

The imaging quality of an optical system such as a metric camera is generally defined by its resolving power. The optical resolving power, or optical resolution, states how many dark lines per millimetre can just be distinguished in the image from the equally wide gaps between them. Figure 3.1-17 shows a test card used to determine resolving power which is expressed in lines per millimetre ([L/mm]) or line pairs per millimetre

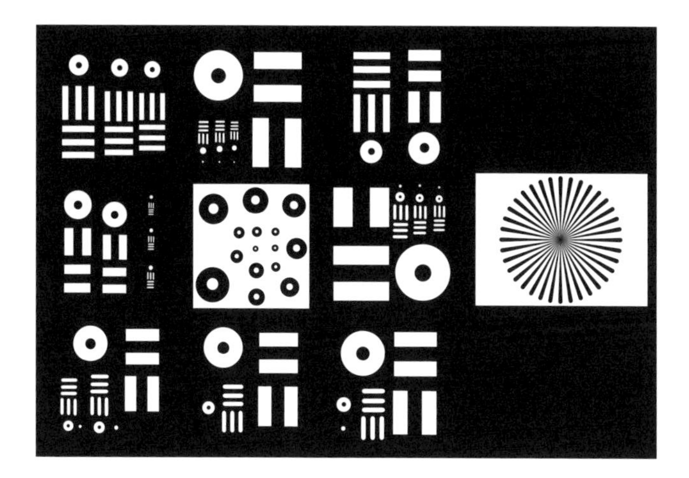

Figure 3.1-17: Resolution test chart (Leica Geosystems)

([Lp/mm]); both expressions are equivalent.[5]

Based on the presentation above, it is possible to state the influence of diffraction on the resolving power. The diffraction-limited point separation of $\delta_{\text{min}}\,[\mu\text{m}] = k/2$ (Equations (3.1-17) and (3.1-18)) results in a resolution limit for an optical system with photochemical image recording, i.e. film-based photography, of:

$$R_{\text{max}}\,[\text{Lp/mm}] = \frac{10^3\,[\mu\text{m/mm}]}{\delta_{\text{min}}\,[\mu\text{m}]} = \frac{2000}{k} \qquad (3.1\text{-}19)$$

Numerical Example.

$k = 22$ (small aperture) (3.1-19): $R_{\text{max}} = 91\,\text{Lp/mm} \mathrel{\hat{=}} 2300\,\text{dpi}$
Width of line pair $= 1/91 = 0.011\,\text{mm} = 11\,\mu\text{m}$
Width of dark line and gap $= 11/2 = 5.5\,\mu\text{m}$
(Ignoring blooming, which can amount to a few micrometres (Section B 2.2, Volume 2))

$k = 8$ (medium aperture) (3.1-19): $R_{\text{max}} = 250\,\text{Lp/mm} \mathrel{\hat{=}} 6350\,\text{dpi}$
Width of line pair $= 1/250 = 0.004\,\text{mm} = 4\,\mu\text{m}$
Width of dark line and gap $= 4/2 = 2\,\mu\text{m}$
(Ignoring blooming, which can amount to a few micrometres (Section B 2.2, Volume 2))

Diffraction is only one factor which influences the resolution of an optical system. The blurring resulting from spherical and chromatic aberration also impairs the resolution.

[5] Particularly in the printing industry the resolution (R) is expressed not in [L/mm] or [Lp/mm] but in [dpi] (dots per inch). Since 1 inch is equivalent to 25.4 mm, the following relationship is obtained: $1 : 25.4 = R\,[\text{L/mm}] : R\,[\text{dpi}]$.

Aberration and other lens errors can be reduced by using a small aperture, i.e. by eliminating the edge rays, though this procedure increases diffraction blurring.

The optimum resolution is achieved with that aperture at which the sum of the optical and diffraction blurring is a minimum (Figure 3.1-18). We call this the "critical aperture", since any smaller aperture will increase the diffraction blurring.

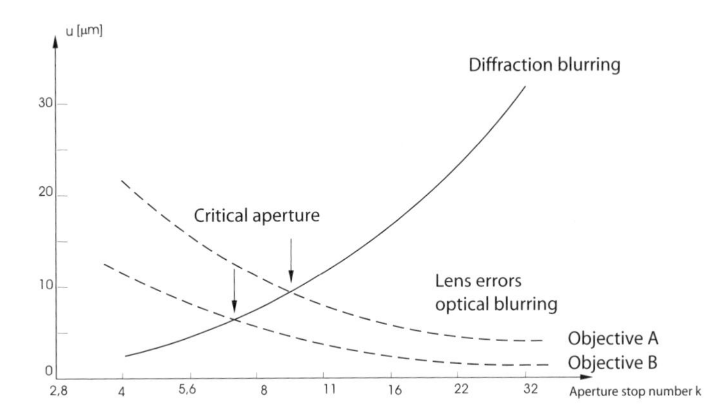

Figure 3.1-18: The definition of critical aperture for two different, good objectives

The optical resolving power depends on the contrast of the line pattern in object space, i.e. from the difference in brightness between the lines and the gaps between them. When object contrast is high, the resolving power can be twice as good as when contrast is low. (Figure 3.1-21 goes into this in more detail.)

The optical resolving power decreases from the centre of the image towards the edge. Reasons for this are the light fall-off (Section 3.1.6) which reduces contrast towards the edge of the image field and the increasing blur from centre to edge as a result of spherical and chromatic aberration.

A practical example will help to illustrate the variation in optical resolving power within the image plane. This example will also demonstrate the derivation of a representative optical resolution using a weighted averaging technique which gives the area weighted average resolution, AWAR. A Wild P31 terrestrial metric camera (Section 3.8.3) was tested with the chart shown in Figure 3.1-17, the optical resolving power determined in a number of annular zones (Figure 3.1-19, left) and the results finally averaged (Figure 3.1-19, right).

Film-based aerial metric cameras currently have an optical resolution of 90 Lp/mm[6]. The resolution is, in fact, determined by exposure of a glass plate supporting a photographic emulsion with a much higher resolution (e.g. 400 Lp/mm).

For users of photogrammetry in particular, the quality in the image plane is not of primary importance. They are instead more interested in the transfer from the image

[6]E.g. Light, D.: PE&RS 58, pp. 185–188, 1992.

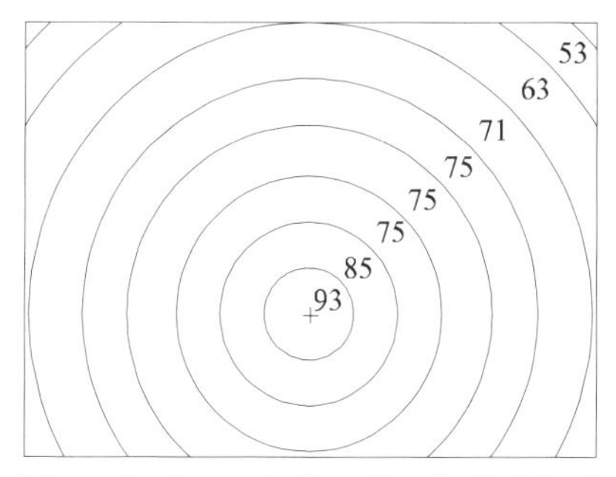

(a) optical resolution for individual annular zones

Area A [mm^2]	Average R [L/mm]
230	93
730	85
1280	75
1660	75
1990	75
2420	71
1640	63
580	53
$\Sigma A = 10530$	$\Sigma(AR) = 759070$
AWAR = $(\Sigma(AR)/(\Sigma A) = 72$L/mm	

(b) calculation of averages to determine AWAR

Figure 3.1-19: Additional details: aperture number 8, principal distance 10 cm, distance of test chart from camera 25 m, high contrast in daylight

plane to the object. The question to be answered is, which details can still be recognized with a metric camera? This question will be answered for the P31 terrestrial camera (Figure 3.1-19, Section 3.8.3) and for aerial metric cameras.

Wild P31:

- optical resolution $= 72\,\text{Lp/mm}$, principal distance $= 10\,\text{cm}$, imaging range $= 25\,\text{m}$
- width of line pair in image $= 1/72 = 0.014\,\text{mm} = 14\,\mu\text{m}$
- width of dark line and gap in image $= 14/2 = 7\,\mu\text{m}$
- width of dark line and gap at object $= 7 \times 25/0.1 = 1750\,\mu\text{m} = 1.7\,\text{mm}$.

Aerial metric camera:

- optical resolution $= 90\,\text{Lp/mm}$, principal distance $= 30\,\text{cm}$, flying height $= 2\,\text{km}$.
- width of line pair in image $= 1/90 = 0.011\,\text{mm} = 11\,\mu\text{m}$
- width of dark line and gap in image $= 11/2 = 5.5\,\mu\text{m}$
- width of dark line and gap at object $= 5.5 \times 2000/0.3 = 36000\,\mu\text{m} = 3.6\,\text{cm}$.

Terrestrial cameras can therefore identify objects in the millimetre range, aerial cameras correspondingly in the centimetre range.

Exercise 3.1-9. Recalculate the estimates for aerial cameras with the following parameters:

a) same flying height but a principal distance of 15 cm, (solution: 7.2 cm),

b) same principal distance but with flying heights of 1 km and 5 km (Solution: 1.8 cm and 9 cm).

3.1.5.3 Definition of contrast

Contrast is a measure of the difference in intensities I between (neighbouring) parts of an object or image (Figure 3.1-20). Intensity differences can either be expressed as a direct ratio or a relative difference. Using the largest and smallest intensities, the contrast as direct ratio is designated K and defined as:

$$K = I_{\text{max}}/I_{\text{min}} \tag{3.1-20}$$

The contrast as relative difference is designated C. The difference between the largest and smallest intensity is usually given with respect to the intensity sum as follows (see also Equations (3.1-24) and (3.1-25):

$$C = \frac{I_{\text{max}} - I_{\text{min}}}{I_{\text{max}} + I_{\text{min}}} \tag{3.1-21}$$

Contrast in digital images is defined by the differences in grey values g. The grey level range, generally 255 (Section 2.2.1), is used to normalize the contrast:

$$C = \frac{g_1 - g_2}{255} \tag{3.1-22}$$

In an analogue photograph (Section 3.2.2.3), the contrast is expressed using the density differences in the exposed and developed film. The density difference ΔD is the logarithm of contrast K above:

$$\Delta D = \log K = \log(I_{\text{max}}/I_{\text{min}}) \tag{3.1-23}$$

For example, if $I_{\text{max}} = 100\, I_{\text{min}}$, then $\Delta D = 2$.

3.1.5.4 Contrast transfer function

The resolving power is an insufficient, although important, measure of imaging quality. The resolving power gives an indication of the smallest details which can be seen in the image. The extent to which differences between bright and dark areas of larger details are transferred can only be answered by the more refined theory of the contrast transfer function.

The theory of the contrast transfer function uses a contrast as defined in Equation (3.1-20). Figure 3.1-20, left, defines the object contrast C as:

$$C = \frac{I_1 - I_2}{I_1 + I_2} \qquad (0 \le C \le 1) \tag{3.1-24}$$

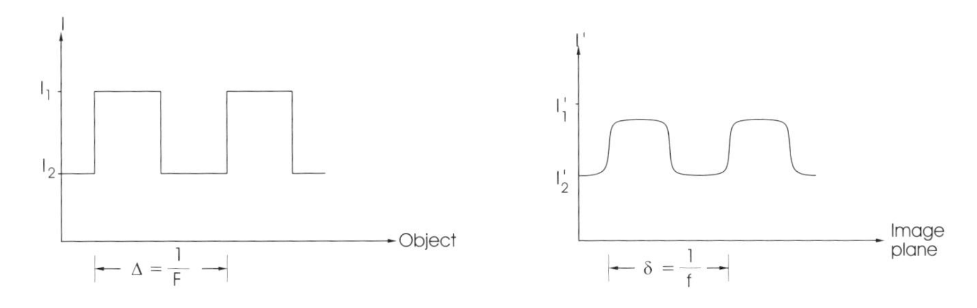

Figure 3.1-20: Left: object contrast. Right: image contrast

In the imaging process the optical system reduces this object contrast. In the image plane a reduced image contrast C' is observed (Figure 3.1-20, right):

$$C' = \frac{I'_1 - I'_2}{I'_1 + I'_2} \qquad (0 \leq C' \leq 1) \tag{3.1-25}$$

The reduction in contrast becomes even larger as the line pattern becomes narrower. The object and image contrast must therefore be defined in terms of the width (wavelength) Δ or δ of the pattern (Figures 3.1-17 and 3.1-20). In the transfer function theory, the reciprocals of these values are used, i.e. the frequencies[7] F and f, instead of the values themselves (Figure 3.1-20). The frequency indicates the number of line pairs (Lp) per unit of length.

The relationship between image contrast C' and object contrast C is, by definition, the contrast transfer function $CTF(f)$ or $CTF(F)$. A dependency on image frequency f is preferred, which in general is stated in line pairs per millimetre ([Lp/mm]):

$$CTF(f) = \frac{C'}{C} \tag{3.1-26}$$

It is common to normalize the contrast transfer function $CTF(f)$ by setting $CTF(0) = 1$. ($CTF(0)$ means the contrast transfer for infinitely wide lines.) Figure 3.1-21 shows the contrast transfer function for a single optical system and three different object contrasts (the various definitions of contrast are investigated in Section 3.1.5.3). The significantly better imaging properties for objects with high contrast can clearly be seen.

The resolution is shown in Figure 3.1-21. It indicates the narrowest line pair which can still be detected, in other words the line pattern with the highest frequency. The resolution therefore indicates the limiting case; it provides no information about the imaged quality of object patterns with lower frequencies. Figure 3.1-22 shows, for example, the contrast transfer function for two optical systems which, by chance, have the same resolving power. System 1, however, is significantly better than system 2.

[7] It is more accurate to speak of spatial frequency rather than temporal frequency. Temporal frequency is used for phenomena dependent on time (e.g. the propagation of electromagnetic radiation) and spatial frequency for effects dependent on spatial location (in other words, from coordinates in a coordinate reference system).

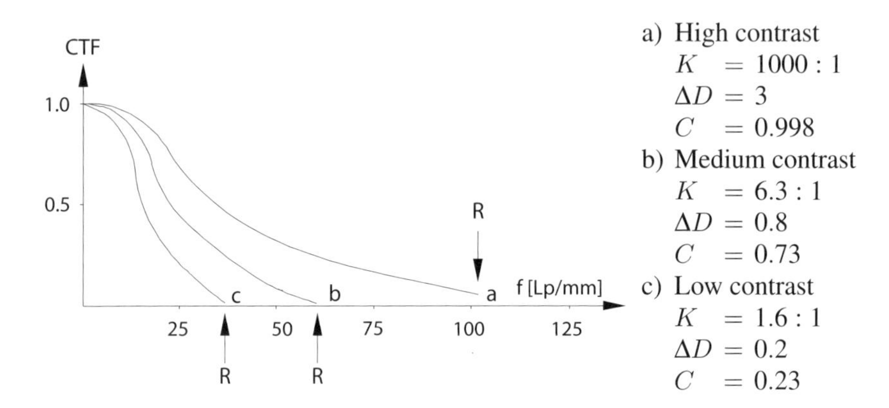

Figure 3.1-21: Contrast transfer function $CTF(f)$ and resolution (R) for three different object contrasts

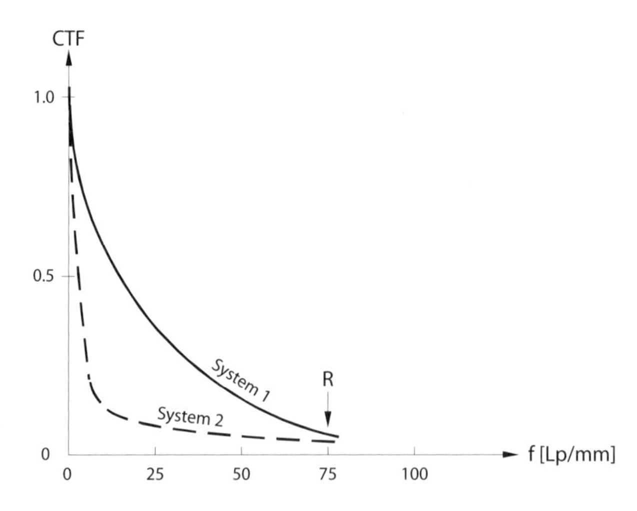

Figure 3.1-22: Contrast transfer functions for two optical systems with the same resolving power R

There is a large amount of literature available on the contrast transfer function and the related modulation transfer function. For a very small selection see: Inglis, A., Luther, A.: Video Engineering. 2nd ed., Mc Graw-Hill, New York, 1996. Graham, R.: Digital Imaging. Whittles Publishing, 1998.

3.1.6 Light fall-off from centre to edge of image

If a camera were constructed using only a single (thin) lens, then the irradiance would reduce from the centre of the image to the edge according to the $\cos^4 \tau$ law (τ is the off-axis angle). This means that areas with the same surface properties (e.g. a newly cropped field), and under the same conditions of illumination, at an off-axis angle of 50 gon (45°), would have only 25% of the image brightness at the edge of the image compared with the centre.

With multiple element lenses (compound lenses), the angle of view can be reduced near the aperture and the light fall-off reduced. Objective lenses in modern metric cameras have a light fall-off which can be described by a $\cos^n \tau$ law where the exponent n lies between 1.5 and 2.5.

Light fall-off can be compensated by the use of a circular graduated filter; this is a grey filter whose density decreases towards the edge. Light fall-off can also be compensated by post-processing, for example using digital image processing (Section 3.5.1.3 or Section C 1.3.1.2, Volume 2).

Numerical Example. A newly cropped field has a grey value of $g = 170$ when imaged at the centre of the field of view by a metric camera with a digital sensor. According to the $\cos^{2.5} \tau$ rule, a newly cropped field at an off-axis angle of 50 gon (45°) has a grey value of $170 \times 0.707^{2.5} = 170 \times 0.42 = 71$. Before cropping, the corresponding values are: image centre: $g = 60$; at an off-axis angle of 50 gon (45°): $g = 60 \times 0.42 = 25$. The contrast reduction, from image centre to image edge, of a field before and after cropping is therefore (Equation (3.1-22)):

$$\text{Image centre: } C = \frac{170 - 60}{255} = 0.43 \qquad \text{Image edge: } C = \frac{71 - 25}{255} = 0.18$$

3.2 Photochemical image recording

A light-sensitive, photographic emulsion is an extremely efficient medium for recording the incoming photons at the image plane of a metric camera. In Section 3.2.1 we examine the issues relating to the use of glass or film for supporting the emulsion. This section also evaluates the correction of film deformation. A wide range of optical and chemical aspects of photography are discussed in Section 3.2.2. Film resolution, contrast and suitability for aerial photography complete the discussion in the final three sections.

3.2.1 Analogue metric image

An image recorded using a film-based metric camera is, after exposure, taken to a darkroom, developed, fixed, washed, dried and then stored under various conditions of temperature and humidity. Also, the diapositives and paper prints of the photographs must be developed and dried. For further photogrammetric analysis it is necessary to take into account the changes of scale and shape of the photographs resulting from these processes.

3.2.1.1 Glass versus film as emulsion carrier

The deformations of photographic materials are, above all, dependent on the emulsion carrier (Figure 3.2-1) which must be chemically inert and highly transparent.

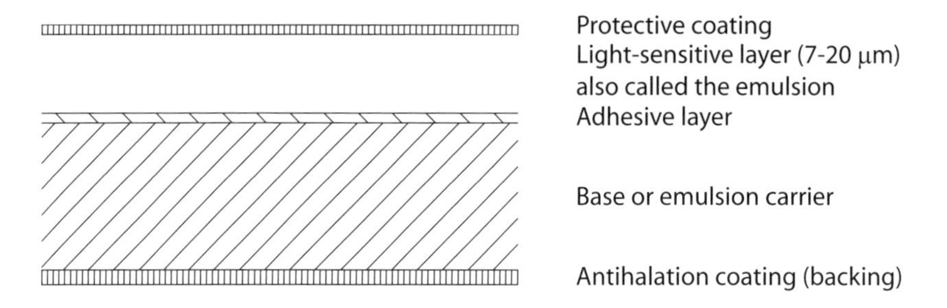

Figure 3.2-1: Cross-section through photographic material

As such, glass plates and films are used in photogrammetry (see Table 3.2-1). Glass plates are no longer used in aerial photogrammetry. Glass is heavy, easily breakable, relatively unflat, but stable. The adhesion of the light-sensitive emulsion is less efficient than with film and therefore deformations of the emulsion tend to be larger[8].

Type of base		Thickness [mm]	Flatness [μm]	Note on flatness
Glass plates (15 × 15 to 24 × 24 cm)		1.3 – 3.0	30 – 50	Flat
		1.3 – 3.0	20 – 30	Ultraflat
		6.0	5 – 10	Mirror glass ground
Films[9] (Width to 24 cm)	Polyester base e.g. Estar thin ... thick	0.06 ± 0.003 to 0.18 ± 0.005	5 – 20	Depending on the vacuum or pressure plate

Table 3.2-1: Thickness and flatness of films and plates (aerial photogrammetry)

Glass plates, usually with a smaller format than used in aerial photogrammetry, are sometimes used in terrestrial photogrammetry since only a few photographs at a time are required. However, a range of terrestrial cameras make use of film cassettes or magazines and a pressure plate for film flattening (see Section 3.8.3).

Film in aerial metric cameras (Section 3.7.2.1) is flattened by means of a vacuum device. Film is less stable than glass. Its deformations are dependent on: temperature, relative humidity, tension in handling, storage and ageing (Table 3.2-2). There is a

[8] Calhoun, J.M. et al.: Photogr. Eng. 29, pp. 661–672, 1960.

[9] Kodak Data for Aerial Photography. Kodak Publ. M-29, Rochester USA, 1982.

distinction between reversible and irreversible deformations. Age deformations are irreversible. Deformations can be conformal, corresponding to a similarity transformation, Equation (2.1-9), or affine, corresponding to an affine transformation, Equation (2.1-8). The linear deformations shown in Table 3.2-2 can be 30% different in the longitudinal direction compared with the lateral direction. Systematic deformations can be corrected. Non-systematic film deformations[10], which cannot be corrected without excessive effort, lie in the range $\pm 3 - 7\,\mu m$.

Base	Unit	Relative change of length caused by				
		Temperature per 1°C	Rel. humidity per 1 % RH		Development and ageing	
			Black/ white film	Colour film	1 week at 50°C & 20% RH	1 year at 25°C & 60% RH
PETP 0.06	‰	0.02	0.035	0.050	0.9	0.4
	$\mu m/20\,cm$	4	7	10	180	80
PETP 0.18	‰	0.02	0.015	0.020	0.2	0.2
	$\mu m/20\,cm$	4	3	4	4	4

Table 3.2-2: Film-base deformations for polyethylene terephthalate (PETP), thin (0.06 mm = 2.5 mil), thick (0.18 mm = 7 mil). (Kodak states film thickness in 1/1000 inch = 1 mil.) RH = rel. humidity[11]

3.2.1.2 Correcting film deformation

This section considers the task of re-creating the imaging bundle of rays from the analogue metric image. For this purpose, fiducial marks are required whose reference coordinates are available in a calibration certificate (Section 3.1.1). These fiducial marks, which are also imaged on the photograph, are measured with a comparator, essentially in an arbitrary coordinate system. This results in measured coordinates for the fiducial marks. Transformation parameters are then calculated which enable the conversion of all measured image points from the current comparator coordinate system of the photograph into the reference coordinate system of the metric camera. Within the scope of this transformation, systematic film deformations are corrected.

Various transformations are available for the correction of systematic film deformation. In order to achieve a reliable result, due regard must be taken with respect to the number and arrangement of fiducial marks (Section B 7.2.2.1, Volume 2). Table 3.2-3 gives the recommended transformation for typical arrangements of fiducial marks:

[10] ASP-Manual of Photogrammetry. 4th ed., p. 335, Falls Church, 1980.

[11] Rüger, W. et al.: Photogrammetrie. 4th ed., VEB Verlag für Bauwesen, Berlin, 1978.

Number and arrangement of fiducial marks	Transformation		Film deformation which can be eliminated (significant part only)
	Number of measurements	Number of unknowns	
4 fiducial marks at the centres of the image edges (Figure 3.1-10)	Plane similarity transform (2.1-9)[12] 8 measurements	4 unknowns	conformal part, i.e. a uniform scale deformation
4 fiducial marks in the corners of the image	Plane affine transformation (2.1-8) 8 measurements	6 unknowns	affine part, i.e. differing scale changes on the longitudinal and transverse axes
8 fiducial marks (Figure 3.7-7)	Bilinear transformation (2.2-1) in both the $\xi-$ and η directions[13] 16 measurements	8 unknowns	affine part and additionally different scale changes on opposite facing image edges

Table 3.2-3: Correction of film deformation by means of plane transformations for different arrangements of fiducial marks (see also Section B 7.1.1, Volume 2)

Exercise 3.2-1. Calibration (camera) and measured (comparator) coordinates of 4 fiducial marks in the centres of the image edges, together with measured coordinates of one of a number of photo points P, are listed in the following table:

Fiducial marks Photo point	Calibration/camera coordinates [mm]		Measured/comparator coordinates [mm]	
	ξ	η	ξ'	η'
1	1113.002	1000.008	1115.133	1000.985
2	1000.001	1112.998	1002.111	1113.956
3	886.994	999.995	889.103	1000.935
4	1000.004	887.002	1002.161	887.971
P	?	?	1024.334	964.847

[12]A plane affine transformation (2.1-8) would be mathematically possible, but it could be dangerous to extrapolate into the image corners. If different scale changes along the axes are still of interest in this configuration of fiducial marks, then differing scale corrections can be made using the calibrated and measured separations of the fiducial marks along the ξ and η directions. (In the solution to Exercise 3.2-3 an adequate 5 parameter transformation is given for this.)

[13]This bilinear interpolation would also be possible with four fiducial marks in the image corners, but there would be no overdetermined solution in this case. A data error would not be detected and the result would be in error. However, if the properties of a bilinear transformation using four fiducial marks remain of interest, the data should first be checked for the presence of gross errors by means of an initial affine or similarity transformation.

Find the coordinates ξ and η of the photo point in the coordinate system of the metric camera. Hint: A similarity transformation is appropriate. Its four unknown transformation parameters can be calculated from the 8 measurement equations using the method of least-squares adjustment of indirect observations (Appendix 4.1-1). The non-linear Equations (2.1-9) should be linearized. A more elegant solution is achieved using a Helmert transformation (Section B 5.1.1.3, Volume 2):

$$\begin{pmatrix} \xi \\ \eta \end{pmatrix} = \begin{pmatrix} -1.6529 \\ -0.5320 \end{pmatrix} + 0.9995493 \begin{pmatrix} 1.0000000 & -0.0000221 \\ 0.0000221 & 1.0000000 \end{pmatrix} \begin{pmatrix} 1024.334 \\ 964.847 \end{pmatrix} = \begin{pmatrix} 1022.198 \\ 963.903 \end{pmatrix}$$

Exercise 3.2-2. Calibration (camera) and measured (comparator) coordinates of 4 fiducial marks in the corners of the image, together with measured coordinates of one of a number of photo points P, are listed in the following table:

Fiducial marks Photo point	Calibration/camera coordinates [mm]		Measured/comparator coordinates [mm]	
	ξ	η	ξ'	η'
1	1106.004	1106.003	1104.412	1104.313
2	893.994	1106.000	892.382	1100.602
3	894.002	894.000	896.107	888.550
4	1106.006	893.993	1108.123	892.241
P	?	?	916.241	1009.421

Find the coordinates ξ and η of the photo point in the coordinate system of the metric camera. Hint: An affine transformation is appropriate. It should only be used for data control in the analysis. The final transformation is made, in this case, by a bilinear transformation. Eight linear equations provide the solution for the eight unknowns. (More exactly, the corresponding four unknowns for each coordinate axis are determined from the four corresponding linear equations.)

Solution:

$$\begin{aligned} \xi &= -17.498 + 0.999804\xi' + 0.017688\eta' - 1.8565 \times 10^{-7}\xi'\eta' \\ \eta &= 21.389 - 0.017239\xi' + 0.999642\eta' - 2.1678 \times 10^{-7}\xi'\eta' \end{aligned}$$

Coordinates for point P: $\xi = 916.246\,\text{mm}$, $\eta = 1014.453\,\text{mm}$

In addition, the corrections to the ξ' and η'-coordinates should be represented graphically. Both translations need not be taken into account in the graphical presentation which, for convenience, can be done separately for each coordinate in the form of lines of equal correction value. The bilinear property should be visible in the graphical result.

Exercise 3.2-3. Re-arrange the affine and similarity transformations such that there are only two translations, a rotation and individual scale factors m_ξ and m_η in both coordinate directions.

Solution:

$$\begin{pmatrix} \xi \\ \eta \end{pmatrix} = \begin{pmatrix} a_{01} \\ a_{02} \end{pmatrix} + \begin{pmatrix} \cos\alpha & -\sin\alpha \\ \sin\alpha & \cos\alpha \end{pmatrix} \begin{pmatrix} m_\xi \xi' \\ m_\eta \eta' \end{pmatrix}$$

In linearized form with small angle $d\alpha$ as well as $m_\xi = 1 + dm_\xi$ and $m_\eta = 1 + dm_\eta$:

$$\xi = a_{01} + \xi' dm_\xi - \eta' d\alpha + \xi'$$

$$\eta = a_{02} + \xi' d\alpha + \eta' dm_\eta + \eta'$$

Exercise 3.2-4. Repeat Exercise 3.2-1 using this 5 parameter transformation.

$$\begin{aligned}\begin{pmatrix} \xi \\ \eta \end{pmatrix} &= \begin{pmatrix} -2.2152 \\ -0.8235 \end{pmatrix} \\ &+ \begin{pmatrix} 1.0000000 & -0.0001858 \\ 0.0001858 & 1.0000000 \end{pmatrix} \begin{pmatrix} 0.9999026 \times 1024.334 \\ 0.9999514 \times 964.847 \end{pmatrix} \\ &= \begin{pmatrix} 1022.198 \\ 963.880 \end{pmatrix}\end{aligned}$$

In addition to film deformation, it may also be possible to account for departures from planarity (Table 3.2-1). Its distorting effect on the bundle of rays is greater with wider angle lenses. With an off-axis ray of 50 gon (45°), a flatness error causes an error in the $\xi\eta$-image plane in the ratio 1 : 1.

In cameras which do not have a vacuum back or pressure plate, lack of film flatness can be corrected using a réseau. Semi-metric cameras which use such réseaus are discussed in more detail in Section 3.8.4.

Exercise 3.2-5. How large is the correction $\Delta\rho$ resulting from the spherical deformation of a glass plate emulsion carrier given by $d = 30\,\mu$m (principal distance $c =$ 152 mm, format = 23 cm × 23 cm)? (Solution: for $\rho = 0, 40, 80, 120$ and 160 mm, $\Delta\rho = 0, 7, 8, -2, -30\,\mu$m.)

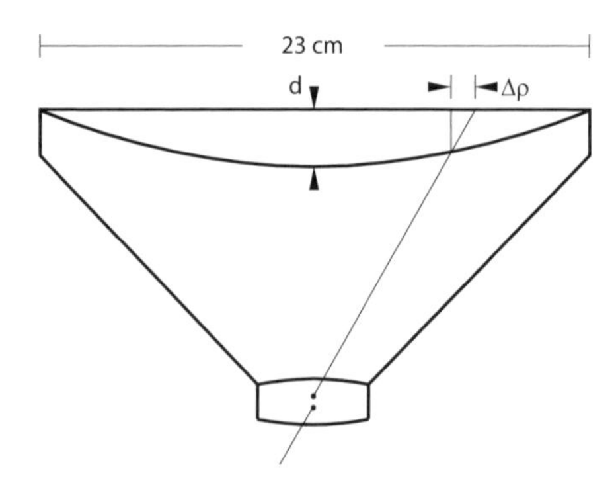

Exercise 3.2-6. Consider a square in object space, placed symmetrically and perpendicularly to the camera axis. Evaluate the deformations resulting from the following distortion of the glass plate: a) spherical towards the object, b) spherical away from the object, c) cylindrical towards the object.

3.2.2 Physical and photochemical aspects

Aerial photographs are taken from moving aircraft (40 – 220 m/s, 80 – 440 knots, 150 – 800 km/h), sometimes from great heights, through the atmosphere. Since only short exposure times can be allowed (1/250 – 1/1000 s), there is often very little light available to form the photograph. The atmosphere causes strong reduction in contrast and change in colours. Practical aerial photography therefore poses many photographic problems, the most important of which are considered below.

3.2.2.1 Colours and filters

The visible part of the electromagnetic spectrum, i.e. the part to which the human eye responds, extends approximately from a wavelength of 400 nm to 700 nm (Figure 3.2-2). The photographic process covers a range from 300 nm to 1000 nm, i.e. includes part of the ultraviolet (UV) and near infra-red (IR) spectra. It does not cover either the medium or the thermal infra-red. If all wavelengths of visible light are uniform in intensity, the eye sees white light. This white light can be split into a large number of monochromatic spectral lines which are seen as saturated "spectral colours".

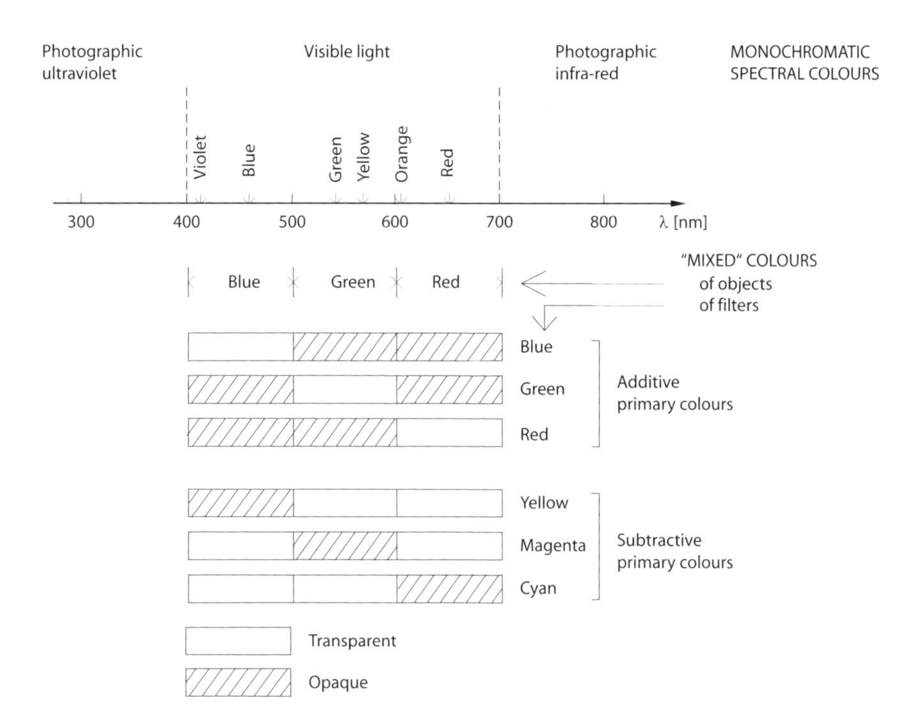

Figure 3.2-2: The spectral range of photography, together with the transmission and absorption characteristics of filters for the additive and subtractive primary colours

If we group these in bandwidths of 100 nm, we obtain the three “mixed” colours blue, green and red, the three additive primary colours. Other colours can be created by adding various proportions of these three primary colours, for example by superimposing them in projection. A uniform mixture of the three primary colours produces white. The addition of pairs of additive primary colours produces the subtractive primary colours cyan, yellow and magenta. These colours are called subtractive since they are created by subtracting one primary colour from white light:

$$\begin{array}{lllll} \text{Cyan} & = \text{Blue} & + \text{Green} & = \text{White} & - \text{Red} \\ \text{Yellow} & = \text{Green} & + \text{Red} & = \text{White} & - \text{Blue} \\ \text{Magenta} & = \text{Red} & + \text{Blue} & = \text{White} & - \text{Green} \end{array} \quad (3.2\text{-}1)$$

Such subtractions are achieved by absorption filters which absorb part of the light reaching them and transmit the remainder. If white light falls on the filter, the transmitted part gives the filter colour, the absorbed part, which if added to the filter colour would give white, is called the complementary colour.

Subtractive and additive primary colours are complementary in pairs. Two colours complementary to each other and superimposed in projection, i.e. added together, produce white.

$$\begin{array}{llll} \text{Blue} & + \text{Yellow} & & = \text{White} \\ \text{Green} & + \text{Magenta} & & = \text{White} \\ \text{Red} & + \text{Cyan} & & = \text{White} \\ \text{Blue} & + \text{Green} & + \text{Red} & = \text{White} \\ \text{Yellow} & + \text{Magenta} & + \text{Cyan} & = \text{White} \end{array} \quad (3.2\text{-}2)$$

Figure 3.2-2 shows the relations of the six primary colours. Each colour filter in a subtractive primary colour absorbs the complementary additive primary colour and transmits the other two additive primary colours. A yellow filter, for example, transmits green and red. A combination of filters of two subtractive primary colours (e.g. cyan and magenta) transmits only the part of white light which is the additive primary colour to both filter colours (blue in this example)[14].

Exercise 3.2-7. Why is a yellow filter also known as a minus-blue filter?

Exercise 3.2-8. Which description of a UV-absorbing filter is more correct: a UV filter or a UV-blocking filter?

Supplementary filters in front of the objective are used in practical photography to prevent unwanted light from reaching the photographic emulsion (e.g. UV-blocking filters in high mountains, yellow filters in hazy weather (Section 3.7.5)). The entire visible spectrum must be blocked with infra-red filters for infra-red photography (Section 3.2.2.6), so that only the infra-red rays form the photograph (Figure 3.2-10).

Filters are not, as shown in Figure 3.2-2, 100% effective. A green filter, for example, can only transmit about 80% of green light, while the other colours are not completely

[14] A detailed treatment of colour theory is given in: Wysecki, G., Stiles, W.: Colour Science. Wiley, 2000.

blocked. About 20% blue and 35% red will be transmitted (Figure 3.2-3). The filter quality must be determined by measuring the intensities of the different transmitted wavelengths. Figure 3.2-4 shows the opacity and transmission of typical filters used in aerial photogrammetric cameras. These highly precise filters have a very steep "cut-off" or edge and therefore block very exactly from a certain wavelength. They absorb as much as possible of the unwanted light and as little as possible of the desired light.

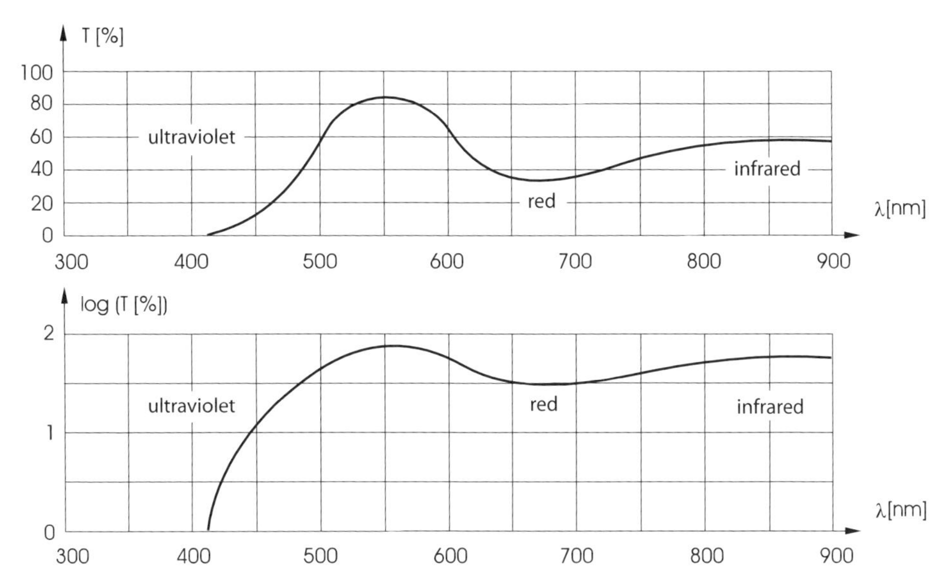

Figure 3.2-3: Transmission characteristics of a green filter. Above: Transmission T in % of light falling on the filter. Below: Similar, but on a semi-logarithmic scale

3.2.2.2 The photochemical process of black-and-white photography

The exposure (see Section 3.2.2.3) produces a latent image in the layer of light-sensitive silver halide crystals (AgBr, AgCl, AgI etc., size 0.2 μm) embedded in the gelatine of the emulsion, which can be made visible by the negative development process (for example with hydroquinone, alkalis and potassium bromide). Here the silver is separated from the bromine. The unexposed silver bromide is converted in a fixing bath (sodium thiosulphate = hypo) into a silver salt easily soluble in water and released. The remainder of about 5% is then removed by washing. Only metallic silver remains finally in the exposed photograph, with the areas of high exposure blacker than those of lower exposure. In the reversal process the exposed silver bromide is released in the predevelopment (bleaching), while the unexposed silver bromide remains. It is then uniformly exposed and finally developed as a positive, fixed and washed. The uniform, intermediate exposure can be replaced by chemical processing with sodium sulphide (instant-photo process!).

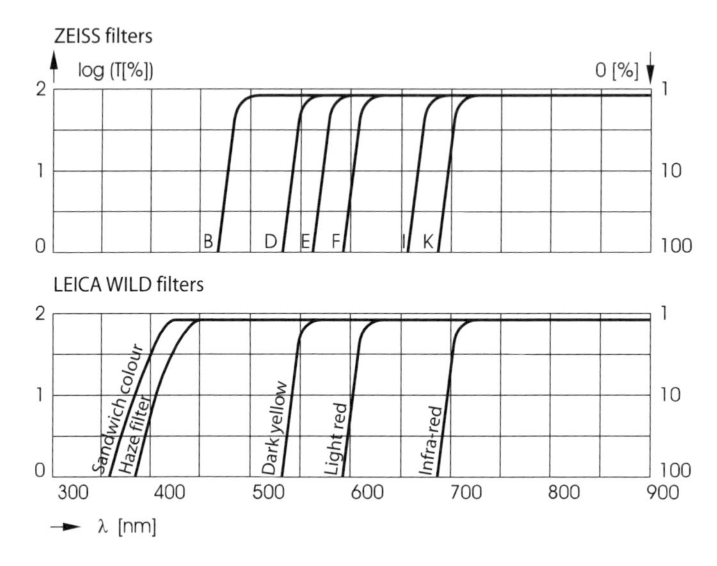

Figure 3.2-4: Spectral transmission/absorption curves of filters for aerial photogrammetric cameras; T ... Transmission, O ... Opacity

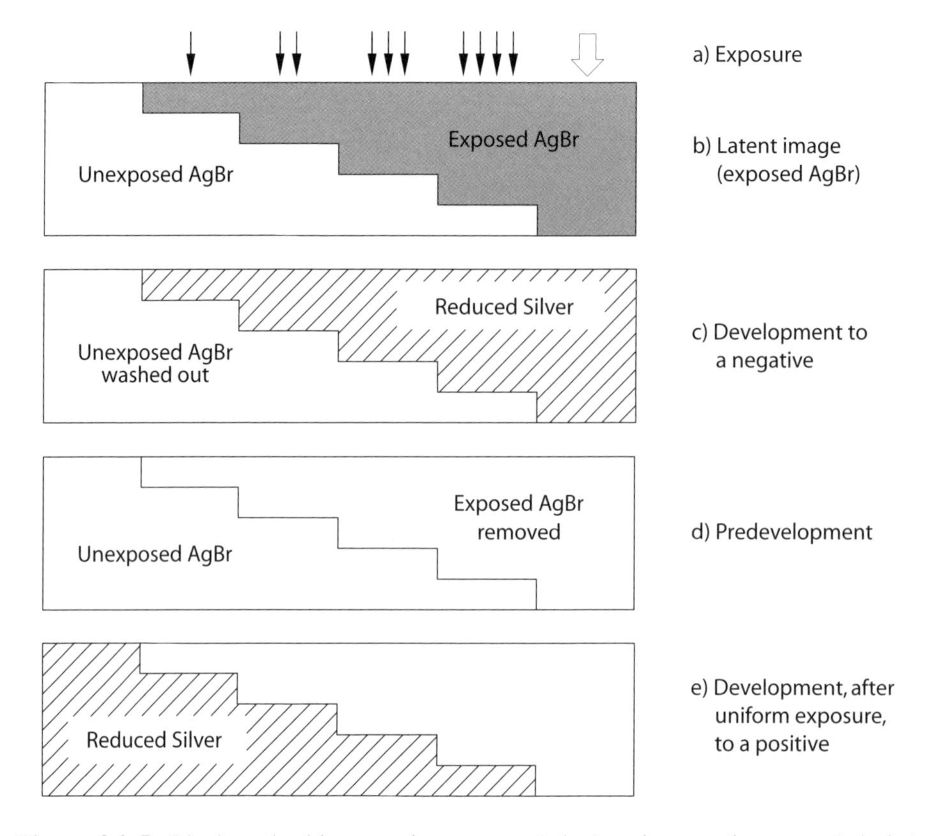

Figure 3.2-5: Black-and-white negative process (a,b,c) and reversal process (a,b,d,e)

3.2.2.3 Gradation

The relation (Figure 3.2-6) between exposure H and density D of a negative or positive is of great practical importance for the judging of photographic materials, processes and results. The density D is measured in a densitometer. A densitometer generates a constant luminous flux Φ_0 (measured in lumen ([lm])) and measures the proportion Φ transmitted through the film. The ratio Φ/Φ_0 is called the transparency τ, the reciprocal $\Phi_0/\Phi = 1/\tau$ is called the opacity O. The logarithm of the opacity is the density D.

$$\tau = \frac{\Phi}{\Phi_0}, \quad (0 \leq \tau \leq 1) \tag{3.2-3}$$

$$D = \log \frac{1}{\tau} = \log O \tag{3.2-4}$$

A density $D = 2$ means therefore that only $1/100$ of the luminous flux Φ_0 falling on the film is transmitted through the film and that $99/100$ is absorbed; for $D = 1$, one tenth of the incident luminous flux is transmitted. A density $D = 0$ indicates complete transparency or $\tau = 1$ (see also Section 3.1.5.3).

The logarithmic definition of density accords with the Weber–Fechner law which states that the sensitivity of human sense organs, at least in the middle range of the usual exciting intensities, is proportional to the logarithm to the base 10 of the exciting intensity. A logarithmic grey wedge, i.e. a series of densities increasing logarithmically, therefore appears as a linear scale to the eye.

The exposure H is defined as the product of the illuminance E falling on the emulsion, measured in [lux] = [lumen/metre^2] ($1\,\text{lx} = 1\,\text{lm/m}^2$), and the exposure time t, and is expressed in lux seconds (lx $\times$ s). The gradation γ is defined as $\gamma = \tan\alpha$, where α is the slope angle of the straight portion of the density curve (also called the characteristic curve) in the range of normal exposure (see Figure 3.2-6). The gradation depends upon the photographic material and its age, the development chemicals as well as the developing temperature and time. If $\gamma > 1$, we have a "hard" photographic material or process which increases contrast: small differences of exposure produce larger differences of density. If $\gamma < 1$, i.e. for "soft" materials or processes, the relation is reversed. If $\gamma = 1$, i.e. "normal", the differences of density reproduce $1:1$ the differences of exposure. The correct exposure for hard films, which is preferred for use in aerial photography, is much more difficult to judge than for soft films. The solarization point takes its name from the photographic effect in which the image of the sun is lighter in a negative than that of the surrounding sky. Apparently an excessive exposure leads to desensitization of the silver halide.

The normal exposure range between points P_2 and P_4 of the density curve (Figure 3.2-6) is known as the dynamic region of the density. In black-and-white films it is around $0.2 - 2.5D$.

Since the maximum resolution in the photograph occurs in the lower third of the normal range of exposure, photogrammetric photographs should be just barely exposed and not too long developed. The practical consequences are that exposures must be carefully measured and, usually, that test photographs must be made.

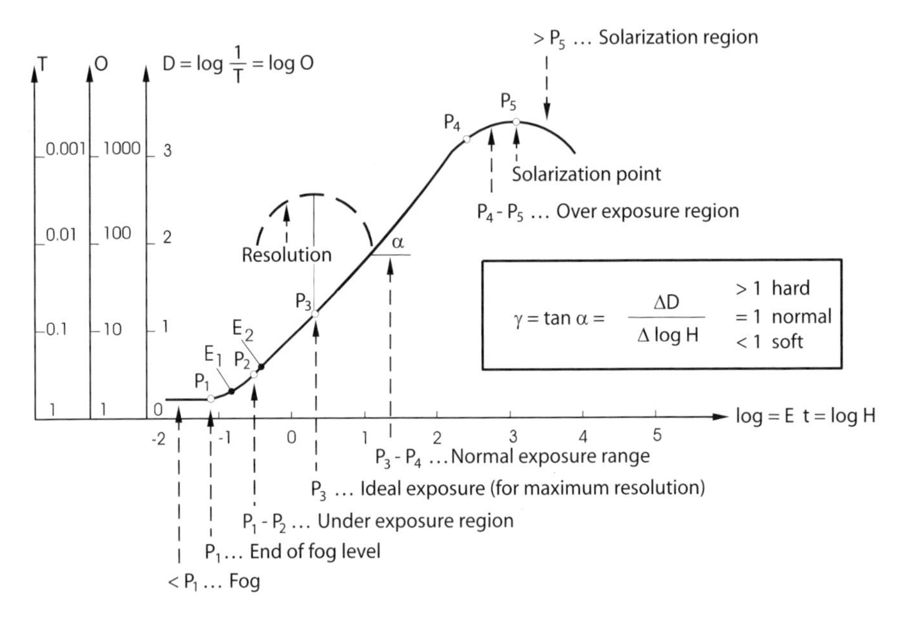

Figure 3.2-6: The characteristic curve
E_1 = Base point for film speed definition (DIN, ASA; ISO), $\Delta D = 0.1$ above fog
E_2 = Base point for film speed definition (AFS, EAFS), $\Delta D = 0.3$ above fog
(see Section 3.2.2.4)

Exposure meters measure the illuminance E as spot or integral measurements. In spot exposure measurement[15], a series of spot measurements is taken along the direction of flight to determine the minimum and maximum illuminances. The series is continually refreshed. A photograph is so exposed that the minimum density lies always at point P_2, but not below it, most of the photograph lies around point P_3 and the maximum density (except for specular reflections such as those of the sun off water surfaces) lies above point P_3, but never above point P_4. Modern exposure meters regulate aperture and exposure time automatically and give warnings if points P_2 or P_4 will be exceeded. In this case, the film and development specifications must be changed. For integral exposure measurements, which can only measure the average object brightness over the entire photograph, one selects the aperture and exposure time for point P_3 and hopes that nothing will be under- or over-exposed.

3.2.2.4 Film sensitivity (speed)

The sensitivity, or speed, of a photographic emulsion is defined as the reciprocal of that exposure $H_{\Delta D}$ which produces a defined density difference ΔD above fog level, under

[15]Zeth, U.: VT 32, pp. 147–151, 1984.

precise conditions of radiation, exposure and development[16]. The German Standards Institute (Deutsches Institut für Normung (DIN)) defines a logarithmic system known as DIN speeds, according to Equation (3.2-5) and based on that exposure $H_{\Delta D}$ which produces a density difference $\Delta D = 0.1$ above fog level (point E_1 in Figure 3.2-6). The American National Standards Institute (ANSI)[17] adopts an arithmetic system rather than logarithmic, according to Equation (3.2-6). The International Standards Organization (ISO) standard adopts both the DIN and ANSI speeds. Table 3.2-4 shows a comparison between the two systems.

$$S_{\text{DIN}} = 10 \log \frac{H_0}{H_{\Delta D=0.1}} \tag{3.2-5}$$

$$S_{\text{ANSI}} = 0.8 \frac{H_0}{H_{\Delta D=0.1}} \tag{3.2-6}$$

H_0 = Unit exposure = 1 lx s (for visible light); $H_{\Delta D}$ in lx s

DIN	ANSI	$H_{\Delta D=0.1}$	DIN	ANSI	$H_{\Delta D=0.1}$
0	0.8	1.0	16	32	0.025
1	1	0.79	17	40	0.020
2	1.2	0.63	18	50	0.016
3	1.6	0.50	19	64	0.013
4	2	0.40	20	80	0.010
5	2.5	0.32	21	100	0.0079
6	3	0.25	22	125	0.0063
7	4	0.20	23	160	0.0050
8	5	0.16	24	200	0.0040
9	6	0.13	25	250	0.0032
10	8	0.10	26	320	0.0025
11	10	0.079	27	400	0.0020
12	12	0.063	28	500	0.0016
13	16	0.050	29	640	0.0013
14	20	0.040	30	800	0.0010
15	25	0.032	40	8000	0.0001

Table 3.2-4: Comparison of DIN and ANSI film speeds. $H_{\Delta D=0.1}$ in lux seconds is that smallest exposure (= illuminance × time) which just produces a density of $\Delta D = 0.1$ above fog level.

There exists in the USA a further standard[18] for black-and-white aerial films, not the same as the DIN and ANSI standards. It is derived according to Equation (3.2-7) from

[16] $\Delta \log H = \log H_B - \log H_{E1} = 1.30$ must yield a $\Delta D = D_B - D_{E1} = 0.80 \pm 0.05$. B is the corresponding point on the characteristic curve between points P_2 and P_4 (Figure 3.2-6).

[17] Formerly the American Standards Association (ASA).

[18] See ANSI Standard PH 2.34-1969.

a unit of $3/2$ times that exposure which produces, for a precisely specified "average" development, a $\Delta D = 0.3$ above fog level (point E_2 in Figure 3.2-6, AFS = Aerial Film Speed).

$$S_{\text{AFS}} = \frac{2}{3} \frac{H_0}{H_{\Delta D=0.3}} \tag{3.2-7}$$

For development conditions different from those of the standard and for colour films, the speed is quoted in EAFS (= Effective AFS). The advantage is that one can develop in any way thought suitable, but must also supply the corresponding information[19].

Exercise 3.2-9. Calculate the film speeds according to the various standards for the density curve in Figure 3.2-6.

3.2.2.5 The colour photographic process

Black-and-white films record only grey tones, but colour films contain further, important information. Colour films are composed of three light-sensitive layers which are so treated in development that each layer becomes a colour filter.

Positive and negative colour films

The three light-sensitive layers are shown diagrammatically in Figure 3.2-7. Each layer is sensitive to a particular spectral range, though each of them is sensitive to blue. Therefore, a yellow filter layer Y is interposed between layers I and II, which prevents blue light (Figure 3.2-2) from passing through to the lower layers and is dissolved in development. Each colour-sensitive layer contains a colour-coupler or dye for the complementary colour, with which in negative films the exposed part and in positive films the unexposed part is dyed. Figures 3.2-7 and 3.2-8 show, in a simplified way, the layers and the effects of development.

Positive colour infra-red film

Colour infra-red, or false-colour, film has layers sensitive to infra-red, green and red rather than to blue, green and red light. These layers are coupled to the colours cyan, yellow and magenta so that in white transmitted light the colours red, blue and green appear, though these do not correspond to the natural colours of the object. This film is therefore also known as false colour film. Since all three layers are also sensitive to blue light, a yellow filter must be used (Figure 3.2-10). A yellow filter layer in the film is therefore not necessary (Figure 3.2-9).

[19]Graham, R., Read, R.E.: Manual of Aerial Photography. Focal Press, London, 1986.

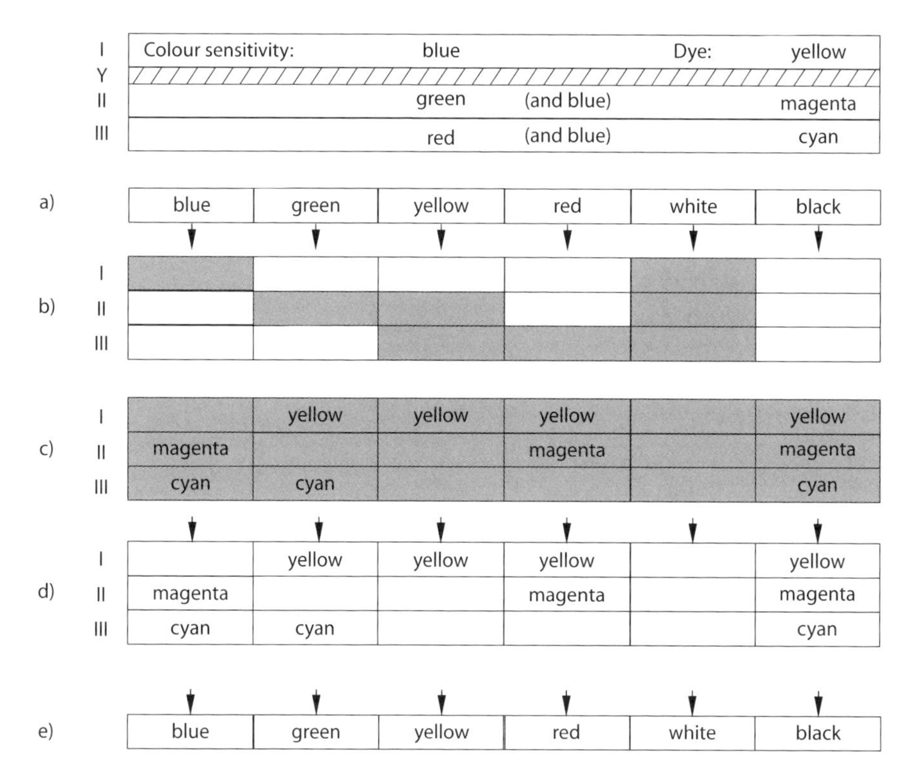

a) object radiation
b) exposure and development without colour-coupling
c) exposure to diffuse white light and development in a colour developer with colour-coupling
d) bleach bath and fixing bath remove reduced silver and silver bromide, the dyes remain
e) observed colours in transmitted white light (subtractive colour mixing

Figure 3.2-7: Positive colour film (colour reversal process). Red, green and blue are the three additive primary colours, yellow is introduced as an example of a mixed colour. Grey areas: reduced silver. Y = Yellow filter layer.

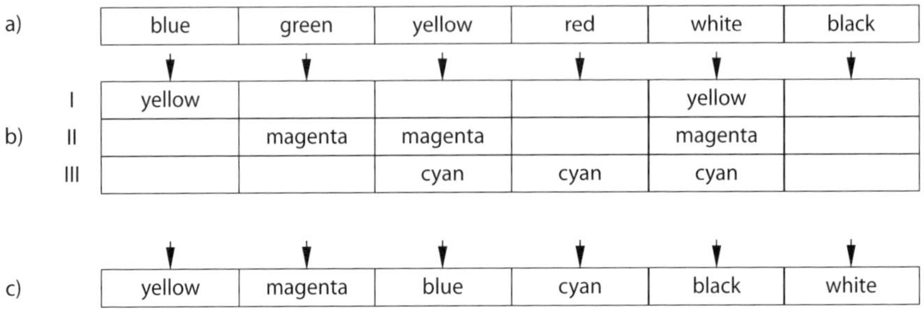

a)		blue	green	yellow	red	white	black
b)	I	yellow				yellow	
	II		magenta	magenta		magenta	
	III			cyan	cyan	cyan	
c)		yellow	magenta	blue	cyan	black	white

a) object radiation
b) exposure and development with colour-coupling developer, followed by bleach bath and fixing bath, produce complementary colours
c) observed colours in transmitted white light

Figure 3.2-8: Negative colour film

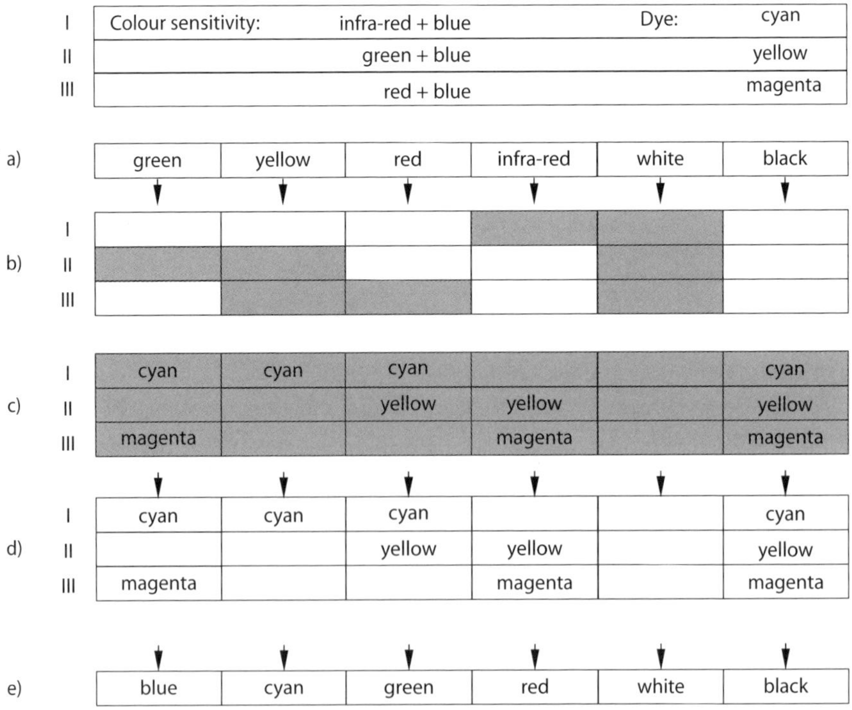

	Colour sensitivity:	Dye:
I	infra-red + blue	cyan
II	green + blue	yellow
III	red + blue	magenta

a)		green	yellow	red	infra-red	white	black
b)	I						
	II						
	III						
c)	I	cyan	cyan	cyan			cyan
	II			yellow	yellow		yellow
	III	magenta			magenta		magenta
d)	I	cyan	cyan	cyan			cyan
	II			yellow	yellow		yellow
	III	magenta			magenta		magenta
e)		blue	cyan	green	red	white	black

Grey areas: reduced silver. a) – e) as in Figure 3.2-7.
Blue light is absorbed by a yellow filter.

Figure 3.2-9: Positive false-colour film

3.2.2.6 Spectral sensitivity

The layers of black-and-white and colour films are sensitized by various chemicals (mostly catalysers) for light of particular ranges of wavelength (Figure 3.2-10). The layers are not, however, equally sensitive to each wavelength. The spectral sensitivity S_λ defines the sensitivity as a function of the wavelength λ. It is, by analogy with the overall sensitivity, expressed as the reciprocal of that exposure (in lx s) which produces under standardized conditions a particular difference of density (usually $\Delta D = 1.0$) above fog level.

$$\log S_{\text{rel}}(\lambda) = \log \frac{H_0}{H_{\Delta D=1.0}(\lambda)} \tag{3.2-8}$$

H_0 is here not the unit exposure (Section 3.2.2.4), but a defined constant reference exposure to which the exposures at all wavelengths are referred.

Orthochromatic emulsions are sensitive to blue-green light and therefore unsuitable for red object details. Since the sensitive layer integrates light of all wavelengths, the loss of the red influence leads to better differentiation in the blue-green range. Orthochromatic emulsions are therefore frequently used in terrestrial photogrammetry. They also have a practical advantage in that they can be developed under a red safety light in often improvised darkrooms in the field.

Panchromatic emulsions reproduce the full visible spectral range in the naturally corresponding grey tones. They must be developed in complete darkness.

Infra-red sensitive black-and-white emulsions are also sensitive to wavelengths from blue to red and must therefore be exposed through infra-red filters if one requires only the infra-red details in the photograph (see also Figure 3.2-4 and Section 3.2.3). The film must also be developed in complete darkness.

(Normal) colour and false-colour films are primarily used for photo-interpretation, though they are all suitable for aerial photogrammetry since they have a sufficiently fine grain and thin layers. Where film is still required, precision photogrammetry nevertheless continues to use the fine-grained, medium-speed black-and-white films.

The human eye can resolve many more tones and hues of colour than it can of grey tones. Colour photographs therefore contain significantly more information than do black-and-white photographs and thus also help to simplify interpretation during stereoscopic observation. The relatively small extra costs of colour photography are well justified for interpretation tasks.

In (normal) colour film, the red sensitive layer receives not only the red object illumination between 600 and 700 nm, as indicated in the idealized Figure 3.2-7, but also part of the green object illumination (500 to 600 nm) and even part of the blue object illumination (Figure 3.2-10). (Normal) colour is also, to some extent, a false colour film.

In false colour film, the colour separation is worse than in colour film (Figure 3.2-10): the infra-red sensitive layer, which appears as red, not only receives the infra-red object illumination but also very large parts of the green and red object illumination.

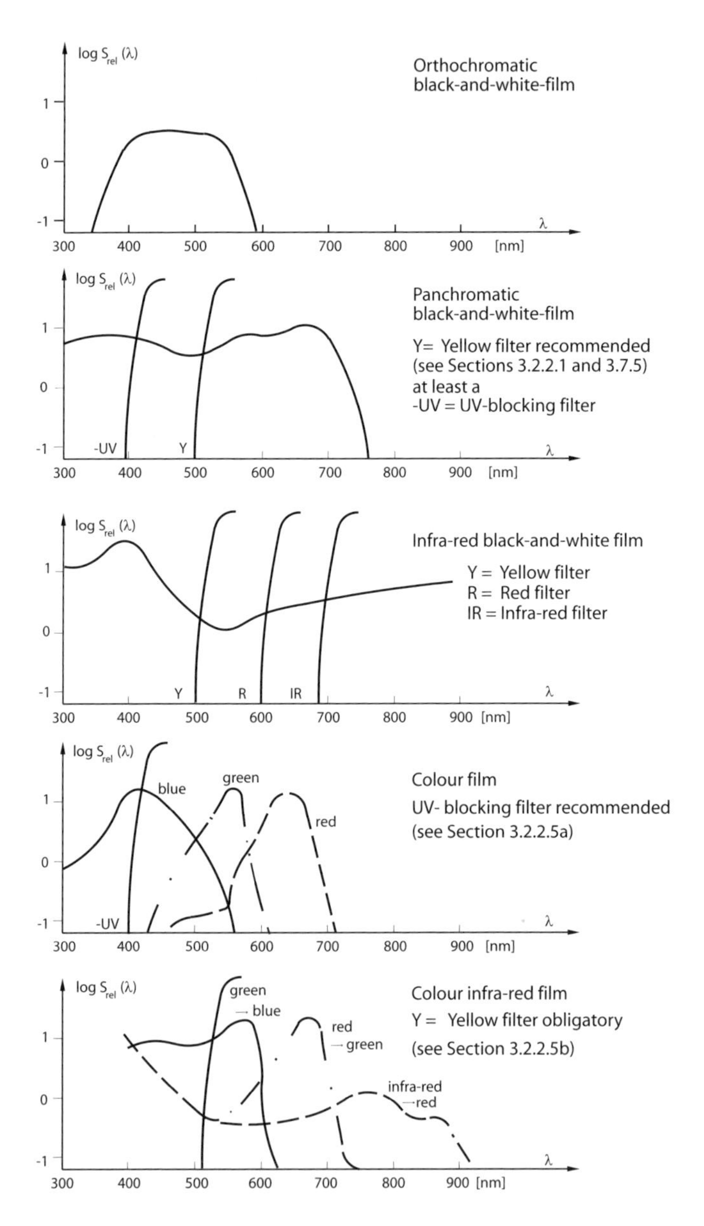

Figure 3.2-10: Spectral sensitivity of various types of film

Practical guidelines concerning the various types of film are given in Section 3.2.3.

3.2.2.7 Resolution of photographic emulsions

The decisive factor in the resolution of a photograph itself is the grain. The "grains" in undeveloped emulsions are the small silver halide crystals embedded in the gelatine. They range in size from a few nm (in emulsions of very low speed, but high resolution) to a few μm (in high speed emulsions). After development, the "grains" are clumps of metallic silver molecules with diameters from 0.5 to 2.0 μm.

Kodak specifies a procedure for measuring granularity indirectly[20]. A microdensitometer with a circular aperture of 48 μm diameter is used to scan part of the film with a relatively even density $D = 1.0$ and to derive the mean density distribution σ_D. This value, multiplied by 1000, is called the Root Mean Square Granularity ($RMSG$).

$$RMSG = 10^3 \sigma_D \tag{3.2-9}$$

A value of $RMSG$ of 13 means that $D = 1.0 \pm 0.013$ under these conditions. The smaller the value of $RMSG$, the higher the resolution of the film can be (see Table 3.2-5). If a section of film with another density D had been used, instead of $D = 1.0$, experience shows that another $RMSG$ would result. According to Diehl[21]:

$$RMSG(D) = RMSG(D = 1.0)\frac{D + 1.5}{2.5} \tag{3.2-10}$$

For the best aerial films (Table 3.2-5) the $RMSG = 6$, i.e. $D = 1.0 \pm 0.006$. According to Equation (3.2-10), for a section of film with $D = 2.0$ then $RMSG(2.0) = 6(2.0 + 1.5)/2.5 = 8$, i.e. $D = 2.0 \pm 0.008$; for a section of film with $D = 0.3$ the corresponding values are $RMSG(0.3) = 6(0.3 + 1.5)/2.5 = 4.3$, d.h. $D = 0.3 \pm 0.0043$.

The resolution of photographic emulsions is determined by test charts (Figure 3.1-17), imaged by a camera onto the emulsion layer under test. This provides the total resolving power R_T of the sensor combination "metric camera with photochemical recording" which depends on the resolving power of the optics R_O and the resolving power of the photographic emulsion R_P as follows[22]:

$$\frac{1}{R_T^2} = \frac{1}{R_O^2} + \frac{1}{R_P^2} \tag{3.2-11}$$

From the observed total resolving power R_T and the known resolving power R_O of the optics, determined for example with a very high resolution emulsion in a separate calibration (Section 3.1.5.2), the resolving power of the photographic emulsion, R_P can be calculated from Equation (3.2-11).

Numerical Example. Observed: $R_T = 72\,\mathrm{Lp/mm}$; known: $R_O = 90\,\mathrm{Lp/mm}$ (Section 3.1.5.2); calculated from Equation (3.2-11): $R_P = 120\,\mathrm{Lp/mm}$.

[20] Kodak Publication M-29. Rochester, USA, 1982.

[21] Diehl, H.: IAPRS XXIX(B1), pp. 1–6, Washington, 1992.

[22] Meier, H.-K.: BuL 52, pp. 143–152, 1984.

With low object contrast, modern films have a resolving power between 40 and 250 Lp/mm; when object contrast is higher, the resolving power is two to three times better.

Numerical Example. The width of a line pair for an aerial metric camera with $R_O = 90\,\mathrm{Lp/mm}$ was determined in Section 3.1.5.2. The question now arises, to what extent do these figures degrade when film with a resolving power $R_P = 100\,\mathrm{Lp/mm}$ is taken into account?

Equation (3.2-11): $1/R_T^2 = 1/90^2 + 1/100^2 \Rightarrow R_T = 67\,\mathrm{Lp/mm}$
Line pair width in image $= 1/67 = 0.015\,\mathrm{mm} = 15\,\mu\mathrm{m}$
Width of dark line (or of gap) in the image $= 15/2 = 7.5\,\mu\mathrm{m}$
Width of dark line (or of gap) at the object (image scale $1 : 6666$) $= 7.5 \times 6666 = 50000\,\mu\mathrm{m} = 5.0\,\mathrm{cm}$ (ignoring blooming, Section B 2.2)

The contrast transfer function, introduced in Section 3.1.5.4, provides more detailed information about imaging quality than the resolving power. Also in this section, the contrast transfer function $CTF(f)_O$ for the optics of a metric camera was introduced. The photographic emulsion has its own contrast transfer function $CTF(f)_P$. Both are combined to give the contrast transfer function $CTF(f)_T$ of the complete system (optics plus photographic emulsion) as follows:

$$CTF(f)_T = CTF(f)_O \, CTF(f)_P \tag{3.2-12}$$

An example is shown in Figure 3.2-11. It shows the contrast transfer function for film, in addition to two contrast transfer functions for the optics, one for the centre of the

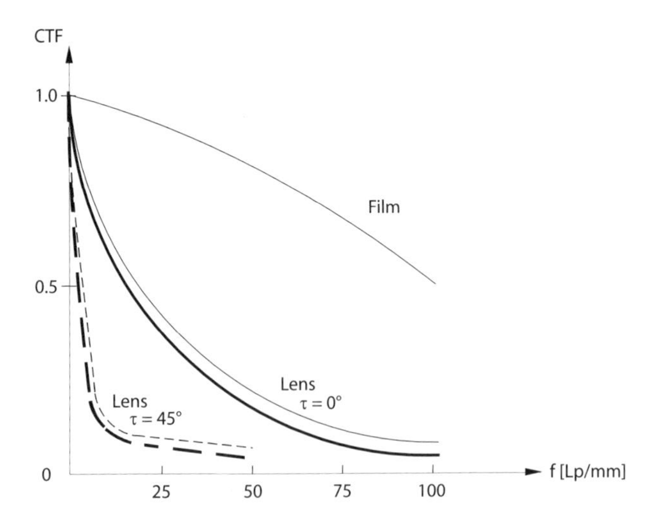

Figure 3.2-11: Contrast transfer function of an objective lens for $\tau = 0°$ and $\tau = 45°$, as well as for film. The thick lines show the resultant contrast transfer functions for a film-based camera.

image (off-axis angle $\tau = 0°$) and one for the edge of the image ($\tau = 45°$). The thick solid line is the total contrast transfer function resulting from the combination of film and optics ($\tau = 0°$) transfer functions according to Equation (3.2-12); the thick dashed line is the corresponding combination at ($\tau = 45°$).

3.2.2.8 Copying with contrast control

The density range of an aerial photograph, $\Delta D = D_{\max} - D_{\min}$, can reach a value as high as $\Delta D = 2.5$ (Section 3.2.2.3) as a result of shadows cast by objects or clouds on the one hand and by snow, rock or fields on the other, or as a result of the gradation of the film. This corresponds to a contrast of 320 : 1 (Section 3.1.5.3). Electronic copying devices permit the global density differences ΔD_g of large areas to be reduced while at the same time increasing the detail contrast ΔD_d (Figure 3.2-12). The negative is scanned by a cathode-ray spot. The light passing through the copy is sensed and either the intensity I or the scan velocity v is instantaneously changed. If copying material with a hard gradation is used ($\gamma = 4$, Figure 3.2-6), the detail contrast within the spot area (variable size between 2 and 144 mm^2) will be greatly increased. The electronic control of density reduction in larger areas is so adjusted as to give densities between about 0.3 and 1.4. A disadvantage is that at the edges of the photograph, as well as at other contrast edges, there is a 2 to 3 mm wide transition zone where illumination is not optimal and detail resolution is impaired.

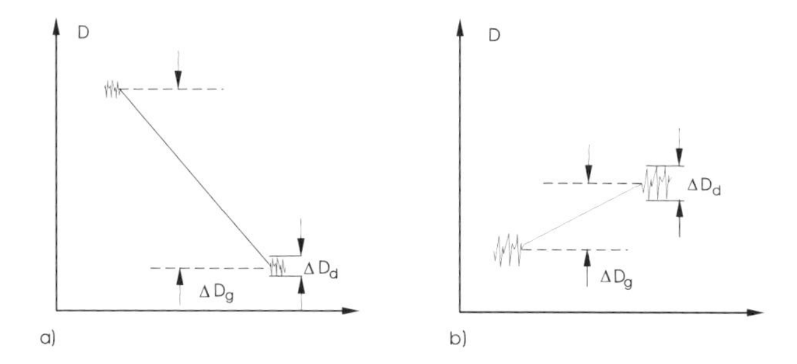

a) density distribution in negative
b) density distribution in positive with reduced global contrast ΔD_g and increased detail contrast ΔD_d

Figure 3.2-12: Copying with contrast control (abscissa = image plane)

3.2.3 Films for aerial photography

Table 3.2-5 shows details of nine typical aerial films. The resolving power of each film is given for two different object contrasts (Section 3.1.5.3). High contrast is represented by $K = I_{\max} : I_{\min} = 1000 : 1$ ($\log K =$ density difference $\Delta D = 3$). Similarly, low

Manu-facturer	Product name and film base	Type	Spectral sensitivity (λ = 400 – 600 – 800 nm; +3, +1, -1)	Thickness of base [mm]	General sensitivity DIN	AFS EAFS x	Recommended Developer	Photographic resolving power [Lp/mm] for object contrast K 1000:1	1:6.1	Gradation	Granu-larity ($RMSG$)
Agfa	Aviphot Pan 80 PE Polyester	Pan +		0.10	18 – 22	50 – 150 x	G 74 c	143	45	1.3 – 2.1	25
Agfa	Aviphot Pan 200 PE Polyester	Pan +		0.10	20 – 25	160 – 500 x	G 74 c	100	50	0.9 – 1.9	28
Kodak	Plus-X-Aerographic 2402 Estar	Pan +		0.10		200 125 – 400 x	Versamat 885	130	55	1.3 0.8 – 2.0	20
Kodak	Panatomic-X Aerographic II 2412 Estar	Pan +		0.10		40 32 – 64 x	Versamat 885	400	125	1.2 – 2.1	9
Kodak	High Definition Aerial 3414 Estar Thin Base	Pan +		0.06		8 4 – 16 x	Versamat 885	800	250	1.6 – 2.5	8
Kodak	Infra-red Aerographic 2424 Estar	IR		0.10		400 200 – 800 x	D-19 Versamat 885	80	40	2.3 1.0 – 2.1	30
Agfa	Aviphot Color X 100 PE1	Color Neg.		0.10	21 – 23	100 – 160 x	Agfacolor AP 70 Process	150	55		6
Kodak	Aerochrome II MS 2448 Estar	Color Pos.		0.10		32 x	AR-5	80	40	2.0	12
Kodak	Aerochrome III Infra-red 1443 Estar	Color IR-Pos.		0.10		40 x	AR-5	100	63		23

Table 3.2-5: Films for aerial photography. The terms "resolution" and "granularity" are treated in Section 3.3.3. Pan + emulsions have an extended spectral sensitivity for red light.

contrast is represented by a contrast $K = 1.6 : 1$ ($\log K$ = density difference $\Delta D = 0.2$). The different sensitivities of black-and-white films are needed to allow photographs to be taken in different seasons, even in late autumn. Panchromatic black-and-white films are used for topographic mapping, i.e. for wide-area recording of ground detail. Black-and-white infra-red films are more suitable when only vegetation is of interest, for example for a forestry inventory (Sections 3.2.2.6 and 3.7.5).

Colour and false colour (colour infra-red) films are primarily used for photo interpretation[23]. The human eye can differentiate many more shades of colour than shades of grey. Colour images contain significantly more information than black-and-white images. Since they have been available with a sufficiently fine grain and thin film base, all are suitable for the analysis of metric aerial photographs. Where film is still required, fine grained, medium sensitivity black-and-white films are still preferred in high precision photogrammetry.

Aerial films have a length from 26 to 140 m and are developed either in film developing machines[24] in a continuous process or in tanks with automatic forward and reverse winding. They are dried after processing in special drying machines in which the film is drawn slowly over a drum while clean air is blown onto it. The handling of the film must impose no mechanical stress and a leader and trailer, each of about 1.5 to 2 m length, must be provided.

3.3 Photoelectronic image recording

It is becoming much less frequent for a photographic emulsion to be used as the sensor in the image plane of a metric camera. Instead, an electronic sensor which directly supplies a digital image (Section 2.2.2) is becoming more common. In Section 3.3.1 the principle of opto-electronic sensors is presented. These devices are suitable for digitally recording optical radiation, that part of the electromagnetic spectrum covering visible and infra-red light (Figure 3.2-2). Sections 3.3.2 and following discuss the quality of image recording by opto-electronic sensors.

3.3.1 Principle of opto-electronic sensors

Solid state sensors are exclusively used nowadays for digital image recording in photogrammetric cameras. They consist of a large number of detectors which register the photons of the incident light falling on the image plane of the camera. The principle of opto-electronic image recording by means of semi-conducting elements (preferentially made of silicon) is illustrated in Figure 3.3-1 for a linear array sensor. In the semi-conducting elements under the electrodes, the photons of the incident light build up an electrical charge by creating electron-hole pairs. This charge is proportional to the number of incident photons (the photo-electric effect).

[23]Literature: Arnold, R.H.: Interpretation of Airphotos and Remotely Sensed Imagery. Waveland PR Inc., 2004.

[24]For example, the Kodak Versamat Ektachrome and Aerochrome RT Processor 1811 (1984).

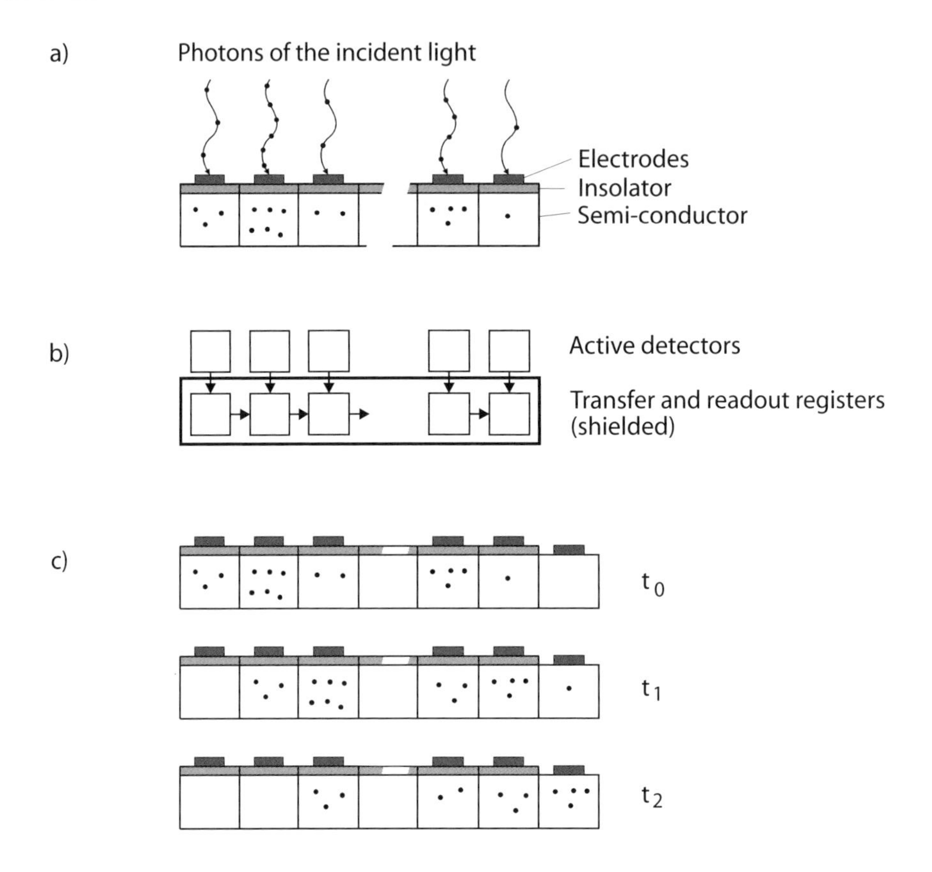

a) generation of charge in semi-conductor according to the intensity (number of photons) of the incident light.
b) transfer and read-out registers arranged next to the active detectors
c) read-out principle by means of CCD (charge coupled device): the charges are shifted at regular time intervals $(t_0, t_1, t_2, \ldots)$ by one element to the right and quantized when they reach the last element.

Figure 3.3-1: Linear array sensor[25]

The varying charges in the individual semi-conducting elements must be read out. For this purpose the charges are passed in parallel to a transfer register and from there read out in series (Figure 3.3-1b). Figure 3.3-1c shows the charges after they have been passed to the transfer register. The quantities of charge here are the ones produced by the photons in the active detectors.

[25]With acknowledgement to Schenk, T.: Digital Photogrammetry. TerraScience, 1999. Figure 3.3-2 is also taken from this source.

The transfer register also consists of semi-conducting elements which are sensitive to light; they must therefore be covered. To determine the amount of each charge, the transfer register has an additional element with its own electrode. At time t_0 this holds no charge. At time t_1 all the charge packets are shifted by one element to the right (Figure 3.3-1c). With the electrode at this additional element, the amount of charge currently resident can be converted into an analogue electrical signal which can subsequently be digitized. At time t_1 this charge is the one generated by the last active detector in the array, at time t_2 it is the charge from the last but one detector, and so on.

This technology based on charges coupled with semi-conducting elements is today dominant. This gives rise to the abbreviation CCD for charge coupled device, with the abbreviation itself in common use, for example in reference to CCD cameras.

Figure 3.3-1 shows a CCD linear array sensor. For photogrammetric requirements, CCD area arrays are more appropriate. Figure 3.3-2 shows such a CCD area array sensor. It consists of:

- a number of active detectors arranged in rows
- transfer registers between the rows
- a serial read-out register

In this case only half the image plane is covered by active detectors; the other half is taken up by the transfer registers. This configuration is known as an interline transfer sensor. Since this development there have been other technical solutions in which the entire image plane is covered with active detectors (full frame sensor).

Some current technical details for CCD sensors prove the high development level of this technology but also show its limits. At present, the largest CCD area arrays have $4096 \times 4096 = 4\text{K} \times 4\text{K} = 16\text{M}$ detectors and CCD linear arrays have up to 20K detectors.

To achieve a high resolution, which is crucially dependent on the number of detectors, mapping flights dispense with CCD area arrays and instead use linear arrays, for example with 20k detectors, in a push broom configuration. This ensures that 20K detectors determine the resolution rather than 4K detectors; resolution is therefore increased by a factor of 5.

A digital line camera with a CCD linear array is shown in Figure 3.3-3, left. For each image line there are individual elements of exterior orientation. In a digital line camera, the six elements of exterior orientation therefore change from imaged line to imaged line. In metric cameras which use CCD area arrays or film, the six elements of exterior orientation only change between each full two-dimensional image (from image plane to image plane).

For the spatial recording of an object it is a condition that each object element is imaged by two rays from different directions. The digital line camera (Figure 3.3-3, left) does not meet this requirement within a flying strip. If instead a 3-line camera is employed

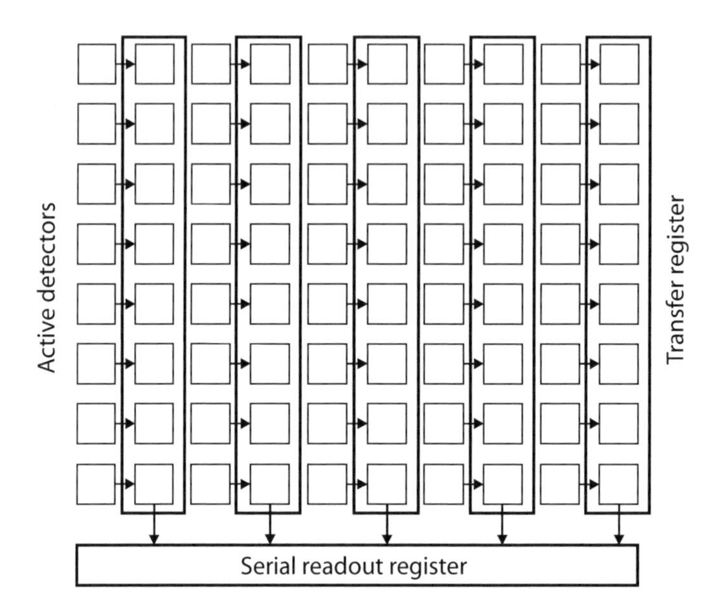

Figure 3.3-2: CCD area array sensor in interline transfer mode

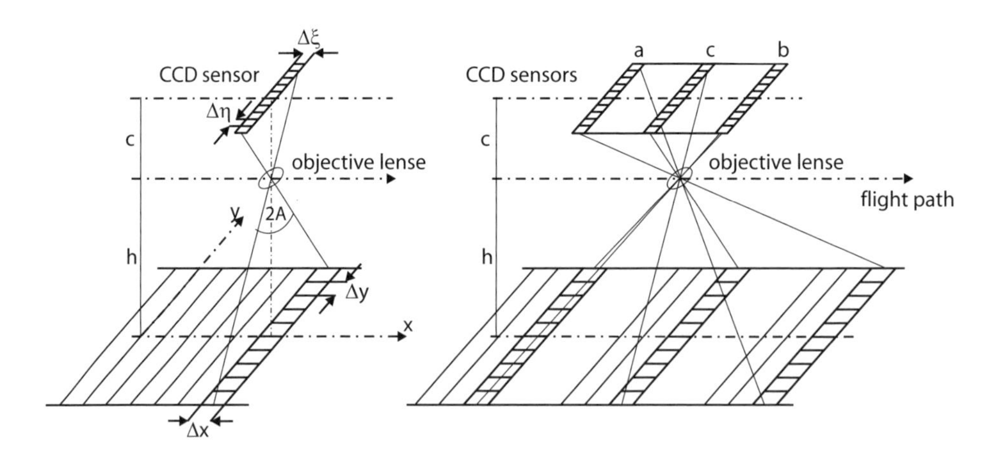

Figure 3.3-3: Line camera (left) and 3-line camera (right)

(Figure 3.3-3, right) then every object element within a flying strip is recorded from not just two but three directions; objects recorded in this way can be spatially reconstructed.

If the time required to read out the charges is significantly longer than the reaction time to build up the charges in the detectors, and if the CCD camera, for example in an aircraft, moves during this time, then a mechanical shutter of the type used in film cameras must be employed to limit the exposure time. A mechanical shutter is not necessary in CCD cameras where the read-out frequency is very high, particularly in line cameras.

Boyle, W.S., Smith, G.E.: Bell Systems Technical Journal, Volume 49, pp. 587–593, 1970 (a key publication relating to CCD technology). Graham, R.: Digital Imaging. Whittles Publishing, 1998. Toth, C.: In Fritsch, D., Spiller, R.: Photogrammetric Week '99, Wichmann Verlag, pp. 95–107, 1999.

3.3.2 Resolution and modulation transfer

The recording quality of opto-electronic image sensors, in terms of their resolution and contrast transfer, is primarily influenced by the size of the pixels $\Delta\xi \times \Delta\eta$. The size of the pixel at the object is given by projecting the surface of an individual detector element through the camera optics onto the surface of the object, for example the Earth's surface. For vertical images taken of flat ground, the size of a pixel $\Delta x \times \Delta y$ on the ground surface, commonly referred to as "the footprint" (Figure 3.3-3), is given by:

$$\Delta x = \frac{h}{c}\Delta\xi \qquad \Delta y = \frac{h}{c}\Delta\eta \tag{3.3-1}$$

$h \ldots$ flying height, $c \ldots$ principal distance or focal length

For images taken with a CCD sensor, the illumination intensity I within a pixel is averaged to a representative value $\bar{I}$. Figure 3.3-4 shows the relationships for an intensity profile along the y direction, i.e. at right angles to the direction of flight.

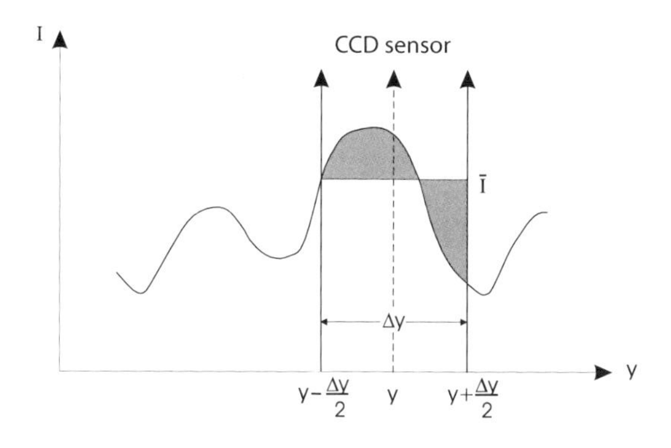

Figure 3.3-4: Average intensity value within a pixel

The illumination intensity function $I(y)$ along a profile can be given as a Fourier series:

$$I(y) = I_0 + \sum_{k=1}^{\infty} C_k \cos(2\pi F_k y - \varphi_k) \tag{3.3-2}$$

I_0 ... Average intensity value along the entire profile length L
F_k ... Spatial frequencies of the individual component waves of the series. They are the reciprocals of the wavelengths L_k, i.e. $L_k = 1/F_k$. The wavelengths derive from $L_k = L/k$, where k runs from 1 to (theoretically) ∞.
C_k ... Amplitudes of the component waves
φ_k ... Phases of the component waves.

A representative mean value $\bar{I}$ within a pixel is given by the following integral taken over the range $[y - \Delta y/2, y + \Delta y/2]$:

$$\bar{I}(y) = \frac{1}{\Delta y} \int_{y-\frac{\Delta y}{2}}^{y+\frac{\Delta y}{2}} I(z)dz \tag{3.3-3}$$

The solution to this integral is:

$$\begin{aligned} \bar{I}(y) &= I_0 + \sum_{k=1}^{\infty} \frac{\sin(\pi F_k \Delta y)}{\pi F_k \Delta y} C_k \cos(2\pi F_k y - \varphi_k) \\ &= I_0 + \sum_{k=1}^{\infty} C'_k \cos(2\pi F_k y - \varphi_k) \end{aligned} \tag{3.3-4}$$

Comparison with the original intensity function $I(y)$ indicates that the amplitudes C_k are attenuated to the values C'_k as a result of imaging with the detectors. In Section 3.3.5.3, the contrast transfer function $CTF(F_k)$ was defined using the ratio of the amplitudes of a square wave function. Since signal analysis is based on the theory of Fourier series, the usage of square waves is replaced by sine waves, and consequently, the term “contrast transfer function” by “modulation transfer function” $MTF(F_k)$. The MTF can be stated mathematically as[26]

$$MTF(F_{k'}\Delta y_D) = \frac{C'_k}{C_k} = \frac{\sin(\pi F_k \Delta y_D)}{\pi F_k \Delta y_D} \tag{3.3-5}$$

Here the pixel size in the ground is no longer designated by Δy but by Δy_D which is the footprint of the detector element D.

In Figure 3.3-5 the modulation transfer function is plotted against $F_k \Delta y_D$ and $L_k/\Delta y_D$. The abscissa gives the intensity as a function of frequency F_k and the lower scale as a function of wavelength L_k.

The modulation transfer function (3.3-5) becomes unity[27], i.e. there is no attenuation of the intensity profile $I(y)$, when either the detector size Δy_D or the frequency F_k becomes zero. The first case implies an infinite number of infinitesimally small detectors, the second case that the intensity profile $I(y)$ has a constant value within the footprint, i.e. an infinitely long wavelength.

[26] The function $\sin x/x$ is known as $\mathrm{sinc}(x)$ function, which resembles a damped sinusoidal oscillation. Figure 3.3-5 shows the first positive lobe of the function.

[27] The limit of $\sin(x)/x$ is unity when $x \rightarrow 0$.

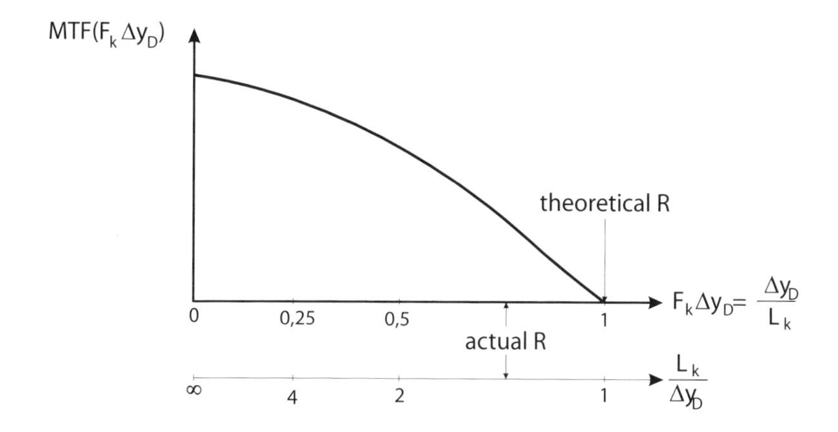

Figure 3.3-5: Modulation transfer function MTF and resolution R of a detector

The modulation transfer function (3.3-5) becomes zero, i.e. the variations in the intensity profile are eliminated in the imaging process, when $F_k \Delta y_D = 1$. This occurs when the wavelength of the intensity variations has the same size Δy_D as the footprint, i.e. $L = \Delta y_D$. Any shorter wavelength, where $L < \Delta y_D$, cannot be recorded correctly, even though non-zero values may be obtained. The limiting case, where MTF becomes zero, corresponds to the point where the resolution can no longer be digitized with that detector. (A related definition of resolution for an analogue photograph is shown by Figure 3.1-21.) This is a theoretical resolution. Due to the limited sensitivity of the detector, the actual resolution lies around the value $F_{\max} \Delta y_D \approx 0.7$ (where the amplitude of function (3.3-5) reaches some 40% of its maximum) and is dependent on the opto-electronic sensor used (Figure 3.3-5). In practice, the width of a line pair $L_{\min}$ which can just be resolved by a detector with a projected extent on the object surface of Δy_D is therefore:

$$\text{Width of line pair } (\hat{=} L_{\min} = 1/F_{\max}) \approx 1.4 \Delta y_D \tag{3.3-6}$$

Between the limits $0 < F_k < F_{\max}$ the intensity amplitudes are attenuated. The higher the frequency (or smaller the wavelength), the larger the attenuation. This attenuation can be corrected by use of an inverse filter (Section 3.5.2.2). The above considerations are not sufficient for deciding if a certain wave pattern, with frequencies higher than the limiting case, can potentially be reconstructed from a digital image, since another parameter, the sampling interval, also plays an important role. One should bear in mind that the considerations in this section presume an infinitesimally narrow sampling interval, i.e. continuous sampling. For a proper assessment of the frequency-related digitization quality, detector size and detector spacing must not be treated independently.

3.3.3 Detector spacing (sampling theory)

The previous section concentrated on the imaging properties within the area of a single detector element, which was continuously moved across the object. The shortest intensity waves which can be resolved by a detector with dimension $\Delta\eta_D$ have a wavelength of $l_{\text{min}} = 1.4\,\Delta\eta_D$, corresponding to a frequency of $f_{\text{max}} = 0.7\,/\Delta\eta_D$. Following Equation (3.3-1), the dimensions Δy_D, L_{min} and F_{max}, which refer to the object, can be replaced by $\Delta\eta_D$, l_{min} and f_{max} which refer to the image. The shortest wave component which can, with certainty, be resolved by a detector of size $\Delta\eta_D$ is represented in Figure 3.3-6.

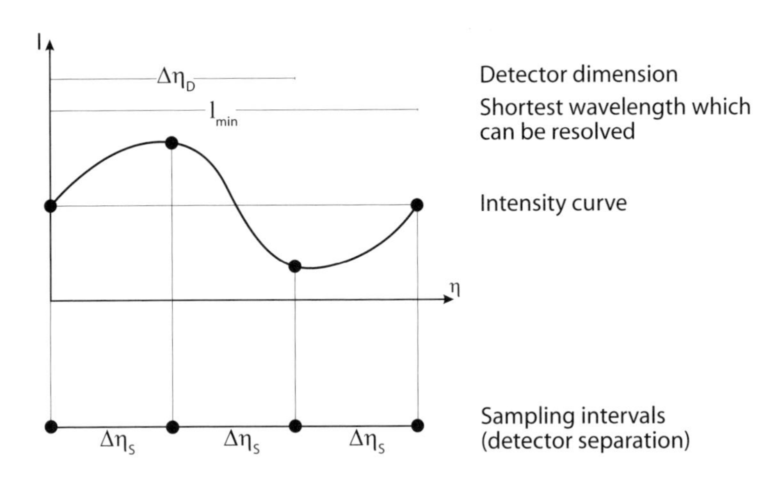

Figure 3.3-6: Detector resolution and corresponding detector separations

In the previous section an infinitesimally narrow spacing $\Delta\eta_s$ of the detector has been presumed. In practice, this spacing is finite. The question now arises as to the spacing required between detectors in order to record the shortest component wavelength in an image. The sampling theorem states that the sampling interval $\Delta\eta_s$ must be smaller than half the wavelength of the sampled wave[28]. Figure 3.3-6 shows an adequate sampling interval $\Delta\eta_s$, which, in the case of a CCD camera, corresponds to the detector spacing. It can be seen that the detector spacing must be significantly smaller than the detector dimension $\Delta\eta_D$[29].

The sampling interval or detector separation $\Delta\eta_s$ can also be related to the resolution R, expressed, for instance, in line pairs per millimetre (Lp/mm). Here the width of a line pair corresponds to the wavelength l_{min} in Figure 3.3-6, yielding the maximum

[28]The sampling theorem is also known as the Shannon-Nyquist Theorem and the limiting frequency as the Nyquist frequency. (Jerri, A.J.: The Shannon Sampling Theorem—Its Various Extensions and Applications: A Tutorial Review. Proc. of the IEEE 65(11), 1977.)

[29]In commercially manufactured CCD sensors, the reverse is the case (Figures 3.3-1 and 3.3-2). By using an array of microlenses in front of the detectors, the effective (and larger) detector area is imaged onto the actual and physically smaller detector surface.

detector spacing $\Delta\eta_s$:

$$\Delta\eta_s\,[\text{mm}] \leq 1/(2R\,[\text{Lp/mm}]) \qquad (3.3\text{-}7)$$

In practice, the numerator 0.7 is chosen[30], as indicated in Figure 3.3-6, i.e.

$$\Delta\eta_s\,[\text{mm}] = 0.7/(2R\,[\text{Lp/mm}]) \qquad (3.3\text{-}8)$$

If the intensity profile is sampled at an interval larger than the values derived from Equation (3.3-8), the result is an undersampling. Artificial (low-frequency) patterns are then generated in digital images, caused by the interference of the higher frequencies in the image with the frequency of the sampling interval. This effect is known as aliasing[31]. A demonstrative example can be seen in Figures 2.2-7 to 2.2-9, where artifacts appear in the window blinds. The sampling interval has a value which clearly cannot fully record the original texture in the slats of the blinds.

Equation (3.3-6) permits a comparison between the resolution of a CCD sensor and photographic film. From Table 3.2-5, a film with a resolution of 100 Lp/mm can be found[32]. This corresponds to a line pair width of 10 μm in the image. In order to achieve the same resolution with an electronic sensor, the pixels must have the following size (derived from Equation (3.3-6): $\Delta\eta_S = 0.7/(2 \times 100) = 3.5\,\mu$m). The analogue photograph provides the given resolution across an image format of 23 cm $\times$ 23 cm. In order for CCD technology to deliver the same quality just by replacing the film plane by a CCD chip, it would require 66000 detector elements ($= 230000/3.5$) along one side of the image. It is difficult to imagine such a number of pixels in area CCD arrays, even in the long term.

In area CCD arrays, sampling interval and detector size are more or less identical and given by the detector arrangement on the sensor plane. In the case of linear array sensors, as they are used in line cameras, the same is valid within the line, while perpendicular to the sensor line, i.e. in the direction of flight, the sampling interval is variable and depends on the relation between flying speed above ground (projected into the image plane) and readout rate. Flight planning has to take this into consideration, particularly in order to avoid undersampling.

Exercise 3.3-1. For a camera with an image format of 23 cm $\times$ 23 cm, there is an equivalence in resolution of 100 Lp/mm between film and a (fictitious) electronic sensor having 66000 $\times$ 66000 electronic detectors. Nowadays, CCD sensors with 10000 $\times$ 10000 elements with 9 μm pixel width are realistic. What is the corresponding resolution in the image plane? If the flying height is kept the same as with a film

[30]Note: the factor 0.7, which has been applied in the previous section to determine the largest detector size, is not related to the factor 0.7 here. While in the first case the detector size has been reduced in order to detect a reasonably high amplitude (noticeably greater than zero), here the detecting interval is reduced in order to guarantee the reconstruction of a certain minimum wavelength (noticeably different from the Nyquist frequency).

[31]Smoothing filters, which suppress high frequencies (prior to digitization) in order to avoid aliasing, are known as anti-aliasing filters.

[32]Note: the specified resolution is valid for the film and not necessarily for an exposed image. A photograph's resolution also depends on negative effects due to the lens, filters and atmospheric disturbances. How the cumulative resolution may be calculated, can be found in Section 3.2.2.7.

camera, how must the principal distance c be changed for the CCD camera in order to achieve the same resolution on ground as film would deliver? What ground area would be covered compared to the film camera? (Solution: c must be 2.57 times longer. Area covered by the digital camera would be just 2.3% of the area covered by the film camera of the same resolution.)

Exercise 3.3-2. Reconsider the situation in the case where the camera optics have a resolving power of 90 Lp/mm (Hint: Make use of Equation (3.2-11)). (Solution: 67 Lp/ mm.)

Further reading for Sections 3.3.2 and 3.3.3: Schenk, T.: Photogrammetry. Terra Science, 1999. Holst, G.C.: CCD Arrays, Cameras and Displays. SPIE Press, JCD Publishing, 1996. Inglis, A.F., Luther, A.C.: Video Engineering. 2nded., McGraw-Hill, 1996.

3.3.4 Geometric aspects of CCD cameras

When CCD cameras are employed for photogrammetric purposes it is, as a rule, necessary to have knowledge of the interior orientation. The required parameters are defined in the calibration certificate (Section 3.1.1) and are:

- the principal distance
- the coordinates of the principal point of autocollimation and principal point of best symmetry
- the objective lens distortion.

In addition, CCD cameras must also provide information on the positions of the active detector elements and the planarity of the recording surface defined by the surfaces of the detectors. Nowadays, detectors can be arranged in an orthogonal grid with a positional accuracy of around 0.2 μm, a deviation which can normally be ignored in photogrammetric processing. In contrast, lack of flatness can be in the order of 10 μm. This should be taken into account in precise analyses, particularly where a wide field of view is employed.

In digital photogrammetry, the correction of lens distortion is similar to that in analytical photogrammetry (Section 3.1.3). The pixels must be "shifted" by the correction values $\Delta\xi$ and $\Delta\eta$ (Equation (3.1-2)). This would disturb the pixels from their arrangement in a strictly orthogonal grid. In order to preserve a strictly orthogonal structure which is convenient for image processing, a resampling is required, as described in Section 2.2.3. However, it is expedient to adopt a virtual image in the solution, as explained in Section 4.5.4.

Correction for lack of flatness is done in a similar way and in the following steps:

- convert the $\Delta\zeta$ (height) deviations of individual detector elements into radial shifts $\Delta\rho$ in the image plane using the formula $\Delta\rho = \Delta\zeta \tan\tau$, where τ is the off-axis angle of the imaging ray on the image plane
- resolve $\Delta\rho$ into component corrections $\Delta\xi$ and $\Delta\eta$ by means of Equation (3.1-2)
- continue the process by means of a virtual image (Section 4.5.4)

In cameras which work with video analogue signals and frame grabbers (Section 3.8.7), pixel positioning is done through line synchronization. Irregularities which may occur in this process can lead to displacements between image lines (line jitter) and also within individual lines (pixel jitter). A typical line shift is illustrated in Figure 3.3-7. The correct image is indicated on the left. The centre diagram shows the image recorded with displaced lines and pixels; the output using pixels in an orthogonal grid (right) results in a distorted image.

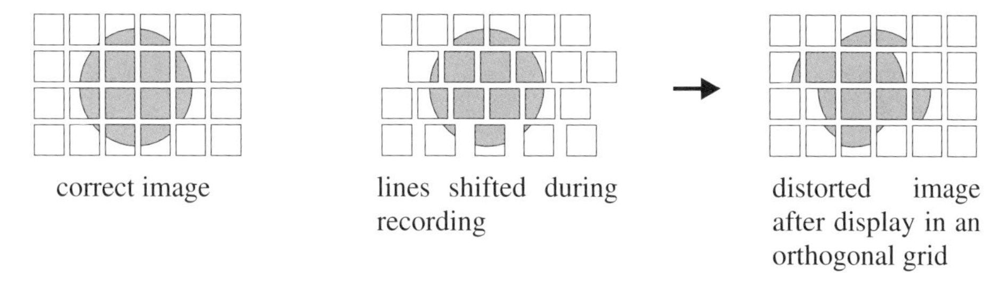

Figure 3.3-7: Effect of line jitter[33]

Finally it should be noted that the electronic signals in analogue videocameras drift by relatively large amounts during the warm-up phase. In practice, therefore, a corresponding time delay should be observed before use.

Further reading: Luhmann, T., Robson, S., Kyle, S., Harley, I.: Close Range Photogrammetry. Whittles Publishing, 2006. Shortis, M., Beyer, H.: In Atkinson (ed.): Close Range Photogrammetry and Machine Vision. Whittles Publishing, Caithness, UK, pp. 106–155, 1996.

3.3.5 Radiometric aspects of CCD cameras

3.3.5.1 Linearity and spectral sensitivity

The electrical signals delivered by CCD sensors are largely proportional to the number of photons incident on the detectors. Departures from linearity can be held to within

[33]Taken from Dähler, J.: Proceedings of the ISPRS Intercommission Conference on "Fast Processing of Photogrammetric Data", pp. 48–59, Interlaken, 1987.

1% of the true value[34] but there are very large departures from linearity when the semiconductor elements are close to saturation. In fact, very large amounts of radiation cause an "overflow" into the neighbouring detector elements. The result is blooming, similar to the effect which also occurs with very bright objects imaged onto film.

The spectral sensitivity (Section 3.2.2.6) of the silicon sensors normally used ranges from 400 to around 1100 nm, i.e. from visible light into near infra-red, thus considerably wider than that with analogue photography.

3.3.5.2 Colour imaging

Imaging in multiple spectral regions, i.e. colour imaging, is possible in several ways with CCD cameras. One variant, which is preferred when employing CCD area arrays, involves the use of a cluster of several cameras. The individual cameras in the cluster are provided with identical sensors, each sensitive in the range 400 to 1100 nm. However, different filters which select different parts of the spectral region are attached to the front of the objective lenses. For example, if normal colour images are required then there are three cameras in the cluster. The first camera has a blue filter, the second a green filter and the third a red filter (Figure 3.2-2). If a colour infra-red image is required then the first camera has a green filter, the second a red filter and the third an infra-red filter (Figures 3.2-2 and 3.2-10). To display a colour infra-red image, for example on a colour monitor, the image matrix from the first camera is represented in blue, the second in green and the third in red (see Figure 3.2-9).

A cluster solution is only realistic for satellite and aerial photography. In these cases the imaging distances are so large that, despite the small relative displacements of the cameras, identical images are obtained. At close ranges an alternative solution is required. One solution makes use of a single camera with a single detector field. The imaging for the three separate colours takes place sequentially. For each exposure a different filter from one of the sets mentioned above is positioned in front of the lens. Between each exposure, the object and the camera's interior and exterior orientation must remain fixed.

Another solution also employs a single camera with a single detector field. This solution achieves colour separation with a filter mask placed in front of the detectors. There are various arrangements of filters. Figures 3.3-8 shows two arrangements of (R)ed, (G)reen and (B)lue filters. For natural objects, the distribution of the green intensities comes close to the brightness distribution. For capturing images with a wide range of brightness levels, the right-hand arrangement has a certain advantage. Full image arrays for each colour are then created by interpolation between the pixels of the individual colours.

There is a very elegant solution for colour images from digital line cameras. The rays from every object point are divided into three separate and displaced components by means of a beam splitter. Before each component ray then reaches one of three

[34]Dierickx, B.: OEEPE-Publication 37, p. 66, 1999. This publication also discusses CMOS technology which may one day replace CCD technology.

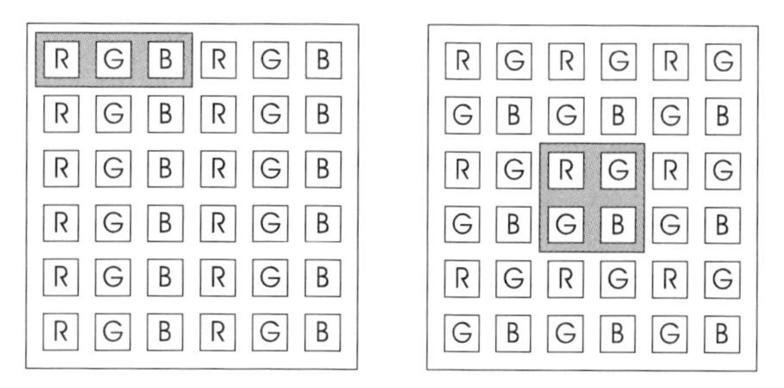

Figure 3.3-8: Filter masks for CCD colour imaging

parallel linear CCD arrays, it passes through a colour filter. The type of filter used is an interference filter which provides a better colour separation than the individual layers of colour film whose spectral sensitivities have a relatively significant overlap (see Figure 3.2-10).

All the techniques of colour separation presented here can be applied beyond the three basic colours. Particular importance is attached to imaging in the four spectral regions blue, green, red and near infra-red. In contrast, analogue photography can only simultaneously record three colour regions.

3.3.5.3 Signal-to-noise ratio

The radiometric sensitivity across the whole spectrum, comparable to the general sensitivity of film (Section 3.2.2.4), is defined for CCD sensors by the signal-to-noise ratio, SNR:

$$SNR = \frac{S}{\sigma_s} \tag{3.3-9}$$

S ... electric signal intensity, σ_s ... standard deviation of system noise

An SNR value of unity means that the signal S, which contains the relevant information, has the same magnitude as the noise σ_s: in this case the signal disappears amongst the noise and no relevant information can be extracted from it.

CCD sensors can reach a maximum SNR of approximately $1000 : 1$ and sometimes more. The maximum SNR is the dynamic range DR_S of the sensor. In order to quantize the recorded illumination intensity, a step size interval of $2\sigma_s$ should be chosen[35], i.e. in this example the dynamic range of the digital image $DR_I = DR_S/2 = 1000/2 = 500$; a pixel must therefore be represented by 9 Bits ($2^9 = 512$).

Noise σ_s has various sources. Amongst these are dark current noise in which photons are generated by thermal effects rather than incident light from the object, charge transfer noise due to inefficiencies in transferring charges between neighbouring detectors during the read-out process, transmission noise in which signals are deformed during

[35] Baltsavias, E. in Fritsch/Spiller: Photogrammetric Week '99, Wichmann Verlag, pp. 155–173, 1999.

transmission and before analogue-digital conversion and quantization noise caused by rounding errors in the analogue-to-digital conversion.

A test card with homogeneous surface properties and illumination is used for the determination of detector noise, a method comparable with the determination of film granularity (Section 3.2.2.7). The noise σ_s can be determined from the variation in grey level. Variations in illumination and surface properties can falsify results. It is possible to allow for these effects by making a series of exposures with the same exterior orientation and, from the spread of results for individual pixels then deduce the standard deviation σ_s.

The signal-to-noise ratio *SNR* depends on various parameters. In the following discussion, only the principal parameters for the *SNR* as a general function $f(\,)$ will be evaluated:

$$\mathit{SNR} = f(\Delta\eta, \Delta t, \dots) \tag{3.3-10}$$

$\Delta\eta$ ... Detector size (Section 3.3.2)
Δt ... Available integration or exposure time for recording photons

The parameters of function (3.3-10) interact with one another. The *SNR*, which can also be called radiometric resolution, improves under the following conditions:

- with larger $\Delta\eta$, which means that geometric resolution is worse
- with a longer exposure time

Both types of resolution (radiometric and geometric) must be carefully balanced according to priorities in the construction of imaging sensors. In photogrammetry, geometric resolution has the higher priority. When resolution is discussed in photogrammetry, then geometric resolution is implied.

The resolution types mentioned are also relevant to analogue photography. Radiometric resolution is expressed by the general sensitivity (Section 3.2.2.4) or the granularity (Section 3.2.2.7). Geometric resolution is given in the form of line pairs/millimetre (Lp/mm).

3.4 Digitizing analogue images

The explanations above have shown that the geometric resolution of analogue photography is significantly higher than the geometric resolution of current opto-electronic (area array) sensors. For many years to come, therefore, objects will be recorded by light-sensitive film in metric cameras when executing the most accurate photogrammetric tasks. However, in order to apply the methods of digital photogrammetric analysis, film can be digitized after development.

3.4.1 Sampling interval

The sampling interval, which equates to the distance between each pixel in an orthogonal grid, must be compatible with the resolution in the analogue image in order to avoid loss of information. Section 3.2.2.7 discussed the geometric resolution R_T of a "photographic emulsion" sensor. For current films it varies between 50 and 100 Lp/mm. In a similar way to the sampling theorem, the digitization sampling intervals $\Delta\xi_s$ and $\Delta\eta_s$ in both coordinate directions should be chosen as follows[36]:

$$\Delta\xi_s\,[\text{mm}] = \Delta\eta_s\,[\text{mm}] = 0.7/(2R) \tag{3.4-1}$$

Numerical Example. Adequate sampling intervals for the analogue images mentioned above are between 3.5 and 7 μm (e.g.: $\Delta\xi_s = \Delta\eta_s = 0.7/(2 \times 50) = 0.007\,\text{mm} = 7\,\mu\text{m}$).

The sampling interval should not be confused with the size of the detector elements. In an ideal situation there is no difference between the detector dimensions and the sampling interval. This concept lies at the basis of Figure 2.2-3 and many related diagrams. However, in Section 3.3.3 it was shown that size of the detectors could certainly be larger than the sampling interval. A larger detecting surface also improves the radiometric resolution (Section 3.3.5.3).

The sampling interval required on the basis of the geometric resolution of an analogue image, generates a very large amount of data. To store a pixel generally requires 8 bits corresponding to 1 byte. At a sampling interval of 7 μm, a 23 cm $\times$ 23 cm metric image requires a storage capacity of approximately 1 gigabyte, where 1 GB $\approx (230000/7)^2$. For a colour image it would be 3 GB. This quantity of data only applies to a single image. In digital photogrammetry, data is required from many images, although only a few must be simultaneously resident in computer memory. From these considerations and many others (e.g. Section C 1.1.1.1, Volume 2), when scanning images a sampling interval between 7 and 30 μm is chosen in practice.

3.4.2 Grey values and colour values

The discussion will initially be limited to the digitizing of an analogue black-and-white photograph. The grey values are measured by means of a densitometer or several densitometers (Section 3.2.2.3). A densitometer transmits a constant stream of light φ_0 and uses a CCD sensor to measure the component not absorbed by the film. The light source and CCD sensor are illustrated in Figure 3.4-1 for a flat-bed scanner, a widely used design. The film carrier is moved with respect to the light source and sensor. The right hand side of Figure 3.4-1 indicates that a diffuse light source is employed and that several CCD detectors receive the component of the light which is not absorbed.

[36] It is interesting to note the connection between the sampling interval, i.e. pixel size Δ, and the parameter dots per inch [dpi], employed in the printing industry. Making use of the footnote in Section 3.1.5.2, the relationship is: $R\,[\text{dpi}] = 25.4 \times 1000/\Delta\,[\mu\text{m}]$. A 7 μm pixel therefore corresponds to 3629 dpi $= 25.4 \times 1000/7$.

[37] Taken from Kölbl, O.: OEEPE-Publication 37, pp. 135–150, 1999.

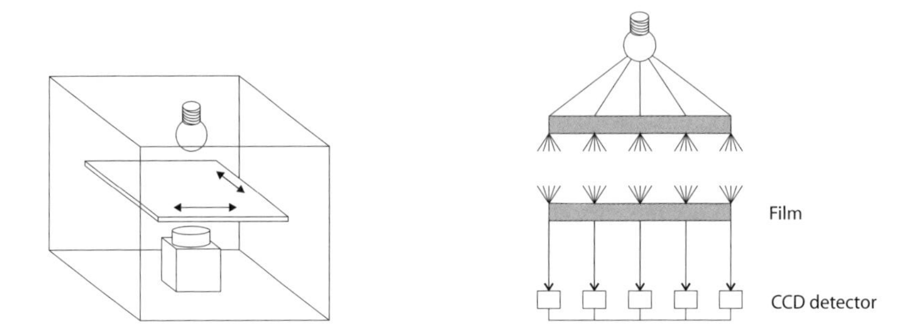

Figure 3.4-1: Flat-bed scanner with diffuse light source[37]

The CCD detectors measure the light which is not absorbed and there exists a linear relationship between this quantity of light and the charges developed in the semiconductors (Section 3.3.1). The charges are quantized and converted by the scanner into grey values g. The logarithms of the grey values g must have a linear relationship with the densities of the corresponding elements of the film (Section 3.2.2.3).

Figure 3.4-2 shows the relationship between the density D of the exposed and developed film and the logarithm of the quantized grey values g for a specific scanner. There is good linearity up to a density of 2.0 but not in darker regions. The steep curve in very dark regions indicates that here the scanner does not satisfactorily reproduce the density gradations.

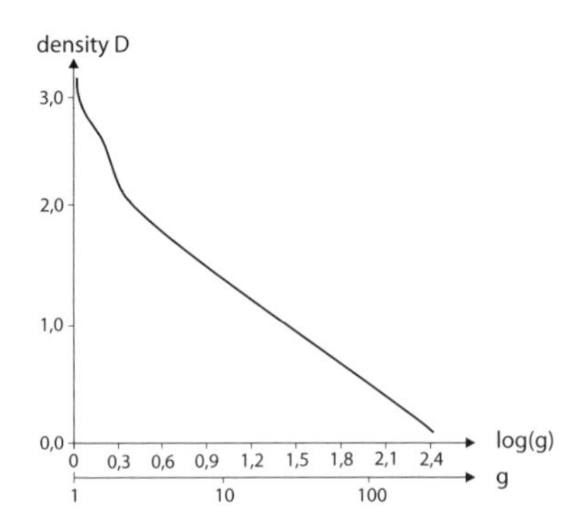

Figure 3.4-2: Density of a reference pattern with respect to grey value g (Scanner: DSW 300 from LH Systems, pixel size 12.5 μm)[38]

Practical tip: Film negatives often generate high densities. To create the conditions for optimal digitizing, a contact copy onto transparency film (Section 3.2.2.8) is recommended. The layers in positive colour film are generally highly saturated. When digitizing colour films, the radiometric demands on the scanner are therefore greater

[38]Taken from Baltsavias, E., Kaeser, Ch.: OEEPE-Publication 37, pp. 111–134, 1999.

than when digitizing black-and-white films. Figure 3.4-2, in fact, results from calibration with a colour film.

Black-and-white images have a density range of $0.1 - 2.0D$, positive colour films a range of $0.3 - 3.5D$ (source: see footnote). The density dispersion σ_D of developed film lies around ± 0.006 in areas with a density of $1D$ (Table 3.2-5 shows the $RMSG$ which is 1000 times greater than σ_D; see also Table E 4.2-1, Volume 2). For less dense regions, say $0.3D$, σ_D lies around ± 0.0043. If a quantization step corresponding to twice the standard deviation of the density dispersion is chosen for the detector charges (Section 3.3.5.3), then the following dynamic ranges DR_I for grey and colour levels are obtained:

- black-and-white film: $DR_I = 2.0/0.0086 \approx 223$
- colour film: $DR_I = 3.5/0.0086 \approx 407$

8 bits ($2^8 = 256$) are therefore just sufficient to represent grey values in black-and-white films. For colour films the 8 bits normally used for image data are clearly too few.

The interaction between geometric and radiometric resolution, mentioned in Section 3.3.5.3, also applies to scanning. For example, sampling with large pixels delivers a better radiometric quality, i.e. a better signal-to-noise ratio, than sampling with small pixels (Table E 4.2-1, Volume 2, provides empirical values).

3.4.3 Technical solutions

In addition to flat-bed scanners (Figure 3.4-1), there also exist drum scanners where the film is mounted onto a cylindrical drum. Drum scanners are widely used in the printing industry. In photogrammetry, flat-bed scanners take precedence; they generally have a smaller format than drum scanners.

With regard to the arrangement of light-sensitive detectors, there are three different principles of construction:

- single detector, which is driven line by line across the film, usually by means of rotation
- linear array, which is moved in strips across the film. Several thousand detectors are arranged in a single array
- area array, which records partial images. These partial images are combined into a complete image by means of a réseau whose crosses have known coordinates. The réseau can be marked on a glass plate which is placed on the film (Figure 3.4-3). The réseau can also be introduced into the image during exposure. This requires a réseau to be incorporated into the recording film camera (Section 3.8.4).

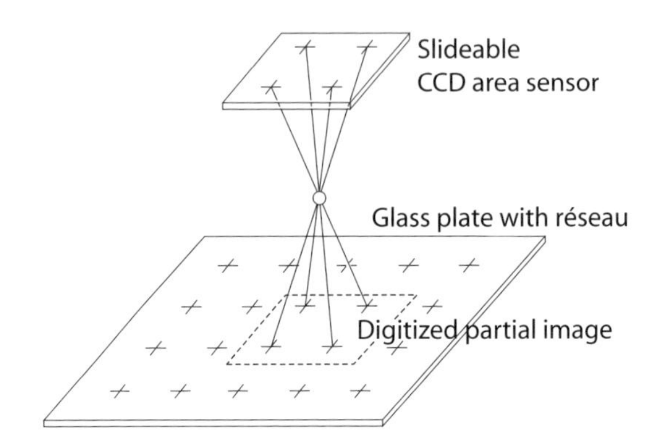

Figure 3.4-3: Principle of réseau scanning

Radiometric calibration of a scanner is particularly difficult with linear and area array scanners. Each detector in the array must have the same sensitivity, otherwise patterns of stripes (linear arrays) or blocks (area arrays) are produced. However, to a large extent these patterns can be removed in digital post-processing.

In photogrammetry, the positional accuracy of the pixels resulting from the digitizing process is of considerable importance. For currently available photogrammetric scanners, the average accuracy of an image coordinate lies around $\pm 1.3 - 2.3\,\mu\text{m}$[39]. A precondition here is that the scanner's calibration certificate provides reference coordinates for the detectors, or at least for a few evenly spaced detectors (further details in Section E 4.2.1, Volume 2). In contrast, repro scanners have significantly larger positional errors but, in general, provide a better radiometric quality than photogrammetric scanners.

Among others, photogrammetric scanners are currently offered by the following companies (product names in brackets): Leica Geosystems (DSW), Microsoft-Vexcel (UltraScan), Wehrli (RM) and Intergraph (PhotoScan).

3.5 Digital image enhancement

Digital images, either directly recorded in a metric camera or produced indirectly by digitizing an analogue image, can be enhanced through the methods of digital image processing. These improvements are made with regard to the optimal conditions for both visual interpretation of the images and automated analysis of the images.

Image enhancement is the central theme of this section; digital image analysis (extraction of 2D and 3D data) will be discussed later, particularly in Section 6.8. The following discussion is limited to black-and-white images (colour images are handled in Section C 1.3.4, Volume 2) and starts with contrast and brightness enhancement.

[39] Baltsavias, E., Kaeser, Ch.: OEEPE-Publication 37, pp. 111–134, 1999.

3.5.1 Contrast and brightness enhancement

In these procedures, the grey values g_{ij} of the original image matrix **G** are transformed to grey values $\bar{g}_{ij}$ in the enhanced matrix. The simplest transform function is:

$$\bar{g}_{ij} := c g_{ij} + d \tag{3.5-1}$$

$c\ldots$ parameter responsible for contrast, $d\ldots$ parameter responsible for brightness

With this transform function, the grey levels of individual pixels are altered independently of one another. Viewing all the grey levels together in an image results in a change of contrast for which parameter c is responsible. Parameter d alters the image brightness. Figure 3.5-1 shows how contrast and brightness of a digital image can be altered by the choice of c and d.

Contrast enhancement can also be applied differentially (Figure 3.5-2). Other functions can also be chosen in place of the linear transform function. Figure 3.5-3 suggests two general functions.

By making contrast enhancement a function of grey level, the sensitivity of the human eye can also be taken into account. CCD sensors generate grey levels which are proportional to the number of photons falling on the individual detectors. To the human eye, grey value sequences appear linear when they follow a logarithmic curve to base 10 (see Section 3.2.2.3 and also Sections 3.3.5 and 3.4.2). Conversion using a logarithmic transform function is therefore a standard procedure in digital image processing and extremely efficiently implemented in many systems.

In order to save computing time, transformations such as the one in (3.5-1) are solved by the use of look-up tables, abbreviated to LUT. To this end, transformation values $\bar{g}_{ij}$ are computed for all grey levels g_{ij}, usually in the range $0-255$, and stored in a table. When an image is processed it is then only necessary to "look up" the position g_{ij} in the LUT and extract the corresponding grey value $\bar{g}_{ij}$; in this way the modified grey levels are obtained without computation.

When grey values are manipulated as described, care must be taken to ensure that the new grey values $\bar{g}_{ij}$ continue to lie within the available value and dynamic ranges (for example from $\bar{g}_{\min} = 0$ to $\bar{g}_{\max} = 255$). For values lying outside the ranges the following conditions are applied:

$$\begin{aligned} &\text{if} \quad \bar{g}_{ij} > \bar{g}_{\max} \quad \text{("too bright"), then set} \quad \bar{g}_{ij} = \bar{g}_{\max} \\ &\text{if} \quad \bar{g}_{ij} < \bar{g}_{\min} \quad \text{("too dark"), then set} \quad \bar{g}_{ij} = \bar{g}_{\min} \end{aligned} \tag{3.5-2}$$

By a suitable choice of transformation parameter it is possible to avoid exceeding the limits of the value range. For example, parameters c and d in function (3.5-1) can be chosen as follows:

$$c = \frac{255}{g_{\max} - g_{\min}} \qquad d = -c\, g_{\min} \tag{3.5-3}$$

The extent to which a human observer judges the quality of an image is dependent on the frequency distribution of the grey values; in other words, the human eye expects a "balanced" image containing grey values across the entire range. Figure 3.5-4

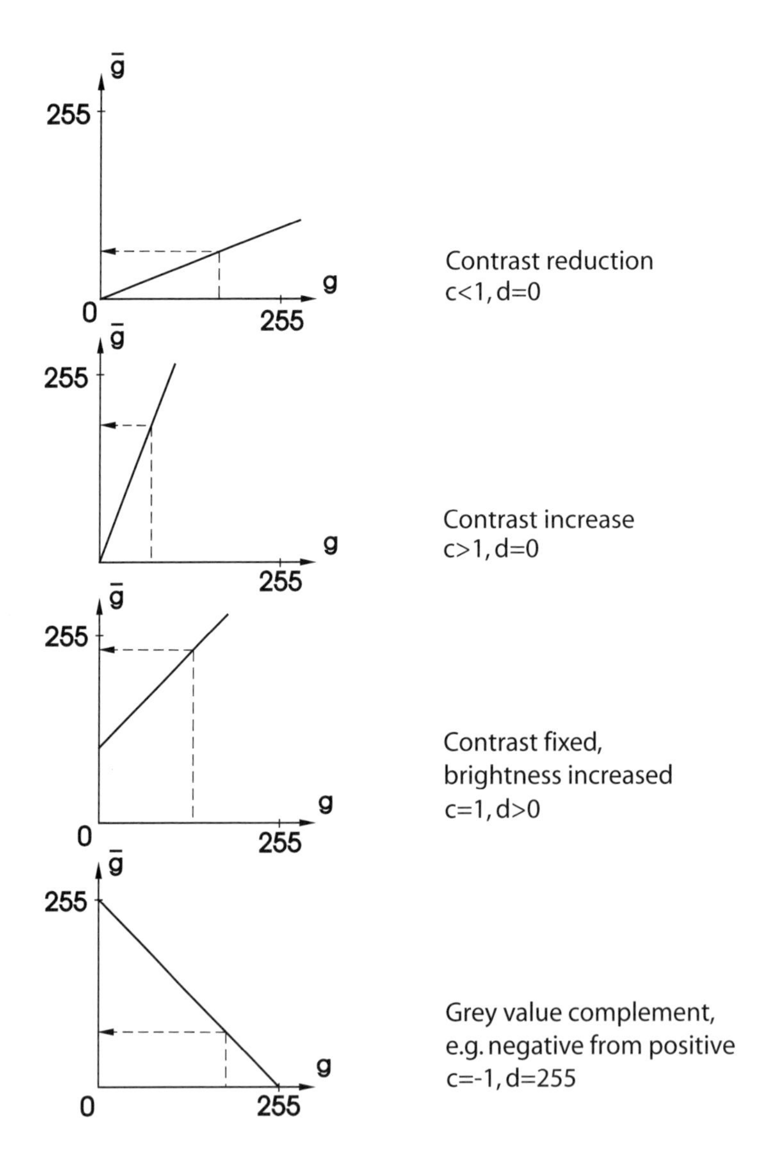

Figure 3.5-1: Contrast and brightness enhancement with a linear transform function

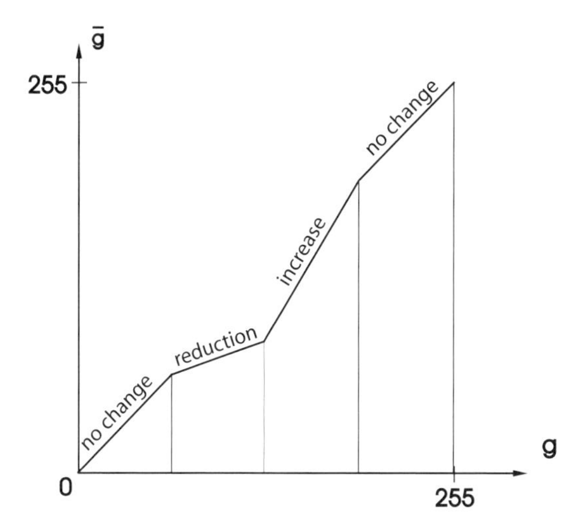

Figure 3.5-2: Contrast enhancement with different linear transform functions applied to different grey level regions

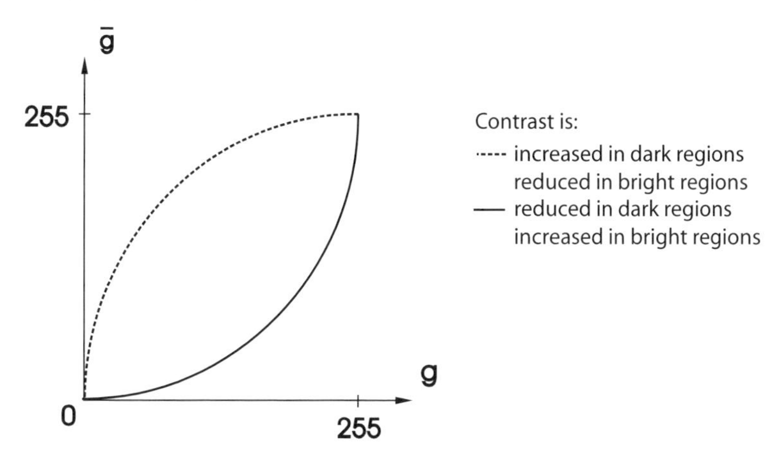

Figure 3.5-3: Contrast enhancement dependent on grey level

shows the frequency $h(g)$ with respect to grey value g. This representation is called a histogram.

An interesting possibility for altering the grey level distribution in an image lies in modifying its grey level histogram. Although many changes can be implemented in this way, two types of histogram modification have achieved particular importance. Both offer the observer a balanced distribution of grey levels. One of the modifications is histogram equalization and the other is histogram normalization (Figure 3.5-4).

The discussion will first look at histogram equalization and then at histogram normalization.

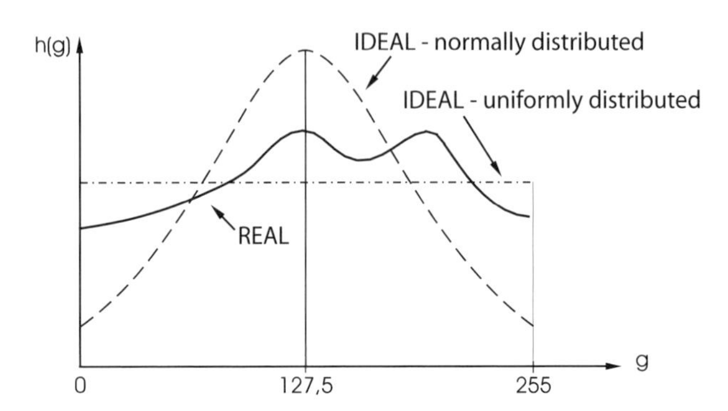

Figure 3.5-4: Frequency distribution (histogram) of a real image (REAL) and ideal image (IDEAL-uniformly distributed, IDEAL-normally distributed)

3.5.1.1 Histogram equalization

To a first approximation this represents a balanced image although, from a statistical point of view, the objective is an equal distribution of grey levels. The procedure is called histogram equalization because the frequency distribution is "levelled". As is generally the case with histogram modifications, the required transformation is based on cumulative frequency distributions, here using $S_h(g)$ for the cumulative frequency distribution of the real image **G** being processed and $S_h(\bar{g})$ for the ideal modified image $\bar{g}$. For histogram equalization, the cumulative frequency function of the modified image is a sloping line. Figure 3.5-5 shows the functions for the real and ideal images whose frequency distributions are shown in Figure 3.5-4. These functions are also known as cumulative histograms.

Figure 3.5-5 also indicates how to access the look-up table, i.e. the tabulated relationship between real grey level g and ideal grey level $\bar{g}$. For any grey level g, the cumulative frequency $S_h(g)$ ($\uparrow$ in Figure 3.5-5) indicates how often this grey level and lower values appear in the real image. By setting this cumulative value $S_h(g)$ equal to the cumulative frequency $S_h(\bar{g})$ ($\leftarrow$ in Figure 3.5-5), the ideal grey value corresponding to g is directly obtained on the $\bar{g}$ axis ($\downarrow$ in Figure 3.5-5). In this way the corresponding ideal value $\bar{g}$ is found for every grey level g and the transformation function for histogram equalization is thereby defined through the look-up table. The histogram equalization requires no additional outside information; it can therefore run automatically.

3.5.1.2 Histogram normalization

An image with equally distributed grey levels (derived from histogram equalization, Section 3.5.1.1) is not (yet) visually balanced. The human eye can differentiate medium grey levels better than bright or dark regions. It is possible to take account of this characteristic by introducing a normal distribution of frequencies as the ideal target function. In this case the required transformation function is given by a look-up table based

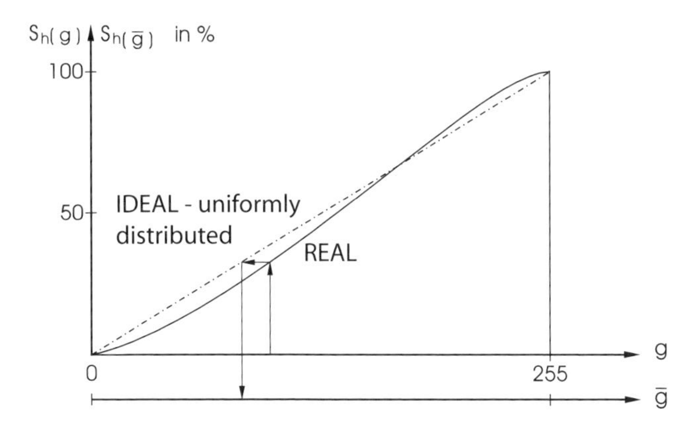

Figure 3.5-5: Cumulative frequency distributions of a real (REAL) and ideal image (IDEAL-uniformly distributed $\hat{=}$ Histogram equalization), whose frequencies are given in Figure 3.5-4

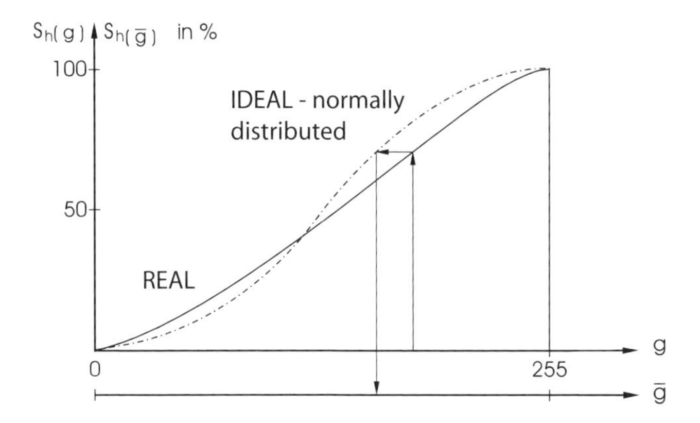

Figure 3.5-6: Cumulative frequency distributions of a real (REAL) and ideal image (IDEAL-normally distributed $\hat{=}$ Histogram normalization), whose frequencies are given in Figure 3.5-4

on the cumulative frequency distribution $S_h(\bar{g})$ of the required normal distribution and the cumulative frequency $S_h(g)$ of the real image. Figure 3.5-6 shows both these cumulative distributions which relate to the frequency distributions of Figure 3.5-4.

The corresponding values g and $\bar{g}$ in this look-up table are found by the same procedure as used for histogram equalization. The necessary cumulative frequency distribution $S_h(\bar{g})$ for the normal distribution can be found in textbooks on statistics and least squares adjustment. The user can determine the form of the normal distribution by defining the standard deviation and average value. For digital images with a grey level spread of $0-255$, values of ± 60 for standard deviation and 127 for the average have proven themselves in practice. When these values are used internally in software, histogram normalization can take place fully automatically.

Numerical Example (of histogram normalization). Until now the methods have been explained on the basis of continuous frequency and cumulative frequency distributions. Using the discrete distributions generated by real images, the target objective can only be achieved approximately. The limitations are even more significant when the grey level range is very small. For reasons of clarity this numerical example will be limited to an image with only 16 grey levels.

The frequency distribution will be stated in percentages. Columns (1) and (2) in Table 3.5-1 provide the starting information. The cumulative distribution $S_h(g)$ is calculated from the frequencies $h(g)$ by successive addition. Relative frequencies are shown in Figure 3.5-7 and cumulative frequencies in Figure 3.5-8. After histogram normalization a relative frequency $h(\bar{g})_{\text{targ}}$ is expected which corresponds to a normal distribution. Such a normal distribution, in this case with standard deviation of ± 3, is shown in column (4) of Table 3.5-1 and has been drawn in Figure 3.5-7. The cumulative distribution corresponding to this normal distribution is given in column (5)

g (1)	$h(g)$ % (2)	$S_h(g)$ % (3)	$h(\bar{g})_{\text{targ}}$ % (4)	$S_h(\bar{g})_{\text{targ}}$ % (5)	$\bar{g}$ (6)
0	1	1	1	1	0
1	1	2	1	2	1
2	2	4	3	5	2
3	3	7	4	9	3
4	6	13	7	16	4
5	7	20	9	25	5
6	7	27	12	37	6
7	8	35	13	50	7
8	6	41	13	63	8
9	6	47	12	75	9
10	7	54	9	84	10
11	9	63	7	91	11
12	10	73	4	95	12
13	8	81	3	98	13
14	9	90	1	99	14
15	10	100	1	100	15
	100		100		

Table 3.5-1: Histogram normalization

$\bar{g}$ (1)	$h(\bar{g})$ % (2)	$S_h(\bar{g})$ % (3)
0	1	1
1	1	2
2	2	4
3	3	7
4	13	20
5	7	27
6	14	41
7	13	54
8	9	63
9	10	73
10	8	81
11	9	90
12	0	90
13	0	90
14	0	90
15	10	100
	100	

Table 3.5-2: Result of histogram normalization

(1) and (2) ... given grey levels with relative frequencies
(3) ... cumulative frequencies calculated from (2)
(4) ... required relative frequency
(5) ... cumulative frequencies calculated from (4)
(6) ... which together with (1) form the look-up table

of Table 3.5-1. Figure 3.5-8 is the corresponding diagram. Following the procedure sketched in Figure 3.5-6, Figure 3.5-8 shows how, for a grey level $g = 9$, the corresponding target grey level is found (and in this example $\bar{g} = 7$). The entire look-up table is given in columns (1) and (6) of Table 3.5-1. Using this look-up table the new image can be created. The result of the histogram normalization, expressed as relative frequencies $h(\bar{g})$ and cumulative frequencies $S_h(\bar{g})$is presented in Table 3.5-2 and Figures 3.5-9 and 3.5-10.

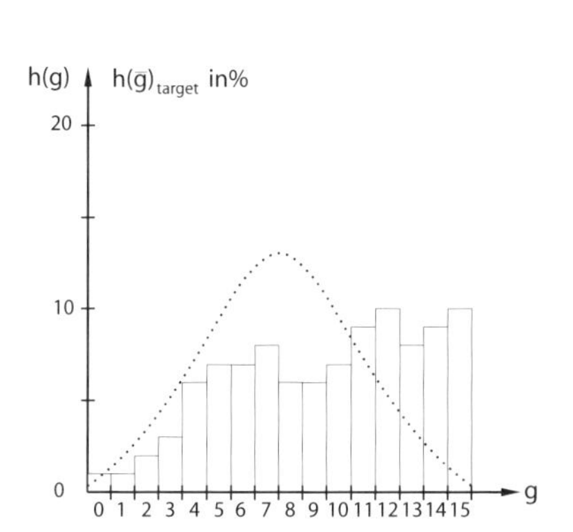

Figure 3.5-7: Given relative frequencies and target relative frequencies (dotted line) for a histogram normalization

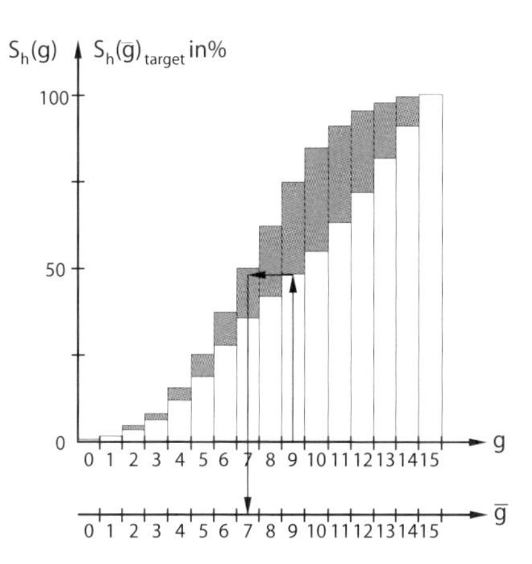

Figure 3.5-8: Given cumulative frequencies and required cumulative frequencies (shaded) of a histogram normalization, together with the construction of the look-up table

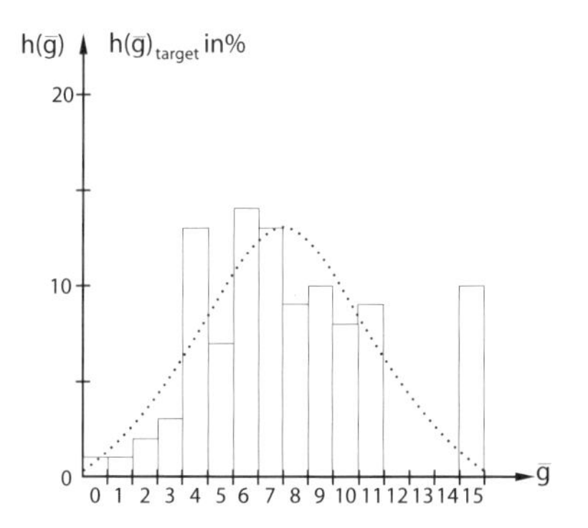

Figure 3.5-9: Result of histogram normalization (relative frequencies) compared with target result

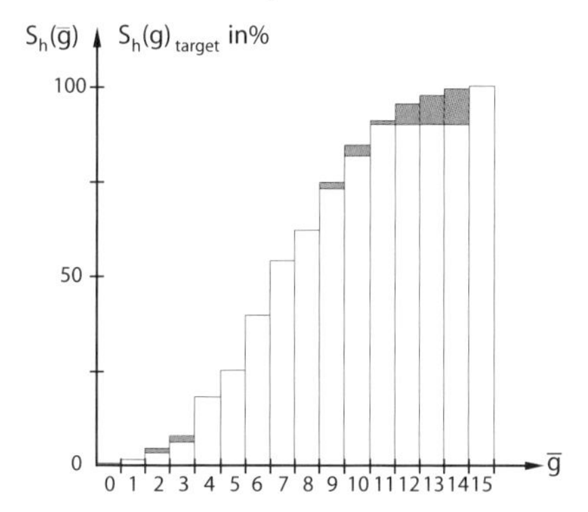

Figure 3.5-10: Result of histogram normalization (cumulative frequencies) compared with target result

Exercise 3.5-1. Perform a histogram equalization on the data contained in columns (1) and (2) of Table 3.5-1. (Solution: For given grey levels $g = 0, 1, 2, 3, 4, 5, 6, 7, 8, 9, 10, 11, 12, 13, 14, 15$ the corresponding grey levels $\bar{g} = 0, 0, 0, 0, 1, 2, 3, 5, 6, 7, 8, 9, 11, 12, 13, 15$). Draw the corresponding diagram. (Details of the solution can be found in Section C 1.3.1.1, Volume 2).

Comment on the result:

- the objective of histogram normalization is only approximately achieved with digital images
- histogram normalization of digital images normally involves loss of information. The original pixels with grey levels $g = 4, 5$ are converted to grey level $\bar{g} = 4$, grey levels $g = 7, 8$ to $\bar{g} = 6$ and grey levels $g = 9, 10$ to $\bar{g} = 7$.
- there are normally gaps in the grey level sequence after histogram normalization of digital images. There are no pixels with grey levels $\bar{g} = 12, 13, 14$

In spite of the negative comments, histogram normalization plays a significant role in practice, particularly when the starting images have a very skewed frequency distribution. An impressive example from practice is presented further below (Figure 3.5-12). Because of the disadvantages, which occur with both histogram normalization and equalization, often only the simpler method of expansion, commonly called stretching, is employed. Here the given grey level range is mapped onto the available grey level range by distributing the original values, without grouping, evenly across the available grey level scale. Naturally this also produces gaps in the grey level sequence. An expansion is only possible when the frequency distribution of the given image has places where there are no grey levels, which often occurs at the beginning or end of the distribution. Isolated grey levels, particularly near the boundaries of the distribution, can also be grouped into a single grey level in order to create “space” for the expansion. An expansion method implemented in many image processing systems first calculates the average value m and standard deviation σ of the grey levels. The following assignments are then made:

- average value m (given image) corresponds to 127 (new image).
- grey level range -2σ to m (given image) corresponds to 1 to 127 (new image).
- grey level range m to $+2\sigma$ (given image) corresponds to 127 to 255 (new image).

Grey levels outside the range $m \pm 2\sigma$ are ignored. If there are too many in this class, the factor 2 can be replaced by a larger one.

The following section explains two further contrast and brightness enhancements typically employed in photogrammetry.

3.5.1.3 Compensation for light fall-off from centre to edge of image

Light fall-off follows a $\cos^n \tau$ law (n lies between 1.5 and 2.5; τ is the off-axis angle, see Section 3.1.6). If there is no graduated filter in front of the objective lens at the time of exposure, light fall-off from image centre to edge is significant (see numerical example in Section 3.1.6).

A relative contrast and brightness adjustment is discussed next, which derives its transformation parameters directly from the image content. For this purpose a reasonably large circular patch surrounded by concentric annular rings is defined at the centre of the image. Within the central area and one of the surrounding rings, several image regions are located which should have equal grey levels (for example, similar ground types). Grey levels g_{Ring} and $\bar{g}_{\text{Centre}}$ for the corresponding regions are then related by means of a regression line; the result provides the transformation function (3.5-1) for the annular ring under evaluation.

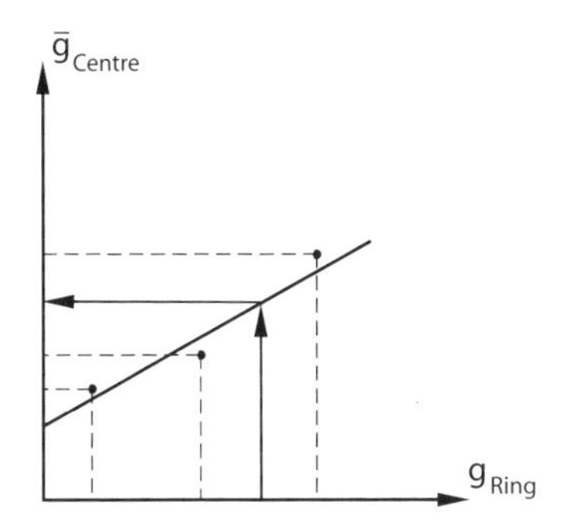

Figure 3.5-11: Transformation function to compensate for light fall-off from image centre to image edge

Figure 3.5-11 shows grey levels for 3 corresponding image regions, used to define the regression line. With the aid of the transformation function, every grey level g_{Ring} within a ring can be converted to a grey level $\bar{g}_{\text{Centre}}$ which is free of the effects of light fall-off.

This procedure is executed for each annular ring. Parameters c and d of the corresponding transformation function (3.5-1) can be further modified by the physically based $\cos^n \tau$ law.

Exercise 3.5-2. The following grey levels for two corresponding image regions at the image centre and outer ring are found in the original image:

	g_{Ring}	$\bar{g}_{\text{Centre}}$
Open fields on flat ground	25	50
Highway	120	200

Derive the transformation function and the corrected grey level for a pixel at the image edge with a value of $g = 100$. (Solution: $\bar{g}_{\text{Centre}} = 168$.)

3.5.1.4 Histogram normalization with additional contrast enhancement

The second example is related to electronic film copying with contrast control, presented in Section 3.2.2.8. In particular, the copying devices increase contrast in local areas. During digital image enhancement it is also the case that such a local improvement in contrast should take place together with a global improvement.

In digital photogrammetry, histogram normalization with additional contrast enhancement is used in place of a copying device providing contrast improvement. Figure 3.5-12a shows a digitized aerial film positive with a very limited contrast range. A histogram normalization is first performed. Figure 3.5-12b shows the result which makes use of a normal distribution with a standard deviation of ± 60 and average value of 127. Figure 3.5-13a shows the frequency distribution $h(g)$ before histogram normalization and the corresponding distribution $h(\bar{g})$ after normalization. The transformation function F_1 for this global contrast adjustment is shown in Figure 3.5-13b.

For additional local contrast enhancement, Kalliany[40] has developed a filter whose effectiveness has been proven in use at the Vienna University of Technology's Institute for Photogrammetry and Remote Sensing (Institut für Photogrammetrie und Fernerkundung)[41]. Here grey levels in a 3×3 or 5×5 neighbourhood are averaged and the average value m compared with the grey value g of the central pixel. The difference is transformed by function F_2 (Figure 3.5-13b). The steep rise at the beginning of the curve has the effect of a contrast enhancement. The remaining part of curve F_2 is not of significance; it would effect an attenuation in contrast if there were very large differences between the average value and the grey level of the central pixel. The enhanced difference, with appropriate sign, is added to the average value and assigned to the central pixel. Figure 3.5-12c shows part of Figure 3.5-12b which has been enhanced in this way (result in Figure 3.5-12d).

Both modifications to contrast, global and local, do not need to be executed sequentially; they can be carried out in a single processing step by the following operator:

$$\bar{g} = F_1(m + (\text{sign}(g - m)F_2|g - m|)) \qquad (3.5\text{-}4)$$

g ... grey level of the central pixel in the original image matrix
m ... average value of the grey levels in a 3×3 or 5×5 neighbourhood
F_1 ... global transform function from histogram normalization
F_2 ... local transform function which can be selected externally
$\bar{g}$ ... grey level of the result for the central pixel

It is appropriate to indicate the size of the region within which the grey levels are averaged: in small scale aerial photos with short-range changes in texture, a 3×3 region should be chosen; at medium to large scales a 5×5 region has proven to work. Figure 3.5-12d was generated using a 5×5 region.

[40] Ecker, R., Kalliany, R., Otepka, G.: In Fritsch/Hobby (Eds.): Photogrammetric Week '93. Wichmann Verlag, pp. 143–155, 1993. The examples in this section are taken from this publication.

[41] A filter with a similar effect has been developed by Wallis and, amongst other places, published in: Pratt, W.: Digital Image Processing. A Wiley-Interscience publication, New York, 1991.

a) photo as delivered by photo scanner (see Figure 3.5-13(a) for histogram)
b) result of histogram normalization with transform function F_1 in Figure 3.5-13 (b) (see Figure 3.5-13 (a) for histogram)
c) part of Figure 3.5-12b, before applying a local contrast operator
d) result of local contrast enhancement with transform function F_2 of Figure 3.5-13(b)

Figure 3.5-12

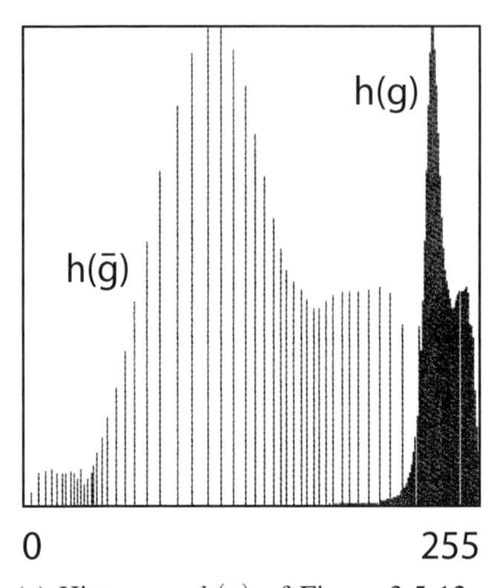

(a) Histogram $h(g)$ of Figure 3.5-12a and histogram $h(\bar{g})$ after histogram normalization (result shown in Figure 3.5-12b)

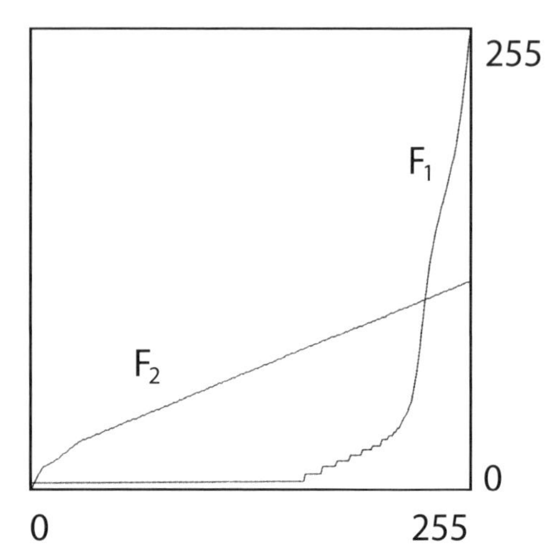

(b) Transform function F_1 for global contrast enhancement and function F_2 for local contrast enhancement

Figure 3.5-13

3.5.2 Filtering

The purpose of filtering is to separate the wanted from the unwanted. Undesirable image content, for example, is the noise from the electronic recording sensors (Section 3.3.5.3). However, filtering is also employed in order to enhance indistinct image information. Human observers, for example appreciate strongly pronounced edges in an image.

Filtering can take place in the spatial domain (Section 3.5.2.1) or in the frequency domain (Section 3.5.2.2).

3.5.2.1 Filtering in the spatial domain

A very simple filter is the moving average. In this case the grey levels $\bar{g}_{ij}$ of the new image $\bar{\mathbf{G}}$ are derived from an averaging of the grey levels g_{ij} in the original image matrix $\mathbf{G}$. It is formulated as follows:

$$\bar{g}_{ij} = \frac{1}{(2n+1)^2} \sum_{k=-n}^{n} \sum_{l=-n}^{n} g_{i+k,j+l} \qquad n \geq 1 \tag{3.5-5}$$

Numerical Example (for $n = 1$). Given matrix $\mathbf{G}$ (Figure 3.5-14, left). By application of Equation (3.5-5) it is possible to obtain, for example, image element $\bar{g}_{22}$:

$$\bar{g}_{22} = \frac{1}{9}(10 + 14 + 2 + 11 + 13 + 10 + 9 + 25 + 8) = 11$$

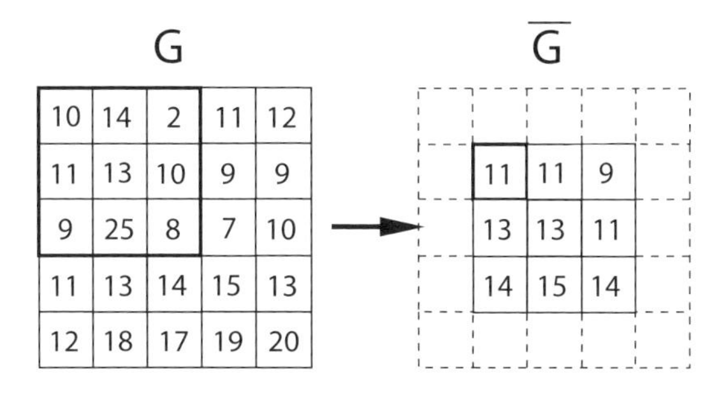

Figure 3.5-14: Moving average

By means of a filter matrix $\mathbf{W}$, Equation (3.5-5) can be alternatively written:

$$\bar{g}_{ij} = \sum_{k=-n}^{n} \sum_{l=-n}^{n} g_{i-k,j-l} w_{n+1+k,n+1+l} \qquad (3.5\text{-}6)$$

Equation (3.5-6) describes an operation known as convolution and the filter matrix $\mathbf{W}$ is therefore known as a convolution operator. Equation (3.5-6) can be expressed more simply as:

$$\bar{\mathbf{G}} = \mathbf{G} * \mathbf{W} \qquad (3.5\text{-}7)$$

For the moving average, the convolution operator or filter matrix $\mathbf{W}$ is:

$$\mathbf{W} = \begin{pmatrix} w_{11} & w_{12} & w_{13} \\ w_{21} & w_{22} & w_{23} \\ w_{31} & w_{32} & w_{33} \end{pmatrix} = \frac{1}{9} \begin{pmatrix} 1 & 1 & 1 \\ 1 & 1 & 1 \\ 1 & 1 & 1 \end{pmatrix} \qquad (3.5\text{-}8)$$

The central pixels can be assigned a higher weight in the filter matrix $\mathbf{W}$ which creates a general arithmetic mean from the neighbouring grey levels. For example, such a matrix known as a binomial filter can be formed with the aid of binomial coefficients:

$$\mathbf{W} = \frac{1}{16} \begin{pmatrix} 1 & 2 & 1 \\ 2 & 4 & 2 \\ 1 & 2 & 1 \end{pmatrix} \qquad (3.5\text{-}9)$$

When the binomial filter is enlarged, it approximates to the Gaussian filter. This name expresses the fact that the elements of filter matrix $\mathbf{W}$ correspond to a two-dimensional Gaussian bell curve. It should also be noted that the sum of all elements w_{ij} always equals unity.

Exercise 3.5-3. Filter Figure 3.5-14, left, using operator (3.5-9). Comment on the grey levels in the new image compared with the grey levels after convolution by means of a moving average (Figure 3.5-14, right). (Solution: e.g. $\bar{g}_{22} = 13$.)

Comment: The negative signs in front of k and l in Equation (3.5-6) have the effect of opposing directions for the indices in filter matrix $\mathbf{W}$ and image matrix $\mathbf{G}$. However, with symmetrical filter matrices such as the moving average, this is of no importance.

Filtering with a moving average results in an attenuation of (high frequency) noise. Admittedly, (high frequency) image information, i.e. edges, is also attenuated. The moving average with operator (3.5-8) has reduced the grey level difference in Figure 3.5-14 from 18 (= 25 – 7) to 6 (= 15 – 9).

The opposite effect, i.e. an increase in grey level differences, can be achieved with a Laplacian operator. It is obtained through a double differencing of neighbouring grey levels:

First difference in ξ direction:

$$\left(\frac{\partial g}{\partial \xi}\right)_{i,j} = g_{i+1,j} - g_{i,j} \tag{3.5-10a}$$

First difference in η direction:

$$\left(\frac{\partial g}{\partial \eta}\right)_{i,j} = g_{i,j+1} - g_{i,j} \tag{3.5-10b}$$

Second difference in ξ direction:

$$\left(\frac{\partial^2 g}{\partial \xi^2}\right)_{i,j} = \left(\frac{\partial g}{\partial \xi}\right)_{i,j} - \left(\frac{\partial g}{\partial \xi}\right)_{i-1,j} = g_{i+1,j} + g_{i-1,j} - 2g_{i,j} \tag{3.5-11}$$

Second difference in η direction:

$$\left(\frac{\partial^2 g}{\partial \eta^2}\right)_{i,j} = \left(\frac{\partial g}{\partial \eta}\right)_{i,j} - \left(\frac{\partial g}{\partial \eta}\right)_{i-1,j} = g_{i,j+1} + g_{i,j-1} - 2g_{i,j} \tag{3.5-12}$$

Finally by addition of (3.5-11) and (3.5-12):

$$\nabla^2 g_{i,j} := \left(\frac{\partial^2 g}{\partial \xi^2}\right)_{i,j} + \left(\frac{\partial^2 g}{\partial \eta^2}\right)_{i,j} = g_{i+1,j} + g_{i-1,j} + g_{i,j+1} + g_{i,j-1} - 4g_{i,j} \tag{3.5-13}$$

However, the algorithm (3.5-13) can also be interpreted as a convolution with the following operator:

$$\mathbf{W} = \begin{pmatrix} 0 & 1 & 0 \\ 1 & -4 & 1 \\ 0 & 1 & 0 \end{pmatrix} \tag{3.5-14}$$

This operator (∇^2) is called the Laplacian operator. A related operator which also includes the diagonals of the convolution matrix is:

$$\mathbf{W} = \begin{pmatrix} 1 & 1 & 1 \\ 1 & -8 & 1 \\ 1 & 1 & 1 \end{pmatrix} \tag{3.5-15}$$

The convolution of the photo in Figure 3.5-15 with this operator produces Figure 3.5-16 as a result. Both the operator (3.5-15) and the Laplacian operator (3.5-14) generate positive and negative pixel values; prior to visualization they must therefore be shifted into "positive".

The Laplacian operator, and all others of similar construction, extract the edges in an image. This property is of great importance for automated image processing and will be discussed in further detail in Sections 6.8.3.4 and 6.8.7.2. Subtracting the image of extracted edges from the original image produces an image with enhanced edges (details can be found in Section C 1.3.3, Volume 2). Figure 3.5-17 shows the result of this subtraction, i.e. Figure 3.5-15 minus Figure 3.5-16.

In practice the intermediate result of an image of extracted edges can be avoided and the original image directly convolved with the following operator:

$$\mathbf{W} = \begin{pmatrix} 0 & 0 & 0 \\ 0 & 1 & 0 \\ 0 & 0 & 0 \end{pmatrix} - \begin{pmatrix} 0 & 1 & 0 \\ 1 & -4 & 1 \\ 0 & 1 & 0 \end{pmatrix} = \begin{pmatrix} 0 & -1 & 0 \\ -1 & 5 & -1 \\ 0 & -1 & 0 \end{pmatrix} \tag{3.5-16}$$

If operator (3.5-15) is used in place of the Laplacian operator (3.5-14) in the combination operator (3.5-16), the following operator for an edge enhanced image is obtained:

$$\mathbf{W} = \begin{pmatrix} 0 & 0 & 0 \\ 0 & 1 & 0 \\ 0 & 0 & 0 \end{pmatrix} - \begin{pmatrix} 1 & 1 & 1 \\ 1 & -8 & 1 \\ 1 & 1 & 1 \end{pmatrix} = \begin{pmatrix} -1 & -1 & -1 \\ -1 & 9 & -1 \\ -1 & -1 & -1 \end{pmatrix} \tag{3.5-17}$$

Note: the elements of operators which enhance edges have a sum equal to unity; elements of operators which extract edges have a sum equal to zero.

Exercise 3.5-4. Create the convolution of Figure 3.5-14, left, with operator (3.5-16). Comment on the grey level differences in the new image compared with the grey level differences after convolution with a moving average (Figure 3.5-14, right).

3.5.2.2 Filtering in the frequency domain

Filtering in the frequency domain starts with a spectral analysis. Here the grey level image is reduced to its component waves, characterized by amplitude and wavelength or frequency. The grey levels g_{ij} in the original image represent a function in the spatial domain, dependent on indices i and j or image coordinates ξ and η. By means of a spectral analysis, achieved through a Fourier transformation, the amplitude spectrum, dependent on frequency f, is obtained.

Figure 3.5-15: Original photo

Figure 3.5-16: Convolution of Figure 3.5-15 with operator (3.5-15)

Figure 3.5-17: Photo with enhanced edges

The relationship between spatial and frequency domains will be demonstrated with a simple example. Imagine a (continuous) grey level profile g(η), consisting of three wave components $g_1(\eta) = 50\sin(2\pi\,0.01\,\eta)$, $g_2(\eta) = 16.7\sin(2\pi\,0.03\,\eta)$ and $g_3(\eta) = 10\sin(2\pi\,0.05\,\eta)$. The three waves are shown in Figure 3.5-18, left. The sum of these component waves is the grey level profile $g(\eta)$, represented in the spatial domain of the cube (Figure 3.5-18, right). The related frequency domain shows the three component waves with their amplitudes C as a function of frequency f.

Figure 3.5-19 illustrates the filtering in the frequency domain. By means of a Fourier transform, the original image is converted into its amplitude spectrum $C(f)$. Using a freely chosen modulation transfer function $MTF(f)$, the amplitudes can be transformed into new amplitudes $\bar{C}$ (see also Equation (3.1-26)). The result is a filtered

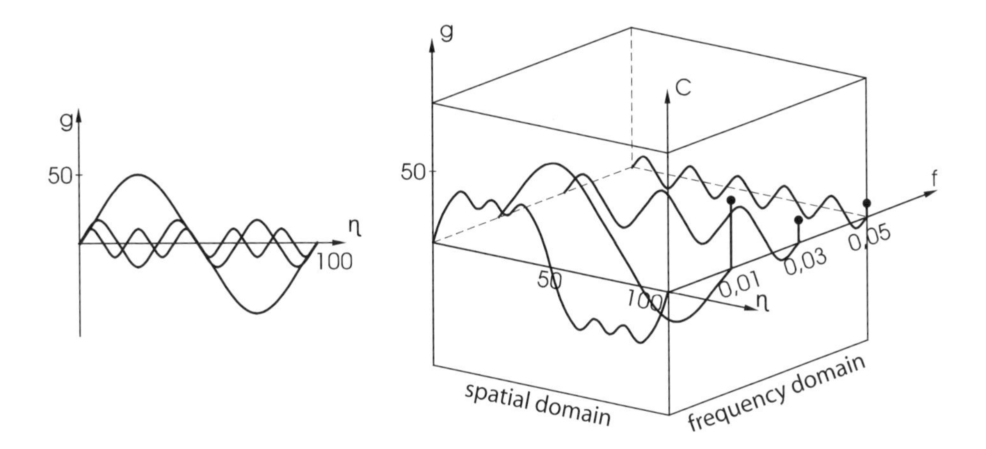

Figure 3.5-18: Relationship between spatial and frequency domains

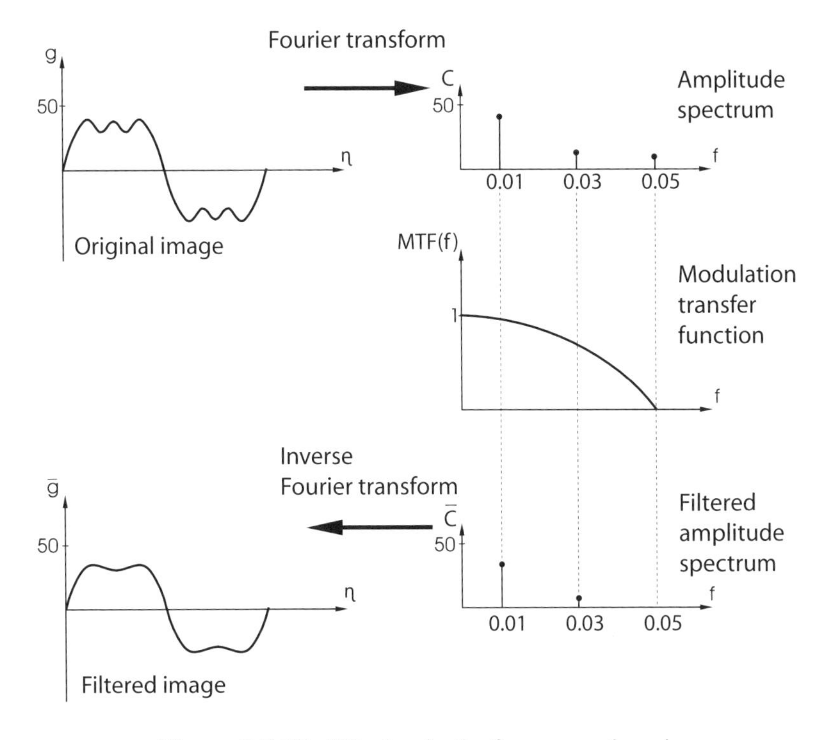

Figure 3.5-19: Filtering in the frequency domain

amplitude spectrum $\bar{C}$(f) from which the filtered image can be generated by means of an inverse Fourier transform. The equation for the inverse Fourier transformation has already been given (Equation (3.3-2)).

In the filtered image of Figure 3.5-19, the amplitude with the smallest frequency ($f = 0.01$) was somewhat attenuated, the amplitude with the mid frequency ($f = 0.03$) more strongly attenuated and the component wave with the highest frequency was eliminated. Since the high frequency parts of an image generally contain the noise, this can therefore be reduced by using the modulation transfer function in Figure 3.5-19. In this regard it is usual to refer to a low-pass filter which suppresses the high frequencies and allows the low frequencies to pass through.

The opposite is a high-pass filter. This can be employed to reverse the effects of a reduction in contrast caused by a detector (Section 3.3.2). For this purpose the reciprocal of the function (3.3-5), i.e. $1/MTF$, must be used for filtering in the frequency domain.

Exercise 3.5-5. For filtering in the frequency domain, sketch the modulation transfer function which removes the contrast reduction caused by image recording with a detector. The transfer function causing contrast reduction during image recording is shown in Figure 3.3-5.

Filtering in the frequency domain can be converted into filtering in the spatial domain and vice versa. A simple comparison indicates:

- averaging (Equations (3.5-8) and (3.5-9)) correspond to low-pass filtering.
- edge-enhancement operators (3.5-16) and (3.5-17) correspond to high-pass filtering.

Further reading: Albertz, J., Zelianeos, K.: Phia, pp. 161–174, 1990. Gonzalez, R.C.: Digital Image Processing. Prentice Hall, 2002. Schenk, T.: Digital Photogrammetry, Volume 1. TerraScience, 1999.

3.6 Image pyramids/data compression

Image pyramids enable efficient management of image data (Section 3.6.1) and compression methods reduce the amount of image data (Section 3.6.2).

3.6.1 Image pyramids

Figure 3.6-1 shows an image pyramid. The geometric resolution in the rows and columns is reduced by a factor of 2 from level to level. The pixels at the next lower resolution can be determined in various ways, the simplest method being the elimination of every second row and column of the current resolution. However, in order to keep the information loss smaller when changing to a coarser resolution, the grey levels at the coarser resolution are often derived from the grey levels of the finer resolution by means of a Gaussian or binomial filter, e.g. using the filter matrix (3.5-9).

Figure 3.6-1: Image pyramid with 512×512, 256×256, 128×128 and 64×64 pixels

Exercise 3.6-1. Create two reduced images from Figure 3.5-14, left, one by eliminating every second row and column and the other by application of the binomial filter (3.5-9).

Compared with the original image, the storage space required, for all pyramid levels, increases by a third.

Exercise 3.6-2. Prove this statement with an image pyramid based on an original image with 512×512 pixels.

This additional storage requirement offers, however, many advantages which will be evaluated later in further discussions of metric image analysis.

Further reading: Ackermann, F., Hahn, M.: in Ebner, H., Fritsch, D., Heipke, C.: Digital Photogrammetric Systems, pp. 43–58, Wichmann Verlag, 1991. Li, M.: PE&RS 57, pp. 1039–1047, 1991. Pan, J.-J., Li, S.-T.: IAPR XXIX(B2), pp. 474–478, Washington, 1992. Kropatsch, W., Bischof, H., Englert, R.: in Kropatsch, W., Bischof, H.: Digital image analysis. Selected techniques and applications, pp. 211–230, Springer, 2000.

3.6.2 Image compression

A reminder about the storage requirement for high resolution, metric images serves as an introduction to this section. According to Section 3.4.1, the storage requirement for a $23\,\text{cm} \times 23\,\text{cm}$ metric image with $7\,\mu\text{m} \times 7\,\mu\text{m}$ pixel size is around 1 GB. The processing of a large array of images, which can consist of several hundred, would necessitate a storage capacity which current hard disks do not offer in medium priced computers. Data transfer between hard disk and memory, and over a computer network, would also be too time-consuming.

It is for this reason that compression methods are in demand. Here it is necessary to distinguish between lossless and lossy compression methods. With a lossless method the original image can be reconstructed from the compressed image without loss of information; with lossy methods a certain loss of information occurs during reconstruction, which may also be called decompression. As would be expected, with lossy compression methods a significantly higher compression factor can be achieved. The compression factor is the ratio of the number of bits in the original to the number of bits in the compressed image.

The method of run-length encoding is a lossless compression. It utilizes the redundancy between neighbouring pixels. Here, for example row by row, successive pixels with identical grey levels are grouped together and only the one representative grey level, together with the repetition number, is stored. Run-length encoding achieves only a small compression factor with photographs. In small-scale photographs there is almost no reduction in storage, in large-scale photographs, particularly for close range work, a compression factor of 3 : 1 can be achieved in the best case.

Lossy compression methods are considerably more interesting. Here a small loss of information can be accepted in return for high compression. The JPEG compression method[42] is very well known and can reduce photographic storage to a third without appreciable information loss. In JPEG compression the digital image is broken down into tiles of 8×8 pixels. By means of a Discrete Cosine transform (DCT, closely related to the Fourier transform) there is a transition from the spatial to the frequency domain within each tile (Section 3.5.2.2). The amplitude spectra are then low-pass filtered, and saved in compressed form. The degree to which the amplitude spectra are filtered is controlled by the "quality factor" which also influences the loss which occurs on reverse transformation into the spatial domain. Loss of grey level information occurs with JPEG compression, as well as losses and shifts of grey level edges, which is very important from a photogrammetric perspective. Figure 3.6-2 illustrates such losses and shifts. On the left are edges extracted from the original image (perhaps with operator (3.5-15)). In the middle are edges from a decompressed JPEG image which was made from an original image compressed by a factor 10. As a result of image shifts caused by this procedure, smoothed features appear, as well as artifacts which develop in particular at the edges of the independently filtered 8×8 tiles.

The disadvantages of JPEG compression can be compensated, to some extent, by use of a wavelet transform instead, allowing a more flexible sub-division of the original image based on frequency properties[43]. The high quality of wavelet-based compression can be seen in the extracted edges shown on the right of Figure 3.6-2. Although the same compression factor as in the JPEG compression has been chosen, there are hardly any shifts or artifacts compared with the original image. In practice there are a number of

[42] Joint Photographic Experts Group, `http://www.jpeg.org/`

[43] Literature: Pennebaker, W.B., Mitchell, J.L.: JPEG Still Image Data Compression Standard. Springer, 1993. Usevitch, B.E.: A Tutorial on Modern Lossy Wavelet Image Compression: Foundations of JPEG 2000. IEEE Signal Processing Magazine, Volume 18(2), pp. 22–35, 2001.

from the original image

from a decompressed image which had previously been compressed by a factor of 10 using the JPEG algorithm

from a decompressed image which had previously been compressed by a factor of 10 by a wavelet-based method

Figure 3.6-2: Extracted edges[44]

wavelet-based compression methods in use and the JPE group has also announced the use of such a method.

A totally different approach to compression lies in extracting relevant image information rather than storing the grey values. Where images contain small, elliptical object features as a result, for example, of applying circular or concentric ring targets to object surfaces, then the relevant information, in this case the image coordinates of the target centre points, should be extracted as soon as possible, for example using algorithm (6.8-21). In this way, the image matrix is reduced to the image coordinates of the critical features. This high reduction method is currently offered by cameras equipped with the corresponding "intelligence" (microcomputer and software, also known as "smart cameras").

3.7 Aerial cameras and their use in practice

Before discussing these cameras in detail, a section on flight planning is appropriate.

3.7.1 Flight planning

The main task of aerial surveying is the three-dimensional recording of natural and man-made features on the ground. It is a requirement of stereophotogrammetry that every point on the ground is imaged in at least two metric photographs (Section 2.1.3). This requirement is met when the images within a strip of photographs overlap by 50%. In single image photogrammetry (Section 2.1.4 and, in particular, Chapter 7) it is sufficient when every point on the ground appears in only one metric image.

[44]Taken from Schiewe, J.: PFG 1998(1), pp. 17–25.

Our attentions will now be turned to the stereophotogrammetry for which metric cameras with a square image format are employed[45]. The area to be mapped is imaged in a sequence of photographs forming parallel strips which together create a block of photographs. The geometrical relationships (Formulae (3.7-1)) involved in the flight planning are illustrated in Figure 3.7-1 for flat ground.

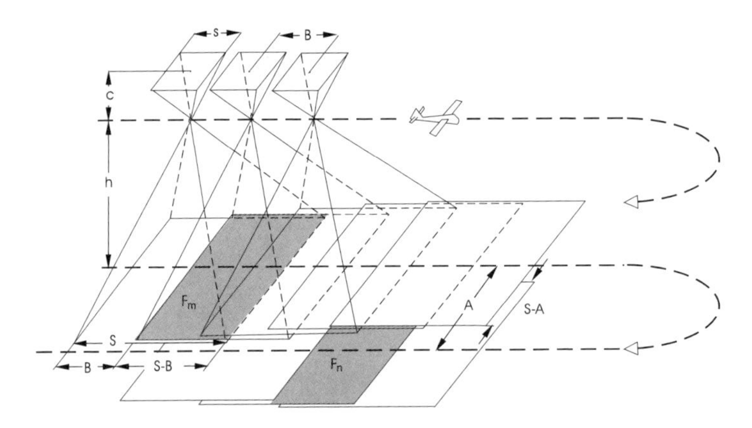

Figure 3.7-1: The geometry of flight planning for flat ground[46]

In Figure 3.7-1 and in Formulae (3.7-1), symbols have the following meaning:

A ... Distance between flight lines/strips
B ... Base (between consecutive images)
c ... Principal distance
s ... Image side (to edge!)
h ... Flying height above ground
Z ... Ground height
v ... Flying speed over ground
L ... Length of a strip/block
Q ... Width of block

[45] Although current aerial cameras such as the UltraCam or DMC provide rectangular images, in the following text only square images are considered for reasons of simplicity.

[46] Taken from Albertz/Kreiling: Photogrammetric Guide, Wichmann, 1980.

Photo scale number	$m_b = h/c$	
Image side in ground units	$S = s\, m_b$	
Base in photograph	$b = B/m_b$	
Flying height above ground	$h = c\, m_b$	
Absolute flying height	$Z_0 = h + Z$	
Forward overlap (%)	$l = \frac{S-B}{S} 100 = \left(1 - \frac{B}{S}\right) 100$	
Side overlap (%)	$q = \frac{S-A}{S} 100 = \left(1 - \frac{A}{S}\right) 100$	
Ground area of one photograph	$F_b = S^2 = s^2 m_b^2$	
Baselength for l% forward overlap	$B = S\left(1 - \frac{l}{100}\right)$	
Distance between strips for q% side overlap	$A = S\left(1 - \frac{q}{100}\right)$	(3.7-1)
Number of models in a strip (length L)[47]	$n_m = \lfloor \frac{L}{B} + 1 \rfloor$	
Number of photographs in a strip	$n_b = n_m + 1$	
Number of strips in a block (width Q)[47]	$n_s = \lfloor \frac{Q}{A} + 1 \rfloor$	
Area of a stereoscopic model	$F_m = (S - B)S$	
New area for each model in a block	$F_n = AB$	
Time between photographs (see Section 3.7.2)	$\Delta t\,[\mathrm{s}] = \frac{B\,[\mathrm{m}]}{v\,[\mathrm{m/s}]} \geq 2.0$	

Comments:

a) every pair of overlapping images in a strip should correspond to the "normal case" for stereopairs (Section 2.1.5). In practice, this "normal case" is never exactly achieved in aerial surveying and it is necessary to deal with a near "normal case". Individual images do not have their camera axes exactly vertical but only near vertical. Assuming there are no extensive navigation aids in use and no stabilization devices for the camera mounting, these departures from ideal amount to the following in practice:

[47] $\lfloor\ \rfloor$ = Largest integer number

- rotation about the longitudinal (fore and aft) axis of the aircraft (roll angle) ± 5 gon $(4.5°)$,
- rotation about the transverse (wing tip to wing tip) axis (pitch angle) ± 3 gon $(2.7°)$,
- deviation from nominal heading (yaw angle): ± 15 gon $(13.5°)$,
- variation in flying height $\pm 2\%$,
- deviation from flight path around ± 200 m.

b) even for an ideal flight plan, a forward overlap $l = 50\%$ and a side overlap $q = 0\%$ would not be sufficient. To process the entire block of photographs (see references to aerotriangulation in Chapter 5), neighbouring strips, as well as stereomodels within a strip, must overlap to some extent. For this reason, and also because of the departures from an ideal imaging configuration listed in paragraph (a), a forward overlap $l = 60\%$ and a side overlap $q = 25 - 30\%$ are planned.

   For single image photogrammetry, equal forward and side overlaps are chosen with $l = q = 25 - 30\%$ (Section 2.1.4 and, in particular, Chapter 7).

c) Figure 3.7-1 and Formulae (3.7-1) assume flat ground. Where there are large height differences, the recommended forward and side overlaps should be designed for the highest ground. At lower heights the overlaps are then larger. For flight planning over mountainous terrain, good knowledge of height variations is necessary.

Exercise 3.7-1. Take a topographic map, 1 : 50000, of a mountainous region and draw in the edges of a strip flown in a direction that crosses the ridges and valleys roughly at right angles. Smallest photo scale 1 : 30000, $c = 150$ mm. Then draw the axis of a second strip, parallel to the first, which has a minimum side overlap of 30%. Draw, in a different colour, the edges of this second strip and finally measure the maximum side overlap of the two strips.

Exercise 3.7-2. What overlap results when photographs are flown along a course down the centre of a U-shaped valley when the forward overlap at the edge of the strip is to be 60%? Flying height above the valley floor is 1200 m, the highest points to be photographed lie 400 m above the valley floor, and the principal distance is 152 mm. (Solution: 73.3%.)

d) the required flying height and photo scale depend on the required resolution on the ground (Sections 3.1.5, 3.2.2.7, 3.3.2) and the required accuracy of the photogrammetric end product (Sections 4.6 and 6.7).

e) for the various applications in photogrammetry, metric cameras with different objective lenses are required, i.e. lenses with different fields of view or principal distances. The lenses vary from normal to super wide angle and their parameters are listed in Table 3.7-1. The standard format for aerial film cameras is 23 cm $\times$ 23 cm. The base-to-height ratios, stereomodel areas and flying height

	Angle			
	Normal	Intermediate	Wide	Super wide
Field of view: Diagonal	62 gon (56°)	85 gon (77°)	100 gon (90°)	140 gon (126°)
Field of view: Image side	47 gon (42°)	64 gon (58°)	83 gon (75°)	117 gon (105°)
Principal Distance c [cm]	30	21	15	9
Ratio of Principal Distance to image diagonal	1 : 1	2 : 3	1 : 2	1 : 4
Base-to-height ratio for 60% forward overlap	1 : 3	1 : 2	2 : 3	1 : 1
Stereomodel area (in A) (h = const.)	$A/4$	$A/2$	A	$2.9A$
Flying height above ground (in z), area = const. scale = const.	$2z$	$1.5z$	z	$0.6z$
Applications and evaluation criteria	Single image photogrammetry High mountain regions Interpretation Orthophotos Cities		Overview flights Height accuracy Lower flight costs Stereophotogrammetry (Table 2.1-1 and Section 4.6)	

Table 3.7-1: Most common objective lenses used in aerial surveying (standard image format 23 cm × 23 cm, h = flying height above ground)

above ground shown in the table have been derived with the aid of this format, the various principal distances and the relationships in (3.7-1). The lower part of the table indicates the applications in which cameras with longer principal distances and those with shorter principal distances tend to be used.

Exercise 3.7-3. Check the fields of view given in Table 3.7-1 using the image format and the corresponding principal distances.

f) the depth-of-field problem (Section 3.1.4) does not exist with aerial photographs due to the relatively large flying height. Aerial metric cameras are generally focused at 1000 m, i.e. according to Equation (3.1-4) flying heights between 500 m and infinity lie within the acceptable depth of field.

Exercise 3.7-4. An aircraft flies with a metric camera at an absolute flying height of 3000 m. The ground height varies between 500 m and 1200 m. State the greatest image blur (diameter of cricle of confusion) for exposures with a principal distance of 15 cm and an aperture of 11. (Solution: $1.2\,\mu$m.)

g) if defined map sheet edges must be taken into account within a strip, there are two possible solutions: use of aimed single images or dense image sequences. For flights requiring aimed single images (also called a point flight) very good navigational aids are required in order to fly to, and "hit", the selected target area. GPS (Global Positioning System) significantly eases the problem of a point flight. Flights which adopt a dense image sequence, from which the best placed images can be selected, require a forward overlap of around 90%.

In addition to the flight parameters mentioned above, particular attention shoud be given to flying speed v when flight planning with digital line cameras. To ensure that individual image lines can be sequenced together without gaps, the following condition must be met:

$$v = \Delta x f_s \tag{3.7-2}$$

v ... flying speed
Δx ... detector dimension projected onto the ground (Equation (3.3-1))
f_s ... scan frequency.

Numerical Example. A digital line camera with a scan frequency of 750 Hz (750 lines are read out per second), a pixel size of $6\,\mu$m and a focal length ($\approx$ principal distance) of 80 mm, should be flown at the following speed when at a flying height of 2 km:

$$v = \frac{6 \times 2000}{80000} 750 = 112.5\,\text{m/s} \mathrel{\hat{=}} 400\,\text{km/h}$$

In practice a speed of around 300 km/h would be set in order to deal with any pitch angle changes (paragraph a) above) and to achieve some overlap of pixels in the direction of flight (see Section 3.4.1). A digital line camera (Figure 3.3-3) provides records for single image photogrammetry. In contrast, the 3-line camera provides images suitable for stereophotogrammetry. The elements of a 3-line camera can be chosen so that, for example, the combination of the a and b line correspond to a wide angle view and the combination of the a and c lines, or c and b lines, correspond to a normal angle of view (Figure 3.3-3, right).

As part of flight planning, a navigation plan is produced. This can either be an enlarged aerial photograph or, better, a good topographic map. It contains:

- the area of interest without any rounding off, i.e. the region which absolutely must be recorded, normally stereoscopically
- obstacles in the flight path

- restricted areas which on no account must be overflown (e.g. army training grounds, foreign countries)

The region marked on the navigation plan must be sufficiently large to accomodate a 5 to 10 km extension at both ends of the strips which allows for navigation manoeuvres. At the end of the strip an incremented numbering, absolute flying height and heading (in sexagesimal degrees) is added. An additional sheet contains the following information, for each strip if required: project identification, purpose, date; m_b, c, Z_0, l, q; minimum film requirement, film type; organizational details such as agreed targeting of points (control points, new points, etc.), proximity of international borders, flights only in overcast weather, etc.

Standards exist for quality control of survey flights, e.g. DIN standards: Schwebel, R.: PFG 2001(1), pp. 39–44.

3.7.2 Metric aerial cameras

Large format, aerial film cameras, which are presently most commonly used in practice, will be discussed in more detail in the first Section 3.7.2.1. Digital aerial cameras, which record the illumination intensity electronically using a CCD area array, are then discussed in Section 3.7.2.2. Finally, digital 3-line cameras are presented in Section 3.7.2.3.

3.7.2.1 Large format, metric film cameras

Figure 3.7-2 is a schematic diagram of the most important components of an aerial film camera. They are designed for serial photography in which a sequence of related

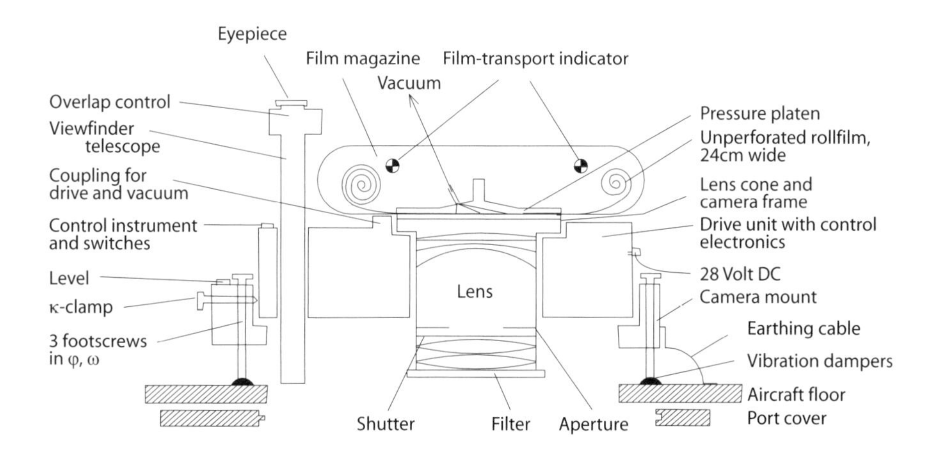

Figure 3.7-2: Schematic diagram of an aerial film camera

metric images is recorded. (German information sources may refer to these as "Reihenmesskameras" (RMK), i.e. serial metric cameras.) A film cassette contains around 400 exposures. Above the port is the camera mount which, in the simplest versions, can be levelled manually by footscrews and has rubber or steel spring fixings for vibration damping. The mount carries the drive unit which can be rotated about a vertical axis and which contains all the elements required for operating the imaging cycle. The lens cone, containing the required lens type, is inserted in the drive unit. The film magazine is placed on the focal-plane frame of the lens cone and contains the pressure platen and the mechanisms for creating the vacuum and pressing the platen against the focal-plane frame.

After every exposure command, the imaging cycle runs automatically as follows:

- expose.
- in modern cameras, shift the image plane during exposure (image motion compensation, to be discussed in more detail in Section 3.7.4).
- raise the pressure platen and release the vacuum.
- transport the film and advance the photo number counter.
- apply the vacuum again (also known as pneumatic flattening).
- press the platen against the photo frame plane.

The camera is now ready for another exposure and awaits the next exposure command. The operator can generate this manually by pressing a button for "aimed single photographs" or it can be generated automatically for serial photographs. The imaging cycle requires about 1.6 to 2.0 s, a time which defines the shortest possible interval between photographs.

In the viewfinder the camera operator sees the ground moving past. Combined with the viewfinder telescope is the overlap control which sets the required forward overlap when the set of moving curved lines appear to move at the same speed as the ground. If the lines move faster than the image of the ground, the operator must slow them down manually, or accelerate them if they are moving too slowly. When the lines move at the same speed as the image of the ground, the photographs will be released at the proper intervals to generate the required forward overlap.

The drift angle is also adjusted via the viewfinder. A ground point imaged on the central line must appear to move along this line if the camera is correctly oriented. If it moves off at an angle α to this line, the camera must be rotated by this angle α and the aircraft heading correspondingly adjusted (Figure 3.7-4) to ensure the correct ground track.

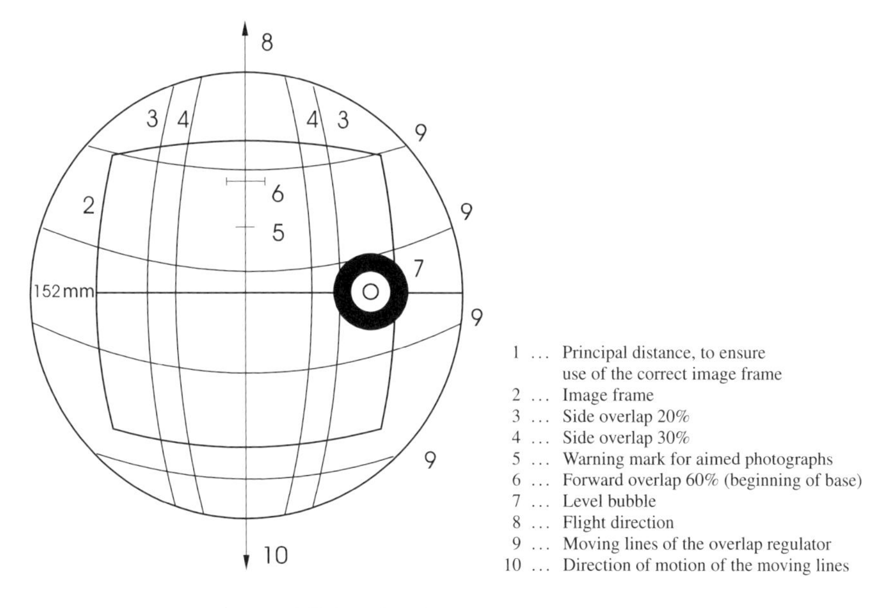

Figure 3.7-3: Viewfinder image (Wild RC10)

The following information can also be seen in the viewfinder (Figure 3.7-3):

- the image frame for the selected objective lens
- the base end points, for example for 60% forward overlap
- the side overlap, for example for 20% and 30%
- horizontal control by means of an image of the level bubble
- warning mark for aimed single photographs

The requirement for a photogrammetric image to be close to a central projection ensures the need for a central shutter instead of a focal-plane shutter and a very short exposure time ($1/250 - 1/1000$ s). Figure 3.7-5 shows a central shutter used by the manufacturer Zeiss.

Shorter exposure times require a compensating increase of light. Therefore, during exposure time the shutter needs to be fully open as long as possible. The conditions are indicated in Figure 3.7-6. The required exposure time is determined by an exposure meter (Section 3.2.2.3), taking into account to the sensitivity of the photographic material and the prior choice of aperture. The reverse procedure is sometimes required, i.e. the exposure time is specified and, from the exposure reading the camera system determines the appropriate aperture. The moment of exposure is the mean time: $t_m = 0.5(t_0 + t_1)$. If precise positioning and orientation sensors are built into the

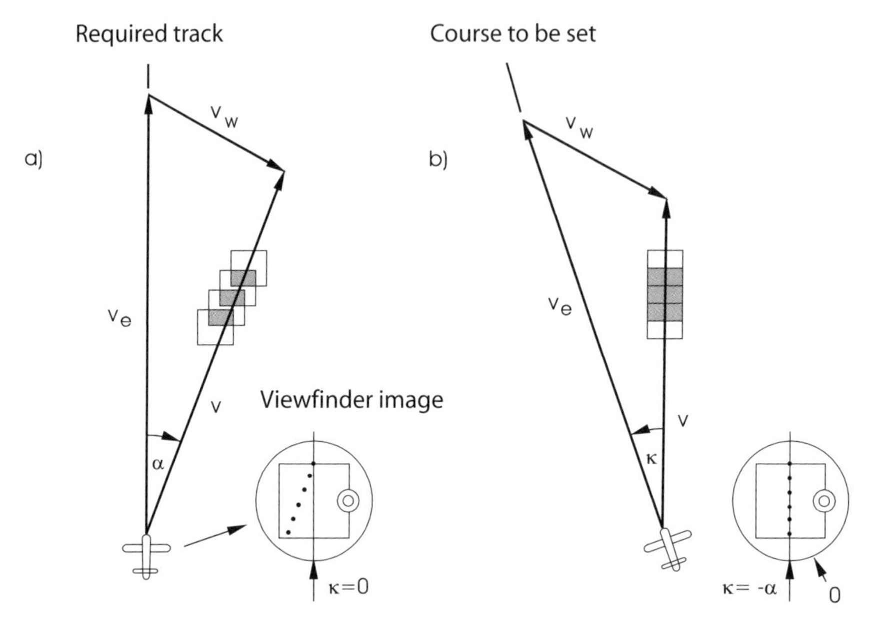

a) If the aircraft flies along its nominal course, a side wind will drive it along a different course or track over the ground. The drift angle is determined by checking the apparent motion of a selected point on the ground.

b) The course is corrected by the angle $\kappa \doteq -\alpha$. The camera is rotated by $\kappa \doteq +\alpha$. The true ground track of the aircraft now lies in the required direction and a selected ground point will appear to move along the central line.

Figure 3.7-4: Principle of drift correction

aircraft (Section 3.7.3), then the moment of exposure t_m must be synchronized with these. In a metric camera, the flash exposure of the fiducial marks must be made at the moment of exposure t_m.

An analogue metric image not only contains the image itself in the 23 cm × 23 cm format but an auxiliary image displaying extensive additional data (Figure 3.7-7) which can be supplied in analogue or digital form. The auxiliary image or, more accurately, the data it contains, can be divided into two groups. One contains strictly necessary information, the other contains optional data.

Strictly necessary auxiliary information:

- the number of symmetrically distributed fiducial marks should be 8 (previously 4 were common) in order to support sufficient redundancy for the transformation of fiducial marks. Identifiers for fiducial marks, such as numbers, are necessary in order to identify them automatically in the digitized images.

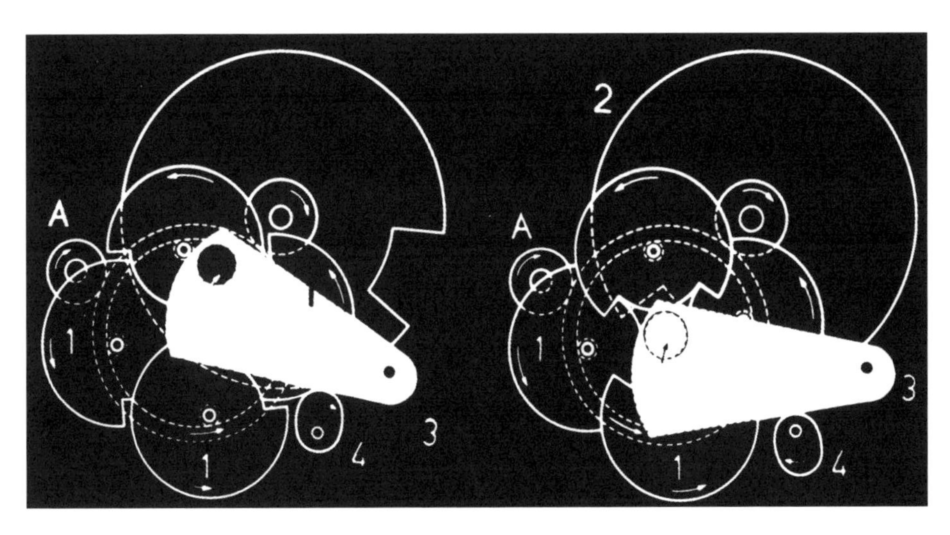

Figure 3.7-5: Example of a rotary shutter (Zeiss Oberkochen), left closed, right open. A is the drive. The lamellae 1 rotate faster than lamella 2. The sword lamella 3, controlled by the overlap regulator, is swung aside relatively slowly by the eccentric 4 while lamella 2 is not yet in the open position. Lamella 2 opens while lamellae 1 are still closed, before opening very rapidly. The speed of rotation of the lamellae determines the exposure time. The closing procedure is the same in reverse.

Figure 3.7-6: The shutter efficiency is an optimum when the cross-hatched area is as small as possible, i.e. the shutter is fully open as long as possible. The shutter must therefore be opened and closed as fast as possible.

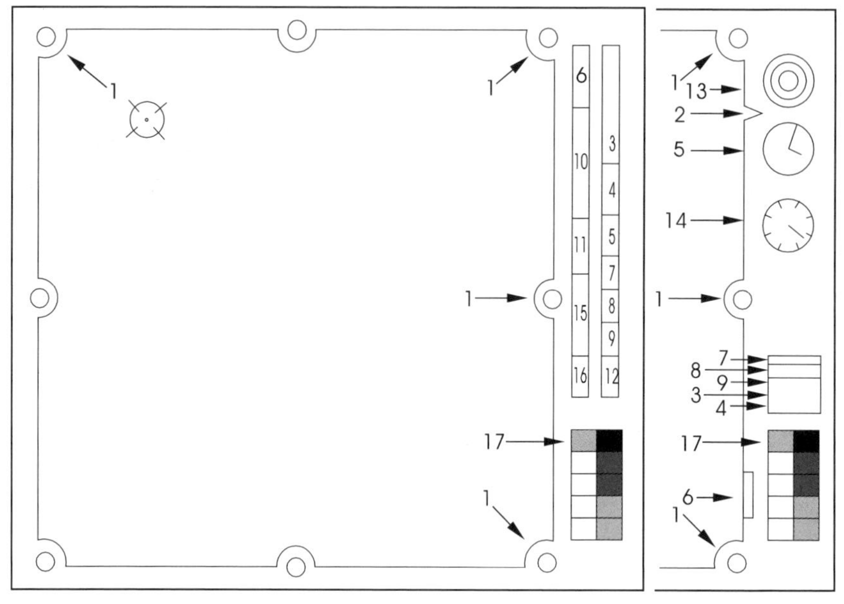

1) eight fiducial marks (with number)
2) ninth, asymmetric fiducial mark
3) brief name of project (data pad)
4) date (data pad)
5) time
6) photo number
7) camera (lens) number
8) principal distance
9) magazine number (data pad)
10) data of exterior orientation
11) overlap
12) photo scale
13) circular level
14) coarse height indicator
15) exposure time, aperture
16) image motion (length)
17) grey wedge

Figure 3.7-7: Auxiliary data imaged with a photograph; left: digital, right: analogue miniplots of instruments

- if the fiducial marks do not have identifiers then a ninth, asymmetrically placed fiducial mark, is required in order to identify automatically the fiducial marks in a digitized image, as well as to determine automatically the orientation of the image[48] (laterally correct, mirror reversed, rotated). If the additional mark is missing then the level bubble or centre of the clock (Figure 3.7-7) can be used instead.
- the principal distance for re-creating the interior orientation and a possible determination of scale
- the date, camera number and film cassette or pressure platen number in order to have a link to both the flight plan and the calibration certificate
- project name and incrementing image number for managing larger image blocks

[48]Ellenbeck, K.H., Waldhäusl, P.: BuL 52, pp. 70–71, 1984.

Optional auxiliary information:

- grey wedge or colour chart for contrast control and colour adjustment during photographic post-processing and for brightness and contrast enhancement, as well as colour adjustment of digitized metric images

- current exposure data: aperture, exposure time and image motion compensation (Section 3.7.4)

- clock to record moment of exposure so that the agreed (midday) period for flying can be checked and, for example, the direction of north in the image can be determined from time of day and direction of shadows

- information relevant to exterior orientation:

  - level bubble to detect a poorly levelled aircraft. Centrifugal forces will also, of course, affect the bubble.

  - Coarse height measurement of absolute flying height to an accuracy of around ± 50 m. Combined with a coarse ground height, this enables an average image scale to be derived.

  - GPS position and IMU orientation which are only available in expensive multi-sensor systems (Section 3.7.3). Since post-processing is generally required to provide good quality values, they maybe exposed on film later on.

Leica Geosystems (formerly Wild, LH Systems) currently offers the RC30 camera system (Figure 3.7-8). Technical details include: Interchangeable lens cones with principal distances of 8.8, 15 and 30 cm, largest aperture of $f/4$, 100 Lp/mm resolution, exposure times between 1/1000 and 1/100 s. For further details: `http://www.leica-geosystems.com/corporate/en/ndef/lgs_57632.htm` (November 2006).

Intergraph (formerly: Z/I Imaging, Carl Zeiss Jena, Carl Zeiss Oberkochen) currently offers the RMK TOP camera system (Figure 3.7-9). RMK TOP15 (15 cm principal distance, aperture between $f/4$ and $f/22$), RMK30 (30 cm principal distance, aperture between $f/5.6$ and $f/22$), Objective lens distortion $\leq 3\,\mu$m, rotary shutter (Figure 3.7-5), exposure times between 1/500 and 1/50 s, two different circular graduated filters (Section 3.1.6) with a transmission in the centre of 35% and 60%, 8 numbered fiducial marks, shortest interval between photographs 1.5 s. Further details: `http://www.intergraph.com/rmktopascs/default.asp` (November 2006).

Both camera systems can be combined with GPS and IMU (Section 3.7.3). In addition there are components for image motion compensation (Section 3.7.4).

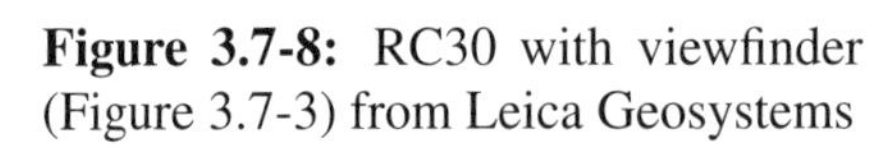

Figure 3.7-8: RC30 with viewfinder (Figure 3.7-3) from Leica Geosystems

Figure 3.7-9: RMK TOP from Intergraph

3.7.2.2 Digital cameras with CCD area sensors

There are now aerial cameras with CCD area sensors which significantly fulfil the quality requirements demanded by photogrammetry. One problem is the number of pixels which are currently available in arrays up to 7K $\times$ 4K; aerial photogrammetry requires a multiple of this (Section 3.3.1). A second problem is the read-out time: this must lie within the shortest interval between photographs, between 1 and 2 s.

With its DMC (Digital Mapping Camera), Intergraph has solved these problems by mounting four CCD cameras in the camera housing in place of a single CCD camera, each with 7K $\times$ 4K pixels and synchronous exposure. The four images overlap only to a limited extent (Figure 3.7-10, top). During post-processing a single metric image is created with around 13500 (normal to flight path) $\times$ 8000 (along flight path) pixels. The pixel size is 12 μm and field of view is 82 gon (74°) (normal to flight path) and 49 gon (45°) (along flight path). In other words, the DMC corresponds to a wide angle camera normal to the flight path and to a normal angle camera along the flight path (Table 3.7-1). Image recording uses 12 bits per pixel.

Exercise 3.7-5. The following data are redundant: principal distance, field of view, pixel size and number of pixels. Consider to what extent the DMC data are free of contradiction.

In addition to the four panchromatic camera heads, the DMC has three cameras for the spectral regions blue, green and red, and a fourth camera for near infra-red (Section 3.3.5.2). These last four cameras have a resolution which is worse by a factor of around 4 compared with the panchromatic camera, i.e. approximately the same area is recorded by a multi-spectral image (3000 $\times$ 2000 pixel) as is covered by 4 images in

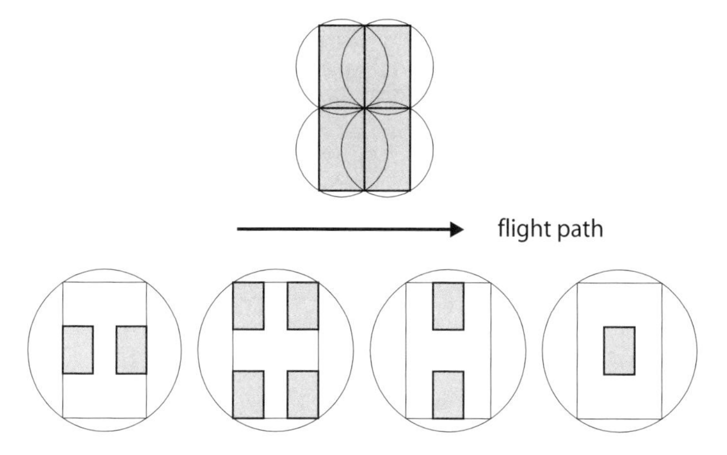

Figure 3.7-10: The DMC from Intergraph (top) and UltraCam-D from Microsoft-Vexcel (bottom)

panchromatic mode (Figure 3.7-10, top).

Like the RMK TOP (Figure 3.7-9), the DMC can be coupled with a GPS and IMU (Section 3.7.3). The DMC likewise has a facility for image motion compensation (Section 3.7.4.2).

Further reading and information: Hinz, A., Dörstel, C., Heier, H.: IAPRS XXXIII(B2), pp. 164–167, 2000. `http://www.intergraph.com/rmktopascs/default.asp` (November 2006).

Microsoft-Vexcel has also presented a digital photogrammetric camera, named the UltraCam-D, with very similar technical features to the DMC. In addition it has the following specifications:

- pixel size on the ground at a flying height of 300 m: 3 cm (at this extremely low altitude the system requires 50 pixel for image motion compensation, Section 3.7.4.2)
- minimum interval between images: 0.75 s
- geometric accuracy in image plane: $< \pm 2\,\mu\text{m}$

A prominent difference between the UltraCam-D and DMC is that the four panchromatic CCD cameras are arranged in a row along the direction of flight and are not synchronously exposed. Exposures are delayed such that the perspective centres of the four images are approximately at the same location in the ground coordinate system ("syntopic exposure"). Each camera has one, two or four CCD area arrays in the image plane, arranged such that the final composite image is built up of 3×3 tiles (Figure 3.7-10, bottom). A further difference between the cameras is that the objective lenses of the four panchromatic cameras in the UltraCam-D each cover the entire image format of 11500 (normal to flight path) $\times$ 7500 (along flight path) pixels. In contrast, each lens in the DMC covers only the format of its corresponding image tile of 7K $\times$ 4K pixels (Figure 3.7-10, top). The second camera in the UltraCam-D cluster, with 4 CCD area arrays (one in each corner), defines the geometric extent of the final composite image, created by post-processing from the nine image tiles each with 4K $\times$ 2.7K pixels (Figure 3.7-10, bottom).

Further reading and information: Leberl, F., Gruber, M., Ponticelli, M., Bernoegger, S., Perko, R.: ASPRS Meeting, Anchorage, Alaska, 2003. www.vexcel.com (November 2006).

3.7.2.3 Digital 3-line cameras

The concept of the digital 3-line camera (Figure 3.3-3, right) is due to Hofmann[49]. Digital 3-line cameras were initially designed for space missions[50] and were then adapted for aerial use, in some cases requiring significant additional development work. The HRSC-A (High Resolution Stereo Camera-Airborne) from the German Aerospace Centre (Deutsches Zentrum für Luft- und Raumfahrt (DLR)) is an adopted model for airborne sensing and has been successfully used for photogrammetric sensing[51].

The following text will deal in more detail with the ADS (Airborne Digital Sensor) from Leica Geosystems. In contrast to other multispectral line cameras, the ADS40 uses a sensor arrangement called tetrachroid which splits up the incoming light beam into the four spectral regions of blue, green, red and near infra-red. By this principle, the exact co-registration of all colour ranges from one object detail can be guaranteed. There is no need for any colour fusion or resolution merging for generating true colour and colour infra-red images. In addition to the tetrachroid, three panchromatic line sensors (covering the range from blue to red) are responsible for the forward (40°), backward (16°) and nadir looking data acquisition. Two models are available: the first model has one nadir looking tetrachroid sensor, which is mainly intended for colour orthophoto production. The second model has one nadir looking and one backwards looking tetrachroid for more universal applications including colour stereo compilation. The nadir looking panchromatic line is complemented by a second line staggered by half a pixel. This reduces the sampling interval which leads to a more reliable reconstruction of object frequencies. All sensor lines consist of 12000 individual sensor elements.

The combination with GPS and IMU is a valuable addition to digital cameras with CCD area arrays, but is not strictly necessary. However, with digital line cameras such as the ADS, such a combination is extremely important (Section 3.7.3). ADS with image motion compensation will be considered in Section 3.7.4.3.

[49]Hofmann, O.: BuL 54(3), pp. 105–121, 1986.

[50]Wewel, F., Scholten, F., Neukum, G., Albertz, J.: PFG 1998(6), pp. 337–348. Reulke, R., Scheele, M.: PFG 1998(3), pp. 157–163. Sandau, R., Eckhardt, A.: IAPRS XXXI(B1), pp. 170–175, 1996.

[51]Neukum, G.: in Fritsch/Spiller: Photogrammetric Week '99, Wichmann Verlag, pp. 83–88, 1999.

Further reading and information: Fricker, P., Sandau, R., Walker, S.: OEEPE-Publication No. 37, pp. 81–90, 1999. Sandau, R. et al.: IAPR 33(B1), pp. 258–265, 2000. http://www.leica-geosystems.com/corporate/en/ndef/lgs_57627.htm (November 2006).

3.7.3 Satellite positioning and inertial systems

Use of the well established GPS (Global Positioning System)[52] for navigation purposes during flying missions will be discussed first. For this purpose a modest accuracy is sufficient. In the subsequent sections the use of satellite and inertial systems for exterior orientation control will be presented.

3.7.3.1 Use of GPS during photogrammetric flying missions and image exposure

With regard to its original concept, GPS can be advantageously used for navigating photogrammetric flying missions. An additional application for either aerial film cameras or CCD area array cameras makes use of the GPS to release the shutter for individual exposures. For this purpose the aircraft's position must be continuously monitored in real time by the GPS and compared with the planned positions of the perspective centres given in the coordinate system of the GPS system currently in use. On the basis of this comparison, individual exposures can be made under computer control. Such an application of GPS can provide aimed single images (Section 3.7.1). The overlap regulator (Section 3.7.2) is then largely redundant.

For this application, GPS is used in kinematic mode with the C/A (Coarse/Acquisition) Code (wavelength $= 293.1$ m). In this mode, the distance of the GPS receiver from the satellite is determined by electronic distance measurement using the transmission time t of the signal and its known speed of propagation $c = 299762\,\mathrm{km/s}$. The C/A code is repeated after one millisecond (ms) and unambiguous range measurement with this code is possible up to 300 km.

The method of trilateration applied to the ranges measured simultaneously from the aircraft receiver to three GPS satellites, enables the calculation of receiver coordinates in WGS84 (World Geodetic System 1984), in which GPS satellite parameters are currently provided. Since the receiver clock cannot be exactly synchronized with the satellite clocks, an unknown time difference Δt must be taken into account. It is therefore necessary to measure range to at least four GPS satellites simultaneously. (Ranges which incorporate the unknown time difference are known as pseudo ranges.)

The position of individual imaging locations, initially provided in the geocentric, Cartesian coordinate system of WGS84, can be converted into geographical coordinates φ and λ and ellipsoid height h, for example using the formulae given in Section B 5.3.1, Volume 2. In modern aerial survey systems, the geographical coordinates φ and λ and

[52] In addition to the American GPS there is also the comparable Russian GLONASS. GALILEO is a third satellite positioning system currently being deployed by the European Union. A general term for such systems is GNSS (Global Navigation Satellite Systems). In this textbook "GPS" is also used as a synonym for these other systems.

ellipsoid height h are recorded during the flight onto the individual metric images (see Figure 3.7-7). Accuracy is at the C/A level, i.e. some tens of metres. The accuracy of GPS-supported navigation of the flying mission lies around $\pm 50\,\mathrm{m}$[53]. If it is assumed that, on average, an acceptable deviation between planned and actual imaging location is ± 5 mm in the image, then this accuracy is sufficient for flying missions which deliver an image scale smaller than $1 : 10000$ ($= 5/50000$). More accurate navigation is only required for flights delivering very large image scales.

The navigational accuracy of GPS in kinematic mode can be raised to a few metres when differential GPS (DGPS) is employed. In this case two receivers are required: one receiver remains at a (fixed) reference station whose coordiantes are known; the other receiver is located in the aircraft. Both receivers are positioned absolutely using the C/A code. At the (fixed) reference station corrections can be calculated and transmitted by telemetry in real time to the receiver in the aircraft. The function of a fixed reference station is being increasingly taken over by a network of permanently installed GPS stations.

3.7.3.2 Accurate determination of exterior orientation elements by GPS and IMU

The coordinates of the imaging locations can be determined by GPS and the angles which define the attitude of a given image can be determined by IMU (inertial measurement unit). Nowadays with DGPS a positional accuracy of around a decimetre can be achieved "on the fly". Instead of time-of-flight measurement with the C/A code, a phase comparison technique using the relatively short wavelengths of both GPS carrier waves is used ($L1 = 0.1905\,\mathrm{m}$, $L2 = 0.2445\,\mathrm{m}$). With phase measurement only the residual part of a single wavelength is determined and some multiple of the full wavelength (0.1905 m or 0.2445 m) must be added to this in order to calculate the actual range.

The GPS literature list at the end of this section provides several methods to eliminate the ambiguity. After solving for the ambiguity, the GPS receiver counts the increase or decrease of integer multiples of the wavelength compared with a starting value. The measured distances between a GPS receiver and a GPS transmitter are strongly distorted by atmospheric influences. These error sources have a similar effect on neighbouring receivers such that the differences between the receivers are largely free of the influences of the (systematic) errors. The differences between the original distance equations provide the components of a vector between both receivers. Provided one receiver is located at a point with known coordinates (the reference station) then the vector, applied to this station, provides the coordinates of the second receiver which is located in the aircraft.

Cycle Slips and the Multipath effect can, however, interrupt the almost continuous determination of the vector from the (stationary) reference station to the GPS antenna in the aircraft. Bridging techniques are required which will be discussed below.

[53]Becker, R., Barriere, J.: PE&RS 59, pp. 1659–1665, 1993. Arnold, H., Schroth, R.: in Fritsch/Hobbie (Eds.): Photogrammetric Week '93, pp. 53–62, Wichmann Verlag, Karlsruhe, 1993.

Exact positioning of the GPS antennae in the aircraft takes place approximately every second. Depending on aircraft speed, this corresponds to separations in position of around 50 to 100 m. These nodal points can be connected by a spatial curve[54]. A spatial flight path supported by GPS points is illustrated in Figure 3.7-11. Time acts as a parameter along this curve. If the individual exposures follow this time scale, normally GPS time (t_m in Section 3.7.2), then by interpolation it is possible to determine the coordinates of the antenna tips A_i at the times when the images are exposed. This synchronization is possible within about 1 ms, i.e. at an aircraft speed of 250 km/h the error is some 7 cm.

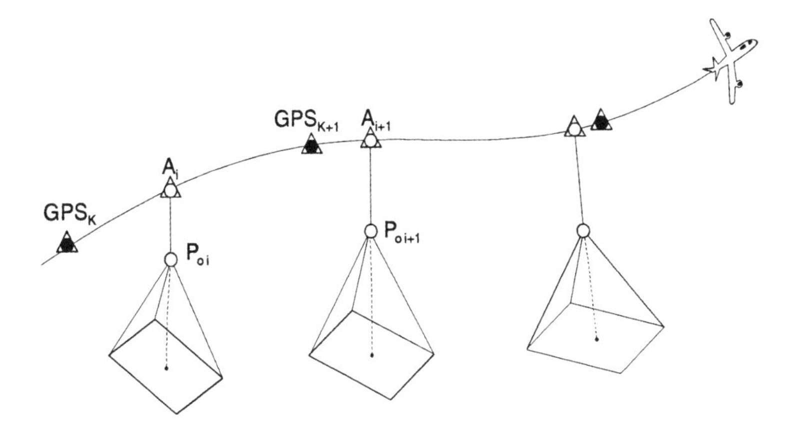

Figure 3.7-11: Photo strips with GPS positioning (A_i: interpolated positions of antenna tips, P_{oi}: perspective centres)

For the transfer to the coordinates of the perspective centres P_{0i} the angular attitude of the sensor, e.g. the aerial film camera, is required at the moment of exposure (Figure 3.7-11). These three orientation parameters are also of considerable interest in completing all six elements of exterior orientation for every image.

However, GPS only provides the positions of the imaging locations A_i and, from the known time intervals between exposures, the speed along the flight path of the GPS antenna mounted in the aircraft. Other measurement systems, which provide the position as well as angular orientation of the sensors are (I)nertial (N)avigation (S)ystems (INS)[55]. These make use of the inertia of a mass in relation to its acceleration $\ddot{a}$. According to Newton's second law of motion, force f is the product of mass m and acceleration $\ddot{a}$, i.e. $f = m\ddot{a}$. (For clarification: a = distance, $da/dt = \dot{a}$ = v = speed, $dv/dt = \ddot{a}$ = acceleration.)

Figure 3.7-12 shows one of the measurement principles. A force, acting along the axis of a cylinder to which a known mass is connected through an elastic spring, causes a

[54] Lancaster, P., Salkauskas, K.: Curve and Surface Fitting. Academic Press Ltd., 1986.

[55] We distinguish between INS and IMU. While the INS is a complete navigation system, the IMU is just one of its components which measures acceleration and attitude changes.

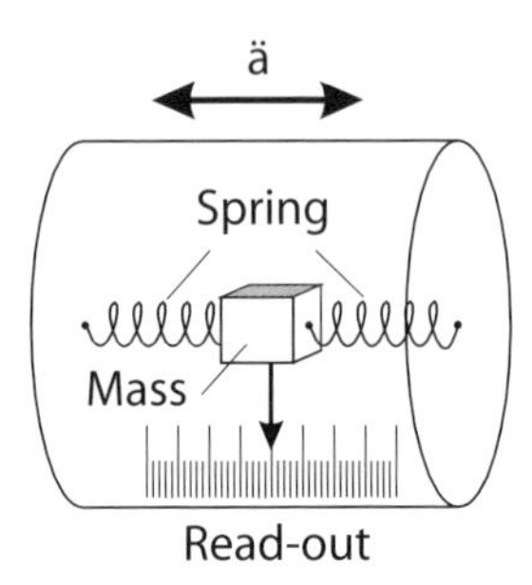

Figure 3.7-12: Deflection method of measuring acceleration

change in position of the mass and a resultant deflection on a scale also attached to the cylinder. This deflection is a measure of the corresponding acceleration $\ddot{a}$ of the mass. From the acceleration determined in this way, speed is obtained by integration over time; a further integration over time provides the distance or, effectively, position.

In order to record the movement of a mobile vehicle, in this case an aircraft, it is necessary to have three accelerometers, arranged orthogonally to one another (Figure 3.7-13). Three accelerometers are insufficient to completely determine the movement of a vehicle and directional information is also required. This is determined continuously by three gyroscopes (gyros), one mounted on each axis of rotation (Figure 3.7-13). Nowadays optical gyroscopes are mostly used. An optical gyroscope with a simple constructional principle is illustrated in Figure 3.7-14.

Two contra-rotating monochromatic beams of light are introduced into a circular optical waveguide. If the device has not rotated then, after the beams have traversed the waveguide, they arrive simultaneously at their starting point, i.e. without any phase difference. However, if the circular waveguide rotates with angular velocity $\dot{\alpha}$, then the beam travelling in the direction of rotation has a greater distance to cover to reach the starting point which has now moved ahead (Figure 3.7-14, left); for the beam travelling in the opposite direction the path is shortened (Figure 3.7-14, right). When the beams are brought together after completing the circular path, a relative phase shift $\Delta\varphi$ is detected from which the angular velocity can be determined as follows:[56]

$$\dot{\alpha} = \frac{C_0 \lambda}{8\pi F} \Delta\varphi \tag{3.7-3}$$

C_0 ... speed of light in vacuum: 299792 km/s
λ ... wavelength of light beam, e.g. 0.6 μm
F ... circular area enclosed by the light guide (fibre optic) which can be multiplied by a factor n through the use of n windings of the fiber

[56]Taken from: Schwarz, W.: Schriftenreihe des Deutschen Vereins für Vermessungswesen, Volume 22, pp. 54–97, 1996. This publication provides a comprehensive presentation of kinematic sensors.

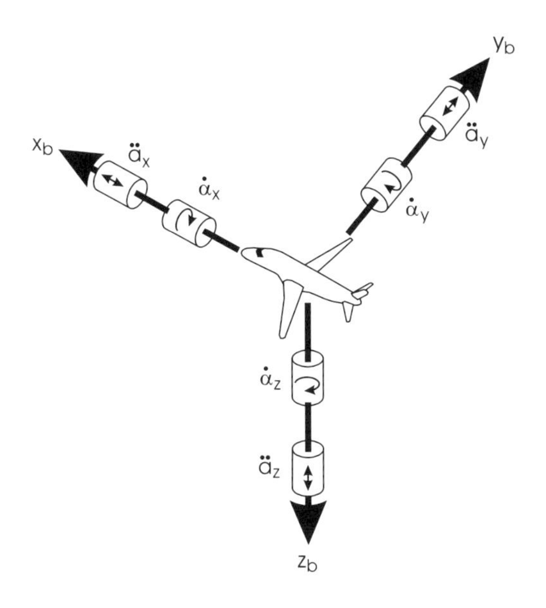

Figure 3.7-13: Body coordinate axes for a strapdown inertial navigation system

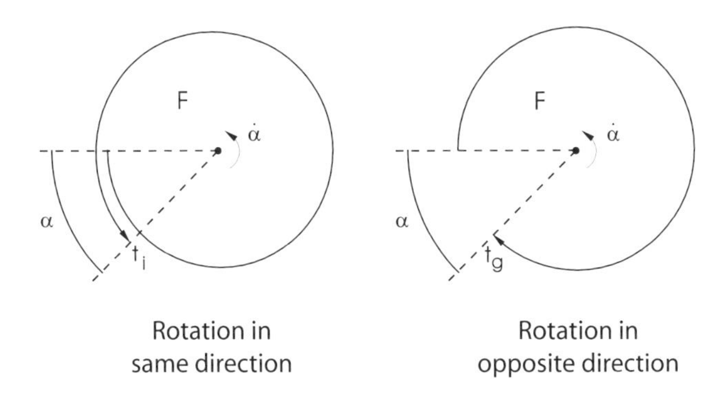

Figure 3.7-14: Optical gyroscope for an intertial system

By integration of the angular velocity $\dot{\alpha}$, for example over the time difference $(t_i - t_g)$, the angle of rotation a can be determined (Figure 3.7-14).

Table 3.7-2 shows the three currently available accuracy classifications for inertial navigation systems[57]. The mid accuracy class is appropriate for determining the exterior

[57]Schwarz. K.-P.: in Fritsch, D., Hobbie, D.: Photogrammetric Week'95, pp. 139–153, 1995. The stated tilt accuracies relate to roll and pitch; yaw angle (heading) is a factor 3 to 5 less accurate. At the Photogrammetric Week 2001 the following accuracies, achieved during practical tests, were reported: GPS positioning between ±5 and ±30 cm, IMU orientation for roll and pitch angle between ±3 and ±5 mgon (10″ and 16″, resp.) and for the heading angle ±8 mgon (26″).

orientation of aerial photographs and is reasonably priced. Over short time periods it provides positional accuracy at the decimeter level. An aircraft's tilt can only be determined by an IMU. Over a period of one minute, an accuracy of 5 mgon (15″) can be achieved which is sufficient for standard requirements. At a flying height of 2000 m, this corresponds to a planimetric error of 15 cm on the ground.

Time interval	System accuracy		
	high	medium	low
Position			
1 s	0.01 – 0.02 m	0.03 – 0.1 m	0.3 – 0.5 m
1 min	0.3 – 0.5 m	0.5 – 3 m	30 – 50 m
1 h	300 – 500 m	1 – 3 km	200 – 300 km
Tilt			
1 s	$< 1''$	$10'' - 15''$	$30'' - 2'$
1 min	$1'' - 2''$	$15'' - 20''$	$10' - 20'$
1 h	$10'' - 30''$	$30'' - 3'$	$1^\circ - 3^\circ$
Price (US$)	≈ 1000000	≈ 100000	≈ 10000

Table 3.7-2: Three accuracy classes for inertial navigation systems

As can be seen from Table 3.7-2, an IMU's positional accuracy is only acceptable over short time periods. A Position and Orientation System (POS) therefore combines GPS with IMU. GPS supplies position and speed and supports the IMU. The IMU provides direction and tilt and densifies the relatively large separations between GPS measurements which, as already mentioned, are between 50 m and 100 m at a recording interval of 1 s. In contrast, the IMU's recording interval is around 0.01 s (100 Hz). Short interruptions in GPS recording (cycle slips, etc.) can also be bridged with the IMU.

Figure 3.7-15 shows the arrangement of the most important sensors in a POS, together with an imaging sensor. Great care must be taken to determine the relative positions of the three illustrated sensor modules. Details of a corresponding calibration can be found in specialist literature[58]. The relationships are relatively simple if all the sensors are fixed to the aircraft:[59]

- from the GPS coordinates of the antenna tips A_i, the perspective centres P_{0i} can be calculated along the spatial directions delivered by the IMU (Figure 3.7-11).
- if the coordinate axes of the imaging sensor and IMU are strictly parallel then the rotation angles of the IMU directly correspond to the rotation angles of the image sensor[60].

[58] E.g. Shin, E., El-Sheimy, N.: ZfV 127(1), pp. 41–49, 2002. Heipke, Ch. et al.: OEEPE Publication No 43, 2002.

[59] If the inertial system is fixed to the aircraft or vehicle, it is known as a strapdown system, as shown in Figure 3.7-13.

[60] This type of configuration cannot generally be made to a sufficient accuracy. Deviations from the

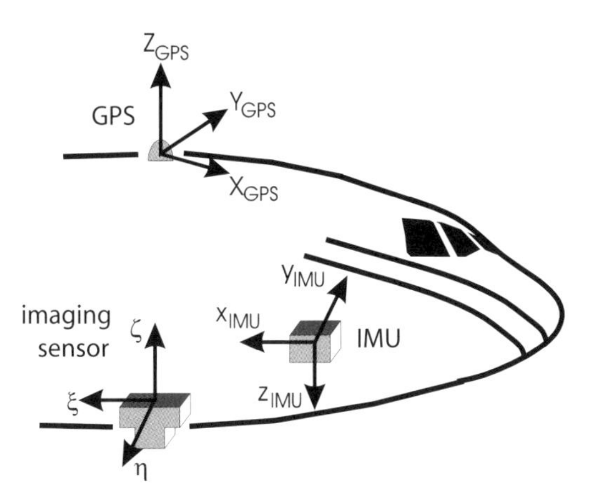

Figure 3.7-15: Imaging sensor together with GPS and IMU in a Position and Orientation System (POS)

The relationships are considerably more complicated when there is continuous banking to compensate for the drift angle and, on the basis of the IMU gyroscope, the imaging sensor platform is continually tilted, in other words stabilized. In this case the rotations must be recorded with timing information in order to establish the relationships between all three sensor groups during post-processing. This procedure is greatly simplified if at least the imaging sensor and IMU are fixed to the same tilting platform so that their relative positions remain unchanged throughout the entire flying mission.

The POS data are valuable for processing aerial images taken with either film or CCD area array cameras but their importance is not so great as when processing images taken with digital line cameras. POS data are also essential for airborne laser scanners, discussed in Section 8.1. This will be discussed in more detail in Section 3.7.3.3.

A position and orientation system is marketed by Applanix under the name POS/AV. Further reading and information: Lithopoulos, E.: in Fritsch/Spiller: Photogrammetric Week '99, Wichmann Verlag, pp. 53–57, 1999. `www.applanix.com`. The German company IGI offers a comparable system (e-mail: `info@igi-ccns.com`).

3.7.3.3 Gyro-stabilized platforms and particular features of line cameras and laser scanners

On flying missions with a digital line camera (Section 3.3.1) and/or a laser scanner (Section 8.1), aircraft navigation along the planned route will also, of course, employ GPS in a similar way to a flying mission with film or CCD area array cameras. This usage was discussed in Section 3.7.3.1.

ideal are known as misalignments. The transformation of IMU rotations into rotations of the imaging sensor or into the coordinate system of the image processing software generally involves the successive multiplication of several rotation matrices.

Gyroscopes provide rotation angles as a function of time. Using this information, an image sensor's platform can be continually stabilized. Compensation can therefore be made for the considerable changes in roll, pitch and yaw angles (Section 3.7.1) during a flight. Such gyro-stabilized platforms are available for the large format, aerial film camears mentioned at the end of Section 3.7.2.1 and the CCD area array cameras discussed in Section 3.7.2.2. However, the mechanical and electronic complexity associated with a gyro-stabilized platform does not, in general, justify their use for such cameras. Roll, pitch and yaw angles can either be extracted from POS information during photogrammetric processing or, if a POS was not employed, by means of aerotriangulation (Chapter 5). This is possible because with these sensors the recording of a single image effectively takes place at a single point in time[61].

The situation is different with dynamic image recording, as is the case when using a single or 3-line camera or a laser scanner. Images taken without the use of a gyro-stabilized platform show large distortions as a result of the flight dynamics, and these are much worse with line cameras than with laser scanners. Certainly such images can be processed with the aid of POS information. However, conditions for processing are significantly improved if line cameras, and possibly also laser scanners, are mounted on gyro-stabilized platforms. The IMU should also be mounted on this platform. An accurate IMU need then only record small "residual" tilts, which is a considerable advantage in the design and construction of such a device.

However, the constant rotations of the gyro-stabilized platform continually alter the relationship between the GPS measurements and the imaging sensor or laser scanner, as can be appreciated in Figure 3.7-15. In order to provide a continuous "eccentricity vector" between GPS antenna and the reference point of the imaging sensor, continuous recording of the gyro-stabilized platform rotations is necessary. A separate gyroscope system is required to stabilize the mounting platform and this must be:

- fixed to the aircraft
- able to measure large rotations
- less accurate than the gyroscopes in the IMU which is fixed to the imaging sensor

In summary it can be said that the exterior orientation of a line camera alters from imaged line to imaged line and that therefore the direct recording of the six orientation elements by GPS and IMU is obligatory. For laser scanners, the exterior orientation even alters for each range measurement. With film and CCD area array cameras the POS information is, in contrast, only an interesting option, as will be seen in Section 4.1. A gyro-stabilized platform is here not strictly necessary; however, it improves the quality of the image and makes processing easier, particularly for dynamically recorded images.

[61]More exactly: during exposure there should be no large shifts or rotations of the camera. To some very reasonable extent, a gyro-stabilized platform improves image quality, particularly for longer exposure times (Section 3.7.4.1).

Further reading for Section 3.7.3 (selected examples only): Hofmann-Wellenhof, B., Lichtenegger, H., Collins, J.: GPS—Theory and Practice. Springer, 1997. Hutton, J., Lithopoulos, E.: PFG 1998(6), pp. 363–370.

3.7.4 Image motion and its compensation

The blur which an optical system produces in an image arises from the following sources:

- from the spread in depth of a three-dimensional object, which cannot be sharply focused onto a single image plane (Section 3.1.4)
- from the wave nature of light, which results in diffraction at the aperture stop (Section 3.1.5.1)
- from the resolving power of the optics (Section 3.1.5.2)
- from the resolution of the emulsion in film cameras (Section 3.2.2.7)
- from the size of the opto-electronic sensor elements and their geometric arrangement in digital cameras (Section 3.3.2) and film scanners (Section 3.4.1)

Further blurring of the image occurs as a result of image motion of the object or camera during the time the shutter is open. The following discussion concentrates on aerial photogrammetry.

The theoretical image motion u_{th} of an object in the image plane is (Figure 3.7-16):

$$u_{th} = vt\frac{c}{h} = \frac{vt}{m_b} \tag{3.7-4a}$$

$$u_{th} = \frac{10^3 vt}{3.6 m_b} \tag{3.7-4b}$$

v = velocity in [km/h] u = image motion in [mm]
t = exposure time in [s] m_b = image scale number

During the exposure time t the image moves over the image plane and the energy in the illumination is distributed. Only in places where sufficient light energy falls on the chemical or electronic sensor will the image become visible. A single object point becomes a "line" in the image. If the chemical or electronic sensors could react to the smallest possible level of illumination energy, then the length of the "lines" would be given by Equation (3.7-4a). In fact, the sensors react only to a certain minimum energy, so that the "lines" in the image are shortened. (When a camera is subject to vibration, particularly in a helicopter, double images can appear.) On the basis of current film

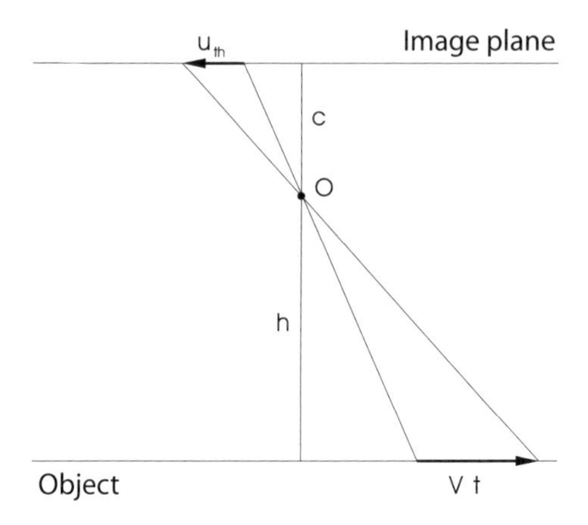

Figure 3.7-16: Theoretical image motion

sensitivities used in photogrammetry, the "shortened" image movement u amounts to around half the theoretical image movement:

$$u \approx 0.5u_{th} = \frac{140vt}{m_b} = 140ct\frac{v}{h} \tag{3.7-5}$$

c in [mm] v in [km/h] t in [s]
h in [m] u in [μm]

Numerical Example (to estimate the image motion).

	from linear motion	from rotation
Velocity v	300 km/h	2 gon/s $\hat{=}$ 170 km/h
Exposure time t	1/300 s	1/300 s
Principal distance c	150 mm	150 mm
Flying height h	1500 m	1500 m
Image scale $1 : m_b$	1 : 10000	1 : 10000
Theoretical image motion u_{th}	28 μm	16 μm
Image motion u visible in practice	14 μm	8 μm

The blurring of the image caused by image motion significantly reduces the resolution. Meier[62] shows that a theoretical image motion of 1.5 times the reciprocal of the resolution is just tolerable in practice.

$$u_{th,\max} \leq 1.5R^{-1} \tag{3.7-6}$$

[62]Meier, H.K.: BuL, pp. 65–77, 1960.

Numerical Example (of the maximum permissible exposure time for which the image motion remains tolerably small).

Given: $R = 80\,\mathrm{Lp/mm}$
Equation (3.7-6): $u_{th} = 19\,\mu\mathrm{m}$
Equation (3.7-5): $u \approx 10\,\mu\mathrm{m}$
Given: $v = 300\,\mathrm{km/h}$, $c = 150\,\mathrm{mm}$, $h = 1500\,\mathrm{m}$,
$\Rightarrow$ $m_b = 10000$
Equation (3.7-5): $t_{\max} = 1/440\,\mathrm{s} \approx 1/500\,\mathrm{s}$

Whether an exposure time of 1/500 s is acceptable depends upon the overall sensitivity of the film to be used and the object brightness (season, time of day, weather, light intensity...).

Exercise 3.7-6. What is the largest acceptable image scale for a flying speed of 300 km/h and an exposure time of 1/300 s ($R = 40\,\mathrm{Lp/mm}$)? (Solution: $1 : m_{\min} = 1 : 7400$.)

Compensation for image motion is particularly required at low flying heights and large image scales. There are a number of solutions.

3.7.4.1 Compensation of image motion in aerial film cameras

In modern aerial cameras, the image motion due to the aircraft's forward movement can be compensated during exposure by a computer controlled displacement of the film's pressure plate with a velocity of $v' = u_{th}/t = v\,c/h$ (see Equation (3.7-4a)). This is known as Forward Motion Compensation (FMC).

This type of motion compensation eliminates the most significant part of the image blurring, but it does not eliminate the effects of:

- the three rotations of the aircraft; gyro-stabilized camera mounts are necessary to compensate for this effect (see also the ends of Sections 3.7.3.2 and 3.7.3.3).
- the variations of flying speed, unless the actual flying speed v for image motion compensation is taken on line from the overlap control (Figure 3.7-3), or the navigation system.
- differences of ground heights. To compensate for these, the image motion compensation must be at least partially separated from the overlap control system. The overlap must be controlled so as to ensure overlap in the highest parts of the ground (see Section 3.7.1c); the image motion compensation, on the other hand, must be set to achieve sharp imaging for the most important parts of the ground (the valleys, up to the average ground height).

Exercise 3.7-7. With the data of the numerical example above, compute the remaining image motions: for the three rotations of the aircraft of about 2 gon/s, for height differences of $\pm 500\,\mathrm{m}$ and for variations of flying speed of $\pm 20\,\mathrm{km/h}$. (Solution in [μm]: $u_\omega = u_\varphi = 13$, $u_\kappa = 9$, $u_{\Delta h} = 7$, $u_{\Delta v} = 1$.)

Exercise 3.7-8. What is the loss in resolution of the film camera introduced in the numerical example of Section 3.2.2.7 ($R_T = 67\,\text{Lp/mm}$, which according to Equation (3.2-11) derives from $R_O = 90\,\text{Lp/mm}$ and $R_P = 100\,\text{Lp/mm}$), if a theoretical image motion of $u_{th} = 28\,\mu\text{m}$ (see first numerical example of this section) is taken into account? (Solution: From Equation (3.7-6), which only has an empirical basis: $R_{\text{ImageMotion}} = 54\,\text{Lp/mm}$; applying this to an expanded version of Equation (3.2-11): $R_T = 42\,\text{Lp/mm}$; there is therefore a reduction in resolution from $67\,\text{Lp/mm}$ to $42\,\text{Lp/mm}$.)

Exercise 3.7-9. What is the deterioration in the modulation transfer function $MTF(f)$ of the full system introduced in Figure 3.2-11 when the above image motion $u_{th} = 28\,\mu\text{m}$ is taken into account?

Tip: As modulation transfer function for $MTF(f)_{\text{ImageMotion}}$ use the sinc function (3.3-5) with values $MTF(0)_{\text{ImageMotion}} = 1$ and $MTF(54\,\text{Lp/mm})_{\text{ImageMotion}} = 0$. To evaluate Equation (3.2-12) use $MTF(f)_O$ and $MTF(f)_P$ from Figure 3.2-11.

Another solution for image motion compensation is to "nod" the camera by a small angle very quickly during exposure.

Exercise 3.7-10. Calculate the angular velocity of nodding (pitch angle velocity) in order to compensate for the image motion of $28\,\mu\text{m}$ in the above numerical example. (Solution: $3.5\,\text{gon/s} \mathrel{\hat{=}} 3.15°/\text{s}$).

3.7.4.2 Image motion compensation for digital cameras with CCD area arrays

For as long as the camera shutter is open, an image of the ground moves continuously across the focal plane in the direction of flight at a speed of $w = vc/h$. At the moment the shutter opens to take a photograph ($t = 0$, say) an image of a particular narrow strip of ground at right angles to the direction of flight will start to be imaged on the sensor elements of row n; in general the exposure will be insufficient. At time $(t + \Delta t)$, where Δt can be calculated from w and the pixel dimensions, an image of the same strip of ground will fall on, and start to be recorded on, row $(n + 1)$; again the exposure will be insufficient. Immediately before this, however, the charge packets from row n were moved to row $(n + 1)$. This process can be repeated several times, row by row, at intervals of Δt; the image of that particular strip of ground is therefore integrated over a number of rows of sensor elements, resulting in an image with an adequate signal-to-noise ratio. Finally, readout of all rows is initiated; that is to say, readout of the whole array. In this way, during the period when the shutter is open, the whole image has been shifted electronically to keep pace with the moving image of the ground, rather than being shifted mechanically as in analogue cameras.

This method of FMC is called Time Delay Integration. Very high demands are made of the readout process when the CCD arrays are large since, within exposure times which are a fraction of a second, all the pixels in the array must be read several times (see also Section 3.3.1). The limitations on FMC for film cameras mentioned at the end of Section 3.7.4.2 above also apply to motion compensation with TDI.

The DMC from Intergraph and the UltraCam-D from Microsoft-Vexcel (Section 3.7.2.2) incorporate TDI for motion compensation.

3.7.4.3 Image motion compensation for digital line cameras

A compensation for image motion, caused by forward movement of the aircraft, would only be possible with TDI in digital line cameras if every linear array was composed of several parallel lines of detectors. The ADS from Leica Geosystems has not made use of this possibility. In consideration of the very high read-out rate of 800 Hz, achieved with very sensitive detectors, motion compensation at mid to high altitudes is not required.

3.7.5 Effective illumination in aerial photography

In Section 3.2.2.8 the exposure H was defined[63] as the illuminance E falling on the emulsion, or opto-electronic detector, times the exposure time t. But what is E for an aerial photograph? Of the total radiation from the sun reaching the atmosphere about one-third is immediately reflected back into space by the atmosphere. The remainder is subject to extinction (absorption plus scattering) in the atmosphere, caused by air molecules and aerosols (suspended particles from 0.01 to 100 μm in diameter such as dust, water droplets in haze, fog and clouds, and ice crystals). Short wavelengths are scattered much more strongly than long wavelengths, hence the blue sky we see. The remainder of the sun's direct radiation, S_d, (emitted at zenith angle Z_I, Figure 3.7-17), together with the skylight, S_s, form the global radiation which finally illuminates the ground. The two components of this total incident light on an object vary in proportion from $S_d : S_s = 3 : 1$ at a solar altitude of 50° and with a cloudless sky, to $1 : 30$ in light haze.

Photographs can be taken without difficulty under a sky completely covered by a high layer of cloud. The greater degree of scattering lightens the shadows of objects but also reduces the contrast (see Section 3.1.5), thus reducing measuring accuracy. In order to increase the contrast various yellow filters are used according to the haze conditions.

An object on the ground absorbs a part of the total incident light; the remainder is diffusely reflected from the natural and artificial rough surfaces forming the ground, some at a zenith angle Z_R towards the camera. This reflected component mixes with the scattered light which does not reach the surface (haze light), and is further scattered by the atmosphere and aerosols. The strong contrast reduction caused by the total amount of blue scattered light is minimized by placing a yellow filter in front of the objective. The 12 – 15 individual lens elements are also coated with an anti-reflex coating.

[63] In this textbook on photogrammetry we use photometric units. In remote sensing, radiometric units are common. In place of illuminance E the irradiadiance is used; in place of luminous intensity I the radiant intensity is used, etc.

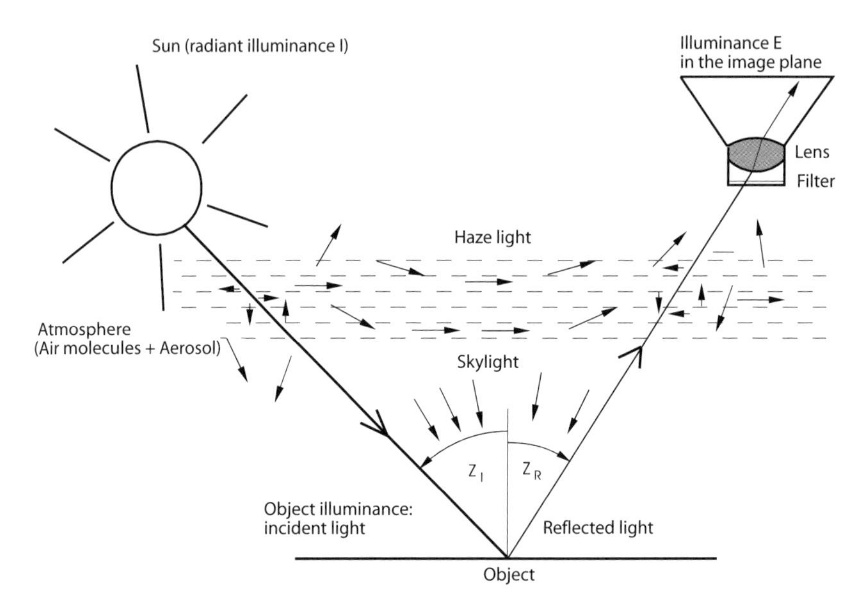

Figure 3.7-17: Raypath from sun to image plane

The result of the superimposition of haze light on the light reflected from the object is that contrast decreases with increasing height. Films with greater gradation ($\gamma = 1.4$ – 1.8) are therefore used for small-scale photographs from greater flying heights.

The non-absorbed, reflected part of the white (sun) light is an important information carrier. Studies of the spectral reflectance of various types of ground objects show:

a) that the reflected light covers strongly differentiated parts of the spectrum, so permitting object colours to be differentiated (Figure 3.7-18, upper left);

b) that even at a uniform spectral distribution, the reflected light can vary greatly in intensity;

c) that the reflectance of vegetation is almost constant in visible light (in the green part of the spectrum somewhat more than in the blue and red), but that significant differences in intensity occur in the near infra-red (Figure 3.7-18, lower part) and for this reason colour infra-red films (Section 3.2.2.5b) or black-and-white infra-red films (Section 3.2.3) are used for classifying vegetation types.

The object brightness, now defined as the integrated intensity over all wavelengths of the reflected visible light, can lead to extreme differences of density in the photograph (Table 3.7-3). The scattered light (Figure 3.7-17) causes, however, a significant reduction of the overall contrast.

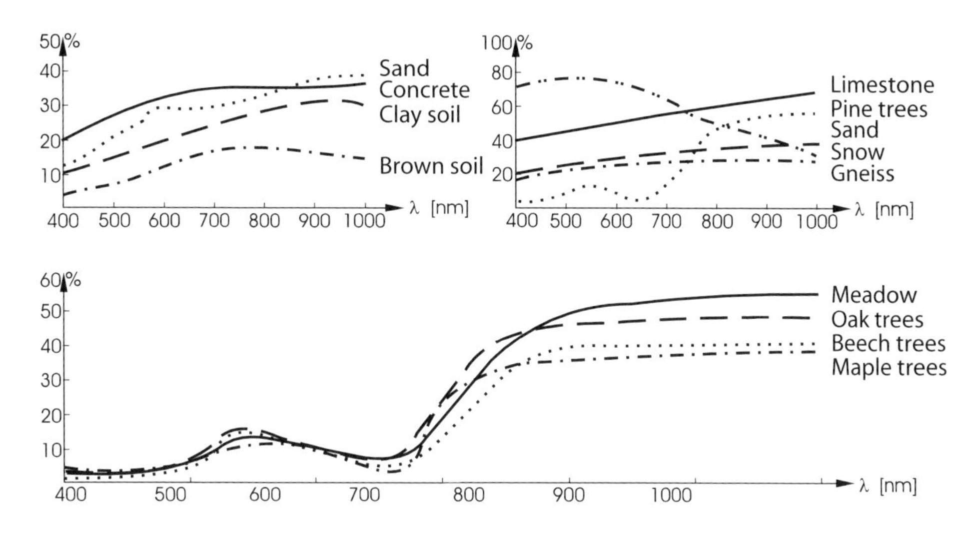

Figure 3.7-18: Spectral reflectance of various ground objects as a percentage of the incident light

Numerical Example. The ratio of object brightnesses of a tarred road and new snow, from Table 3.7-3, is $8 : 80 = 1 : 10$. If we add a haze brightness of 20, the ratio above is reduced to $28 : 100 = 1 : 3.6$. For the same haze brightness, the contrast of a coniferous forest to new snow is reduced from $3 : 80 = 1 : 27$ to $23 : 100 = 1 : 4.3$ and the contrast of a coniferous forest to a tarred road from $3 : 8 = 1 : 2.7$ to $23 : 28 = 1 : 1.2$! (Contrast is defined here as a ratio, see Section 3.1.5.3.)

Coniferous forest	1...3
Water	3
Tarred road	8
Dry meadow	7...14
Wet sand, deciduous forest	18
Yellow, dry sand	31
Light, dry concrete	35
Old snow	42...70
New snow	80...85

Table 3.7-3: Total reflectance as a percentage of object illuminance

3.7.6 Survey aircraft

Naturally, it is possible to convert almost any type of aircraft for aerial photography by cutting a port for the camera in the bottom of the fuselage. The selection of a type must satisfy certain performance characteristics, however. Relatively slow aircraft are preferred for large-scale photographs (currently permitted minimum flying height is around 300 m above open ground and around 1 km over densely populated areas). However, for rapid flight to and from the area of interest and to the required flying height a fast speed[64] and rate of climb are desirable. Since the flights to and from the area of interest are expensive, it is desirable to complete the whole job in one flight and therefore the maximum possible flight endurance is demanded. Projects in developing countries often require the use of short landing strips, i.e the use of STOL aircrafts (short take-off and landing). Oxygen masks are needed above a height of 4000 m. A pressurized cabin is much more comfortable, if only from the point of view of temperature. The camera port must then be sealed by an expensive plate of optical glass which must cause no additional distortion, but which must be capable of withstanding the stresses caused by pressure and temperature differences (ΔT can easily reach 50 Kelvin). An alternative solution here is a remotely operated camera. The navigation instrument with overlap regulator is then placed inside the cabin over a smaller sealed port while the camera is outside the pressurized part of the fuselage above an open port.

Camera ports in the centre of the width of the floor have under them an axially symmetric air cushion which, seen as an optical component, cause less asymmetric distortion by refraction than do ports at the side of the floor. It must be possible to close the port at take-off and landing in order to prevent mechanical damage to the port glass or the camera.

High-wing aircraft give the camera operator a good view downwards, an important aid in navigation during the approach flight, particularly for large-scale photographs. An intercom system is essential for communication between crew members. Navigation, including exposure release, is increasingly done with GPS and for large scale work with DGPS (Section 3.7.3.1).

Aerial-photography flights in central Europe must take full advantage of every day of good weather. IFR[65] (blind flying) equipment is therefore essential for take-off and landing through ground mist. The time available for photography is limited not only by the weather, however, but also by such factors as shadow lengths, which should usually not exceed 3.5 times the object heights. Flights under full, high cloud cover are possible in summer; they may even be desirable, for example for photographs of cities, so as to reduce the loss of detail in shadows. The resulting photographs have relatively weak contrast, however, and the height accuracy may suffer as a result of loss of detail contrast[66]. Sometimes a flight may be expressly required before trees are in leaf, although stereoscopic acuity in the maze of shadows may be so poor that the forest floor cannot be measured. These factors, taken together, produce a maximum flying time, in central Europe, of about 300 flying hours with about 120 flights, per

[64]Airspeed is still given in knots (nautical miles per hour). The conversion formula is: 1knot = 0.515 m/s = 1.852 km/h.

[65]IFR = Instrument Flight Rules

[66]Waldhäusl, P.: Results of the Vienna Test of OEEPE Commission C. Institute for Applied Geodesy, Frankfurt, pp. 13–41, 1986.

year and machine. The result is that each hour of photographic flight is relatively expensive[67].

3.8 Terrestrial metric cameras and their application

The photographs for terrestrial photogrammetry are usually taken with the cameras in fixed positions. Photographs at large distances, camera to object, are only used in special cases, for example for topographic surveys by expeditions and for glaciological research. In general, the objects are much closer and the camera must therefore be focused on finite distances and the depth of field (Section 3.1.4) has to be considered.

The most important applications in close-range photogrammetry are:

- architecture and civil engineering (surveys of old and new buildings, documentation of buildings and damage to buildings, etc.)
- computer Aided Facility Management (internal building surveys, etc.)
- conservation of monuments (preserving cultural assets by photographic records, which can be used at any time for restoration or reconstruction of the monuments)
- archaeology (documentation and surveys of excavations)
- biophotogrammetry (measurements of living creatures)
- forensic photogrammetry (scenes of crimes, most often used for documentation and reconstruction of the details of traffic accidents)
- industrial photogrammetry (automobile design, production control, assembly control)
- mobile mapping (surveys along traffic routes)
- computer-controlled vehicle navigation (robotics)
- support and control for medical operations

The "normal case" of terrestrial photogrammetry will be outlined in the following Section 3.8.1, after which terrestrial cameras will be presented. This presentation starts with cameras based on chemical sensors. These fulfill the rigorous conditions of metric cameras but in the concluding discussion the strict demands of photogrammetry are relaxed and the discussion moves to off-the-shelf (amateur) cameras. Electronic sensors then take the place of chemical sensors.

[67] A review of the more common survey aircraft can be found in Albertz/Kreiling: Photogrammetric Guide, 4th ed., Wichmann, Karlsruhe, pp. 112–117, 1989.

3.8.1 "Normal case" of terrestrial photogrammetry

The "normal case" of stereophotogrammetry was presented in Section 2.1.5. In the "normal case", both camera axes are perpendicular to the base and parallel to one another. In Section 2.1.5 the object coordinate system is configured for aerial photogrammetry, i.e. the direction of photography is the opposite direction of the Z axis (Figure 2.1-13). In terrestrial photogrammetry, the direction of photography is the direction of the y axis (Figure 3.8-1)[68].

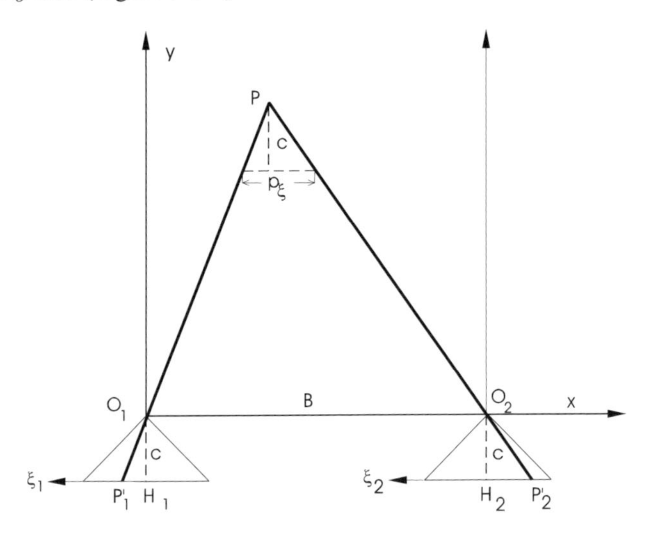

Figure 3.8-1: The horizontal "normal case"

The horizontal "normal case" is derived from the "normal case" of aerial photogrammetry (Figure 2.1-13) by substituting[69] $\omega_1 = \omega_2 = 100\,\text{gon} \mathrel{\hat{=}} 90°$. Equations (2.1-32) and (2.1-34) become, for the "normal case" of terrestrial photogrammetry:

$$
\begin{aligned}
x &= y\frac{\xi_1}{c} = B\frac{\xi_1}{p_\xi} \\
y &= B\frac{c}{p_\xi} \\
z &= y\frac{\eta_1}{c} = B\frac{\eta_1}{p_\xi} = y\frac{\eta_2}{c}
\end{aligned}
\tag{3.8-1}
$$

[68] The XYZ coordinates have been replaced by xyz coordinates. This expresses the fact that the object coordinate system in the normal configuration is a local xyz system which can be arbitrarily oriented in 3D space, for example with vertical base B or with tilted camera axes.

[69] In order to avoid singularities in the conversion of rotation matrix elements into rotation angles, the angles α, ν and κ are often used in terrestrial photogrammetry instead of the angles ω, φ, κ (Section B 3.4.2, Volume 2). Here α (primary) is the azimuth of the camera axis, and ν (secondary) is the angle between the camera axis and the vertical (nadir distance). In terrestrial photogrammetry the rotation matrix B (3.4-6) from Volume 2 is therefore used instead of the rotation matrix of Equation (2.1-13).

$$\sigma_x = \sqrt{\left(\frac{\xi_1}{c} m_b \frac{y}{B} \sigma_{p_\xi}\right)^2 + (m_b \sigma_\xi)^2}$$

$$\sigma_y = m_b \frac{y}{B} \sigma_{p_\xi} = \frac{y^2}{cB} \sigma_{p_\xi} \tag{3.8-2}$$

$$\sigma_z = \sqrt{\left(\frac{\eta_1}{c} m_b \frac{y}{B} \sigma_{p_\xi}\right)^2 + (m_b \sigma_\eta)^2}$$

3.8.2 Stereometric cameras

To reconstruct spatial objects at close ranges, it can be a great convenience to obtain the required two metric images simultaneously using a single camera system known as a stereometric camera. Such a stereometric camera is illustrated in Figure 3.8-2. In most systems the base is fixed and very accurately known. A common base length is 120 cm and these devices are often known as fixed base cameras. When using a fixed base, which has a metric camera at each end and is configured for the "normal case", the 12 exterior orientation elements of both cameras are known with respect to the local xyz coordinate system. With this type of photogrammetry, no control points are required, a considerable advantage in practical use. Control points, or other control elements such as measured lengths, are recommended only for checking purposes. However, control

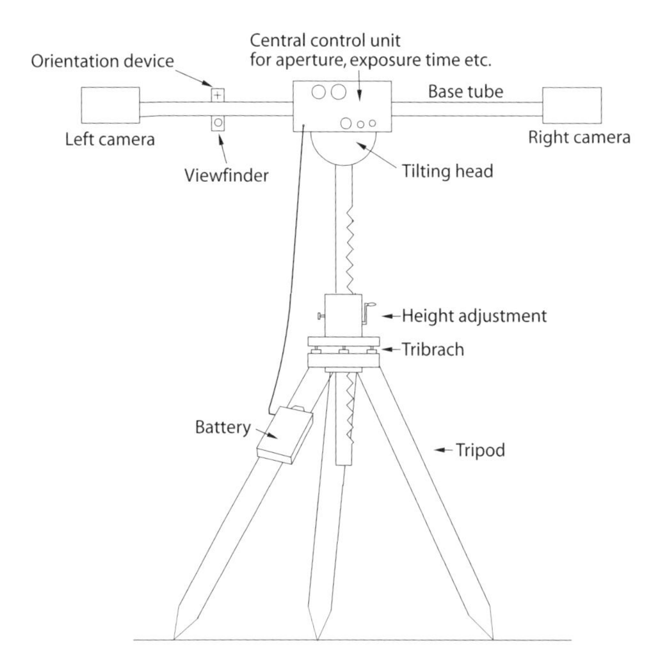

Figure 3.8-2: Schematic diagram of a stereometric camera

points are necessary if the photogrammetric results must be supplied in a global XYZ system rather than a local xyz system.

Despite the great advantages of stereometric cameras, they play a secondary role in practice. The fixed base has the effect that the average error in the direction of the camera axes, i.e. in the y direction, increases with the square of the distance (Equation (3.8-2b)). Stereometric cameras should therefore be available with different base lengths; the most flexible solution is a rail along which a variable base can be individually set.

Numerical Example. The accuracy behaviour of stereometric cameras will be examined on the basis of data from Zeiss and Wild cameras. These stereometric cameras were constructed with base lengths of 120 cm and 40 cm. Evaluation of Formula (3.8-2b) results in the following accuracies ($c = 60\,\text{mm}$, $\sigma_{P_\xi} = \pm 7\,\mu\text{m}$):

Base 120		Base 40	
y [m]	σ_y [mm]	y [m]	σ_y [mm]
3	0.9	2	1.2
6	3.5	3	2.6
10	10	6	11
20	39	10	29

Table 3.8-1: Depth accuracy of stereometric cameras with a base of 120 cm and 40 cm ($c = 60\,\text{mm}$, $\sigma_{P_\xi} = \pm 7\,\mu\text{m}$)

Exercise 3.8-1. The Zeiss and Wild stereometric cameras are fixed focus cameras. The Zeiss camera with a base of 120 cm is focused at 9 m and the 40 cm base is focused at 4 m. For these cameras calculate the circles of confusion for the distances given in Table 3.8-1 (aperture setting $f/11$). For both camera systems also calculate the depth of field for a maximum circle of confusion of 40 μm diameter and give recommendations for the use of the Zeiss stereometric camera with respect to expected accuracies (Table 3.8-1).

Exercise 3.8-2. Give the 12 elements of exterior orientation for a stereometric camera with respect to the local xyz system. (Solution: Corresponds to details given on Figure 2.1-13.)

3.8.3 Independent metric cameras

Independent metric cameras are preferable to stereometric cameras when the base should be increased in order to achieve optimal accuracy at increased object ranges. Figure 3.8-3 shows a universal independent metric camera. To determine elements of exterior orientation there are levelling bubbles and horizontal and vertical angle settings. By means of forced centering the camera can be replaced by another measuring instrument such as a theodolite. These components for determing the elements of exterior orientation are, however, used less and less (Section 3.8.8).

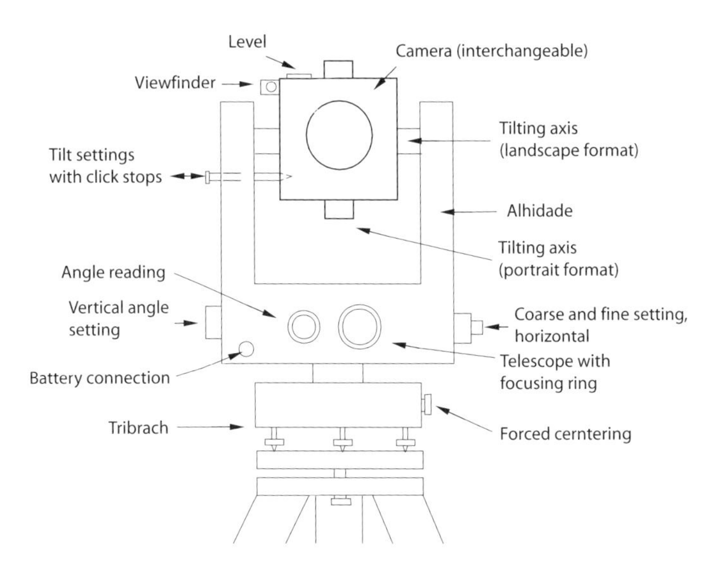

Figure 3.8-3: Schematic diagram of a terrestrial metric camera

The camera illustrated in Figure 3.8-3 has a rectangular format which can be set in either landscape or portrait orientation. To further accommodate a particular imaging situation, for example photographing tall buildings, Wild cameras incorporated an offset principal point. Three typical configurations of the image field are shown in Figure 3.8-4.

The technical data for the Wild P31 Universal Terrestrial Camera, a system comprising three interchangeable precision metric cameras, are summarized in Table 3.8-2. Optical distortion is smaller than 4 μm. Glass plates or cut-film in special holders can be used (see also Table 3.2-1). Calibrated intermediate focusing rings control the depth of field (Section 3.1.4 and in particular Figure 3.1-14). The optical resolving power of the P31 has already been given in Section 3.1.5, in particular with Figure 3.1-19.

The smaller version of the P31 is the Wild P32. It is designed to be mounted on a theodolite telescope. Principal distance $c = 64$ mm. The P32 can also use glass plates (64 mm $\times$ 89 mm) or roll film. The principal point is offset by 10 mm. It has a fixed focus distance of 25 m (see numerical example b in Section 3.1.4).

Jenoptik Jena has manufactured the universal terrestrial camera UMK with a format of 13 cm $\times$ 18 cm. There are five different objective lenses from narrow angle to super wide angle. The UMK's construction corresponds largely to Figure 3.8-3. Calibrated focusing controls the depth of field (Section 3.1.4, and in particular Figure 3.1-15).

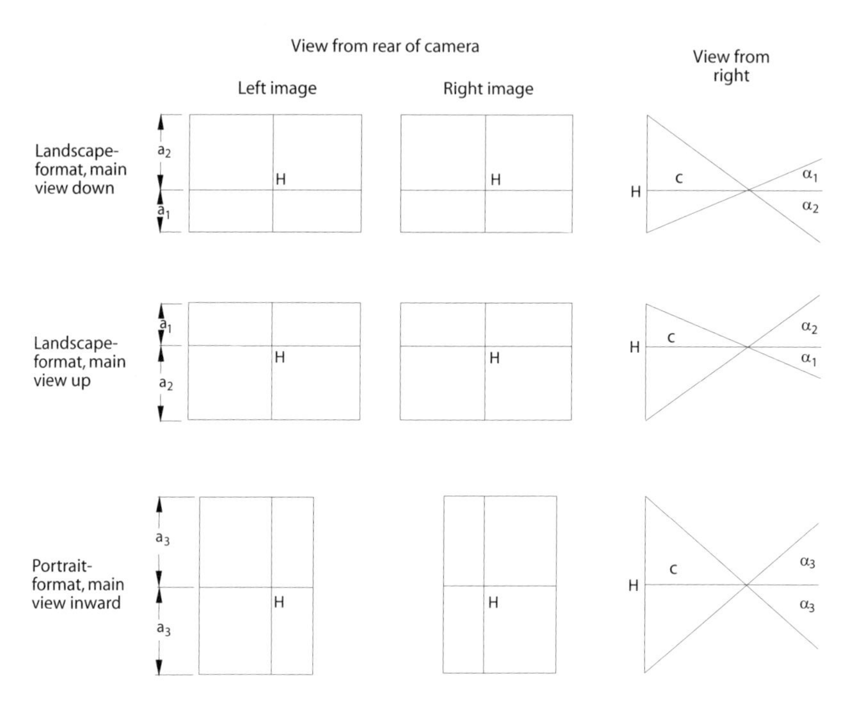

Figure 3.8-4: The various image-field arrangements for Wild cameras with offset principal point

Field of view	Super wide angle	Wide angle	Normal angle
Principal distance c [mm]	45	100	202
Aperture stops	$5, 6 \ldots 22$	$8 \ldots 22$	$8 \ldots 22$
Exposure times [s]	$B, 1, \ldots, 1/500$		
Usable format to edges of image[70] [mm]	92×118	84×117	$(90 - 83) \times 118$
Principal point displacement [mm]	0	19	12
Standard focusing [m]	7	25	35
Intermediate rings for other distances	no	yes	yes
Plates (P) or film (F)	P	P,F	P,F

Table 3.8-2: Technical data for Wild P31

[70]The object extent which can be measured can be seen in Figure 3.8-9.

3.8.4 Semi-metric cameras

Semi-metric cameras occupy a position between metric and amateur cameras. Figure 3.8-5 contrasts a semi-metric camera with a metric camera. The key component of a semi-metric camera is the réseau. This is a (stable) glass plate on which an array of crosses at intervals of several millimetres is marked. The réseau is exposed on every image. The réseau crosses have known coordinates for their centres which are determined prior to mounting the plate in the camera. Measured values for these points are acquired when individual images taken with the camera are processed. A transformation[71] onto reference coordinates significantly eliminates distortion and unflatness in the film. In a sense, this transformation provides a metric image in the plane of the stable glass plate on which the réseau is marked.

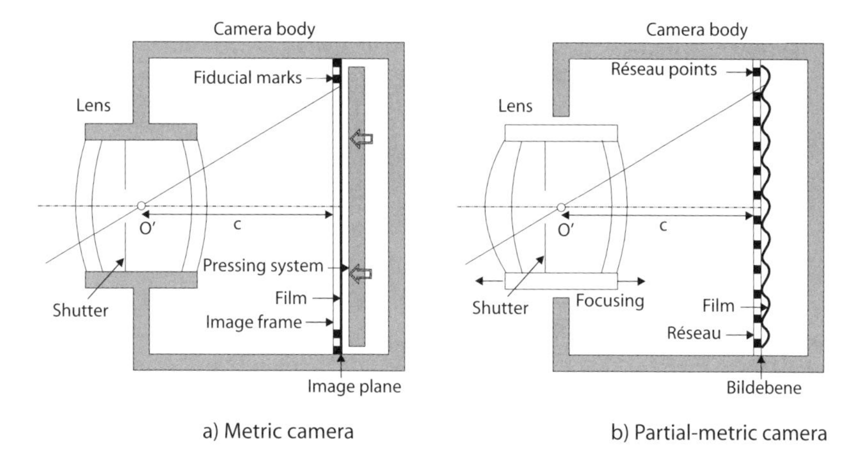

Figure 3.8-5: Metric camera (left) and semi-metric camera (right)[72]

The following comments can be made with regard to calibration and the elements of interior orientation:

- the optical distortion of semi-metric cameras is usually significantly greater than in metric cameras (Figure 3.1-9 shows a typical distortion curve for a semi-metric camera). Numerical correction of optical distortion (Section 3.1.3) is therefore absolutely necessary for high accuracy applications.

[71]The bilinear transformation (Section 3.2.1.2) is a suitable transformation for each grid square. Since the 8 unknowns of the bilinear transformation are derived without any overdetermination from the 4 réseau crosses in the corners of the square, a prior check of the data using a similarity or affine transformation is recommended (see Table 3.2-3). An interesting alternative is the derivation of a closed interpolation function for the entire image by simultaneous use of all measured réseau points. An appropriate method is based on interpolation by least squares, which can also largely eliminate (filter out) random measurement errors (Section B 9.5.1.3, Volume 2).

[72]Taken from Luhmann, T., Robson, S., Kyle, S., Harley, I.: Close Range Photogrammetry. Whittles Publishing, 2006. This textbook discusses semi-metric cameras in some detail, since semi-metric cameras play a significant role in close range measurement.

- semi-metric cameras can be focused to almost any distance. Calibration data exists for marked focal positions. The location of the perspective centre over time is not as stable as in a metric camera (indicated in Figure 3.8-5); the position of the perspective centre is therefore only approximately known.
- semi-metric cameras are often calibrated using test fields. In applications demanding high accuracy there would also be an on-the-job calibration (Sections 3.1.2 and E 3.4, Volume 2).

The image format of semi-metric cameras varies from small format up to full metric camera format. Medium format cameras (6 cm $\times$ 6 cm) are most widely represented. Commercially available lenses vary from super wide angle to telephoto.

Semi-metric cameras using the réseau principle are available from Geodesign, Hasselblad, Leica, Linhof, Pentax and Rollei. A list is available in Luhmann, T., Robson, S., Kyle, S., Harley, I.: Close Range Photogrammetry. Whittles Publishing, 2006 (see Table 3.8, p. 143). Wester-Ebbinghaus (Ph.Rec. 76, pp. 603–608, 1990) has made major contributions to the development of semi-metric cameras.

3.8.5 Amateur cameras

Amateur cameras can only be used in photogrammetry to a limited extent. The simplicity of the photography is offset by the complexity of analysis and processing. This is discussed in Volume 2 of this series of textbooks. As regards calibration, only on-the-job calibration is a realistic option (Section 3.1.2 and E 3.3, Volume 2). The particular detail of defining the image coordinate system, in which the (initially unknown) perspective centre is determined during on-the-job calibration, is discussed in the following section (Figure 3.8-6).

3.8.6 Terminology and classification

The classification of photogrammetric cameras into terrestrial and aerial cameras derives from practical experience. From a methodological point of view, however, the following classification is relevant:

- metric cameras, specially developed for photogrammetry.
- semi-metric cameras, not originally intended for measuring purposes, but which have been developed for photogrammetry even though not all the elements of interior orientation may be stable.
- non-metric cameras, which can be used for low-accuracy photogrammetric applications, particularly when supported by modern, non-specialist software packages.

Cameras can also be classified according to the internal features used to define interior orientation:

- fiducial-mark cameras, specially developed for photogrammetry, with 4, 8 or more fiducial marks.

- réseau cameras, which have a glass plate supporting a regular grid of crosses placed in front of the image plane. This grid can be used to detect, in particular, film deformation and lack of flatness.

- frame cameras, which have no fiducial marks, but sharply imaged edges. The "indirect" fiducial marks, formed by the corners of the frame, replace the fiducial marks and are located by intersecting lines derived from measurements of points on the frame edges (see Figure 3.8-6).

Figure 3.8-6: Typical 35 mm film. The frame sides are measured at the points shown and the corners determined exactly by intersection of the lines.

- false-frame cameras, e.g. Polaroid Land Camera, in which the image is bounded not by a fixed frame in the camera, but by a paper frame which is then removed. The interior orientation cannot be reconstructed and such photographs can hardly be used for photogrammetry. A similar situation occurs when only an enlarged part of an image is available.

Exercise 3.8-3. An object is to be reconstructed using two images. Consider the number of unknown elements of interior orientation when using the following cameras:

- metric camera (Solution: no unknowns.)

- calibrated and stable réseau camera (Solution: no unknowns.)

- uncalibrated but stable réseau camera (Solution: 3 unknowns and additional polynomial coefficients for the optical distortion)

- partially stable amateur camera or frame camera with the same focus setting for both images (Solution: 3 unknowns and additional polynomial coefficients, both for the optical distortion and for the film deformation.)

- amateur or false-frame camera (e.g. two different image details) (Solution: 6 unknowns and additional polynomial coefficients, both for the optical distortion and for the film deformation.)

3.8.7 CCD cameras

In close range photogrammetry film cameras have been more or less completely replaced by digital cameras. The lower geometric resolution of digital cameras can be replaced, if necessary, by choosing a shorter imaging distance to the object. Of particular interest is the data flow in a complete photogrammetric system when data acquisition is made not with a film camera but with a digital camera. In analytical photogrammetry, considering the need to develop the film, there is inevitably an interruption in the process; this is an off-line technique. In contrast, digital photogrammetry can offer an on-line process from data acquisition through to the presentation of results. By means of this real-time photogrammetry new application areas can be opened up (robotics, computer-controlled medical operations, etc.).

CCD cameras used in close range application normally have a single CCD area array sensor (Section 3.3.1) covering the entire image format. In order to increase the geometric resolution there are digital cameras which sequentially sample the image in the image plane, using the same principle already discussed when digitizing photographs (Section 3.4, in particular Figure 3.4-3). When using a digital scanning camera there must be no appreciable movement between camera and object during the scan. In addition, the real-time property mentioned above is compromised.

Only a few manufacturers have been able to provide small numbers of CCD cameras which, based on their geometric stability, can be classed as digital metric cameras. (e.g. Figure 3.8-7[73]). Integral arrays of 4K $\times$ 4K detectors (and higher) are already commercially available.

In addition to their medium format film cameras, many manufacturers offer as an alternative digital semi-metric cameras. A CCD area array sensor is used in place of the film magazine (e.g. Figure 3.8-8).

WWW addresses for digital metric cameras: `www.rollei.de`, `www.imetric.com`, `www.geodetic.com`.

For moderate accuracy demands there are inexpensive standard CCD video cameras on offer which follow television standards.

Video cameras were used in the past as electronic imaging devices in real time photogrammetry. They operate according to diverse standards (for instance, CCIR/PAL in Europe and EIA/NTSC in the US), which are quite similar with differences mainly in the frame frequency (25 vs. 30 image frames per second) and the number of lines per image (625 vs. 525 lines). It is important to mention that they use an analogue technique. If digital images are required, additional frame grabbers with analogue-to-digital converters have to be employed. Due to the analogue nature of the signal, various perturbing influences can be observed and high quality imaging with an acceptable geometric stability for photogrammetric purposes needs very careful set-ups. Motion pictures have never had significant importance in photogrammetry and, where

[73]The photogrammetric close range systems, to which these cameras belong, reduce the images to the coordinates of targeted points at a very early stage (Section 3.6.2).

Figure 3.8-7: INCA 3 digital metric camera from Geodetic Systems, Inc. (GSI)

Figure 3.8-8: AIC integral from Rollei

they have been employed, video movie cameras have been "misused" as still image cameras.

The predecessors of today's digital cameras were often called "still video cameras". The new generation of digital consumer cameras excels the former still video technology concerning stability, resolution, convenience and price. Cameras with a sensor array of up to 10 Megapixels are becoming standard, some professional models have more impressive specifications. Still, most of these cameras have all the shortcomings of a typical amateur camera: unstable zoom lenses, careless mounting of the sensor plate, automatic settings which can rarely be adjusted manually and often cannot be reset to certain predefined values or cannot even be fixed. Conventional digital cameras are also not suited to certain industrial applications. Some manufactures offer special cameras for very specific applications (e.g. DALSA (`www.dalsa.com`), JAI (`www.jai.com`), TVI (`www.tvivision.com`)). From the photogrammetric point of view they also belong to the category "amateur cameras".

Further reading for Section 3.8.7: Luhmann, T., Robson, S., Kyle, S., Harley, I.: Close Range Photogrammetry. Whittles Publishing, 2006. Shortis and Beyer: In Atkinson (ed.): Close Range Photogrammetry and Machine Vision, Whittles Publishing, UK, pp. 106–155, 1996.

3.8.8 Planning and execution of terrestrial photogrammetry

Guide values for the maximum possible object distance or smallest possible image scale can be derived from the required accuracy of coordinates which, on the object, are parallel to the image plane. This is done by inverting the Formulae (3.8-2a) and (3.8-2c), restricted to the second term (see also Equation (2.1-35)). Using Formula (3.8-2b), the smallest possible base can then be derived from the required accuracy in depth (a relevant example is solved in Exercise 2.1-12). On this basis it can be decided if a stereometric camera with a fixed base can be used or if images must be taken with an independent camera.

After clarifying the issue of accuracy it is then necessary to decide from where the object can be imaged most advantageously for photogrammetric purposes, i.e. the area within which camera stations can be located and where line-of-sight restrictions must be avoided. In this procedure the field of view, or more accurately the focal length and image format, can be decided. Large object distances demand long focal lengths, short ranges not infrequently require super wide angles. The depth of field is also important for planning the optimal object distances and it is necessary to stay within the depth-of-field limits of s_n and s_f (see Section 3.1.4). A variety of cameras and objective lenses is valuable in making these decisions.

Finally, the object extent acquired by the images must be taken into account when planning terrestrial photogrammetry. The width of an object acquired by a stereopair of images can be seen in Figure 3.8-9. The relationship between base B, object range y, object width OW, principal distance c and usable image width s (from edge to edge) is illustrated in Figure 3.8-9 and expressed in Equation (3.8-3) which allows for a 5 mm safety margin at the image edges. (For CCD cameras 100 pixels should be chosen as a safety margin.)

$$\frac{s - 10\,\mathrm{mm}}{c} = \frac{B + OW}{y} \tag{3.8-3}$$

Considerations relevant to the optimal baselength, which is a very critical parameter in planning, can be summarized as follows:

- the required accuracy of depth measurement σ_y (Equation (3.8-2b)),
- the need to view into object recesses, taking into account that spatial location of an object point requires it to be visible from at least two camera stations,
- recording the object with as few images as possible.

The greater the required accuracy of the result, then the greater the base/distance ratio B/y must be; i.e. for a given y the base must be larger. When using independent metric cameras the base/distance ratio B/y can be held constant by scaling the base in proportion to the distance. The range error σ_y then is directly proportional to y, according to Equation (3.8-2b). However, with fixed base stereometric cameras the base cannot be changed; σ_y therefore varies in proportion to y^2. If a photogrammetric error analysis shows that a fixed base stereometric camera will not give the required accuracy in a particular case, an independent camera with freely selectable baselength must be used instead.

Working with independent metric cameras is similar to aerial photogrammetry. Camera stations are not exactly determined and the camera axes are not exactly oriented

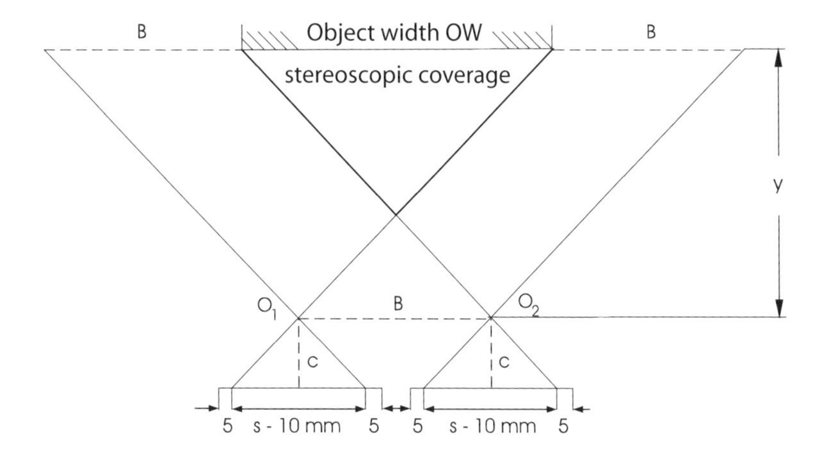

Figure 3.8-9: Planning the horizontal "normal case"

but aligned approximately according to the "normal case". The elements of exterior orientation are therefore derived from control points and tie points[74].

This has a number of advantages:

- photography is accelerated.
- systematic errors in interior orientation, as well as exterior orientation during photography and processing, are largely detected by the control points.
- photography can be made from unstable platforms such as ladders or lifting platforms, for zenith views the camera can be laid on the ground pointing upwards, or it can simply be used hand-held. The most important issue is often to ensure good and accurate ray intersections at the object.

Figure 3.8-10 is intended to encourage some solutions.

Where an object to be recorded is not roughly flat, such as a building façade, but has some significant extent in all directions, such as a complete building, then a configuration of multiple images surrounding the object is required. Four images are sufficient if they are directed at the object corners; in this configuration every object point would appear in at least two images. In contrast, with four images each looking perpendicular to the object sides a three-dimensional measurement of the object is not possible.

[74]When determining control points inside buildings, in narrow streets, on balconies which overhang one another etc., a software package is required which does not separate out plan and height determination but works directly in 3D space (Waldhäusl, P.: Vermessungswesen und Raumordnung, pp. 128–139, 1979).

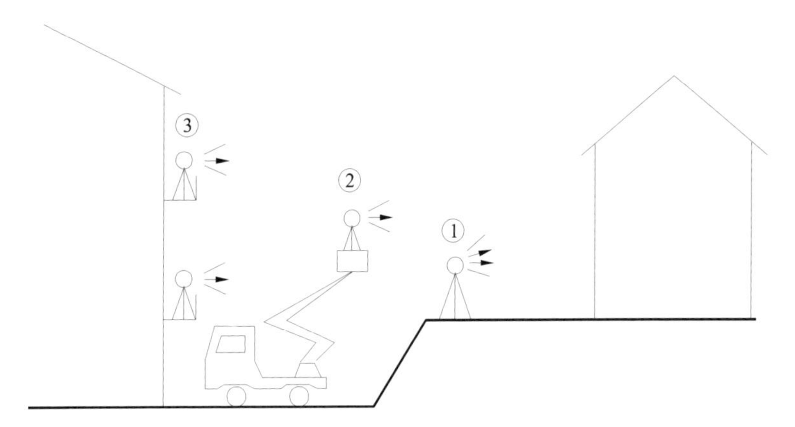

1) horizontal and/or tilted images from the ground, (approximately) horizontal base
2) approximately orthogonal images with an approximately horizontal base, taken from a lifting platform
3) stereoimagery with an approximately vertical base, orthogonal to the object façade

Figure 3.8-10: Various camera configurations

If an all-round configuration of eight images is created, as shown in Figure 3.8-11, the following advantages are obtained:

- the diagonal images contribute to a very stable network.
- every object point is recorded in at least three images; this produces results of high accuracy and reliability.
- the perpendicular images are little distorted; they are very suitable for producing orthophotos.

The imaging configuration shown in Figure 3.8-11 has one disadvantage: there are no image pairs which correspond approximately to the "normal case". The photogrammetric operators, with their stereoscopic vision, would use an approximately "normal case" (Section 6.1.2) for stereoscopic processing, for example of a façade with many three-dimensional details. With due regard for the various processing options, the imaging configuration of Figure 3.8-11 should be modified or extended such that, in place of just one perpendicular image there are two, with an appropriate base. Simple tools exist to aid the approximate alignment of the camera axes perpendicular to the object plane (e.g. a regular grid in the viewfinder, aiming at the camera's reflection in a window or polished surface of the façade).

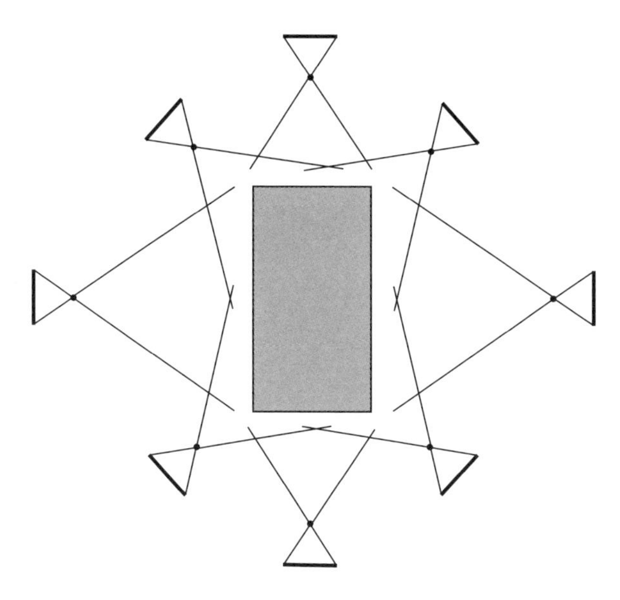

Figure 3.8-11: Recording an object with all-round imaging[75]

This section on planning, which has a continuation in Section 5.7, will be concluded with comments on the application area known as Mobile Mapping. This type of data acquisition makes use of a road vehicle equipped with the following sensors (Figure 3.8-12):

- GPS receiver which determines position with reference to a specific GPS reference station
- inertial measurement unit (IMU) which determines the angular attitude of the camera, as in aerial survey use (Section 3.7.3.2), and which also supports GPS positioning
- odometer which determines distance travelled by counting the rotations of the wheel
- barometer which provides a coarse measurement of height
- digital stereometric camera or single digital camera
- laser scanner which provides polar (spherical) coordinates to points on the road surface (Chapter 8)

[75] Other object forms and imaging configurations have been analysed with respect to accuracy and reliability and are discussed in Section B 4.5, Volume 2.

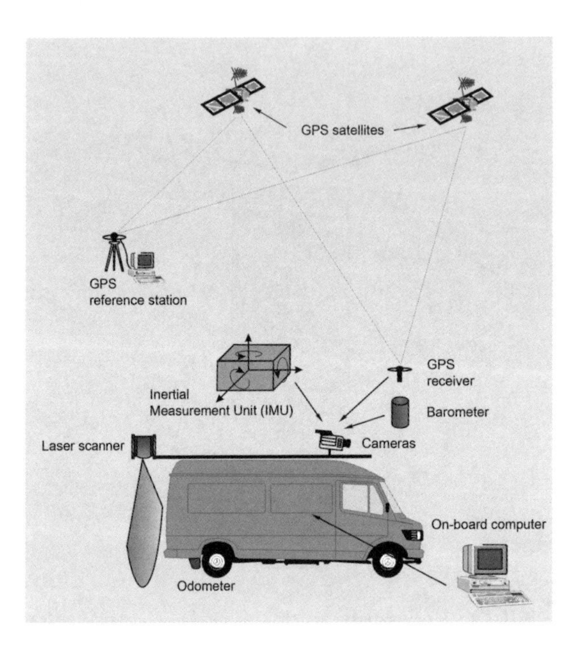

Figure 3.8-12: Mobile Mapping System[76]

Where a single metric camera is used, the base is defined between successive images and is roughly oriented in the direction of travel when the road is straight. Analysis of the images is very difficult because corresponding object elements are imaged in successive images at very different scales.

For sampled object points, such mobile mapping systems achieve an accuracy of around $\pm 30\,\mathrm{cm}$ at a speed of $60\,\mathrm{km/h}$[77].

Further reading on mobile mapping: Novak, K.: ZPF 59, pp. 112–120, 1991. Benning, W., Aussems, T.: ZfV 123, pp. 202–209, 1998. Gajdamowicz, K., Öhman, D., Rise, K.: Journal of the Swedish Society for Photogr. and RS, Nr 2002(1), pp. 103–112, 2002. He, G., Novak, K., Feng, W.: IAPRS 30, Commission II, pp. 480–486, 1992, and Commission V, pp. 139–145, 1992.

Exercise 3.8-4. An object with an extent of $2\,\mathrm{m} \times 3\,\mathrm{m}$ is to be photographed by a Rollei Scanning Camera RSC and the format is to be filled with the object's image. The number of pixels in a full image is 4200×6250 and in the scanning array's partial image 512×512. What is the resolution of the full image? (Solution: width of line pair $= 0.67\,\mathrm{mm}$.) What would be the resolution of the partial image only? (Solution: maximum width of line pair $= 8.2\,\mathrm{mm}$.) Values should be calculated with respect to the object surface.

[76]Provided by the ikv working group at the Institut für Geodäsie der Universität der Bundeswehr, Munich.

[77]Schwarz, K.P., El-Sheimy, N.: IAPRS XXXI(B3), pp. 774–785, Vienna, 1996.

Exercise 3.8-5. Repeat the exercise above for a CCIR video camera. (Solution: maximum width of line pair = 6.7 mm.)

Exercise 3.8-6. Repeat the exercise above for a Wild P32 (Section 3.8.3) with a total resolving power (film and optics) of 75 Lp/mm. (Solution: maximum width of line pair = 0.45 mm.)

Exercise 3.8-7. An 18 m wide and 12 m high, almost flat façade is to be photographed at height of 8 m from the full width balcony of the building opposite. Stereophotography is required which will deliver an accuracy of $\sigma_x = \sigma_z = \pm 0.5$ cm in the plane of the façade and $\sigma_y = \pm 1$ cm normal to the façade. The building opposite is 20 m away. Accuracy within the image can be assumed to be $\pm 10\,\mu$m. Is the Wild P32 sufficient or must the P31 be used (for camera data see Section 3.8.3)? (Solution: The P32 is sufficient.) What baselength should be used? (Solution: 6.25 m.) What is the image scale? (Solution: 1 : 312.) How many photographs are required? (Solution: 2.) What is the base/distance ratio? (Solution: 1 : 3.2.) How many times must an image be enlarged to obtain an orthophoto at a scale of 1 : 50? (Solution: 6.3 times.)

Chapter 4

Orientation procedures and some methods of stereoprocessing

This chapter deals with the reconstruction of shape and position of an object point by point from two images. In much of Chapter 4, the alignment of the two photographs approximates to the "normal case". In the case of aerial survey one speaks of near vertical photographs. The base is generally chosen so that the two pictures overlap each other by about 60% (Figure 4.0-1). Such pictures can be observed stereoscopically, or "in three dimensions" (Section 6.1.2), when one sees a stereoscopic model, sometimes also called an optical model, of the photographed object. In several passages of this chapter we also deal with the general case where the photographs depart from the "normal case".

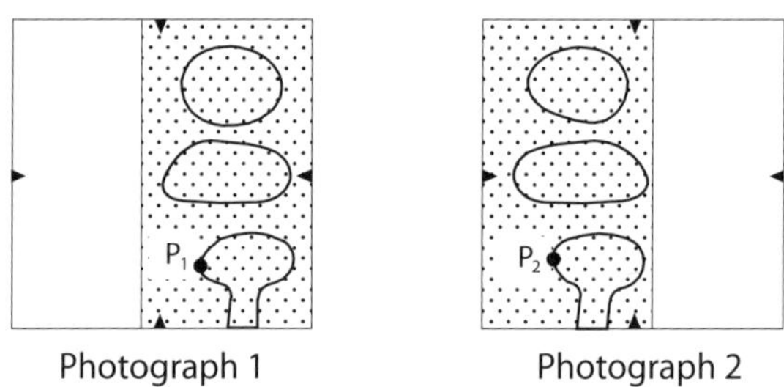

Figure 4.0-1: The "model area" (overlap) in two nearly vertical photographs

Stereoprocessing in the narrow sense, that is, the creation of three-dimensional digital models of the photographed objects, is preceded by a quite extensive preparation phase, which is generally called orientation. This preparatory phase is crucially dependent on whether or not the elements of exterior orientation of the two photographs are known. In principle, photogrammetric processing leads to the computational re-establishment of the geometric relationship of the cameras to the global coordinate system at the time of exposure and, then, to the digital reconstruction of the object from corresponding image points or image elements.

The orientation of the photographs is considerably simplified if the interior orientation of the pictures is known—as, in general, is assumed in this chapter. It is also presumed that we have photographs taken with a metric camera on film or with CCD sensors (Section 3.7.2.2) or by means of three-line digital cameras (Section 3.7.2.3).

4.1 With known exterior orientation

Here we distinguish between the cases of two metric photographs (film or CCD sensor, Section 4.1.1) and that of metric images taken with a three-line sensor camera (Section 4.1.2).

4.1.1 Two overlapping metric photographs

The elements of exterior orientation are available when the pictures were taken:

- with terrestrial stereometric cameras (though only with respect to a local object coordinate system, Section 3.8.2), or
- with a terrestrial metric camera when the position and attitude of the camera have been accurately determined (Section 3.8.8), or
- with aerial survey cameras of which the position and attitude have been accurately established in post-processing of the corresponding GPS/IMU recordings from the photo flight (Section 3.7.3.2).

When using GPS and IMU, one speaks of direct georeferencing. GPS and IMU allow direct transformation into a terrestrial coordinate system, normally the national coordinate system. One speaks of indirect georeferencing when the elements of exterior orientation are determined in a roundabout way using control points.

The photogrammetric processing of a stereopair with known exterior orientation starts with the measured image coordinates ξ_1, η_1 and ξ_2, η_2 of the corresponding points P_1 and P_2 (Figure 4.0-1). The following relationships (4.1-1), derived from Equations (2.1-20), define intersection of rays in three dimensions; they may be used to find the object coordinates X, Y, Z of the point P:

$$
\left.
\begin{aligned}
X &= X_{01} + (Z - Z_{01})k_{x1} \\
Y &= Y_{01} + (Z - Z_{01})k_{y1}
\end{aligned}
\right\} \text{Image 1}
\qquad
\left.
\begin{aligned}
X &= X_{02} + (Z - Z_{02})k_{x2} \\
Y &= Y_{02} + (Z - Z_{02})k_{y2}
\end{aligned}
\right\} \text{Image 2}
\tag{4.1-1}
$$

The quantities k are derived from the elements of the interior and exterior orientation together with the four measured image coordinates. Since the coordinates X_{0i}, Y_{0i} and Z_{0i} of the perspective centres are known, there are four linear equations in the three unknown object coordinates X, Y and Z.

Provided the Z-coordinate axis is roughly parallel to the direction of photography (as in the case of aerial images), then from the first and the third of the four Equations (4.1-1) we have:

$$Z = \frac{X_{02} - Z_{02}k_{x2} + Z_{01}k_{x1} - X_{01}}{k_{x1} - k_{x2}} \tag{4.1-2}$$

The X coordinate may be found from the first or the third equation, while Y can be found from both the second and the fourth equations (providing a check). The two slightly different Y values are then averaged.

For the "normal case" of photogrammetry (which, as we know, is arrived at directly when using stereometric cameras) the relationships (4.1-1) become especially simple, as one can see with the help of Equations (2.1-32) or (3.8-1).

The equation systems (4.1-1) and (4.1-2) can also be used during the processing of images which depart greatly from the "normal case". The solution for intersection outlined above is, however, not rigorous in the case of such a general camera arrangement. The redundancy cannot really be used by an averaging of the different Y values, but must be dealt with using least squares estimation from the original measurements, that is the image coordinates $\bar{\xi}_1$, $\bar{\eta}_1$ and $\bar{\xi}_2$, $\bar{\eta}_2$ (Appendix 4.1-1). In this estimation problem we have four (non-linear) observation equations in three unknowns. Using the collinearity equations (2.1-19) we derive four linearized observation equations:

$$\begin{aligned} v_{\xi_1} &= \left(\frac{\partial \xi_1}{\partial X}\right)^0 dX + \left(\frac{\partial \xi_1}{\partial Y}\right)^0 dY + \left(\frac{\partial \xi_1}{\partial Z}\right)^0 dZ - (\bar{\xi}_1 - \xi_1^0) \\ v_{\eta_1} &= \left(\frac{\partial \eta_1}{\partial X}\right)^0 dX + \left(\frac{\partial \eta_1}{\partial Y}\right)^0 dY + \left(\frac{\partial \eta_1}{\partial Z}\right)^0 dZ - (\bar{\eta}_1 - \eta_1^0) \\ v_{\xi_2} &= \left(\frac{\partial \xi_2}{\partial X}\right)^0 dX + \left(\frac{\partial \xi_2}{\partial Y}\right)^0 dY + \left(\frac{\partial \xi_2}{\partial Z}\right)^0 dZ - (\bar{\xi}_2 - \xi_2^0) \\ v_{\eta_2} &= \left(\frac{\partial \eta_2}{\partial X}\right)^0 dX + \left(\frac{\partial \eta_2}{\partial Y}\right)^0 dY + \left(\frac{\partial \eta_2}{\partial Z}\right)^0 dZ - (\bar{\eta}_2 - \eta_2^0) \end{aligned} \tag{4.1-3}$$

The partial derivatives $(\ )^0$ are evaluated from the relationships in Appendix 2.1-3 using approximate values for the unknown coordinates. $\xi_{1,2}^0$ and $\eta_{1,2}^0$ are computed image coordinates derived from Equations (2.1-19) using the approximate values for the unknown object coordinates and the known elements of interior and exterior orientation. Approximate values for the unknown object coordinates can be determined by means of relationships (4.1-1) and (4.1-2).

Numerical Example. In the exact "normal case" the approximate solution and the rigorous solution result in identical answers (why?); nevertheless we wish to compute such a case by means of least squares estimation because the computational effort of the least squares solution will can be kept small for the "normal case". $B = 1.20\,\text{m}$, $c = 64.20\,\text{mm}$.

Image coordinates:

	ξ [mm]	η [mm]
Image 1	3.624	34.202
Image 2	-14.697	34.196

Approximate values from (2.1-32): $-Z^0 = 4.205\,\text{m}$, $Y^0 = 2.240\,\text{m}$, $X^0 = 0.237\,\text{m}$.

The design matrix **A** (Appendix 4.1-1) is developed using the partial derivatives which are found in Exercise 2.1-18 (Appendix 2.1-3). The image coordinates $\xi_{1,2}^0$, $\eta_{1,2}^0$ (4.1-3) to be calculated for the vector $\mathbf{l}$ of observations (Appendix 4.1-1) come from the (simple) central perspective equations (2.1-28) to (2.1-31) of the exact "normal case":

$$\mathbf{A} = \begin{pmatrix} 15.268 & 0 & 0.861 \\ 0 & 15.268 & 8.133 \\ 15.268 & 0 & -3.496 \\ 0 & 15.268 & 8.133 \end{pmatrix} \qquad \mathbf{l}\,[\text{mm}] = \begin{pmatrix} 0.00559 \\ 0.00271 \\ 0.00564 \\ -0.00329 \end{pmatrix}$$

After setting up and solution of the normal equations and calculating the accuracy one finally obtains (Appendix 2.1-3):

$$\begin{pmatrix} X \\ Y \\ Z \end{pmatrix} = \begin{pmatrix} 0.23700 \\ 2.24000 \\ -4.20500 \end{pmatrix} + \begin{pmatrix} 0.00037 \\ -0.00001 \\ -0.00001 \end{pmatrix} = \begin{pmatrix} 0.23737\,\text{m} \\ 2.23999\,\text{m} \\ -4.20501\,\text{m} \end{pmatrix}$$

$$\begin{pmatrix} \hat{\sigma}_X \\ \hat{\sigma}_Y \\ \hat{\sigma}_Z \end{pmatrix} = \begin{pmatrix} \pm 0.23\,\text{mm} \\ \pm 0.75\,\text{mm} \\ \pm 1.36\,\text{mm} \end{pmatrix}$$

The solution of a three-dimensional intersection using least squares has the great advantage that, as a by-product of the adjustment computation, the accuracy of the new points is also found. The accuracies calculated in the above numerical example are too optimistic because the elements of exterior orientation were taken to be error-free. In Volume 2 (Sections B 3.5.7 and B 3.5.8), extended adjustment formulae are dealt with for the case of imperfect elements of interior and exterior orientation.

Publication No. 43 of the EuroSDR (European Spatial Data Research, former OEEPE—Organisation Européenne d'Études Photogrammétriques Expérimentales) concerned an international test of direct georeferencing. The results achieved were roughly in accord with those given in Section 3.7.3.2. See also Cramer, M.: GIS 6/2002, pp. 37–42.

In principle, with known exterior orientation one requires no control points. In many cases, nevertheless, control points are incorporated in the processing; on the one hand, this provides checks and, on the other hand, improves the known elements of exterior orientation.

4.1.2 Metric images with a three-line sensor camera

When taking images with a three-line sensor camera, a GPS/IMU recording must be made simultaneously, from which at every point in time the elements of exterior orientation are available with a relatively higher accuracy (Sections 3.7.3.2 and 3.7.3.3). When one identifies a point P, either manually or automatically, then the three time-stamps t_1, t_2 and t_3 are known from the indices of the three rows; these are the times at which the point P was imaged in the three lines one after the other (Figure 4.1-1). Corresponding to these three time-stamps, the elements of exterior orientation available from the GPS/IMU records are taken:

$$\begin{aligned}
&X_0(t_1), Y_0(t_1), Z_0(t_1), \omega(t_1), \varphi(t_1), \kappa(t_1) \\
&X_0(t_2), Y_0(t_2), Z_0(t_2), \omega(t_2), \varphi(t_2), \kappa(t_2) \\
&X_0(t_3), Y_0(t_3), Z_0(t_3), \omega(t_3), \varphi(t_3), \kappa(t_3)
\end{aligned} \tag{4.1-4}$$

As a result, the prerequisites exist for a three-dimensional intersection, using three directions, to determine the XYZ coordinates of the point P. To do so, the equation system (4.1-1) or (4.1-3) is extended to a third picture; that is, there are six observation equations for the three unknowns. In calculating the quantities k, the following calibrated values for the ξ coordinates are used:

- for k_{x1}, k_{y1} the distance a,
- for k_{x2}, k_{y2} the distance 0 and, (4.1-5)
- for k_{x3}, k_{y3} the distance $-b$.

The η coordinates of the three image points P_1, P_2 and P_3 differ only slightly from each other.

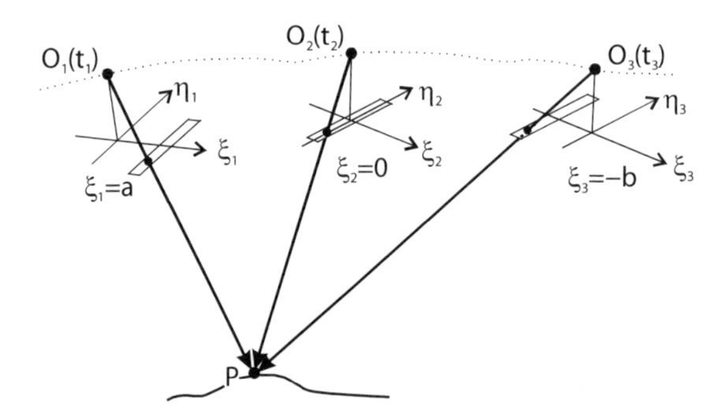

Figure 4.1-1: Object point P in the three images of a three-line sensor camera

In processing the three scan lines, if the elements of exterior orientation obtained from the GPS and IMU are available only with limited accuracy and are to be improved, control points are introduced in the procedure. This matter is discussed in Section 5.5.

4.2 With unknown exterior orientation

In this case, using control points, we must find the following twelve elements of orientation for two images, whether on film or on CCD chip:

$$\begin{aligned}
&\text{Picture 1: } X_{01}, Y_{01}, Z_{01}, \omega_1, \varphi_1, \kappa_1 \\
&\text{Picture 2: } X_{02}, Y_{02}, Z_{02}, \omega_2, \varphi_2, \kappa_2
\end{aligned}$$

The different methods of solution may be classified in three groups:

- separate orientation of the two images (see Section 4.2.1)
- combined, single-stage orientation of the two images (see Section 4.2.2)
- two-step combined orientation of a pair of photographs (see Section 4.2.3)

4.2.1 Separate orientation of the two images

Here it is assumed that for each image at least three control points are known (Figure 4.2-1). For each image at least six equations (2.1-19) can be written in the six unknowns, which are underlined below:

$$\begin{aligned}\bar{\xi}_i &= f(\xi_0, c, \underline{X}_0, \underline{Y}_0, \underline{Z}_0, \underline{\omega}, \underline{\varphi}, \underline{\kappa}, X_i, Y_i, Z_i) \\ & \qquad\qquad\qquad\qquad\qquad\qquad i = 1, 2, 3 \qquad (4.2\text{-}1) \\ \bar{\eta}_i &= f(\eta_0, c, \underline{X}_0, \underline{Y}_0, \underline{Z}_0, \underline{\omega}, \underline{\varphi}, \underline{\kappa}, X_i, Y_i, Z_i)\end{aligned}$$

After linearizing them using approximate values the six equations can be solved (iteratively, Newton's method) for the six unknowns. The procedure is known resection in three dimensions.

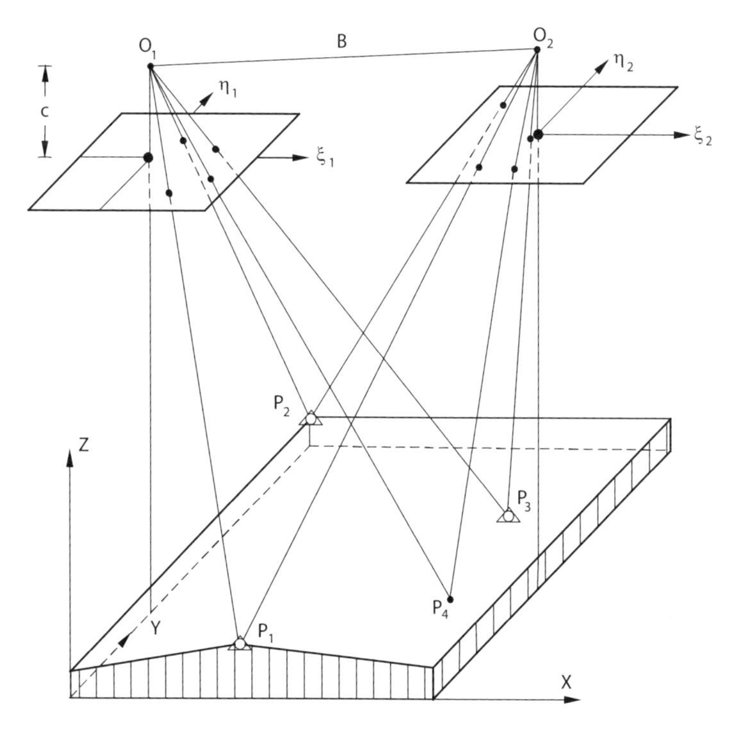

Figure 4.2-1: Stereomodel with three control points

Overdetermined resection in three dimensions is solved using the method of least squares estimation (Appendix 4.1-1). Each point P_i provides two linearized observation equations for the observations $\bar{\xi}_i$ and $\bar{\eta}_i$ (instead of writing out the partial derivatives $(\,)^0$, which are calculated using the approximate values for the unknowns, we use the quantities a and b introduced in Appendix 2.1-3):

$$\begin{aligned} v_{\xi i} &= a_2 dX_0 + a_3 dY_0 + a_4 dZ_0 + a_5 d\omega + a_6 d\varphi + a_7 d\kappa - (\bar{\xi}_i - \xi_i^0) \\ v_{\eta i} &= b_2 dX_0 + b_3 dY_0 + b_4 dZ_0 + b_5 d\omega + b_6 d\varphi + b_7 d\kappa - (\bar{\eta}_i - \eta_i^0) \end{aligned} \tag{4.2-2}$$

ξ_i^0 and η_i^0 are the image coordinates computed from Equations (2.1-19) using the known elements of interior orientation, the known control point coordinates X_i,Y_i,Z_i and the approximate values for the unknown elements of exterior orientation. For the first iteration, in the case of near-vertical photographs taken in a West-East flying direction, one can use the following approximations: $\omega^0 = \varphi^0 = \kappa^0 = 0$. Values introduced from a GPS-supported navigation system in the aircraft normally provide sufficiently accurate first approximations for X_0^0, Y_0^0, Z_0^0 (Section 3.7.3.1).

Numerical Example. We are given image coordinates and ground coordinates of the following points:

	ξ [mm]	η [mm]	X [m]	Y [m]	Z [m]
1	−86.15	−68.99	36589.41	25273.32	2195.17
2	−53.40	82.21	37631.08	31324.51	728.69
3	−14.78	−76.63	39100.97	24934.98	2386.50
4	10.46	64.43	40426.54	30319.81	757.31

The elements of interior orientation are: $c = 153.24\,\text{mm}$, $\xi_0 = \eta_0 = 0$.

Since we are concerned here with a near-vertical photograph, we can use simple proportion as for a truly vertical photograph to find approximate starting values for the camera position. From two control points P_i and P_j one gets:

$$\begin{aligned} Z_0^0 &= (c(X_j - X_i) + \xi_j Z_j - \xi_i Z_i)/(\xi_j - \xi_i) \\ X_0^0 &= X_i - \xi_i (Z_0^0 - Z_i^0)/c \\ Y_0^0 &= Y_i - \eta_i (Z_0^0 - Z_1^0)/c \end{aligned}$$

The two control points P_i and P_j should lie at approximately equal ground heights Z_i and furthermore their ξ-coordinates should not be too close together. One would therefore use either points P_1 and P_3 or points P_2 and P_4.

Solution:

$$\begin{aligned} X_0 &= 39795.45\,\text{m} \\ Y_0 &= 27476.46\,\text{m} \\ Z_0 &= 7572.69\,\text{m} \end{aligned} \qquad \mathbf{R} = \begin{pmatrix} 0.99771 & 0.06753 & 0.00399 \\ -0.06753 & 0.99772 & -0.00211 \\ -0.00412 & 0.00184 & 0.99999 \end{pmatrix}$$

Calculation and presentation of the accuracies are omitted.

The computation of resection is frequently required in photogrammetry. For this reason there are many solutions; efforts have been concentrated on linear equations which work without initial approximations. A successful linear solution exists for four control points as fully described in Volume 2 (Section B 4.1.2). It is not, however, a least squares adjustment so the accuracy is limited. If the highest accuracy is demanded one will regard the results of this method as approximations and then complete a precise computation with the observation equations (4.2-1).

On many flying missions nowadays the position of the camera station is already very accurately determined by GPS (Section 3.7.3.2). The attitude of the photograph, however, is not always recorded with IMU. In such a case the missing orientation angles can be found using the equation system (4.2-2), without the terms for dX_0, dY_0 and dZ_0.

An alternative for finding the elements of orientation is the Direct Linear Transformation (DLT). The DLT establishes the relationship between the two-dimensional image coordinates and the three-dimensional object coordinates by means of a projective transformation, with which we are already acquainted for the case of plane objects (Sections 2.1.4 and 2.2.3b) and which for three-dimensional objects may be extended in the following form (an elegant derivation using homogeneous coordinates is to be found in Appendix 4.1-1):

$$\xi = \frac{a_1X + a_2Y + a_3Z + a_4}{c_1X + c_2Y + c_3Z + 1}$$
$$\eta = \frac{b_1X + b_2Y + b_3Z + b_4}{c_1X + c_2Y + c_3Z + 1} \tag{4.2-3}$$

These transformation equations are linear with respect to the transformation parameters a_i, b_i, c_i (Section 2.1.4 contains a numerical example for a plane object; the computational steps outlined there are no different from those for a three-dimensional object). Each control points leads to two equations. Therefore, for the determination of the eleven transformation parameters a_i, b_i, c_i a minimum of six control points is needed (more accurately, for the sixth control point only one of the two image coordinates needs to be measured). The fact that there are eleven transformation parameters is surprising. In Section 2.1.3 we established that a central projection of three-dimensional space is defined by nine independent parameters (three for the interior orientation and six for the exterior orientation). The additional two parameters in the DLT can be interpreted as an extension of the interior orientation to allow a scale difference between the two image coordinate axes and non-orthogonality of these axes. Projective photogrammetry, therefore, usually discards a Cartesian coordinate system.

The parameters of the DLT cannot be interpreted physically. While the eleven parameters a_i, b_i, c_i can indeed be successfully expressed in terms of the parameters of interior and exterior orientation, accounting for a known interior orientation within a computation is difficult. Hence, the DLT is most suitable for the processing of non-metric images, such as amateur photographs, video images and so on (Section 3.8.6). Nonetheless, DLT is also used for the evaluation of metric images. The results of the

DLT, which are easy to obtain on account of the linearity of the DLT equations, are used as approximations for a subsequent adjustment based on central perspective relations. Valuable information concerning the interior orientation can be extracted from this rigorous adjustment, its accuracy considerably enhanced. In Volume 2, Section B 4.7.1, the DLT is dealt with in detail.

Separate orientation of the two images, for example by means of resection, making use of the relationships (4.2-2), has disadvantages which can be formulated as follows with reference to Figure 4.2-1:

- no use is made of the information that the homologous rays intersect each other (as, for example, in the new object point P_4).
- one needs at least three full control points (with known X, Y, Z) in the stereopair as opposed to the procedures described below; for these later methods the minimum requirement is just two plane control points (with known X, Y) and three height control points (with known Z).

4.2.2 Combined, single-stage orientation of the two images

The image coordinates of the control points and some new points are measured (Figure 4.2-1). For each control point there are four equations (2.1-19) in the twelve unknowns:

$$\left.\begin{aligned} \xi_{i1} &= f(\xi_0, c, \underline{X}_{01}, \underline{Y}_{01}, \underline{Z}_{01}, \underline{\omega}_1, \underline{\varphi}_1, \underline{\kappa}_1, X_i, Y_i, Z_i) \\ & \qquad\qquad\qquad\qquad\qquad\qquad \text{Image 1} \\ \eta_{i1} &= f(\eta_0, c, \underline{X}_{01}, \underline{Y}_{01}, \underline{Z}_{01}, \underline{\omega}_1, \underline{\varphi}_1, \underline{\kappa}_1, X_i, Y_i, Z_i) \\ \xi_{i2} &= f(\xi_0, c, \underline{X}_{02}, \underline{Y}_{02}, \underline{Z}_{02}, \underline{\omega}_2, \underline{\varphi}_2, \underline{\kappa}_2, X_i, Y_i, Z_i) \\ & \qquad\qquad\qquad\qquad\qquad\qquad \text{Image 2} \\ \eta_{i2} &= f(\eta_0, c, \underline{X}_{02}, \underline{Y}_{02}, \underline{Z}_{02}, \underline{\omega}_2, \underline{\varphi}_2, \underline{\kappa}_2, X_i, Y_i, Z_i) \end{aligned}\right. \qquad (4.2\text{-}4)$$

For each new point there are in fact three further unknowns (with double underlining below), but also four additional equations as follows (2.1-19):

$$\left.\begin{aligned} \xi_{i1} &= f(\xi_0, c, \underline{X}_{01}, \underline{Y}_{01}, \underline{Z}_{01}, \underline{\omega}_1, \underline{\varphi}_1, \underline{\kappa}_1, \underline{\underline{X_i}}, \underline{\underline{Y_i}}, \underline{\underline{Z_i}}) \\ & \qquad\qquad\qquad\qquad\qquad\qquad \text{Image 1} \\ \eta_{i1} &= f(\eta_0, c, \underline{X}_{01}, \underline{Y}_{01}, \underline{Z}_{01}, \underline{\omega}_1, \underline{\varphi}_1, \underline{\kappa}_1, \underline{\underline{X_i}}, \underline{\underline{Y_i}}, \underline{\underline{Z_i}}) \\ \xi_{i2} &= f(\xi_0, c, \underline{X}_{02}, \underline{Y}_{02}, \underline{Z}_{02}, \underline{\omega}_2, \underline{\varphi}_2, \underline{\kappa}_2, \underline{\underline{X_i}}, \underline{\underline{Y_i}}, \underline{\underline{Z_i}}) \\ & \qquad\qquad\qquad\qquad\qquad\qquad \text{Image 2} \\ \eta_{i2} &= f(\eta_0, c, \underline{X}_{02}, \underline{Y}_{02}, \underline{Z}_{02}, \underline{\omega}_2, \underline{\varphi}_2, \underline{\kappa}_2, \underline{\underline{X_i}}, \underline{\underline{Y_i}}, \underline{\underline{Z_i}}) \end{aligned}\right. \qquad (4.2\text{-}5)$$

First, it is necessary to linearize the Equations (4.2-4) and (4.2-5) using suitable first approximations (Appendix 2.1-3); these systems of equations are usually highly overdetermined. Subsequently the linearized equations are to be solved by means of least squares estimation (Appendix 4.1-1). The results of the computation are the twelve elements of exterior orientation and the object coordinates X, Y, Z of the new points. For this combined, two-image resection[1] in the following example, a tally of the observations and unknowns involved may be given:

Given:	2 metric images with
	3 control points and
	3 new points
Observations:	12 coordinates, 1st image
	12 coordinates, 2nd image
	$\Rightarrow$ 24 observations in all
Unknowns:	6 rotations of the two images
	6 coordinates X_0, Y_0, Z_0 of the two perspective centres
	9 coordinates X, Y, Z of the three new points
	$\Rightarrow$ 21 unknowns in all
Redundancy:	$24 - 21 = 3$

Exercise 4.2-1. Revise the above summations assuming that both images have the same interior orientation, but that it is unknown. Give your critical opinion of this exercise especially from the viewpoint of comprehensive checking of the computation. (Answer: In this case, there are 24 observations and an equal number of unknowns. That is to say, the system of equations is soluble but gross errors cannot be discovered. Note that for this special, degenerate, case of least squares one still needs the partial derivatives a_1 and b_1 of Appendix 2.1-3.)

Since arbitrarily many control points and new points may be introduced in this method and since the adjustment is based on the indirect observation equations (4.2-4) and (4.2-5) (which equations establish the direct connection between the unknowns and the original observations, the image coordinates), this method of orientation is the most accurate. It is a one-step solution, in contrast to the following procedure which splits the orientation into two steps and ignores the correlations between the results of the first step in solving the second. A disadvantage of the one step solution is that additional operations are unavoidable in order to find approximate values, as described in Section 4.2.1.

4.2.3 Two-step combined orientation of a pair of images

The orientation procedure shown in Figure 4.2-2 works in two steps. In the first step a stereomodel in an arbitrary three-dimensional xyz coordinate system is created from the two photographs. In the second step this model is transformed into the XYZ coordinate system. The procedure is most easily understood if we begin with the second step:

[1] There is a single, special word in German for this procedure: "Doppelbildeinschaltung", literally "double image insertion". As is frequently the case, there is no word in English for this German word. One might call it "two image resection" but, more frequently, one would simply say that the problem was solved using the "bundle method" (Section 5.3).

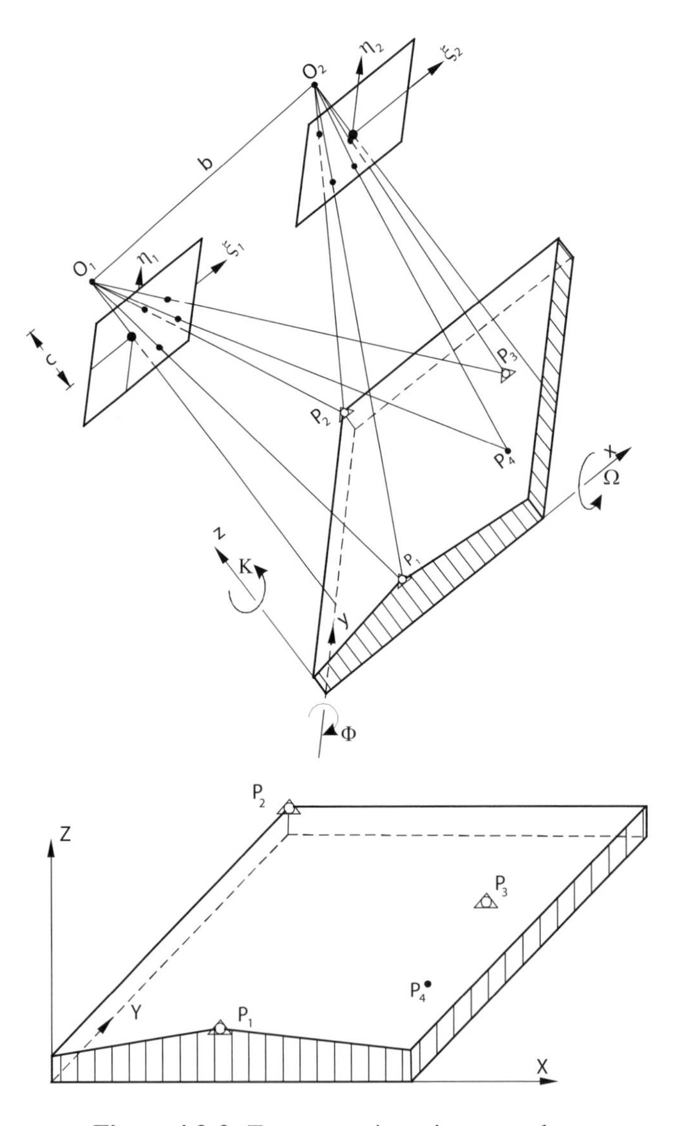

Figure 4.2-2: Two-step orientation procedure

Second step: The relation between the model coordinates x, y, z and the object coordinates X, Y, Z can be expressed by the equations (see also Figure 4.2-2):

$$\begin{pmatrix} X \\ Y \\ Z \end{pmatrix} = \begin{pmatrix} X_u \\ Y_u \\ Z_u \end{pmatrix} + m\mathbf{R} \begin{pmatrix} x \\ y \\ z \end{pmatrix} \tag{4.2-6}$$

in which:

X_u, Y_u, Z_u ... are the object coordinates of the origin of the xyz system
m ... is the scale number of the xyz system
$\mathbf{R}$ ... is the matrix of the three-dimensional rotation of the xyz system into the XYZ system defined in terms of the three rotations Ω, Φ, K See Equations (2.1-11) and (2.1-13) or Appendix 2.1-1.

The seven parameters, X_u, Y_u, Z_u, Ω, Φ, K, m, are called the elements of absolute orientation. Equation (4.2-6) represents a three-dimensional similarity transformation (2.1-18).

At least seven equations are required for the computation of the seven elements. Equation (4.2-6) gives us:

- three equations for a full control point (X, Y, Z all known)
- two equations for a horizontal, or plan, control point (X, Y both known)
- one equation for a height control point (Z known).

The absolute orientation requires at least two plan control points and three (non-collinear, in plan) height control points; alternatively, it requires two full control points and a height control point not collinear in plan with the full control points. Further details of absolute orientation, especially in the over-determined case, are dealt with in Section 4.4.

Exercise 4.2-2. On the basis of the matrix (2.1-13) write out the individual equations for a full control point, for a plan control point and for a height control point.

First step: When the photographs were taken, each individual point on the object could be thought of as having given rise to two rays, one towards each camera, and hence to two image points. If the positions and orientations of the images with respect to each other are correctly restored, all such pairs of homologous rays will once again intersect, in points which define the surface of the model in the xyz system.

Since seven of the twelve unknown elements of exterior orientation can be determined by the absolute orientation, five unknowns remain for the first step. Therefore it is reasonable to assume that a necessary condition for relative orientation is that the homologous rays from at least five well distributed points intersect and that if this is achieved, all other pairs of homologous rays will intersect, thus forming a complete photogrammetric model[2], see Section 4.3.

The procedure required to achieve the above state is called relative orientation, since only the relative positions and orientations of the two bundles of rays are determined. There is no reference to the XYZ coordinate system and no control points are required for relative orientation.

The condition for correct relative orientation, the intersection of each of five pairs of homologous rays in a model point, may be formulated using the scalar triple product of the three vectors $\mathbf{b}$, $\mathbf{p}_{1i}$, $\mathbf{p}_{2i}$ (Figure 4.2-3 and Equation (4.2-7a)). The scalar triple product, which is composed of the scalar product of one of three vectors with the vector product of the other two vectors, gives the volume of the parallelepiped of which three concurrent edges have the lengths and directions of the three vectors concerned. If the three vectors are coplanar their scalar triple product is zero, because the vector product of two of them is then perpendicular to the third and thus their scalar product vanishes. Equation (4.2-7b), which is frequently called the coplanarity condition and which is the condition that homologous rays intersect at each of five points, expresses the basic condition for relative orientation.

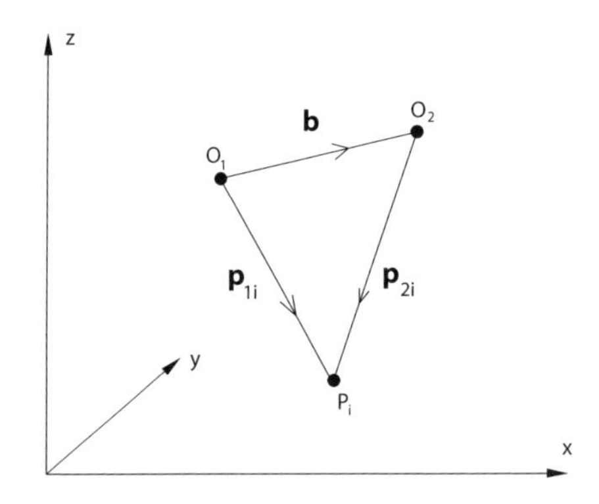

Figure 4.2-3: Condition for intersection, or coplanarity

$$\mathbf{b}^\top(\mathbf{p}_{1i} \times \mathbf{p}_{2i}) = 0 \qquad i = 1, \ldots, 5 \tag{4.2-7a}$$

The scalar triple product can also be written using the following determinant:

$$D = \begin{vmatrix} b_x & p_{1i,x} & p_{2i,x} \\ b_y & p_{1i,y} & p_{2i,y} \\ b_z & p_{1i,z} & p_{2i,z} \end{vmatrix} = 0 \qquad i = 1, \ldots, 5 \tag{4.2-7b}$$

[2]This two stage process brings to mind the definition of photogrammetry given at the very start of Section 1.1. In relative orientation a model is created which reproduces the form of the object; absolute orientation determines the scale, position and orientation of the object with respect to the ground coordinate system. Instead of absolute orientation one could use the term georeferencing.

Sections 4.3 and 4.4 are devoted, respectively, to relative orientation and to absolute orientation.

4.3 Relative orientation

4.3.1 Relative orientation of near-vertical photographs

The coplanarity condition (4.2-7) is considerably simplified if it can be assumed that the photographs are almost vertical. It is first necessary to set out the relationship between the image coordinates and the model coordinates under the assumption of near vertical photographs. In this case in the relationships (2.1-20), instead of the object coordinates X, Y, Z, the model coordinates x, y, z are introduced and, instead of the orientation elements, increments to the orientation elements. The quantities used can be seen in Figure 4.3-1:

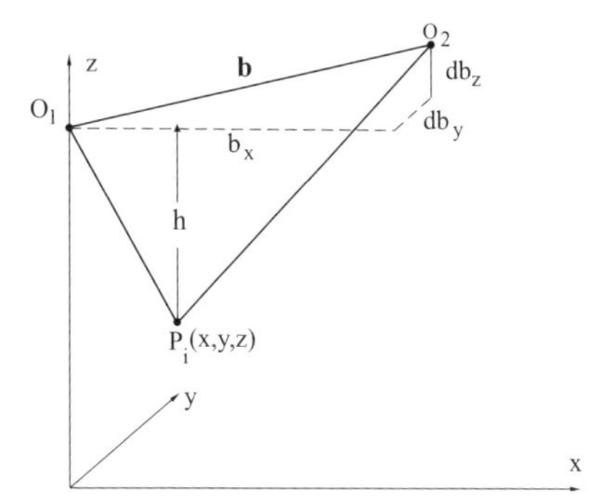

Figure 4.3-1: Orientation elements in the case of near-vertical images

$$\begin{aligned} \omega &= d\omega, \\ \varphi &= d\varphi, \\ \kappa &= d\kappa, \\ \xi_0 &= \eta_0 = x_{01} = y_{01} = 0, \\ x_{02} &= b_x, \\ y_{02} &= db_y \\ z_{02} &= z_{01} + db_z \\ h &= z_{01} - z. \end{aligned} \tag{4.3-1}$$

For small rotations the matrix $\mathbf{R}$ (2.1-13) simplifies to:

$$d\mathbf{R} = \begin{pmatrix} 1 & -d\kappa & d\varphi \\ d\kappa & 1 & -d\omega \\ -d\varphi & d\omega & q \end{pmatrix} \tag{4.3-2}$$

For photograph 1, therefore, from Equation (2.1-20):

$$x = (-h)\frac{\xi_1 - \eta_1 d\kappa_1 - c d\varphi_1}{-\xi_1 d\varphi_1 + \eta_1 d\omega_1 - c}$$

Division of both numerator and denominator by $-c$ gives:

$$x = (-h)\frac{-\frac{\xi_1}{c} + \frac{\eta_1}{c}d\kappa_1 + d\varphi_1}{1 + \frac{\xi_1}{c}d\varphi_1 - \frac{\eta_1}{c}d\omega_1}$$

Expanding the expression on the right-hand side in terms of the series $1/(1+x) = 1 - x + x^2 - \dots$ and ignoring second and higher order products of small quantities gives:

$$x = (-h)\frac{\xi_1 - \eta_1 d\kappa_1 - c d\varphi_1}{-\xi_1 d\varphi_1 + \eta_1 d\omega_1 - c}$$
$$= (-h)\left(-\frac{\xi_1}{c} + \frac{\xi_1^2}{c^2} d\varphi_1 - \frac{\xi_1\eta_1}{c^2} d\omega_1 + \frac{\eta_1}{c} d\kappa_1 + d\varphi_1\right)$$

An expression for the y coordinate may be found in a similar way:

$$y = (-h)\frac{\xi_1 d\kappa_1 + \eta_1 + c d\omega_1}{-\xi_1 d\varphi_1 + \eta_1 d\omega_1 - c}$$
$$= (-h)\left(-\frac{\eta_1}{c} - \frac{\eta_1^2}{c^2} d\omega_1 + \frac{\xi_1\eta_1}{c^2} d\varphi_1 - \frac{\xi_1}{c} d\kappa_1 - d\omega_1\right)$$

Rearrangement then gives us the following relationship between the model coordinates x, y, h (h instead of z), the image coordinates of a point P, the orientation elements and the principal distance for picture 1:

$$\begin{aligned} x_1 &= h\left(\frac{\xi_1}{c} - \left(1 + \frac{\xi_1^2}{c^2}\right) d\varphi_1 + \frac{\xi_1\eta_1}{c^2} d\omega_1 - \frac{\eta_1}{c} d\kappa_1\right) \\ y_1 &= h\left(\frac{\eta_1}{c} - \frac{\xi_1\eta_1}{c^2} d\varphi_1 + \left(1 + \frac{\eta_1^2}{c^2}\right) d\omega_1 + \frac{\xi_1}{c} d\kappa_1\right) \end{aligned} \quad (4.3\text{-}3)$$

Taking into account a shift of origin to O_2, a similar process gives corresponding relationships for picture 2:

$$\begin{aligned} x_2 &= b_x + (h + db_z)\left(\frac{\xi_2}{c} - \left(1 + \frac{\xi_2^2}{c^2}\right) d\varphi_2 + \frac{\xi_2\eta_2}{c_2} d\omega_2 - \frac{\eta_2}{c} d\kappa_2\right) \\ y_2 &= db_y + (h + db_z)\left(\frac{\eta_2}{c} - \frac{\xi_2\eta_2}{c^2} d\varphi_2 + \left(1 + \frac{\eta_2^2}{c^2}\right) d\omega_2 + \frac{\xi_2}{c} d\kappa_2\right) \end{aligned} \quad (4.3\text{-}4)$$

Setting y_1 equal to y_2 gives the condition for intersection of two rays, (4.2-7), in the case of near-vertical photographs:

$$\begin{aligned} 0 = db_y + \frac{\eta_2}{c} db_z + h\Bigg(&\frac{\eta_2 - \eta_1}{c} + \frac{\xi_1\eta_1}{c^2} d\varphi_1 - \left(1 + \frac{\eta_1^2}{c^2}\right) d\omega_1 \\ &- \frac{\xi_1}{c} d\kappa_1 - \frac{\xi_2\eta_2}{c^2} d\varphi_2 + \left(1 + \frac{\eta_2^2}{c^2}\right) d\omega_2 + \frac{\xi_2}{c} d\kappa_2\Bigg) \end{aligned}$$

In Section 2.1.5 we introduced the definition of y-parallax, $p_\eta = \eta_1 - \eta_2$. We may therefore rewrite the condition for intersection as:

$$\begin{aligned} p_\eta =& \frac{c}{h} db_y + \frac{\eta_2}{h} db_z + \frac{\xi_1\eta_1}{c} d\varphi_1 - \left(c + \frac{\eta_1^2}{c}\right) d\omega_1 - \xi_1 d\kappa_1 - \frac{\xi_2\eta_2}{c} d\varphi_2 \\ &+ \left(c + \frac{\eta_2^2}{c}\right) d\omega_2 + \xi_2 d\kappa_2 \end{aligned} \quad (4.3\text{-}5)$$

Equation (4.3-5) expresses the η parallax, in the case of near-vertical photographs, as a function of eight orientation elements. Consideration of the effects of small physical translations and rotations of the photographs shows that not all eight elements can, however, be computed from measured η parallaxes. Equal but opposite small rotations, $d\kappa_1$ and $d\kappa_2$, of the left and right photographs about their (near-vertical) axes will have the same relative effect on the η parallax as a small change, db_y, in the y position of the right-hand photograph.

Clearly, not all three of $d\kappa_1$, $d\kappa_2$ and db_y may be computed simultaneously from measured η-parallaxes; it is possible to compute the following pairs of elements from measured η-parallaxes: $d\kappa_1$, $d\kappa_2$; $d\kappa_1$, db_y; $d\kappa_2$, db_y. Likewise, either but not both of $\mathrm{d}\varphi_1$ and db_z can be found since they have the same relative effect; and either but not both of $d\omega_1$ and $d\omega_2$ can be found. Three of the elements of orientation included in Equation (4.3-5) must be excluded from the computation; if, for instance, both $d\omega_1$ and $d\omega_2$ were to have been included, the matrix of observation equations, corresponding to that found in the following numerical example, would have been singular or almost singular. The implication in Section 4.2.3 that only five elements of orientation were to be found in relative orientation is confirmed.

Of the possible choices of five orientation elements for the computation of numerical relative orientation of two near-vertical photographs, the following two variants are usual:

Relative orientation using rotations only. The photographs are rotated only, their positions meanwhile remaining unchanged; this is known as independent relative orientation and in the US as the "swing-swing method".

$$p_\eta = -\xi_1 d\kappa_1 + \xi_2 d\kappa_2 + \frac{\xi_1\eta_1}{c} d\varphi_1 - \frac{\xi_2\eta_2}{c} d\varphi_2 + \left(c + \frac{\eta_2^2}{c}\right) d\omega_2 \tag{4.3-6}$$

For approximately vertical photographs the image coordinate η_1 is approximately equal to η_2; the coefficients of $d\omega_1$ and $d\omega_2$ are almost identical, apart from their signs. One can choose to determine either $d\omega_1$ or $d\omega_2$, but not both; $d\omega_2$ has been chosen in Equation (4.3-6).

Relative orientation using only elements of the second photograph. One photograph remains fixed while the other is translated and rotated; this is frequently known as dependent relative orientation and often as the "one-projector method".

$$p_\eta = \frac{c}{h} db_y + \frac{\eta_2}{h} db_z - \frac{\xi_2\eta_2}{c} d\varphi_2 + \left(c + \frac{\eta_2^2}{c}\right) d\omega_2 + \xi_2 d\kappa_2 \tag{4.3-7}$$

It can be seen that, of the six elements of orientation of the right-hand photograph, only five occur in Equation (4.3-5). For near vertical photographs, within the limits of approximation applied in this section, translation b_x of the right-hand photograph does not influence the η-parallaxes. This may be seen as further confirmation that the measurement of five η-parallaxes, or the intersection of five pairs of homologous rays, suffices for relative orientation.

If more than five η-parallaxes have been measured, a least squares solution may be used (Appendix 4.1-1), in which case Equation (4.3-6) or (4.3-7), augmented by the residual v_{p_η} is regarded as observation equation with the substitution of the residual v_{p_η} as the constant term.

Numerical Example (of relative orientation of independent images). We are given the image coordinates (in [mm]) of eight points and the principal distance $c = 152.67$ mm.

Point	ξ_1	η_1	ξ_2	η_2	$p_\eta = \eta_1 - \eta_2$
1	93.176	5.890	6.072	5.176	0.714
2	−27.403	6.672	−112.842	1.121	5.551
3	83.951	107.422	−4.872	105.029	2.393
4	−11.659	101.544	−99.298	95.206	6.338
5	110.326	−97.800	34.333	−99.522	1.722
6	−12.653	−87.645	−96.127	−93.761	6.166
7	37.872	40.969	−48.306	37.862	3.107
8	41.503	−37.085	−42.191	−40.138	3.053

We wish to compute the elements of relative orientation and their accuracies. From Equation (4.3-6) the observation equations are of the form:

$$v_{p_\eta} = -\xi_1 d\hat{\kappa}_1 + \xi_2 d\hat{\kappa}_2 + \frac{\xi_1\eta_1}{c} d\hat{\varphi}_1 - \frac{\xi_2\eta_2}{c} d\hat{\varphi}_2 + \left(c + \frac{\eta_2^2}{c}\right) d\hat{\omega}_2 - p_\eta$$

The observation equations in matrix form are (Appendix 4.1-1):

$$\mathbf{v} = \begin{pmatrix} -93 & 6 & 4 & 0 & 153 \\ 27 & -113 & -1 & 1 & 153 \\ -84 & -5 & 59 & 3 & 225 \\ 12 & -99 & -8 & 62 & 212 \\ -110 & 34 & -71 & 22 & 218 \\ 13 & -96 & 7 & -59 & 210 \\ -38 & -48 & 10 & 12 & 162 \\ -42 & -42 & -10 & -11 & 163 \end{pmatrix} \begin{pmatrix} d\hat{\kappa}_1 \\ d\hat{\kappa}_2 \\ d\hat{\varphi}_1 \\ d\hat{\varphi}_2 \\ d\hat{\omega}_2 \end{pmatrix} - \begin{pmatrix} 0.714 \\ 5.551 \\ 2.393 \\ 6.338 \\ 1.722 \\ 6.116 \\ 3.107 \\ 3.053 \end{pmatrix}$$

which we may write as:

$$\mathbf{v} = \mathbf{A}\hat{\mathbf{x}} - \mathbf{l}$$

The normal equations have the form (Appendix 4.1-1):

$$\mathbf{A}^\top\mathbf{A}\hat{\mathbf{x}} = \mathbf{A}^\top\mathbf{l}; \qquad \hat{\mathbf{x}} = (\mathbf{A}^\top\mathbf{A})^{-1}\mathbf{A}^\top\mathbf{l}$$

The solution of the normal equations gives:

$$\begin{aligned} d\kappa_1 &= 1.73\,\text{gon} = 1^\circ 33' & d\kappa_2 &= -0.82\,\text{gon} = -44' \\ d\varphi_1 &= -0.34\,\text{gon} = -18' & d\varphi_2 &= 0.05\,\text{gon} = 3' \\ d\omega_2 &= 1.40\,\text{gon} = 1^\circ 16' \end{aligned}$$

The weight coefficient matrix is (Appendix 4.1-1): $\mathbf{Q} = (\mathbf{A}^\top \mathbf{A})^{-}1$

$$\mathbf{Q} = 10^{-5} \begin{pmatrix} 70 & 63 & 05 & 00 & 29 \\ 63 & 62 & 06 & 00 & 28 \\ 05 & 06 & 13 & 03 & 02 \\ 00 & 00 & 03 & 13 & 00 \\ 29 & 28 & 02 & 00 & 13 \end{pmatrix}$$

Standard deviation of an observation p_η:

$$\sigma_{p\eta} = \sqrt{\frac{\mathbf{v}^\top \mathbf{v}}{n-u}} = \pm 0.009\,\text{mm}$$

n ... number of observation equations (= 8)
u ... number of unknowns (= 5)

Standard deviation of an element of orientation: $\hat{\sigma}_k = \hat{\sigma}_{p\eta}\sqrt{q_{kk}}$

$$\hat{\sigma}_{\kappa_1} = 0.009\sqrt{0.00070} = \pm 15\,\text{mgon}\,(49'')$$
$$\hat{\sigma}_{\kappa_2} = 0.009\sqrt{0.00062} = \pm 14\,\text{mgon}\,(45'')$$
$$\hat{\sigma}_{\varphi_1} = \hat{\sigma}_{\varphi_2} = \hat{\sigma}_{\omega_2} = 0.009\sqrt{0.00013} = \pm 7\,\text{mgon}\,(23'')$$

While the procedure put forward in this Section 4.3.1 is exceedingly instructive, it is inadequate, especially for error analysis, when the photographs have large tilts. For photographs with arbitrarily large tilts, relative orientation and error analysis are dealt with in Sections 4.3.2 and 4.3.6.

Exercise 4.3-1. The rotation matrix (4.3-2) is orthogonal only to a first order of accuracy. If the matrix is to meet the conditions for orthogonality with errors no larger than 10^{-6}, how large can the angles ω, φ and κ be? (Answer: 0.6 gon (32′)).

Exercise 4.3-2. There is an obvious and very strong correlation between some of the unknowns of relative orientation as can be seen from the matrix of weight coefficients, $\mathbf{Q}$, given above. Among other things this leads to the fact that accuracies of functions of the orientation elements should be found by applying the general rules of error propagation. How large is the standard error of the difference $\Delta\kappa = \kappa_1 - \kappa_2$? (Answer: ± 4.4 mgon $= \pm 14''$)

4.3.2 Relative orientation and model formation using highly tilted photographs

We restrict ourselves initially to relative orientation using rotations only. In deriving linearized equations for the case of highly tilted photographs we cannot, as we did in the preceding section (see Equations (4.3-1) and (4.3-2)), adopt the unit matrix for a first approximation to the attitude of the pictures; we must use general rotation matrices. Figure 4.3-2 represents the relative orientation of tilted pictures. The three vectors

$\mathbf{b}$, $\mathbf{p}_1$ and $\mathbf{p}_2$ must be coplanar for a minimum of five points. We wish first to consider relative orientation using rotations only and we set $\omega_1 = 0$. The ζ_1 axis is thus at right angles to the y axis of the model coordinate system (Figure 4.3-2).

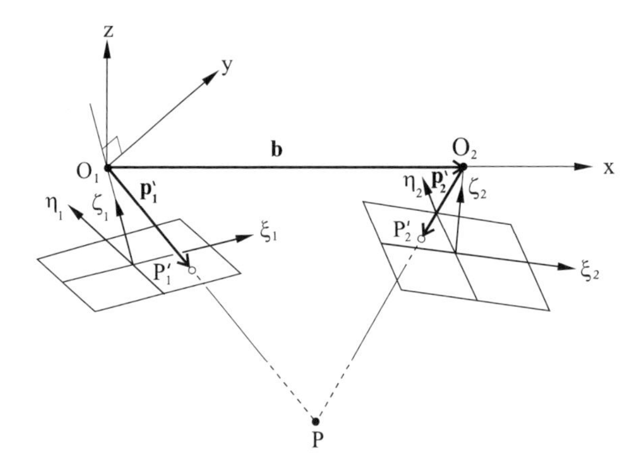

Figure 4.3-2: Relative orientation of tilted photographs using rotations only

With respect to their separate image coordinate systems the two vectors $\mathbf{p}_1$ and $\mathbf{p}_2$ have the following components:

$$\mathbf{p}'_1 = \begin{pmatrix} \xi_1 \\ \eta_1 \\ -c \end{pmatrix} \qquad \mathbf{p}'_2 = \begin{pmatrix} \xi_2 \\ \eta_2 \\ -c \end{pmatrix} \tag{4.3-8}$$

We transform these vectors into a common coordinate system, the xyz model coordinate system, using the two rotation matrices $\mathbf{R}_1$ and $\mathbf{R}_2$.

$$\mathbf{p}_1 = \mathbf{R}_1(\omega_1 = 0, \varphi_1, \kappa_1) \begin{pmatrix} \xi_1 \\ \eta_1 \\ -c \end{pmatrix} \qquad \mathbf{p}_2 = \mathbf{R}_2(\omega_2, \varphi_2, \kappa_2) \begin{pmatrix} \xi_2 \\ \eta_2 \\ -c \end{pmatrix} \tag{4.3-9}$$

Each of the two rotation matrices $\mathbf{R}_1$ and $\mathbf{R}_2$ can be written as the product of one matrix $\mathbf{R}^0$ rigorously computed but based on approximate rotation angles and a matrix $d\mathbf{R}$ representing incremental rotations (corresponding to that of Equation (4.3-2)):

$$\mathbf{p}_1 = d\mathbf{R}_1\mathbf{R}_1^0 \begin{pmatrix} \xi_1 \\ \eta_1 \\ -c \end{pmatrix} \qquad \mathbf{p}_2 = d\mathbf{R}_2\mathbf{R}_2^0 \begin{pmatrix} \xi_2 \\ \eta_2 \\ -c \end{pmatrix} \qquad \mathbf{b} = \begin{pmatrix} 1 \\ 0 \\ 0 \end{pmatrix} \tag{4.3-10}$$

The third vector shown in Equation (4.3-10) is the base vector $\mathbf{b}$; it lies in the direction of the x axis of the model with an arbitrarily chosen unit length. From Equation (4.2-7b) coplanarity of the three vectors $\mathbf{p}_1$, $\mathbf{p}_2$ and $\mathbf{b}$, corresponding to point P_i, is expressed by Equation (4.3-11) in which, as in Figure 4.3-2, the index i is omitted

$$D = \begin{vmatrix} 1 & p_{1,x} & p_{2,x} \\ 0 & p_{1,y} & p_{2,y} \\ 0 & p_{1,z} & p_{2,z} \end{vmatrix} = p_{1,y}p_{2,z} - p_{1,z}p_{2,y} = 0 \tag{4.3-11}$$

From the coplanarity equation (4.3-11) we may derive the following equation in the increments to the five unknowns for relative orientation using rotations only:

$$\begin{aligned} v_D = & \left(\frac{\partial D}{\partial \varphi_1}\right)^0 d\varphi_1 + \left(\frac{\partial D}{\partial \kappa_1}\right)^0 d\kappa_1 + \left(\frac{\partial D}{\partial \omega_2}\right)^0 d\omega_2 \\ & + \left(\frac{\partial D}{\partial \varphi_2}\right)^0 d\varphi_2 + \left(\frac{\partial D}{\partial \kappa_2}\right)^0 d\kappa_2 + D^0 \end{aligned} \tag{4.3-12}$$

The partial derivatives $(\,)^0$ are evaluated as in Appendix 4.3-1 using the approximate values for the unknown angles. The determinant D^0 is evaluated as follows:

a) working out the rotation matrices $\mathbf{R}_1^0$ and $\mathbf{R}_2^0$ using Equation (2.1-13) and the approximate values for the unknown rotation angles φ_1^0, κ_1^0, ω_2^0, φ_2^0 and κ_2^0

b) computing the components of the vectors $\mathbf{p}_1^0$ and $\mathbf{p}_2^0$ from Equation (4.3-9) using the measured image coordinates ξ_1,η_1 and ξ_2,η_2 of homologous points, the principal distance c and the rotation matrices $\mathbf{R}_1^0$ and $\mathbf{R}_2^0$

c) from Equation (4.3-11) evaluating the determinant $D^0 = p_{1,y}^0 p_{2,z}^0 - p_{2,y}^0 p_{1,z}^0$

Following the first iteration which results in increments, $d\varphi_1^0$, $d\kappa_1^0$, $d\omega_2^0$, $d\varphi_2^0$ and $d\kappa_2^0$, respectively, to the approximate values, the steps a), b) and c) above are repeated and the partial derivatives $(\,)^0$ of Equation (4.3-12) are computed anew as in Appendix 4.3-1. After stopping the iterations we have the elements of relative orientation, so that formation of the model can proceed, using intersection (Section 4.1) with the values: $x_{01} = y_{01} = z_{01} = \omega_1 = y_{02} = z_{02} = 0$, $x_{02} = 1$.

Numerical Example (Continuation of the Numerical Example of Section 4.3.1). The results obtained in that example were $\omega_1 = 0$, $\varphi_1 = -0.34\,\text{gon} = -18'$, $\kappa_1 = 1.73\,\text{gon} = 1°33'$, $\omega_2 = 1.40\,\text{gon} = 1°16'$, $\varphi_2 = 0.05\,\text{gon} = 3'$, $\kappa_2 = -0.82\,\text{gon} = -44'$. These are used as approximate values for the second iteration, applying Equation (4.3-12). After some iterations we arrive at the following results which do indeed depart considerably from the simplified solution of Section 4.3.1:

$$\begin{aligned} \omega_1 &= 0 & \omega_2 &= 1.387\,\text{gon} = 1°14'54'' \\ \varphi_1 &= -0.455\,\text{gon} = -24'34'' & \varphi_2 &= -0.096\,\text{gon} = -5'11'' \\ \kappa_1 &= 1.708\,\text{gon} = 1°32'14'' & \kappa_2 &= -0.838\,\text{gon} = -45'15'' \end{aligned}$$

Model formation is performed using these final values.

Exercise 4.3-3. Compute the xyz model coordinates of points 1 to 8 of the numerical example in Section 4.3.1 and of both perspective centres, using the elements of relative orientation from the above numerical example and adopting the value for the base $b_x = 100\,\text{mm}$. Solution (from Equations (4.1-3)):

Point	x [mm]	y [mm]	z [mm]
1	107.236	9.563	−173.269
2	−30.721	6.888	−177.348
3	96.141	128.340	−178.034
4	−15.472	117.838	−177.894
5	140.093	−116.509	−186.622
6	−10.627	−101.529	−176.316
7	44.222	49.029	−178.058
8	50.827	−41.946	−177.727
O_1	0.000	0.000	0.000
O_2	100.000	0.000	0.000

For the dependent relative orientation (using elements of the right-hand photograph only) the three vectors required for the coplanarity equation are as follows ($\mathbf{R}_1 = \mathbf{I}$):

$$\mathbf{p}_1 = \mathbf{I}\begin{pmatrix}\xi_1\\ \eta_1\\ -c\end{pmatrix} \qquad \mathbf{p}_2 = \mathbf{R}_2(\omega_2, \varphi_2, \kappa_2)\begin{pmatrix}\xi_2\\ \eta_2\\ -c\end{pmatrix} \qquad \mathbf{b} = \begin{pmatrix}1\\ b_y\\ b_z\end{pmatrix} \tag{4.3-13}$$

The coplanarity equations (4.2-7b) are set up using these three vectors; subsequently the observation equations (corresponding to Equation (4.3-12)) and the partial derivatives (as in Appendix 4.3-1) have to be evaluated, and so on.

This section should be concluded with a note of criticism. In the strict sense of adjustment by the method of least squares, the corrections (often called residuals) should apply to the original observations, in this case to the image coordinates ξ_1, η_1 and ξ_1, η_2, and should not, as in the Equation system (4.3-12), relate to the "pseudo-observation" the function[3] D. This imprecision can be remedied in one of two ways: one is by changing to the Gauss–Helmert model of adjustment by least squares and the other is by adapting the combined, single-stage orientation of Section 4.2.2 to relative orientation.

4.3.2.1 Gauss–Helmert model of relative orientation

The vectors $\mathbf{p}_1$ and $\mathbf{p}_2$ are functions of the observations ξ_1, η_1 and ξ_2, η_2 in (4.3-9). In the coplanarity equation the residuals v and the unknown elements of orientation appear. A strict relative orientation leads to the Gauss–Helmert solution, also known as the general case of least squares estimation.

$$\begin{aligned}&\left(\frac{\partial D}{\partial \xi_1}\right)^0 v_{\xi_1} + \left(\frac{\partial D}{\partial \eta_1}\right)^0 v_{\eta 1} + \left(\frac{\partial D}{\partial \xi_2}\right)^0 v_{\xi_2} + \left(\frac{\partial D}{\partial \eta_2}\right)^0 v_{\eta_2} + \left(\frac{\partial D}{\partial \varphi_1}\right)^0 d\varphi_1\\ &+ \left(\frac{\partial D}{\partial \kappa_1}\right)^0 d\kappa_1 + \left(\frac{\partial D}{\partial \omega_2}\right)^0 d\omega_2 + \left(\frac{\partial D}{\partial \varphi_2}\right)^0 d\varphi_2 + \left(\frac{\partial D}{\partial \kappa_2}\right)^0 d\kappa_2 + D^0 = 0\end{aligned} \tag{4.3-14}$$

[3] "Residuals" when applied to a quantity other than the original measurements are also called "algebraic residuals".

The partial derivatives for the unknowns and for the determinant D^0 have been fully discussed in connection with Equation (4.3-12). These derivatives can be derived from the relationships (4.3-9) and (4.3-11):

$$\begin{aligned} \frac{\partial D}{\partial \xi_1} &= r_{1,21}\, p_{2,z} - r_{1,31}\, p_{2,y} & \frac{\partial D}{\partial \eta_1} &= r_{1,22}\, p_{2,z} - r_{1,32}\, p_{2,y} \\ \frac{\partial D}{\partial \xi_2} &= r_{2,31}\, p_{1,y} - r_{2,21}\, p_{1,z} & \frac{\partial D}{\partial \eta_2} &= r_{2,32}\, p_{1,y} - r_{2,22}\, p_{1,z} \end{aligned} \tag{4.3-15}$$

The indices in $r_{1,32}$, for example, refer to the element in the third row and the second column of rotation matrix $\mathbf{R}_1$. Details of the solution of the estimation problem (4.3-14) are to be found in the adjustment literature[4]. The partial derivatives of (4.3-15) have to be re-calculated from iteration to iteration, especially when one has begun from very poor approximations for the rotation matrices $\mathbf{R}_1^0$ and $\mathbf{R}_2^0$.

4.3.2.2 A combined, single-stage relative orientation

The single-stage solution for the 12 elements of external orientation of two pictures was described in Section 4.2.2; for relative orientation this takes on a special form. Since there are no control points but just new points, Equations (4.2-5), only, are applicable, i.e. two homologous points leading to four observation equations. The unknowns are the model coordinates, x_i, y_i, z_i, for each pair of homologous points together with the five unknowns of dependent relative orientation, $d\varphi_1$, $d\kappa_1$, $d\omega_2$, $d\varphi_2$ and $d\kappa_2$. Of the original 12 unknowns in Equations (4.2-5), the following are known: $x_{01} = y_{01} = z_{01} = y_{02} = z_{02} = \omega_1 = 0$, $x_{02} = 1$ (compare Figures 4.2-1 and 4.3-1).

Literature relevant to Sections 4.3.1 and 4.3.2: Rinner, K.: Phia, pp. 41–54, 1942. Schut, G.: Phia XIV, pp. 16–32, 1957/58. Thompson, E.H.: Phia 23, pp. 67–75, 1968. Blais, J.A.R.: Can. Surv. 26, pp. 71–76, 1972. Stefanovic, P.: ITC-J (1973), pp. 417–448. Molnar, L.: Geow. Mitt. der TU Wien, vol. 14, 1978. Mikhail, E., Bethel, J., McClone, C.: Introduction to Modern Photogrammetry. John Wiley, 2001.

4.3.3 Alternative formulation of relative orientation

Since the orientation of two overlapping pictures is a central problem in photogrammetry and in computer vision, it is hardly surprising that many different formulations have been developed. In Section 4.3.2 we learnt of a method typically used in photogrammetry, a technology which normally works with metric photographs. The formulation presented in what follows is more frequently associated with computer vision, in which the use of images without known interior orientation predominates. This alternative formulation also offers very interesting insights into the relative orientation of metric images.

[4]See, for example, Mikhail, E., Observations And Least Squares. IEP-A Dun-Donnelley, New York, 1962.

We start with the basic equation (4.2-7a) of relative orientation with three coplanar vectors $\mathbf{b}$, $\mathbf{p}_1$ and $\mathbf{p}_2$. The value of a scalar triple product is independent of the ordering of its terms. We choose the following form for the coplanarity condition:

$$\mathbf{p}_1^\top (\mathbf{b} \times \mathbf{p}_2) = 0 \tag{4.3-16}$$

It can be shown that the cross product $\mathbf{b} \times \mathbf{p}_2$ can be written as in Equation (4.3-17) in which the elements of the skew-symmetric matrix $\mathbf{B}$ come from the vector $\mathbf{b}$:

$$\mathbf{b} \times \mathbf{p}_2 = \mathbf{B}\mathbf{p}_2 = \begin{pmatrix} 0 & -b_z & b_y \\ b_z & 0 & -b_x \\ -b_y & b_x & 0 \end{pmatrix} \begin{pmatrix} p_{2,x} \\ p_{2,y} \\ p_{2,z} \end{pmatrix} \tag{4.3-17}$$

Exercise 4.3-4. Expand the cross product $\mathbf{b} \times \mathbf{p}_2$ and show that the result is the same as that found from the matrix multiplication of (4.3-17).

All skew-symmetric matrices are singular; in particular $\det \mathbf{B} = 0$, as is easily shown: $\det \mathbf{B} = 0(b_x^2) + b_z(-b_x b_y) + b_y(b_x b_z) = 0$.

Here we limit ourselves to the case of independent relative orientation (Figure 4.3-2). Taking the base vector as $\mathbf{b}^\top = (1, 0, 0)$ the coplanarity equation (4.3-16), as expressed in the form of (4.3-17) becomes:

$$\mathbf{p}_1^\top \mathbf{B}\mathbf{p}_2 = \mathbf{p}_1^\top \begin{pmatrix} 0 & 0 & 0 \\ 0 & 0 & -1 \\ 0 & 1 & 0 \end{pmatrix} \mathbf{p}_2 = 0 \tag{4.3-18}$$

The two vectors $\mathbf{p}_1$ and $\mathbf{p}_2$ (see Equations (4.3-9)) are derived from the image coordinates ξ_1, η_1 and ξ_2, η_2, the principal distance c and the rotation matrices $\mathbf{R}_1$ and $\mathbf{R}_2$, which describe the attitudes of the two pictures with respect to the model coordinate system. The matrices $\mathbf{C}_1$ and $\mathbf{C}_2$, describing the interior orientation of the two images, may be introduced into Equations (4.3-9), generalized by the inclusion of the coordinates of the principal point, ξ_0 and η_0, and usually identical for metric photographs:

$$\begin{aligned} \mathbf{p}_1 &= \mathbf{R}_1 \begin{pmatrix} \xi_1 - \xi_{0,1} \\ \eta_1 - \eta_{0,1} \\ 0 - c_1 \end{pmatrix} = \mathbf{R}_1 \begin{pmatrix} 1 & 0 & -\xi_{0,1} \\ 0 & 1 & -\eta_{0,1} \\ 0 & 0 & -c_1 \end{pmatrix} \begin{pmatrix} \xi_1 \\ \eta_1 \\ 1 \end{pmatrix} \\ &= \mathbf{R}_1 \mathbf{C}_1 (\xi_1, \eta_1, 1)^\top \\ \mathbf{p}_2 &= \mathbf{R}_2 \begin{pmatrix} \xi_2 - \xi_{0,2} \\ \eta_2 - \eta_{0,2} \\ 0 - c_2 \end{pmatrix} = \mathbf{R}_2 \begin{pmatrix} 1 & 0 & -\xi_{0,2} \\ 0 & 1 & -\eta_{0,2} \\ 0 & 0 & -c_2 \end{pmatrix} \begin{pmatrix} \xi_2 \\ \eta_2 \\ 1 \end{pmatrix} \\ &= \mathbf{R}_2 \mathbf{C}_2 (\xi_2, \eta_2, 1)^\top \end{aligned} \tag{4.3-19}$$

With this representation, the coplanarity equation (4.3-18) reads as follows:

$$(\xi_1, \eta_1, 1) \underbrace{\mathbf{C}_1^\top \mathbf{R}_1^\top \mathbf{B} \mathbf{R}_2 \mathbf{C}_2}_{\mathbf{F}} \begin{pmatrix} \xi_2 \\ \eta_2 \\ 1 \end{pmatrix} = 0 \tag{4.3-20}$$

The matrix $\mathbf{F}$ is known as the fundamental matrix of relative orientation or the correlation matrix. Provided that we are dealing with metric images, the fundamental matrix $\mathbf{F}$ can be computed using the parameters of relative orientation, that is to say, the two rotation matrices $\mathbf{R}_1$ and $\mathbf{R}_2$ (Section 4.3.2), the known matrix $\mathbf{B}$ (4.3-18) and the known matrix $\mathbf{C}$ of the interior orientation. Although the relationship (4.3-20) describes a relationship between the image coordinates ξ_1, η_1 and ξ_2, η_2 of corresponding points in two images, it is not possible, given image coordinates in one image, to find the corresponding image coordinates in the other image. This becomes clear from the following expansion of the coplanarity equation (4.3-20):

$$\xi_1\xi_2\bar{f}_{11}+\xi_1\eta_2\bar{f}_{12}+\xi_1\bar{f}_{13}+\xi_2\eta_1\bar{f}_{21}+\eta_1\eta_2\bar{f}_{22}+\eta_1\bar{f}_{23}+\xi_2\bar{f}_{31}+\eta_2\bar{f}_{32}+\bar{f}_{33}=0 \quad (4.3\text{-}21)$$

With known matrix elements $\bar{f}_{ik}$ and, for example, both image coordinates ξ_1 and η_1, one has a single linear equation in the two unknowns ξ_2 and η_2. Adopting arbitrary coordinates η_2 one can, therefore, compute matching coordinates ξ_2.

As a consequence, we know that corresponding to a point in one picture there is a straight line in the other picture on which the homologous point must lie. (This result is pursued further in Section 6.8.5.5, as illustrated in Figure 6.8-9.).

An interesting application of the coplanarity equation (4.3-21) lies in the relative orientation of images for which the interior orientation is unknown. Since Equations (4.3-19) contain the coordinates of the principal point (ξ_0,η_0) it is possible for the image coordinates ξ_1,η_1 and ξ_2,η_2 of homologous points to be referred to an arbitrary coordinate system. Division by one of the 9 elements $\bar{f}_{ik}$ of the matrix of Equation (4.3-21) reduces their number to 8. As the divisor one should select the largest possible element, which may be a different element $\bar{f}_{ik}$ in different cases. We choose $\bar{f}_{33}$; thus the nine elements become $f_{ik}=\bar{f}_{ik}/\bar{f}_{33}$. In this way, for each pair of homologous points in two pictures, we get the following linear equation for a least squares estimation (Appendix 4.1-1):

$$v=\xi_1\xi_2 f_{11}+\xi_1\eta_2 f_{12}+\xi_1 f_{13}+\xi_2\eta_1 f_{21}+\eta_1\eta_2 f_{22}+\eta_1 f_{23}+\xi_2 f_{31}+\eta_2 f_{32}=-1 \quad (4.3\text{-}22)$$

Since Equation (4.3-22) has eight unknowns f_{ik} and since each pair of points provides only one such equation, it would seem that this kind of relative orientation requires eight homologous points. The solution using Equations (4.3-22) does not involve the elements of interior orientation, even when they are known, but has more than five unknowns. It was shown in the postscript to Exercise 4.3-4 that the matrix $\mathbf{B}$ is singular; therefore the $\mathbf{F}$ matrix, which contains $\mathbf{B}$ as a factor, is also singular and, thus, not all eight of the f_{ik} are independent; in fact the number of independent unknowns is seven. The two additional unknowns, beyond the five of conventional relative orientation, represent general (unknown) parameters of interior orientation. The question of the conversion of the elements f_{ik} resulting from a relative orientation using the above Equations (4.3-22) into the parameters of conventional relative orientation will not be considered here. It should certainly be mentioned here that, if the interior orientation is known but not is employed for relative orientation using Equations (4.3-22), it is possible to derive the five standard elements of relative orientation from the values f_{ik}.

In conclusion, it remains for the link with projective geometry to be presented. The coplanarity equation (4.3-20) contains eight matrix elements f_{ik}:

$$(\xi_1, \eta_1, 1) \begin{pmatrix} f_{11} & f_{12} & f_{13} \\ f_{21} & f_{22} & f_{23} \\ f_{31} & f_{32} & 1 \end{pmatrix} \begin{pmatrix} \xi_2 \\ \eta_2 \\ 1 \end{pmatrix} = 0 \qquad (4.3\text{-}23)$$

This representation may be compared with the projective relationship between an object plane and an image plane, especially when it is expressed using homogeneous coordinates (see Equation (2.2-1-1) of Appendix 2.2-1). The two image planes in a relative orientation are projectively related to each other, though not "point to point" but "point to straight line", as already mentioned above.

How the unfamiliar elements of relative orientation f_{ik} may be introduced into photogrammetric processing will be discovered in Section 6.8.5.5.

Exercise 4.3-5. In Section 4.3.1 the image coordinates ξ_1, η_1 and ξ_2, η_2 of eight corresponding points in two metric images are given. Determine the relative orientation using Equations (4.3-22).

Solution:

$$\begin{pmatrix} f_{11} & f_{12} & f_{13} \\ f_{21} & f_{22} & f_{23} \\ f_{31} & f_{32} & 1 \end{pmatrix} = \begin{pmatrix} -0.000018 & 0.000284 & -0.005009 \\ -0.000302 & 0.000071 & -0.576408 \\ -0.017098 & 0.550187 & 1 \end{pmatrix}$$

Exercise 4.3-6. This solution, however, takes no account of the fact that the determinant $\det \mathbf{F}$ must vanish. If this additional constraint is introduced in the least squares estimation, one is dealing with an adjustment by the method of least squares with a constraint equation[5].

$$\begin{pmatrix} f_{11} & f_{12} & f_{13} \\ f_{21} & f_{22} & f_{23} \\ f_{31} & f_{32} & 1 \end{pmatrix} = \begin{pmatrix} 0.000001 & -0.000040 & -0.008565 \\ 0.000031 & 0.000039 & -0.261240 \\ -0.002044 & 0.263777 & 1 \end{pmatrix}$$

The additional constraint equation markedly influences the result in this example.

Exercise 4.3-7. Write the **F** matrix for the "normal case" of photogrammetry (Section 2.1.6) and derive the coplanarity equation from it.

Solution:

$$\mathbf{C}_1 = \mathbf{C}_2 = \begin{pmatrix} 1 & & \\ & 1 & \\ & & -c \end{pmatrix} \qquad \mathbf{R}_1 = \mathbf{R}_2 = \mathbf{I}$$

If unit baselength is chosen ($\|\mathbf{b}\| = 1$), $\mathbf{B} = \begin{pmatrix} 0 & 0 & 0 \\ 0 & 0 & -1 \\ 0 & 1 & 0 \end{pmatrix}$.

[5]The solution may be found, for example, in Mikhail, E., Observations And Least Squares. IEP-A Dun-Donnelley, New York, 1962.

(4.3-20): $$(\xi_1, \eta_1, 1) \begin{pmatrix} 0 & 0 & 0 \\ 0 & 0 & c \\ 0 & -c & 0 \end{pmatrix} \begin{pmatrix} \xi_2 \\ \eta_2 \\ 1 \end{pmatrix} = -c\eta_2 + c\eta_1 = \eta_1 - \eta_2 = 0$$

The coplanarity equation for the "normal case" of a stereopair leads to the statement that there are no η-parallaxes, as has already been stated in Section 2.1.6. Further note: Because $\bar{f}_{33} = 0$ in the exact "normal case", relative orientation by means of Equation (4.3-22) will fail.

Literature: Brandstätter, G.: Mitt. der TU Graz, Folge 87, 2000. Haggren, H., Niini, I.: The Photogr. Journal of Finland 12, pp. 22–33, 1990. Hartley, R., Zisserman, A.: Multiple View Geometry in Computer Vision, Cambridge University Press, 2000. Mayer, H.: Journal of the Swedish Society for Photogrammetry and Remote Sensing 2002(1), pp. 129–141, 2002. Förstner, W.: IAPR 33(B3/1), pp. 297–304, Amsterdam, 2000, and PFG (2000), pp. 163–176. Ressl, C.: IAPR 34(3A), pp. 277–282, Graz, 2002.

4.3.4 Relative orientation of near-vertical photographs by y-parallaxes

Following the excursion into an alternative relative orientation, in particular for non-metric photographs, we turn again to relative orientation of metric images. In the following we assume approximately vertical photographs, just as in Section 4.3.1 in which the relative orientation was solved in terms of η-parallaxes, that is differences in image coordinates $(\eta_1 - \eta_2)$. For analogue photogrammetric instruments (which receive no attention in this new edition of the textbook) a relative orientation using y-parallaxes was commonly used, the y-parallaxes being measured in the stereomodel. Although orientation with y-parallaxes is no longer of any practical significance, the formulae are well suited for assessment of accuracy and for consideration of "critical surfaces".

When carrying out relative orientation in an analogue instrument, any existing x-parallax should be removed before the observation of y-parallax; x-parallax is of no significance in relative orientation (see Figures 4.3-3 and 4.3-4).

For a relative orientation using model parallaxes p_y rather than image parallaxes p_η, the relationships (4.3-6) and (4.3-7) must be transformed such that model-related values x, y and p_y replace the values ξ, η and p_η which refer to images. The following relations are valid for near-vertical photographs (see Figure 4.3-1):

$$p_y \approx \frac{h}{c} p_\eta; \quad y \approx \frac{h}{c}\eta_1 \approx \frac{h}{c}\eta_2, \quad x \approx \frac{h}{c}\xi_1 \approx \frac{h}{c}\xi_2 + b \tag{4.3-24}$$

Substituting these results in Equation (4.3-6) gives:

$$p_y = -x d\kappa_1 + (x-b) d\kappa_2 + \frac{xy}{h} d\varphi_1 - \frac{(x-b)y}{h} d\varphi_2 + h\left(1 + \frac{y^2}{h^2}\right) d\omega_2 \tag{4.3-25}$$

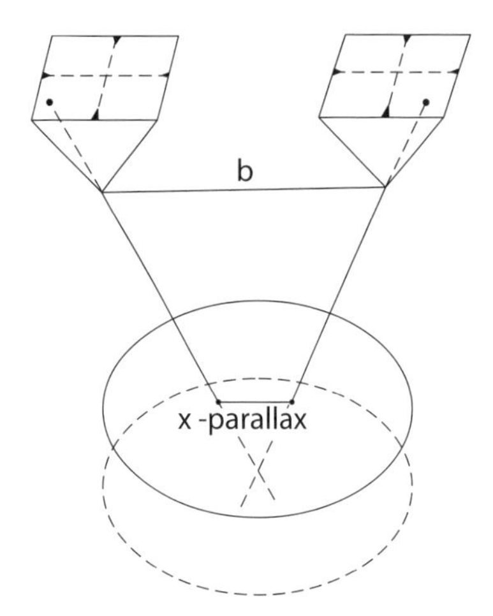

Figure 4.3-3: x-parallax after the removal of y-parallax in a relatively oriented model

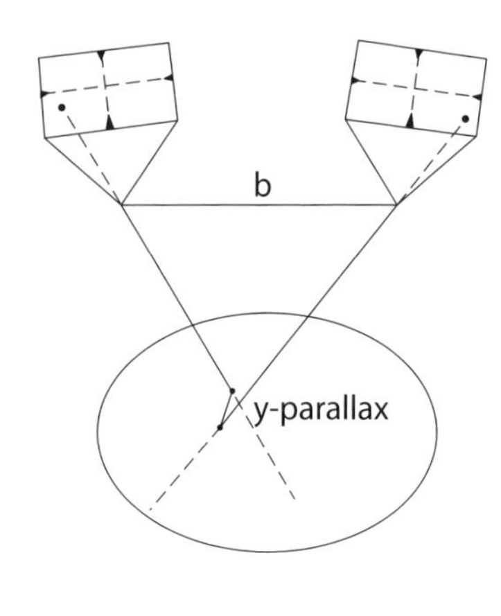

Figure 4.3-4: y-parallax after removal of the x-parallax in a model before relative orientation

Exercise 4.3-8. Derive corresponding equations for dependent relative orientation.

The elements of relative orientation can be estimated by least squares using observed y-parallaxes and the associated model coordinates x, y and h (instead of z) in Equation (4.3-25). If the y-parallaxes are measured in specially chosen positions, explicit expressions can be derived for the required five unknowns. With mountainous or relatively flat ground, however, different procedures are necessary.

4.3.4.1 Mountainous country (after Jerie)

In this procedure one chooses six corresponding points, at which to measure the y-parallax, and which are constrained to lie (Figure 4.3-5)

- on two parallel lines in the model, separated in the x direction by a distance b and
- on three parallel lines, separated in the η direction in the image planes by distances δ (y direction in the model)

Under these conditions the following ratio remains constant for point i where $i = 3, \ldots, 6$.

$$\frac{|y_i|}{h_i} = \frac{\delta}{c} = r \tag{4.3-26}$$

Let

$$R = 1 + r^2 \tag{4.3-27}$$

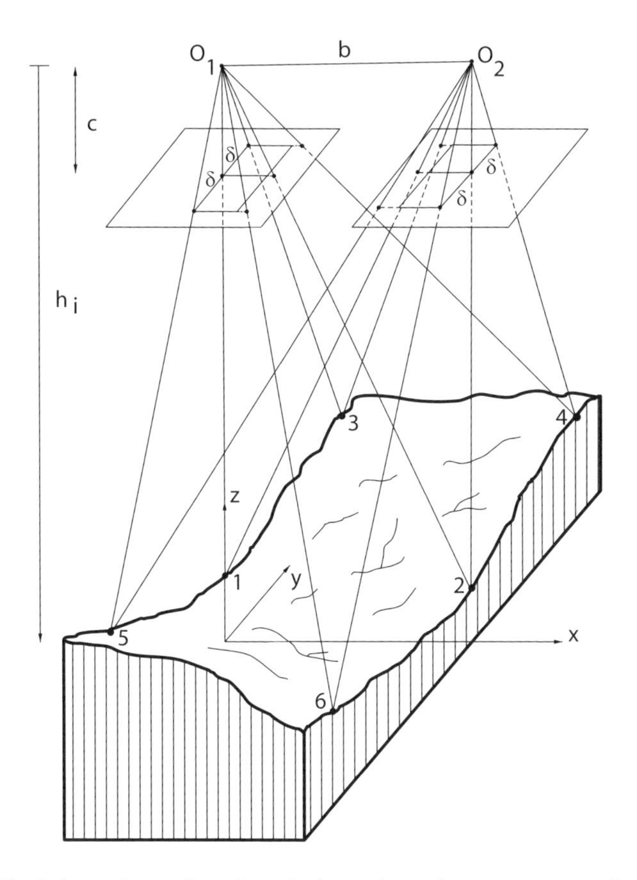

Figure 4.3-5: Orientation points for relative orientation over mountainous country

Defining the constant R as in (4.3-27) and introducing it in (4.3-25) we get the following observation equations for independent relative orientation:

	$d\kappa_1$	$d\kappa_2$	$d\varphi_1$	$d\varphi_2$	$d\omega_2$	
v_1	0	$-b$	0	0	h_1	$-p_{y_1}$
v_2	$-b$	0	0	0	h_2	$-p_{y_2}$
v_3	0	$-b$	0	br	h_3R	$-p_{y_3}$
v_4	$-b$	0	br	0	h_4R	$-p_{y_4}$
v_5	0	$-b$	0	$-br$	h_5R	$-p_{y_5}$
v_6	$-b$	0	$-br$	0	h_6R	$-p_{y_6}$

(4.3-28)

Forming the normal equations gives (Appendix 4.1-1):

$d\kappa_1$	$d\kappa_2$	$d\varphi_1$	$d\varphi_2$	$d\omega_2$	
$3b^2$				$-bh_2 - bh_4R - bh_6R$	$b(p_{y_2} + p_{y_4} + p_{y_6})$
	$3b^2$			$-bh_1 - bh_3R - bh_5R$	$b(p_{y_1} + p_{y_3} + p_{y_5})$
		$2b^2r^2$		$brh_4R - brh_6R$	$-br(p_{y_4} - p_{y_6})$
			$2b^2r^2$	$brh_3R - brh_5R$	$-br(p_{y_3} - p_{y_5})$
				$h_1^2 + h_2^2 +$ $R^2(h_3^2 + h_4^2 + h_5^2 + h_6^2)$	$-h_1p_{y_1} - h_2p_{y_2} -$ $h_3Rp_{y_3} - h_4Rp_{y4} -$ $h_5Rp_{y_5} - h_6Rp_{y_6}$

The solution of the normal equations is:

$$
\begin{aligned}
d\omega_2 &= -\big((-2h_1 + h_3R + h_5R)(2p_{y_1} - p_{y_3} - p_{y_5}) + \\
&\quad (-2h_2 + h_4R + h_6R)(2p_{y_2} - p_{y_4} - p_{y_6})\big)\,/ \\
&\quad \big((-2h_1 + h_3R + h_5R)^2 + (-2h_2 + h_4R + h_6R)^2\big) \\
d\varphi_2 &= \frac{1}{2br}(p_{y_3} - p_{y_5}) + \frac{R}{2br}(-h_3 + h_5)d\omega_2 \\
d\varphi_1 &= \frac{1}{2br}(p_{y_4} - p_{y_6}) + \frac{R}{2br}(-h_4 + h_6)d\omega_2 \\
d\kappa_2 &= -\frac{1}{3b}(p_{y_1} + p_{y_3} + p_{y_5}) + \frac{1}{3b}(h_1 + h_3R + h_5R)d\omega_2 \\
d\kappa_1 &= -\frac{1}{3b}(p_{y_2} + p_{y_4} + p_{y_6}) + \frac{1}{3b}(h_2 + h_4R + h_6R)d\omega_2
\end{aligned}
\tag{4.3-29}
$$

Explicit expressions for the orientation elements of dependent relative orientation can be derived in a similar manner[6].

Numerical Example (of Equation (4.3-29)).

Given: Principal distance $c = 152.64$ mm
Model base $b = 170.00$ mm
Point separation in image $\delta = 80.00$ mm

Observed: Parallaxes in the model p_y and vertical distances h from the six orientation points in the model to the perspective centres:

P	h [mm]	p_y [mm]
1	245	0.2
2	232	–0.70
3	225	0.14
4	225	–0.99
5	240	0.82
6	250	0.70

[6] Jerie, H.G.: Phia (1953/54), pp. 22–30.

Substitution in expressions (4.3-29) gives the following values for the elements of orientation result:

$$
\begin{aligned}
d\kappa_1 &= 0.82\,\text{gon} = 44' & d\kappa_2 &= 0.55\,\text{gon} = 30' \\
d\varphi_1 &= -0.53\,\text{gon} = -29' & d\varphi_2 &= -0.20\,\text{gon} = -11' \\
& & d\omega_2 &= 0.43\,\text{gon} = 23'
\end{aligned}
$$

4.3.4.2 Flat ground (after Hallert)[7]

For approximately flat ground one can assume that $h = \text{const}$. The six orientation points (also known as Gruber points) then lie on a rectangular grid both in the model and in the images (Figure 4.3-6 shows the image space).

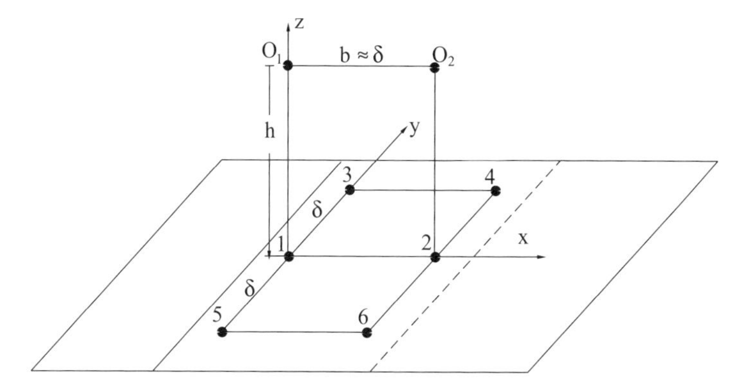

Figure 4.3-6: Orientation points for relative orientation when the ground is flat

In a similar manner to that for Equations (4.3-28) and (4.3-29) one can form the observation equations corresponding to the six orientation points, the normal equations and their solutions in explicit form:

$$
\begin{aligned}
d\kappa_1 &= -\frac{1}{12\delta^3}\left(p_{\eta_1}(6c^2+4\delta^2)+p_{\eta_2}(6c^2+8\delta^2)\right.\\
&\quad \left.-(p_{\eta_3}+p_{\eta_5})(3c^2+2\delta^2)-(p_{\eta_4}++p_{\eta_6})(3c^2-2\delta^2)\right)\\
d\kappa_2 &= -\frac{1}{12\delta^3}\left(p_{\eta_1}(6c^2+8\delta^2)+p_{\eta_2}(6c^2+4\delta^2)\right.\\
&\quad \left.-(p_{\eta_3}+p_{\eta_5})(3c^2-2\delta^2)-(p_{\eta_4}++p_{\eta_6})(3c^2+2\delta^2)\right)\\
d\varphi_1 &= \frac{c}{2\delta^2}(p_{\eta_4}-p_{\eta_6})\\
d\varphi_2 &= \frac{c}{2\delta^2}(p_{\eta_3}-p_{\eta_5})\\
d\omega_2 &= \frac{c}{4\delta^2}(-2p_{\eta_1}-2p_{\eta_2}+p_{\eta_3}+p_{\eta_4}+p_{\eta_5}+p_{\eta_6})
\end{aligned}
\qquad (4.3\text{-}30)
$$

[7] Hallert, B.: Über die Herstellung photogrammetrischer Pläne. Diss. Stockholm 1944, 118 p.

Exercise 4.3-9. Derive Equations (4.3-30) by establishing the observation equations, the normal equations and their solution.

Exercise 4.3-10. In (4.3-29) substitute a constant value for $h(=c)$, the value b in place of δ (Figure 4.3-6) and p_η in place of p_y. Then check the agreement between (4.3-30) and (4.3-29).

4.3.5 Critical surfaces in relative orientation

It is appropriate to ask whether relative orientation procedures lead to a unique and stable result for every imaginable form of object surface and location of exposure stations. We can most simply answer this question by considering the relationships (4.3-29).

A critical situation can arise with respect to the elements of orientation κ_1, κ_2, φ_1 and φ_2 if the base b is zero or very small, since the denominator in the relevant expressions then vanishes or almost vanishes. If the base is very small or, to put it another way, the forward overlap is very large, the two profiles on which the orientation points lie move very close together. It is not surprising then that no solution is possible using Jerie's method. Although such stereomodels are to be avoided on account of their very poor accuracy (in the first of Equations (2.1-34) the base B appears in the denominator) the relative orientation can be solved by applying, for example, the Formula (4.3-6) using corresponding points lying outside the profiles beneath the perspective centres.

If there is a very small value of r or of δ, the elements φ_1 and φ_2 are indeterminate when using the relationships (4.3-29). In this case all the orientation points lie more or less beneath the base (Figure 4.3-5). Such a situation arises in practice if in either the upper or lower margins of the pictures no corresponding points can be found (see also Section 4.3.6). This may happen because of the presence of water surfaces, forested areas, clouds, and so on.

Regarding the orientation element ω_2 it is necessary to consider whether the denominator in the first Equation of the system (4.3-29) can become, or approach, zero. The form of the denominator leads one to suspect that a critical situation may occur if the profiles 3, 1, 5 and 4, 2, 6 (Figure 4.3-5) have the same shape, especially if they are symmetrical about points 1 and 2. Thus we arrive at one of the four identical conditions for a possible critical surface:

$$h_2 = h_4 R \tag{4.3-31}$$

$$\text{Equation (4.3-27):}\quad h_2 = h_4(1 + r^2)$$

$$\text{Equation (4.3-26):}\quad h_2 = h_4\left(1 + \frac{y_4^2}{h_4^2}\right) \tag{4.3-32}$$

$$\text{Re-arranging:}\quad y_4^2 = h_4(h_2 - h_4) \tag{4.3-33}$$

Equation (4.3-33) shows that the triangle whose vertices are the points 2, 4 and O_2 has a right-angle at 4 (Figure 4.3-7); regarding $\overline{O_2 2}$ as the base of the triangle, y_4 is the height while h_4 and $(h_2 - h_4)$ are corresponding segments of the base. Thus O_2 lies

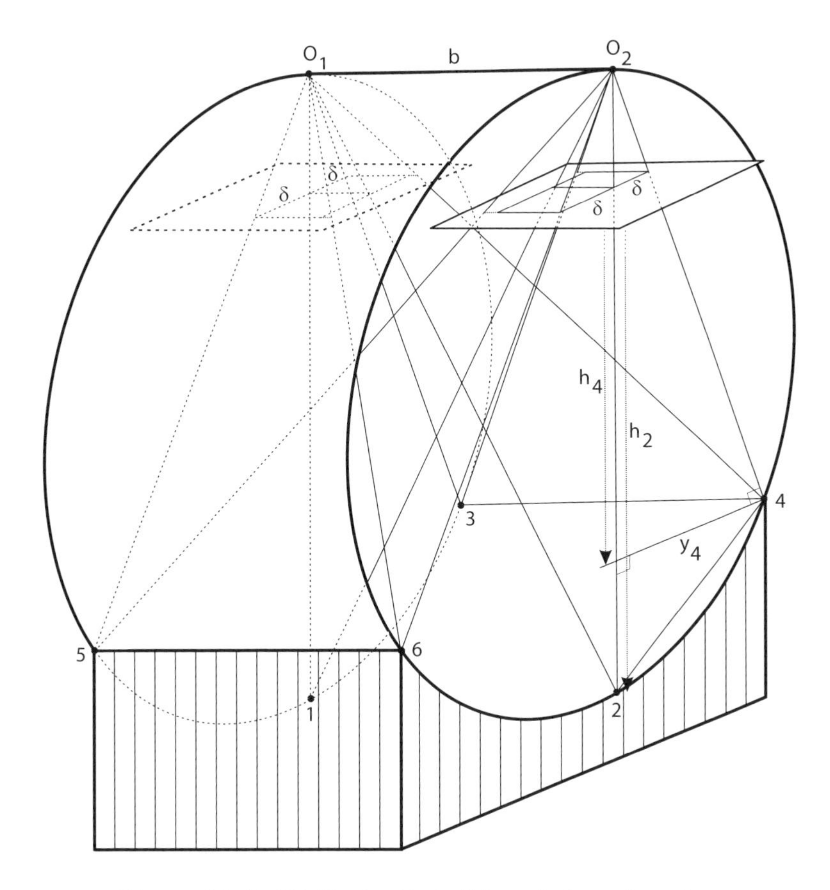

Figure 4.3-7: "Critical cylinder" for relative orientation

on a circle passing through points 2 and 4 (and also 6). Since profiles 3, 1, 5 and 4, 2, 6 have the same shape, the ground surface for which relative orientation cannot be performed is a right circular cylinder containing both camera stations, O_1 and O_2.

If, instead of Equation (4.3-31), we examine the relationships $2h_1 = h_3 R + h_5 R$ and $2h_2 = h_4 R + h_6 R$, we can derive more general conditions for critical surfaces in relative orientation. For example, a circular cylinder whose axis does not lie in the vertical plane containing the base (the line joining the two perspective centres), but which is parallel to that plane, is also a critical surface. Furthermore, the two circular profiles 3, 1, 5 and 4, 2, 6 can have different radii; that is, a conical surface containing O_1 and O_2 is a critical surface. Nor is it necessary for the base, $\overline{O_1 O_2}$, to be a generator of the cone; it is necessary only that the points O_1 and O_2 should lie on a conical surface which also contains the relative orientation points on the object surface. Further details may be found in the technical literature.[8]

[8]Hofmann, W.: DGK, Reihe C, Heft 3, 1953. Rinner, K.: JEK Band IIIa/1. Brandstätter, G.: IAPR

Exercise 4.3-11. Show that the denominator in the $d\omega_2$ of Equation (4.3-29) is zero for the case shown below in which the axis of the circular cylinder is not in the vertical plane containing the base, and thus that the ground surface is a critical surface relative to O.

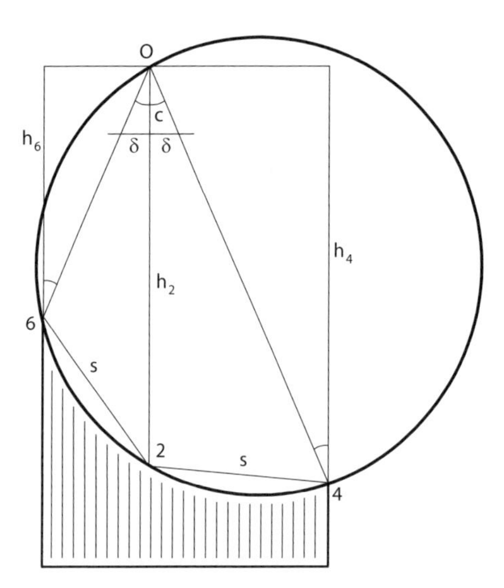

Hint: Use the cosine rule to express the length s in the two triangles and hence show that $2h_2 = (1 + (\delta/c)^2)(h_4 + h_6)$.

In what way do the critical surfaces manifest themselves in relative orientation? Difficulties arise in the solution of the normal equation system because the equations are singular or ill-conditioned. Even when the normal equations can be solved, this leads to very large standard deviations in the ω_2 unknown (see also Section 4.3.6).

Some of the following possibilities may improve the situation:

a) flying with a different type of camera. Figure 4.3-8 illustrates cross-sections of ground forming critical surfaces for each of three camera types (normal angle, wide angle and super-wide angle); the axis of the critical cylinder is taken in the direction of flight. Only when the landscape is very mountainous can it provide a critical surface for super-wide angle cameras. On the other hand, using a normal angle camera the relative orientation can relatively frequently be insoluble or, when the ground surface corresponds approximately to the form of the critical cylinder, simply very unreliable.

31(B3), Vienna, 1996. Brandstätter concerned himself with critical surfaces in relation to alternative methods of relative orientation. The relationship to two-dimensional resection in ground surveying should also be mentioned. As is well known, the position of a new point, A, from which three control points are observed with a theodolite, cannot be determined if the control points and A lie on a circle. The two circles in Figure 4.3-7 and the circle in the figure of Exercise 4.3-11 correspond exactly to this situation.

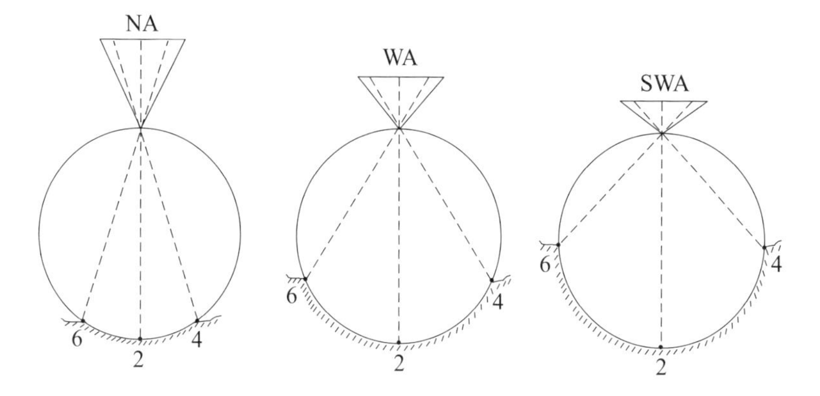

Figure 4.3-8: Critical cylinders in relation to camera type

b) fly across, rather than along, valleys.

c) in the relative orientation include only points which do not lie on or near the critical surface.

d) after absolute orientation of a model which is free of parallax, but which is deformed as a result of errors in relative orientation, corrections to the elements of relative orientation may be derived from residual height errors at check points. (See Section 4.3.6.2 or special literature.)[9]

e) computation of a combined, single-stage orientation of the two images (Section 4.2.2) or computation of a bundle triangulation (Section 5.3) including all the photographs in the block.

4.3.6 Error theory of relative orientation

Standard deviations of the elements of relative orientation are dealt with in the first section while in the following section (Section 4.3.6.2) deformations of photogrammetric models arising from certain errors of relative orientation are discussed.

4.3.6.1 Standard deviations of the elements of orientation[10]

At the end of all orientation procedures using least squares an estimate of the standard deviation of the computed elements is obtained (Appendix 4.1-1). An error computation of this kind is given in the numerical example of Section 4.3.1. In order for such accuracy measures to be evaluated, it is necessary to establish accuracy values

[9]Regensburger, K.: Photogrammetrie. VEB Verlag für Bauwesen, Berlin, 1990.
[10]Literature: Gotthardt, E.: BuL 15, pp. 2–34, 1940.

for the "standard case" of relative orientation, which we define as involving approximately flat, level ground and six orientation points as in Figure 4.3-6. Application of the rules of error propagation to Equations (4.3-30) results in the following expressions for standard deviations of the orientation elements in this standard case:

$$\begin{aligned} \sigma_{\varphi_1} = \sigma_{\varphi_2} &= \frac{c}{\sqrt{2}\delta^2}\sigma_{p_\eta} \\ \sigma_{\omega_2} &= \frac{\sqrt{3}}{2}\frac{c}{\delta^2}\sigma_{p_\eta} \\ \sigma_{\kappa_1} = \sigma_{\kappa_2} &= \frac{1}{\delta}\sqrt{\frac{2}{3} + \frac{c^2}{\delta^2} + \frac{3}{4}\frac{c^4}{\delta^4}}\sigma_{p_\eta} \end{aligned} \tag{4.3-34}$$

It is assumed in the above formulations that the η-parallaxes are measured with equal accuracy σ_{p_η} at all six orientation points and that no correlation exists among the η-parallaxes.

For individual types of cameras and using an assumed value of σ_{p_η}, for example $\sigma_{p_\eta} = \pm 5\,\mu\text{m}$, one can apply Equation (4.3-34) to compute accuracy estimates, as in the following Table 4.3-1. For example, for a normal angle camera:

$$\sigma_\omega = \frac{\sqrt{3}}{2}\frac{300}{92 \times 92}0.005\rho = \pm 10\,\text{mgon}(32'')$$

The other values can be calculated in a similar manner:

Camera type Format and p.d. [cm]	σ_ω [mgon ($''$)]	σ_φ [mgon ($''$)]	σ_κ [mgon ($''$)]
Normal angle (23×23, $c = 30$)	$\pm 10\,(32)$	$\pm 8\,(26)$	$\pm 34\,(110)$
Wide angle (23×23, $c = 15$)	$\pm 5\,(16)$	$\pm 4\,(13)$	$\pm 10\,(32)$
Super-wide angle (23×23, $c = 8.5$)	$\pm 3\,(10)$	$\pm 2\,(6)$	$\pm 5\,(16)$

Table 4.3-1: Accuracy of elements of relative orientation for different camera types

This table has great significance in practical photogrammetry. In any particular case the operator, or the software, assesses the extent to which the accuracy from this table is acceptable or not acceptable. We now carry out such an assessment for the specific results given in the numerical example of Section 4.3.1, in which case a wide angle camera was used. The calculated standard deviations ($\sigma_{\kappa_1} = \pm 15\,\text{mgon}\,(49'')$, $\sigma_{\kappa_2} = \pm 14\,\text{mgon}\,(45'')$, $\sigma_{\varphi_1} = \sigma_{\varphi_2} = \sigma_{\omega_2} = \pm 7\,\text{mgon}\,(23'')$) slightly exceed the target values from Table 4.3-1. This excess stems from the accuracy of the parallax measurements: a standard deviation of $\pm 9\,\mu\text{m}$ was estimated, while Table 4.3-1 is based on $\pm 5\,\mu\text{m}$. The excess can, however, just be tolerated. The values may clearly exceed the targets (a factor of more than two should not be accepted), for the following reasons; in each case a remedy is given:

- insufficient accuracy in measurement of image coordinates or η-parallaxes. Remedy—repeat the measurements.
- a critical surface for the relative orientation. Remedy—the measures quoted in Section 4.3.5.
- poor distribution of homologous points (for example, arising from inability to match points on water surfaces). Remedy—computation of a bundle triangulation (Section 5.3) including all photographs of the block.

The orientation elements constitute an important interim result on the way to the final photogrammetric result, the reconstituted object points. Monitoring these interim results is of great importance in quality control. The accuracy of the object points, which will be considered in Section 4.6, is of still greater interest for the user.

4.3.6.2 Deformation of the photogrammetric model

A relative orientation can be performed with only a specific, limited accuracy and the resulting elements of orientation will therefore contain specific errors. The question arises as to how such errors in orientation result in deformations of the stereomodel. We limit the following discussion to distortion of the z coordinate. In general, model deformations in height are more critical than those in plan (x, y).

Determination of heights in the stereomodel depends on x-parallaxes. It is necessary, therefore, to derive relationships between x-parallaxes, $p_x = x_1 - x_2$ (x_1 and x_2 are model coordinates related to the first and second pictures respectively) and small changes to the orientation elements.

From the first of each of the sets of Equations (4.3-3) and (4.3-4):

$$p_x = x_2 - x_1 = \frac{h}{c}\xi_1 - \left(h + \frac{h\xi_1^2}{c^2}\right) d\varphi_1 + \frac{h\xi_1\eta_1}{c^2} d\omega_1 - \frac{h\eta_1}{c} d\kappa_1 - b_x - \frac{h}{c}\xi_2$$
$$-\frac{\xi_2}{c} db_z + \left(h + \frac{h\xi_2^2}{c^2}\right) d\varphi_2 - \frac{h\xi_2\eta_2}{c^2} d\omega_2 + \frac{h\eta_2}{c} d\kappa_2$$

With reference to the relationships (4.3-24) the image-related quantities can be replaced and instead of the expression $(h\xi_1/c - h\xi_2/c - b_x)$ the small quantity $(-db_x)$ can be introduced:

$$\begin{aligned} p_x = &-db_x - \frac{(x-b)}{h} db_z - h\left(1 + \frac{x^2}{h^2}\right) d\varphi_1 + \frac{xy}{h} d\omega_1 - y d\kappa_1 \\ &+h\left(1 + \frac{(x-b)^2}{h^2}\right) d\varphi_2 - \frac{(x-b)y}{h} d\omega_2 + y d\kappa_2 \end{aligned} \tag{4.3-35}$$

It can be seen from Figure 4.3-9 that:

$$dz = \frac{h}{b} p_x \tag{4.3-36}$$

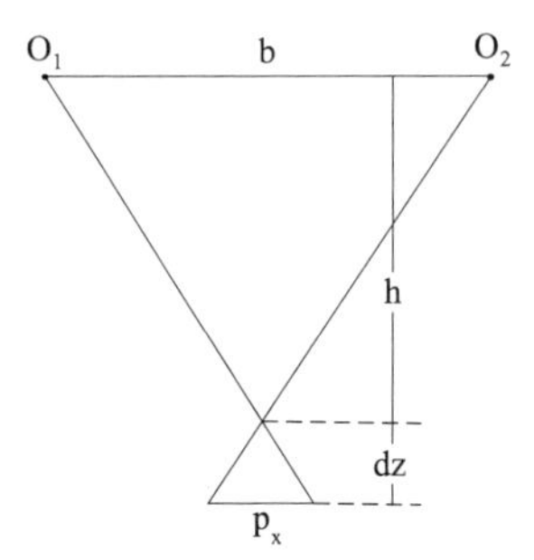

Figure 4.3-9: Relation between the x-parallax and the height error dz in the stereo-model

Substitution in Equation (4.3-35) finally gives us the desired relationship between errors in the orientation elements and the resulting height error in the model, dz.

$$
\begin{aligned}
dz = & -\frac{h}{b} db_x - \frac{(x-b)}{b} db_z - \left(\frac{h^2}{b} + \frac{x^2}{b} \right) d\varphi_1 + \frac{xy}{b} d\omega_1 - \frac{yh}{b} d\kappa_1 \\
& + \left(\frac{h^2}{b} + \frac{(x-b)^2}{b} \right) d\varphi_2 - \frac{(x-b)y}{b} d\omega_2 + \frac{yh}{b} d\kappa_2
\end{aligned}
\tag{4.3-37}
$$

The effects of the most important terms in Equation (4.3-37) are illustrated in Figure 4.3-10[11]; in that figure instead of db_z (see Figure 4.3-1) we have used db_{z2} and we have added the element db_{z1}, equivalent to a negative db_z. We can see from the various diagrams of Figure 4.3-10 that absolute orientation will remove the effects of (small) errors in the relative orientation, except for that arising from error in $d\omega$ and part of that caused by $d\varphi$.

It should be clearly mentioned that these deformations are superimposed on the surface of the model; it is as if the height reference plane, the xy plane, of a model free of errors in relative orientation is distorted into an inclined plane, a cylinder, a paraboloid and so on.

Exercise 4.3-12. In the mathematical literature one usually finds $x^2/a^2 - y^2/b^2 = 2z$ as the equation to a hyperbolic paraboloid. Under what conditions does this reduce to the equation given above, that is, $z = c\,x\,y$ in which c is a constant. (Answer: putting $a = b$ and rotation of the axes by $\pi/4$.)

[11] Albertz and Kreiling: Photogrammetric Guide. 4th ed., 1989.

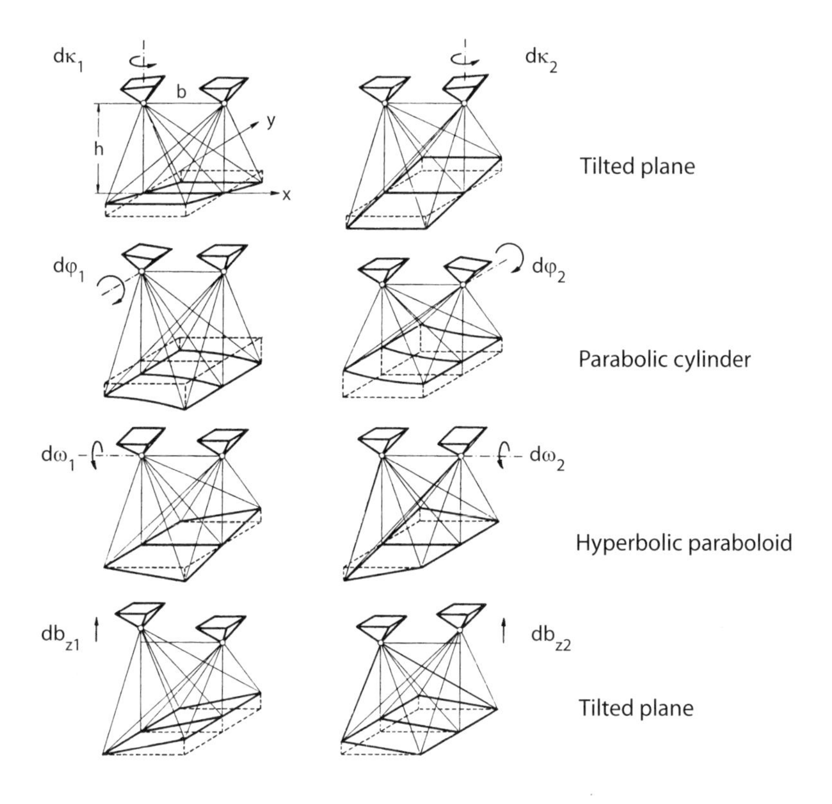

Figure 4.3-10: Model deformations in height

Numerical Example.

Given: Principal distance $c = 85\,\mathrm{mm}$, format $23\,\mathrm{cm} \times 23\,\mathrm{cm}$

Photo scale number $m_b = 10000$

Forward overlap 60%

$\Rightarrow$ base (actual) = 920 m (Equation (3.7-1))

Side overlap 25%

$\Rightarrow$ max. y coordinate (actual) = 862 m (Equation (3.7-1))

a) $d\omega_2$ deformation

Let the error $d\omega_2$ be 30 mgon (or $1'37''$), which is ten times the standard deviation given in Table 4.3-1. We wish to find the maximum height error arising from this ω_2 error; the maximum height error will appear along the line $x = 0$; Equation (4.3-37) gives:

$$dz = -\frac{(x-b)y}{b} d\omega_2 = 862 \frac{30}{\rho} = 0.41\,\mathrm{m}$$

Using four height control points in the corners of the model, this will be reduced during absolute orientation to about ±20 cm.

b) $d\varphi_2$ deformation
Let the error $d\varphi_2$ be 20 mgon (or 1′05″), which is again ten times the standard deviation given in Table 4.3-1. We wish to find the maximum height error arising from this φ_2 error. From Equation (4.3-37):

$$dz = \left(\frac{h^2}{b} + \frac{(x-b)^2}{b}\right) d\varphi_2$$

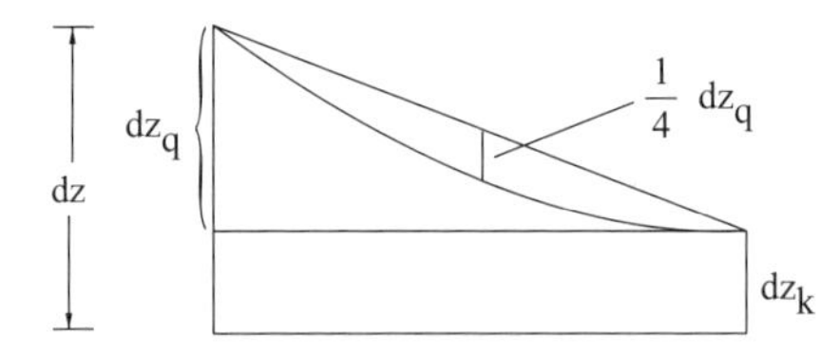

Figure 4.3-11: Bowing of the model as a result of $d\varphi_2$

As well as the bowing of the model, there is a constant vertical displacement, dz:

$$dz_k = \frac{h^2}{b} d\varphi_2 = \frac{850^2}{920}\frac{20}{\rho} = 0.25\,\text{m}$$

This is in addition to a term varying as the square of the x coordinate; this will be a maximum when $x = 0$:

$$dz_q = \frac{(x-b)^2}{b} d\varphi_2 = 920\frac{20}{\rho} = 0.29\,\text{m}$$

Under absolute orientation, the uniform displacement in a vertical direction will be removed completely and the effect of the quadratic term will be reduced to approximately 1/4 of the maximum value (see Figures 4.3-10 and 4.3-11). The maximum remaining error from the bowing of the model will amount to $dz_q/4 = 29/4\,\text{cm} = 7\,\text{cm}$.

Although we have considered very large errors in orientation, $d\omega_2$ and $d\varphi_2$, the resulting height errors in the model are of the same order of magnitude as the standard deviation of height measurement, σ_z. According to Equation (4.6-1) σ_z is about ±10 cm for well defined points using this camera, flying height and overlap. In practice, therefore, the model deformations are of little significance by comparison. Large orientation errors and therefore large model deformations can arise only with an unfavourable distribution of homologous points (choice limited, for example, by water surfaces) or, in some cases, in the presence of a critical surface (Sections 4.3.6.1 and 4.3.5).

Model deformations are, however, caused not only by orientation errors but also by sources of systematic error such as lens distortion and film shrinkage. The effects of these systematic errors on the geometrical correctness of the stereomodel are very complicated. They can manifest themselves as y-parallaxes larger than the measuring accuracy after relative orientation as well as in large residual errors at the control points after absolute orientation. A part of the y-parallaxes caused by these systematic errors is removed during relative orientation, although these parallaxes do not belong to the relative orientation. As a consequence, the orientation elements are wrong and this, therefore, results in model deformations. Such systematic errors produce correlations, especially among the xyz coordinates of neighbouring object points. To a great extent the inclusion of additional parameters in the mathematical model overcomes this problem (see Section B 3.5.6 and B 5.2.4 in Volume 2 as well as the relevant literature[12]).

Exercise 4.3-13. Consider the processing of photographs from a stereometric camera ($c = 60\,\text{mm}$, portrait format 80 mm × 95 mm, Section 3.8.2) in which the right-hand camera makes an angle of 99.9 gon ($89°55'$) with the base. The object distance is 10 m. Find what error arises in the direction of the camera axes (depth error); use formulae for the "normal case". (Answer: The error varies between 13 cm and 18 cm depending on the x coordinate; see Figure 3.8-1.)

Exercise 4.3-14. Repeat the above exercise for the case where the zenith distance of the right-hand camera axis is 100.2 gon ($90°11'$); that is, the axis of the right hand camera is tilted downwards by 0.2 gon ($11'$). (Answer: The error varies between 0 cm and 14 cm depending on the x and z coordinates.)

4.4 Absolute orientation

A procedure for the orientation of a stereopair was described in broad outline in Section 4.2.3; absolute orientation is the second step. In absolute orientation the stereomodel, which is defined with respect to an arbitrary coordinate system (xyz), is brought into a superior, or global, object coordinate system (XYZ) (Figure 4.2-2). The mathematical relationship is given by a three-dimensional similarity transformation. The seven parameters of this transformation are found using control points, points for which both the model coordinates and the object coordinates are known.

4.4.1 Least squares estimation

For a least squares computation by means of indirect observations (Appendix 4.2-1) Equation (4.2-6) has to be linearized. For the time being we assume that initial values are known for the seven parameters of absolute orientation with the model scale number $m \approx 1$. Using these values the original model coordinates $\mathbf{x}$ are transformed into the object coordinate system. These transformed coordinates are denoted by $\mathbf{X}^0$.

This means that:

$$\mathbf{X} \approx \mathbf{x} = \mathbf{X}^0 \Rightarrow \Omega = d\Omega, \Phi = d\Phi, \mathrm{K} = d\mathrm{K}, m = 1 + dm, \mathbf{X}_u = d\mathbf{X}_u \qquad (4.4\text{-}1)$$

[12] Schwidefsky/Ackermann: Photogrammetrie. B.G.Teubner, Stuttgart, 1976.

As a result the linearized rotation matrix $\mathbf{R}$ (2.1-13) is given by:

$$d\mathbf{R} = \begin{pmatrix} 1 & -d\mathrm{K} & d\Phi \\ d\mathrm{K} & 1 & -d\Omega \\ -d\Phi & d\Omega & 1 \end{pmatrix} \tag{4.4-2}$$

Expanding the product $m\mathbf{R}$, where $m = 1 + dm$ and ignoring second order terms gives:

$$\begin{aligned} m\mathbf{R} &= (1 + dm)d\mathbf{R} = \begin{pmatrix} 1 + dm & -d\mathrm{K} & d\Phi \\ d\mathrm{K} & 1 + dm & -d\Omega \\ -d\Phi & d\Omega & 1 + dm \end{pmatrix} \\ &= \mathbf{I} + \begin{pmatrix} dm & -d\mathrm{K} & d\Phi \\ d\mathrm{K} & dm & -d\Omega \\ -d\Phi & d\Omega & dm \end{pmatrix} \end{aligned} \tag{4.4-3}$$

in which $\mathbf{I}$ represents the unit matrix. Thus the linearized form of Equation (4.2-6) for a three-dimensional similarity transformation becomes:

$$\mathbf{X} = d\mathbf{X}_u + (1 + dm)d\mathbf{R}\mathbf{X}^0 \tag{4.4-4}$$

Writing this in ordinary algebraic notation:

$$\begin{aligned} X &= dX_u + X^0 dm \qquad\qquad + Z^0 d\Phi - Y^0 d\mathrm{K} + X^0 \\ Y &= dY_u + Y^0 dm - Z^0 d\Omega \qquad\qquad + X^0 d\mathrm{K} + Y^0 \\ Z &= dZ_u + Z^0 dm + Y^0 d\Omega - X^0 d\Phi \qquad\qquad + Z^0 \end{aligned} \tag{4.4-5}$$

For a least squares estimation by the method of indirect observations these simultaneous linear equations in the seven unknowns are rearranged as follows:

$$\begin{aligned} v_x &= dX_u + X^0 dm \qquad\qquad + Z^0 d\Phi - Y^0 d\mathrm{K} - (X - X^0) \\ v_y &= dY_u + Y^0 dm - Z^0 d\Omega \qquad\qquad + X^0 d\mathrm{K} - (Y - Y^0) \\ v_z &= dZ_u + Z^0 dm + Y^0 d\Omega - X^0 d\Phi \qquad\qquad - (Z - Z^0) \end{aligned} \tag{4.4-6}$$

Equations (4.4-6) may be written in matrix form (Appendix 4.1-1) as:

$$\mathbf{v} = \mathbf{A}\hat{\mathbf{x}} - \mathbf{l} \tag{4.4-7}$$

A full control point results in all three such equations, a plan control point in two and a height control point only in the last of Equations (4.4-6). The computation proceeds in the normal manner (Appendix 4.2-1), with the formation of normal equations and their solution:

$$\mathbf{A}^\top\mathbf{A}\hat{\mathbf{x}} = \mathbf{A}^\top\mathbf{l}; \quad \hat{\mathbf{x}} = (\mathbf{A}^\top\mathbf{A})^{-1}\mathbf{A}^\top\mathbf{l} \tag{4.4-8}$$

Since Equations (4.4-6) are only linear approximations, further iterations are normally required, unless the initial approximations are very good.

Numerical Example. We are given the model coordinates of five points:

[m]	23	24	50	51	45
x	0.303532	0.192638	0.303848	0.204120	0.246931
y	0.595068	0.602834	0.403493	0.434574	0.594227
z	0.034298	0.034116	0.026903	0.036672	0.034676

The first three points are full control points; point 51 is a height control point. The corresponding object points have known coordinates in the control system, frequently the national coordinate system, as follows[13]:

[m]	23	24	50	51
X	3321.65	3402.84	1776.75	
Y	1167.56	2061.10	1196.79	
Z	579.48	579.80	493.19	574.62

In addition, the following approximate values are known, for example from the flight plan (Section 3.7.1) or from one of the methods described in Section 4.4.3:

$$X_u^{(0)} = -1400\,\text{m},\; Y_u^{(0)} = 3600\,\text{m},\quad Z_u^{(0)} = 300\,\text{m}$$
$$m^{(0)} = 8000,\qquad \Omega^{(0)} = \Phi^{(0)} = 0,\; \mathrm{K}^{(0)} = 300\,\text{gon}\,(270^\circ)$$

Using these initial values the original model coordinates $\mathbf{x}$ are transformed into approximate ground coordinates $\mathbf{X}^0$, by means of the following equation:

$$\begin{pmatrix} X^0 \\ Y^0 \\ Z^0 \end{pmatrix} = \begin{pmatrix} -1400 \\ 3600 \\ 300 \end{pmatrix} + 8000 \begin{pmatrix} 0 & 1 & 0 \\ -1 & 0 & 0 \\ 0 & 0 & 1 \end{pmatrix} \begin{pmatrix} x \\ y \\ z \end{pmatrix}$$

In what follows we denote these coordinates by $\mathbf{X}^{(0)}$:

[m]	23	24	50	51	45
$X^{(0)}$	3360.54	3422.67	1827.94	2076.59	3353.82
$Y^{(0)}$	1171.74	2058.90	1169.22	1967.04	1624.55
$Z^{(0)}$	574.38	572.93	515.22	593.38	577.41

[13]The numerical stability of the computation will be compromised if the control coordinates are very large by comparison with the model coordinates; that is, if the origin of the ground coordinate system lies far outside the stereomodel (for relevant literature see, for example, Reinking, J.: ZfV 115, pp. 186–193, 1990). For this reason one should ignore as many of the leftmost digits of the ground coordinate system as are identical for all control points.

The observation equation system (4.4-6) or (4.4-7) then becomes:

$$
\mathbf{v} = \begin{pmatrix}
1 & & & 3.361 & & 0.574 & -1.172 \\
& 1 & & 1.172 & -0.574 & & 3.361 \\
& & 1 & 0.574 & 1.172 & -3.361 & \\
1 & & & 3.423 & & 0.573 & -2.059 \\
& 1 & & 2.059 & -0.573 & & 3.423 \\
& & 1 & 0.573 & 2.059 & -3.423 & \\
1 & & & 1.828 & & 0.515 & -1.169 \\
& 1 & & 1.169 & -0.515 & & 1.828 \\
& & 1 & 0.515 & 1.169 & -1.828 & \\
& & 1 & 0.593 & 1.967 & -2.077 &
\end{pmatrix}
\begin{pmatrix} d\hat{X}_u \\ d\hat{Y}_u \\ d\hat{Z}_u \\ d\hat{m}10^3 \\ d\hat{\Omega}10^3 \\ d\hat{\Phi}10^3 \\ d\hat{\mathrm{K}}10^3 \end{pmatrix}
- \begin{pmatrix} -38.89 \\ -4.18 \\ +5.10 \\ -19.83 \\ +2.20 \\ +3.87 \\ -51.19 \\ +27.57 \\ -22.03 \\ -18.76 \end{pmatrix}
$$

The following unknowns are found from Equations (4.4-8):

$$
\begin{aligned}
d\hat{X}_u &= -82.50\,\mathrm{m} & d\hat{\Omega} &= -0.00238 \mathrel{\hat{=}} -0.152\,\mathrm{gon} &&= -8'12'' \\
d\hat{Y}_u &= 54.29\,\mathrm{m} & d\hat{\Phi} &= -0.01726 \mathrel{\hat{=}} -1.099\,\mathrm{gon} &&= -53'24'' \\
d\hat{Z}_u &= -55.16\,\mathrm{m} & d\hat{\mathrm{K}} &= -0.02083 \mathrel{\hat{=}} -1.326\,\mathrm{gon} &&= -1^\circ 04'25'' \\
d\hat{m} &= 0.008666
\end{aligned}
$$

From Equation (4.4-4), using (2.1-13) to compute a rigorous form of the rotation matrix, the following ground coordinates are found for the new point 45:

$$
\begin{pmatrix} X^{(0)} \\ Y^{(0)} \\ Z^{(0)} \end{pmatrix} = \begin{pmatrix} -82.50 \\ 54.29 \\ -55.16 \end{pmatrix} + 1.008666 \begin{pmatrix} 0.999634 & 0.020825 & -0.017256 \\ -0.020787 & 0.999781 & 0.002380 \\ 0.017302 & -0.002020 & 0.999848 \end{pmatrix}
\times \begin{pmatrix} 3353.82 \\ 1624.55 \\ 577.41 \end{pmatrix} = \begin{pmatrix} 3323.22 \\ 1623.62 \\ 582.39 \end{pmatrix}
$$

In a manner identical to that for point 45, the model (control) points 23, 24, 50 and 51 can be transformed into the ground system:

[m]	23	24	50	51	45
$X^{(1)}$	3320.54	3401.85	1776.20	2042.31	3323.22
$Y^{(1)}$	1166.84	2060.90	1196.29	1995.83	1623.62
$Z^{(1)}$	580.37	578.18	493.96	575.50	582.39

Using the (approximate) ground coordinates $\mathbf{X}^{(1)}$ found in this way for the control points, an absolute orientation can be once again computed. For the second iteration the l-vector of the observation equation systems (4.4-6) or, as the case may be, (4.4-7) is as follows:

$$
\mathbf{l}^\top = (1.11, 0.72, -0.89, 0.99, 0.91, -1.38, 0.50, -0.77, -0.88)
$$

The unknowns derived in the second iteration are:

$$\begin{aligned}
d\hat{X}_u &= 0.05\,\text{m} & d\hat{\Omega} &= -0.00034 \mathrel{\hat{=}} -0.021\,\text{gon} &&= -1'02''\\
d\hat{Y}_u &= 0.23\,\text{m} & d\hat{\Phi} &= 0.00021 \mathrel{\hat{=}} 0.013\,\text{gon} &&= 42''\\
d\hat{Z}_u &= -0.05\,\text{m} & d\hat{\mathrm{K}} &= 0.00010 \mathrel{\hat{=}} 0.007\,\text{gon} &&= 19''\\
d\hat{m} &= 0.000308
\end{aligned}$$

Now the ground coordinates for the new point 45 and the control points 23, 24, 50 and 51 become:

[m]	23	24	50	51	45
$X^{(2)}$	3321.61	3402.86	1776.78	2042.90	3324.25
$Y^{(2)}$	1167.51	2061.15	1196.78	1996.62	1624.44
$Z^{(2)}$	579.40	576.90	493.28	574.52	581.27

At this point we break off the iteration process since, as can be seen from the diminution in the unknowns from the first to the second iteration, a third computation would make no contribution worth mentioning. From a comparison of the coordinates $\mathbf{X}^{(2)}$ with the given ground coordinates $\mathbf{X}$, one obtains the following improvements $\mathbf{v}$.

[cm]	23	24	50	51
v_x	−4	2	3	
v_y	−5	5	−1	
v_z	−8	10	9	−10

The weight coefficient matrix (Appendix 4.1-1): $\mathbf{Q} = (\mathbf{A}^\top \mathbf{A})^{-1}$

$$\mathbf{Q} = \begin{pmatrix}
5.098 & 0.020 & 0.021 & -1.287 & 0.046 & -0.236 & 0.671\\
0.020 & 5.254 & -0.518 & -0.673 & 0.622 & 0.040 & -1.265\\
0.021 & -0.518 & 6.379 & -0.252 & -1.974 & 1.071 & -0.070\\
-1.287 & -0.673 & -0.252 & 0.455 & -0.007 & -0.004 & -0.000\\
0.046 & 0.622 & -1.974 & -0.007 & 1.415 & 0.111 & 0.059\\
-0.236 & 0.040 & 1.071 & -0.004 & 0.111 & 0.473 & 0.009\\
0.671 & -1.265 & -0.070 & -0.000 & 0.059 & 0.009 & 0.458
\end{pmatrix}$$

The standard deviation in an observed coordinate:

$$\hat{\sigma}_0 = \sqrt{\frac{\mathbf{v}^\top \mathbf{v}}{n-u}} = \sqrt{\frac{408}{10-7}} = \pm 12\,\text{cm expressed in the ground coordinate system}$$

± 0.015 mm in the model coordinate system (1 : 8000)

$\pm 12\,\mu$m in the image ($m_B = 10000$)

in which n = number of observation equations (= 10)
u = number of unknowns (= 7)

The root mean square error in the elements of orientation: $\hat{\sigma}_k = \hat{\sigma}_0\sqrt{q_{kk}}$

$$\begin{array}{lll} \hat{\sigma}_{X_u} = \pm 0.28\,\text{m} & & \hat{\sigma}_\Omega = \pm 0.0092\,\text{gon} = \pm 30'' \\ \hat{\sigma}_{Y_u} = \pm 0.28\,\text{m} & \hat{\sigma}_m = \pm 0.000081 & \hat{\sigma}_\Phi = \pm 0.0053\,\text{gon} = \pm 16'' \\ \hat{\sigma}_{Z_u} = \pm 0.30\,\text{m} & & \hat{\sigma}_\mathrm{K} = \pm 0.0052\,\text{gon} = \pm 15'' \end{array}$$

An assessment of these accuracies appears in the following section.

Exercise 4.4-1. Calculate the rotation matrix with which the xyz model coordinates may be transformed directly into the XYZ ground coordinate system. Hint: As in Equation (2.1-14), the rotation matrix describing the resultant of successive rotations is obtained by multiplication of the matrices for the partial rotations. The result is:

$$\begin{pmatrix} -0.020720 & 0.999640 & -0.017045 \\ -0.999783 & -0.020677 & 0.002715 \\ 0.002361 & 0.017098 & 0.999851 \end{pmatrix}$$

Exercise 4.4-2. Repeat the numerical example assuming that point 50 is no longer a control point. Result:

[m]	23	24	50	51	45
X	3321.65	3402.84	1776.95	2042.02	3324.26
Y	1167.56	2061.10	1196.74	1996.53	1624.44
Z	579.48	576.80	493.58	574.62	581.25

$$\begin{array}{lll} \hat{\sigma}_{X_u} = \pm 0.68\,\text{m} & & \hat{\sigma}_\Omega = \pm 0.0113\,\text{gon} = \pm 37'' \\ \hat{\sigma}_{Y_u} = \pm 0.69\,\text{m} & \hat{\sigma}_m = \pm 0.00018 & \hat{\sigma}_\Phi = \pm 0.0074\,\text{gon} = \pm 24'' \\ \hat{\sigma}_{Z_u} = \pm 0.56\,\text{m} & & \hat{\sigma}_\mathrm{K} = \pm 0.0115\,\text{gon} = \pm 37'' \end{array}$$

Since, with this control point arrangement there is no redundancy, one would have been able to solve the problem (iteratively) on the basis of the linear equation system (4.4-5). The long way round using a least squares adjustment leads to the same result, including the error computation (see Equations (4.1-1-11) and (4.1-1-12) in Appendix 4.1-1). Estimation of the standard errors of the "corrections" is, however, not possible. We have adopted $\sigma_0 = \pm 12\,\text{cm}$ from the numerical example above.

Exercise 4.4-3. Taking as a starting point the result of Exercise 4.4-1 and the result of the relative orientation of Section 4.3.1, determine the rotation matrix defining the orientation of each of the images with respect to the ground coordinate system.

Result:

$$\mathbf{R}_1 = \begin{pmatrix} 0.006359 & 0.999836 & -0.016932 \\ -0.999947 & 0.006495 & 0.008057 \\ 0.008167 & 0.016880 & 0.999824 \end{pmatrix}$$

$$\mathbf{R}_2 = \begin{pmatrix} -0.033555 & 0.998674 & -0.039037 \\ -0.999437 & -0.033488 & 0.002387 \\ 0.001077 & 0.039095 & 0.999235 \end{pmatrix}$$

Exercise 4.4-4. Repeat the above numerical example assuming that the xy coordinates are more accurate by a factor two than the z coordinates. (Solution: The ground coordinates differ by about 1 cm at most. The explanation for this slight disparity is to be found in the introduction of Section 5.3.3.)

Further remarks: Every now and then, especially in close range work, the problem arises in which one has to carry out a photogrammetric evaluation without control points, either in a single model or in a block of photographs. In such cases an arbitrary object coordinate system may be adopted. For example, one can measure a distance S and adopt XYZ coordinates $(0,0,0)$ and $(S,0,0)$ for the end points. If one adopts the value $Z = 0$ for an (arbitrary) third point the object coordinate system is defined, although with no redundancy. With an object coordinate system chosen in this very arbitrary manner, a least squares estimation, which is in reality no such thing (see Equations (4.1-1-11) and (4.1-1-12) in Appendix 4.1-1), will deliver (exterior) accuracies for the new points, dependent on the arbitrary selection of the object coordinate system. (On the other hand accuracies for distances—interior accuracy—which can be calculated from coordinate accuracies, taking into account correlations, are independent of the object coordinate system chosen.)

In order to reduce the uncertainty associated with an arbitrary selection of object coordinate system, as hinted at above, the free net adjustment, as it is called, was developed. It is described in detail in Section B 4.6.3 of Volume 2. To set up the theory of free net adjustment one requires the normal equations for an over-determined similarity transformation in three dimensions which can be derived from Equations (4.4-6). Let the model coordinates and the object coordinates referred to origins at their respective centroids be denoted by $\overline{x}$ and $\overline{X}$ respectively; when this is done the translations dX_u, dY_u and dZ_u of Equations (4.4-6) vanish. With the substitutions $\overline{X} - \overline{X}^{(0)} = dX$, $\overline{Y} - \overline{Y}^{(0)} = dY$ and $\overline{Z} - \overline{Z}^{(0)} = dZ$, the normal equation system in the case of n full control points then becomes:

$$\begin{pmatrix} \left[\overline{X}^{(0)2} + \overline{Y}^{(0)2} + \overline{Z}^{(0)2}\right] & 0 & 0 & 0 \\ & \left[\overline{Y}^{(0)2} + \overline{Z}^{(0)2}\right] & \left[-\overline{X}^{(0)}\overline{Y}^{(0)}\right] & \left[-\overline{X}^{(0)}\overline{Z}^{(0)}\right] \\ & & \left[\overline{X}^{(0)2} + \overline{Z}^{(0)2}\right] & \left[-\overline{Y}^{(0)}\overline{Z}^{(0)}\right] \\ \text{symmetric} & & & \left[\overline{X}^{(0)2} - \overline{Y}^{(0)2}\right] \end{pmatrix} \begin{pmatrix} dm \\ d\Omega \\ d\Phi \\ d\mathrm{K} \end{pmatrix}$$

$$= \begin{pmatrix} \left[\overline{X}^{(0)}dX + \overline{Y}^{(0)}dY + \overline{Z}^{(0)}dZ\right] \\ \left[-\overline{Z}^{(0)}dY + \overline{Y}^{(0)}dZ\right] \\ \left[\overline{Z}^{(0)}dX - \overline{X}^{(0)}dZ\right] \\ \left[-\overline{Y}^{(0)}dX + \overline{X}^{(0)}dY\right] \end{pmatrix} \tag{4.4-9}$$

The normal equation system (4.4-9) is also of service in the error theory of absolute orientation which follows.

4.4.2 Error theory of absolute orientation

As a standard case we consider four control points in the corners of the model (Figure 4.4-1a). Based on the normal equations (4.4-9) weight coefficients may be given for the four relevant elements of absolute orientation, that is the three angles and the scale. For this purpose we assume approximately flat level ground, that is $\overline{Z}^{(0)} = 0$, and we take the dimensions of the rectangle at the corners of which the control points lie as 92 mm × 184 mm; under these conditions the matrix of the normal equations (4.4-9) becomes a diagonal matrix:

$$\begin{aligned}
q_{mm} &= \sqrt{1/(4 \times 46^2 + 4 \times 92^2)} = 0.00486\,\text{mm}^{-1}\\
q_{\Omega\Omega} &= \sqrt{1/(4 \times 92^2)} = 0.00543\,\text{mm}^{-1}\\
q_{\Phi\Phi} &= \sqrt{1/(4 \times 46^2)} = 0.01086\,\text{mm}^{-1}\\
q_{\mathrm{KK}} &= \sqrt{1/(4 \times 46^2 + 4 \times 92^2)} = 0.00486\,\text{mm}^{-1}
\end{aligned}$$

Concerning the accuracy of the scale number m and the rotation K, a crucial matter is the accuracy in plan of the photogrammetric coordinates of the control points; the accuracy of the photogrammetric coordinates of the control points in height is decisive for the angles Ω and Φ. For well-measured control points these two accuracies amount to about ±6 μm in the image and ±0.06‰ of the principal distance c (for normal-angle and wide-angle pictures) or ±0.08‰ of the principal distance c (for superwide-angle pictures) (Section 4.6). Using these accuracies for plan and height coordinates and the weight coefficients given above, one obtains root mean square errors in this standard case as compiled in Table 4.4-1.

Camera type Format, p.d. [cm]	σ_m $[10^{-5}]$	σ_ω [mgon ($''$)]	σ_φ [mgon ($''$)]	σ_κ [mgon ($''$)]
Normal angle (23×23, $c = 30$)	±2.9	±6.2(20)	±12.4(40)	±1.9(6)
Wide angle (23×23, $c = 15$)	±2.9	±3.1(10)	±6.2(20)	±1.9(6)
Super-wide angle (23×23, $c = 8.5$)	±2.9	±2.4(8)	±4.7(15)	±1.9(6)

Table 4.4-1: Accuracy of absolute orientation for different camera types

The accuracy of absolute orientation, like that of relative orientation (Table 4.3-1), is independent of the image scale or of the object distance. As is Table 4.3-1, Table 4.4-1 is of considerable significance in practical photogrammetry. On the basis of this table the operator, or the computer program, should judge in a particular case whether or not and to what extent the given accuracy values are acceptable. Such an assessment is carried out below for the numerical example of Section 4.4.1. The root mean square errors arising in this particular case clearly lie above the values prescribed in Table 4.4-1. The excesses require some elucidation:

- for the photo scale number m the unfavourable ratio is 2.8 $(= 0.000081/0.000029)$. A factor 2 stems from the fact that in this particular case an estimated accuracy in plan coordinates of $\pm 12\,\mu$m was used and in the standard case an accuracy of $\pm 6\,\mu$m is assumed. A further cause of the large excess is the different number of plan control points; in the standard case there are four (Figure 4.4-1a) and in this particular case only three.

- for the heading angle K the ratio amounts to 2.7 $(= 0.0052/0.0019)$. The causes of this large excess are the same as for the photo scale number just reviewed.

- it is conspicuous with respect to the inclination angles Ω and Φ that the relationship in the standard case between the higher accuracy in Ω and that in Φ is reversed in this particular example. This apparent contradiction arises from the K rotation of the stereomodel through approximately 300 gon (270°). Taking account of this transposition, the ratios for the two angles are 1.7 $(= 0.0053/0.0031)$ and 1.5 $(= 0.0092/0.0062)$. A factor of 1.3 arises because the particular case under consideration is based on an estimated height accuracy of $\pm 12\,\mu$m while the standard case an accuracy of $\pm 9\,\mu$m $(= 150000 \times 0.00006)$ is assumed. A further cause of the relatively large excess lies in the fact that the figure containing the control points in the image is somewhat smaller than the 92 mm × 184 mm rectangle of the standard case.

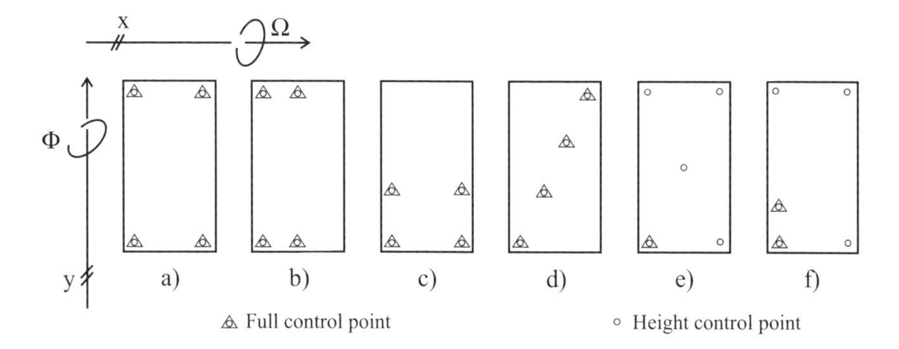

a) standard situation
b) inclination Φ poor
c) inclination Ω poor
d) inclinations Ω and Φ indeterminate, if the control points lie exactly on a straight line.
e) heading angle K and model scale number m indeterminate
f) heading angle K and model scale number m poorly determined

Figure 4.4-1: Assessment of different control point arrangements for absolute orientation

Absolute orientation provides an important interim result in the photogrammetric procedure. For quality control of this important step the excess factor mentioned above should remain below 2. If it exceeds this value there are causes and consequences as follows:

- determination of the model coordinates of the control points insufficiently accurate. To improve matters: repeat the measurements.
- poor distribution of the control points. In Figure 4.4-1 a number of poor control point patterns are presented; Figure 4.4-1d) illustrates a seriously bad arrangement of control points. Remedy: Computation of a bundle triangulation (Section 5.3) with inclusion of all the images in the block.

From the viewpoint of reliability, a topic to be dealt with in Section B 7.2.2.2 (Volume 2), the standard case of absolute orientation is improved by using pairs of control points in the model corners. In this case the prescribed values of Table 4.4-1 reduce by a factor $1/\sqrt{2}$, as one may readily convince oneself with the help of the normal equations (4.4-9).

The assessment above applies in the case of relatively flat models. In close-range photogrammetry, and in aerial photogrammetry among very high mountains, the stereomodels should be surrounded by control points in all three coordinate directions. For these cases specific reference values should be derived as outlined above.

Exercise 4.4-5. With the control point arrangement of Figure 4.4-1d) the angles Ω and Φ cannot be determined. Show that this statement is true. (Solution: Consider the system of normal equations (4.4-9).
For simplicity take $\overline{Z}^{(0)} = 0$. Choose $\overline{X}^{(0)} = aY^{(0)}$. If the second equation is multiplied by the factor $(-a)$ it becomes identical to the third equation. That is, the two rows of the matrix are linearly dependent, meaning that the matrix is singular.)

Exercise 4.4-6 (Continuation of Exercise 4.4-5). Assess the result of this absolute orientation with only three control points in the light of Table 4.4-1. (Answer: The excess factors are 6.2 (m), 2.4 (Ω), 1.8 (Φ) and 6.1 (K). Since three of these factors exceed 2, the absolute orientation must be rejected.

4.4.3 Determination of approximate values

Least squares estimation assumes approximate initial values for the unknowns. Methods to find these values, suitable for automation, are given in this section.

The first procedure assumes at least four full control points. For the connection of the model coordinates (x, y, z) and the object coordinates (X, Y, Z) the (linear) three-dimensional affine transformation (2.1-17), with its twelve unknowns suggests itself. The unknowns are three translations $\mathbf{a}_0$ and nine matrix elements a_{ik}. Each control point provides three linear equations; with four full control points twelve unknowns may be found. With more than four full control points a least squares estimation should be carried out.

The question must still be resolved as to how one gets the seven parameters of absolute orientation (4.2-6) from the twelve transformation parameters of the affine transformation. Comparison of the relationships (2.1-17) with (4.2-6) gives: $a_{10} = X_u$, $a_{20} = Y_u$ and $a_{30} = Z_u$. The other elements a_{ik} of the affine transformation (2.1-17) differ from the elements r_{ik} of the three-dimensional similarity transformation (2.1-18) or (4.2-6) mainly on account of the scale number m. The relationship

$$\sqrt{a_{11}^2 + a_{21}^2 + \cdots + a_{33}^2} = \sqrt{3}m \tag{4.4-10}$$

gives an approximate value for the scale number m. Finally, all the elements a_{ik} can be divided by the scale number m so that by means of the relationships (2.1-1-8) of Appendix 2.1-1 the angles Ω, Φ and K can also be found[14].

Numerical Example. We take up the numerical example of Section 4.4.1, but replacing the height control point with a full control point.

Model coordinates

[m]	23	24	50	51
x	0.303532	0.192638	0.303848	0.204120
y	0.595068	0.602834	0.403493	0.434574
z	0.034298	0.034116	0.026903	0.036672

Object coordinates or national coordinates

[m]	23	24	50	51
X	3321.65	3402.84	1776.75	2043.11
Y	1167.56	2061.10	1196.79	1996.72
Z	579.48	576.80	493.19	574.62

Setting up and solving the equations results in:

$$\begin{pmatrix} X_u \\ Y_u \\ Z_u \end{pmatrix} = \begin{pmatrix} -1425 \\ 3715 \\ 213 \end{pmatrix} \qquad \mathbf{A} = \begin{pmatrix} -166.95 & 8068.09 & -107.88 \\ -8069.40 & -167.42 & 39.82 \\ 20.49 & 137.51 & 8107.24 \end{pmatrix}$$

$$\mathbf{R}^0 = \begin{pmatrix} -0.020685 & 0.999639 & -0.013366 \\ -0.999802 & -0.020743 & 0.004934 \\ 0.002539 & 0.017037 & 1.004490 \end{pmatrix}$$

Equation (4.4-10) gives $m^0 = 8070$. The elements of the approximate rotation matrix $\mathbf{R}^0$ are formed as follows: $\mathbf{R}^0 = (1/m^0)\mathbf{A}$. Using relationships (2.1-1-8) of

[14]Further literature: Schmid, H.H., Heggli, S.: Mitt. des Inst. für Geod. und Photogr. an der ETH Zürich, Nr. 23, 1978.

Appendix 2.1-1 one obtains the following approximate values for the angles: $\Omega^0 = -0.3\,\text{gon} = -16'$, $\Phi^0 = -0.8\,\text{gon} = -43'$ and $\mathrm{K}^0 = 298.7\,\text{gon} = 268°50'$.

The solution using the three-dimensional affine transformation has two disadvantages: on the one hand, the method breaks down when the four control points lie in a plane and, on the other hand, in the presence of only three control points, which is the basic requirement for an absolute orientation, it cannot be employed. A solution free of these disadvantages can be found by creating orthogonal vectors in both coordinate systems from three full control points. Details of this procedure are to be found in Section B 4.1.2.2 in Volume 2. Another method, with a detailed example, is included in Section B 5.1.1.3, again in Volume 2, which solves for the absolute orientation of near-vertical photographs without the need for initial approximate values.

4.5 Image coordinate refinement

Image coordinates taken from photographs must be corrected in different respects. This correction is known as image coordinate refinement. We have already learnt of a number of corrections:

- correction for lens distortion (Equation (3.1-3)).
- correction for film deformation in metric cameras (Section 3.2.1.2) and in semi-metric cameras (Section 3.8.4).
- correction for aplanarity (lack of flatness) of the CCD detector array (Section 3.3.4)
- correction for positional errors of detectors in CCD cameras or in film scanners (Section 3.4.3).

To meet demands for high accuracy one must also eliminate the influence of atmospheric refraction and take into account the curvature of the Earth.

4.5.1 Refraction correction for near-vertical photographs

Because the atmospheric pressure, the temperature and the humidity vary along the path of a ray of light, so too does the index of refraction, which is dependent on these three quantities. This means that the rays which give rise to a photograph are bent. The position of the image point P' must, therefore, be corrected by an amount $\Delta\rho$ in order to get the hypothetical metric image which would have been produced by central projection (Figure 4.5-1).

In image space, the tangent to the curved light ray will appear to represent the direction OP. In the case of a near-vertical photograph the angle $\Delta\tau$ between the light ray and

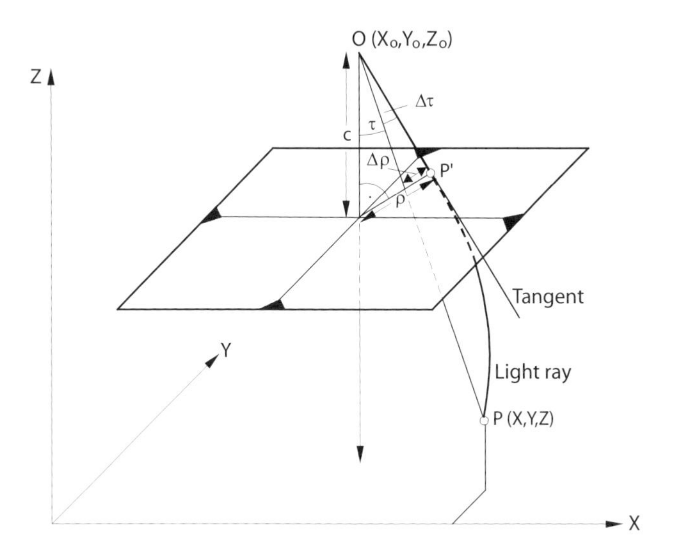

Figure 4.5-1: The effect of atmospheric refraction on a vertical photograph

the straight line OP can be represented with sufficient accuracy by a function of the angle τ and a coefficient K:

$$\Delta\tau = K \tan\tau = K\frac{\rho}{c} \tag{4.5-1}$$

It goes without saying that the coefficient K will vary according to the meteorologic data at the time of a particular photograph and, since the index of refraction varies with the wavelength, also with the wavelength to which the sensor is responsive. For a normal atmosphere stratified parallel to the surface of the Earth and for light in the visible wavelengths, K can be estimated as follows:[15]

$$K = 0.00241\left(\frac{Z_0}{Z_0^2 - 6Z_0 + 250} - \frac{Z^2}{Z_0(Z^2 - 6Z + 250)}\right) \tag{4.5-2}$$

in which Z_0 = flying height above sea level (in [km])
and Z = height of the ground (in [km])

Normally it suffices to use approximate values of Z_0 and Z representative of the whole area being surveyed.

[15] Harris, W.D., Tewinkel, S.C. and Whitten, C.A.: Analytic Aerotriangulation. Coast and Geodetic Survey, Technical Bulletin No. 21, July 1962, corrected July 1963. Schut, G.H.: Ph.Eng. 35, pp. 79–86, 1969.

It remains only to give the relationship using which the radial correction $\Delta\rho$ can be found from $\Delta\tau$ (Figure 4.5-1):

$$\Delta\rho\,\cos\tau \approx \Delta\tau\sqrt{c^2+\rho^2}$$

$$\Delta\rho \approx \frac{c^2+\rho^2}{c}\Delta\tau$$

Taking Equation (4.5-1) into account:

$$\Delta\rho \approx \rho\left(1+\frac{\rho^2}{c^2}\right)K \tag{4.5-3}$$

Finally, using the relationship (3.1-2), the correction $\Delta\rho$ can be resolved in the two coordinate directions ξ and η.

In Table 4.5-1 the expressions (4.5-2) and (4.5-3) are evaluated for particular cameras, photo scales and radial distances.

Photo scale	c [mm]	Z_0 [km]	Correction $\Delta\rho$ [μm]	
			$\rho = 90$ mm	$\rho = 130$ mm
1 : 10000	300	3.5	3	5
	150	2.0	2	4
	85	1.3	2	5
1 : 30000	300	9.5	8	12
	150	5.0	6	11
	85	3.0	6	13
1 : 100000	85	9.0	15	34
1 : 800000	300	240	1	2

Table 4.5-1: Correction of image coordinates for the effect of atmospheric refraction (Ground height $Z = 0.5$ km)

One sees from this table that the corrections reach a sufficient number of micrometers to be taken into account in processing only in small-scale photographs, above all in the case of super-wide photography. In the case of very precise aerotriangulation for larger-scale projects, however, refraction also needs to be considered. The last row shows that satellite images are not adversely affected by refraction.

In conclusion it must be mentioned that the Formulae (4.5-2) and (4.5-3) are valid only in the case of near-vertical photographs. With larger tilts it is necessary to move to more general formulae[16].

Exercise 4.5-1. An aerial photograph ($c = 85$ mm) was taken from a height of 9100 m above sea level. Calculate corrections for the effect of atmospheric refraction at 25 points of a 5 cm × 5 cm rectangular raster. In the left-hand side of the image the ground

[16] See, for example, Rinner, K.: JEK, Band IIIa/1, §§ 22–23.

is flat and level with a height of 600 m. It rises from the mid-line of the photographs with a slope of 30%. (Answer: In two corners of the picture $\Delta\rho$ reaches 37 μm; in the other two corners $\Delta\rho = 42\,\mu$m.)

An important special case concerns horizontal photographs in terrestrial photogrammetry. In this case the influence of atmospheric refraction is usually dealt with together with Earth curvature, although treating it separately is didactically more correct. Refraction is actually a physical error in the formation of the image while Earth curvature is a geometrical "error" in the ground coordinates.

4.5.2 Correction for refraction and Earth curvature in horizontal photographs

As has already been indicated, the effect of Earth curvature is not a photogrammetric error but is related simply to the definition of coordinate systems. Photogrammetry provides coordinates in a three-dimensional Cartesian coordinate system in which the position of an object point is fixed by orthogonal projection on the respective coordinate planes. In the field survey system, on the other hand, heights of ground points are referred to the ellipsoid while the XY coordinates are more or less distorted by the mathematical projection of the ellipsoid onto a plane. For this reason the ground survey coordinates of control points should first be transformed into three-dimensional Cartesian coordinates before the start of the photogrammetric work.

Fortunately, however, the differences between the ground survey coordinate system and a three-dimensional Cartesian coordinate system are small so that in practice the problem can be simplified. In Volume 2, Section B 5.4, however, a full discussion appears. Here and in the following section the discussion is restricted to approximate solutions for the coordinate most strongly affected, namely the ground height.

First, the effect of Earth curvature combined with atmospheric refraction for horizontal photographs is dealt with. The Earth is regarded as a sphere of radius $R = 6370\,\text{km}$. The symbol $\overline{Z}$ is used for heights referred to the sphere as opposed to Z for heights referred to the three-dimensional Cartesian system (Figure 4.5-2).

The effects of Earth curvature ΔE for the point P are given by the well-known formula:

$$\Delta E \approx \frac{A^2}{2R} \qquad (4.5\text{-}4)$$

Standard textbooks on surveying[17] give an approximate equation[18] for trigonometric height measurement which takes account of Earth curvature and the influence of refraction:

[17] For example Kahmen, H., Faig, W.: Surveying. 578 p., Walter de Gruyter, Berlin, 1988.

[18] The approximation in this relation is too crude for large distances and large height differences, conditions which, however, seldom arise in photogrammetric practice.

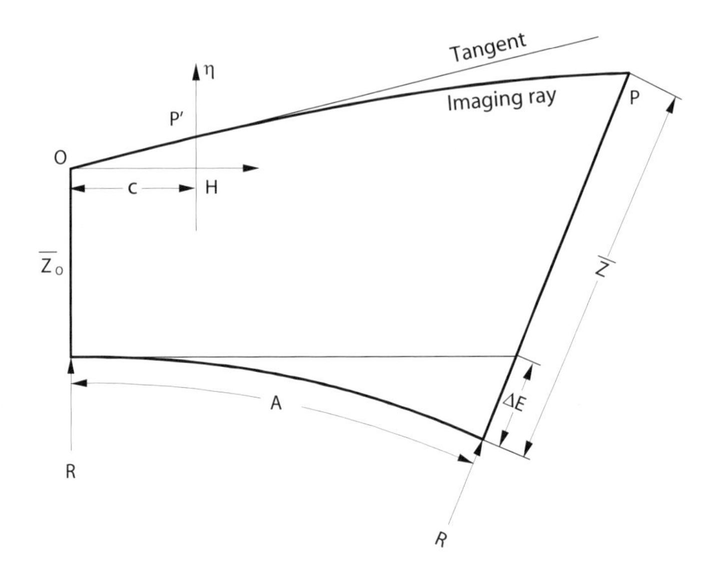

Figure 4.5-2: Atmospheric refraction and Earth curvature for a horizontal photograph

$$\overline{Z} = \overline{Z}_0 + A\tan\alpha + A\tan\alpha\,\frac{\overline{Z}_0 + \overline{Z}}{2R} + (1-k)\frac{A^2}{2R} \tag{4.5-5}$$

in which k ... refraction coefficient, usually taken as 0.13

For photogrammetric point determination with horizontal photographs $\tan\alpha$ may be replaced by the ratio η/c:

$$\overline{Z} = \overline{Z}_0 + \eta\frac{A}{c} + \eta\frac{A}{c}t\frac{\overline{Z}_0 + \overline{Z}}{2R} + (1-k)\frac{A^2}{2R} \tag{4.5-6}$$

What is required, however, for use with the three-dimensional object coordinate system is the significantly simpler relation:

$$\overline{Z} = \overline{Z}_0 + \frac{A}{c}(\eta + \Delta\eta) \tag{4.5-7}$$

To compensate for the use of heights referred to the sphere rather than Z coordinates from the Cartesian coordinate system, the correction $\Delta\eta$ is introduced to the image coordinates η. Subtraction of Equation (4.5-7) from (4.5-6) gives:

$$\frac{A}{c}\Delta\eta = \eta\frac{A}{c}\frac{\overline{Z}_0 + \overline{Z}}{2R} + (1-k)\frac{A^2}{2R}$$

Finally, re-arranging gives the desired formula:

$$\Delta\eta = \eta\frac{\overline{Z}_0 + \overline{Z}}{2R} + (1-k)\frac{Ac}{2R} \tag{4.5-8}$$

If the image coordinates η are corrected by an amount $\Delta\eta$ and if the photogrammetric measurements are made in the usual way, then the height coordinates, whether of control points or of new points, will be referred to the sphere; they will also be free of the influence of refraction. For the computation of the correction $\Delta\eta$, however, the distance A and the value $\overline{Z}$ are necessary. If P is a new point the following procedure, suitable for automation, is recommended:

- first measure the point photogrammetrically without correction for Earth curvature and refraction
- using the result so obtained, compute A, $\overline{Z}$ (with adequate accuracy) and $\Delta\eta$
- finally, repeat the photogrammetric computation using the corrected image coordinates.

Table 4.5-2 gives some typical values for the correction $\Delta\eta$.

A [km]	0.1	0.25	0.5	1	5	20
$\Delta\eta$ [μm]	0.7	1.7	3.4	6.8	34	137

Table 4.5-2: Corrections, $\Delta\eta$, to image coordinates for the influence of refraction and Earth curvature ($c = 100\,\text{mm}$, $k = 0.13$, $R = 6370\,\text{km}$, $\eta = 0$)

It will be seen that, up to a few hundred metres object distance the correction is smaller than the measuring accuracy of a few micrometres, but may not be ignored for distances greater than about 1 km.

Exercise 4.5-2. Compute the correction $\Delta\eta$ and the $\overline{Z}$ coordinate for an extreme case with $A = 5000\,\text{m}$, $\overline{Z}_0 = 1000\,\text{m}$, $\eta = 30\,\text{mm}$, $c = 100\,\text{mm}$. (Result: $\Delta\eta = 42.4\,\mu\text{m}$, $\overline{Z} = 2502.12\,\text{m}$). Additional exercise: Compute the $\overline{Z}$ coordinate using a refraction coefficient of 0.10. (Result: $\Delta\eta = 43.6\,\mu\text{m}$, $\overline{Z} = 2502.18$.)

Exercise 4.5-3. Using the data of Table 4.5-2 compute the standard errors of both the correction $\Delta\eta$ and the ground height Z under the assumption that the refraction coefficient $k = 0.13$ has a standard error of ± 0.03. (Solution: for example, if $A = 5\,\text{km}$: $\sigma_{\Delta\eta} = \pm 1.2\,\mu\text{m}$, $\sigma_{\overline{Z}} = \pm 5.9\,\text{cm}$.)

4.5.3 Earth curvature correction for near-vertical photographs

In this section the influence of and correction for Earth curvature in near-vertical photographs is considered. A stereomodel produced photogrammetrically in the usual manner from two aerial photographs is referred to the three-dimensional Cartesian coordinate system XYZ (Figure 4.5-3). Heights from land survey, however, refer to a sphere with the Earth radius R.

Earth curvature can be taken into account in the photogrammetric process as now described. Control point heights $\overline{Z}$, referred to the curved datum surface $\overline{Z}_0 = 0$, are

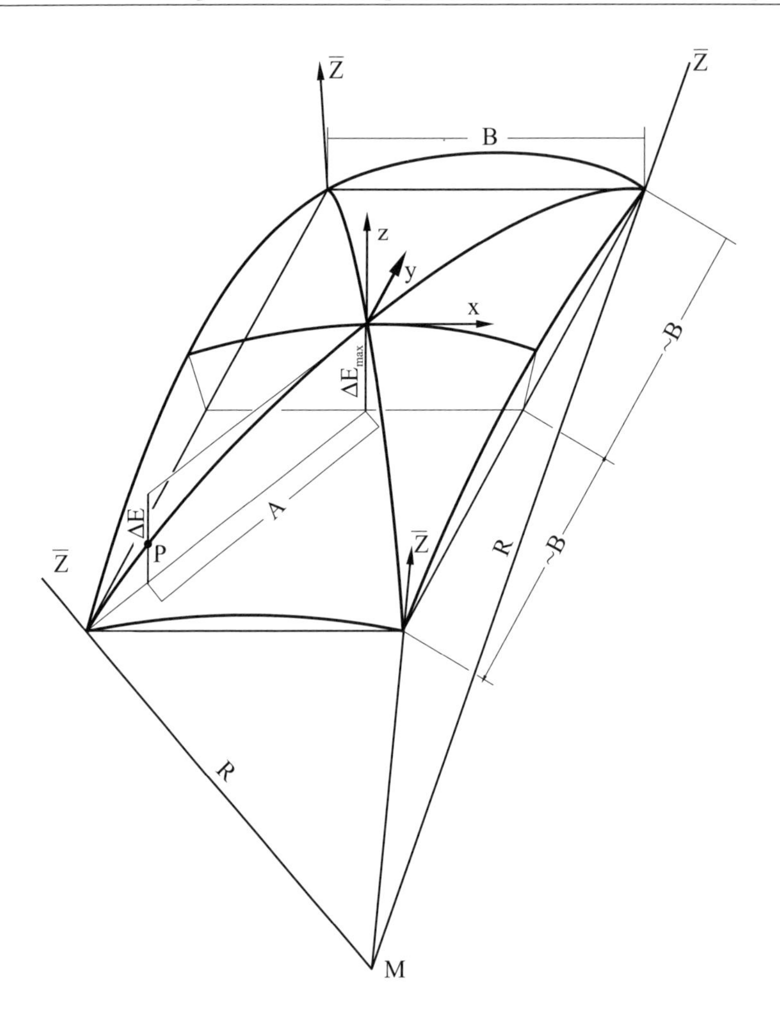

Figure 4.5-3: Earth curvature in a stereomodel from near-vertical aerial photographs

reduced by the correction ΔE from Equation (4.5-4), in which A is the approximate distance of the particular control point from the centre of the stereomodel. The result of this operation is that the control point heights are referred to the tangent plane at the centre of the model, the XY plane of a Cartesian coordinate system. Subsequently all object points are determined in this XYZ system using the computed image coordinates; finally all Z values are increased by the amount ΔE (Equation 4.5-4).

It is also possible to take account of Earth curvature by "correcting" the image coordinates in aerial photographs in a similar way to the correction in horizontal photographs. This "Earth curvature correction" is applied in the same way as the correction for refraction. It has the opposite sign, however, and is significantly larger than the $\Delta\rho$ refraction correction (Equation (4.5-3)). This makeshift Earth curvature correction is dealt with in more detail in Section 5.6.

So that an assessment may be made of the extent to which a correction for Earth curvature is important, the maximum correction $\Delta E_{\max}$ in parts per thousand of the flying height is given in Table 4.5-3 for various photo scales. The value $\Delta E_{\max}$ (Figure 4.5-3) comes from Equation (4.5-4) in which $B\sqrt{5}/2$, the semi-diagonal of the model, is substituted for A:

$$\Delta E_{\max} = \frac{\left(\frac{B}{2}\sqrt{5}\right)^2}{2\ R} = \frac{5\ B^2}{8\ R} \tag{4.5-9}$$

In view of photogrammetric height accuracy of 0.06‰ for a measured point (Section 4.6), Table 4.5-3 shows that Earth curvature must be taken into account:

- for super-wide angle photographs from a photo scale of about 1 : 6000,
- for wide angle photographs from a photo scale of about 1 : 10000 and
- for normal angle photographs from a photo scale of about 1 : 20000.

Photo scale	Maximum Earth curvature correction $\Delta E_{\max}$			
	[m]	(‰ of the flying height)		
		$c = 85$ mm	$c = 150$ mm	$c = 300$ mm
1:10000	0.08	0.10	0.06	0.03
1:50000	2.1	0.49	0.28	0.14
1:100000	8.3	0.98	0.56	0.28

Table 4.5-3: Maximum Earth curvature correction $\Delta E_{\max}$ at the corners of a stereomodel with 60% forward overlap

4.5.4 Virtual (digital) correction image

In digital photogrammetry every completed correction to be implemented in the context of image coordinate refinement entails a resampling, that is a restructuring of the pixels in a new orthogonal raster (Section 2.2.3). As a rule every resampling is accompanied by loss of information. In view of the relatively large number of corrections which mount up in the course of image coordinate refinement a solution is sought which, if possible, manages with one single resampling.

With a virtual correction image[19] one achieves this aim. The procedure is explained with reference to Figure 4.5-4; it is a typical example which can readily be generalized. Let the digitized photograph have an arbitrary $\xi'\eta'$ coordinate system (Section 3.4). A corrected image matrix is required in the $\xi\eta$ system, the coordinate system of a digital metric camera or a digital metric image (Figures 4.5-4 and 3.1-10). As in indirect resampling (Figure 2.2-5), an orthogonal raster is first defined in the digital image

[19]The idea was adopted from T. Schenk (Digital Photogrammetry. TerraScience, Volume I, 1999).

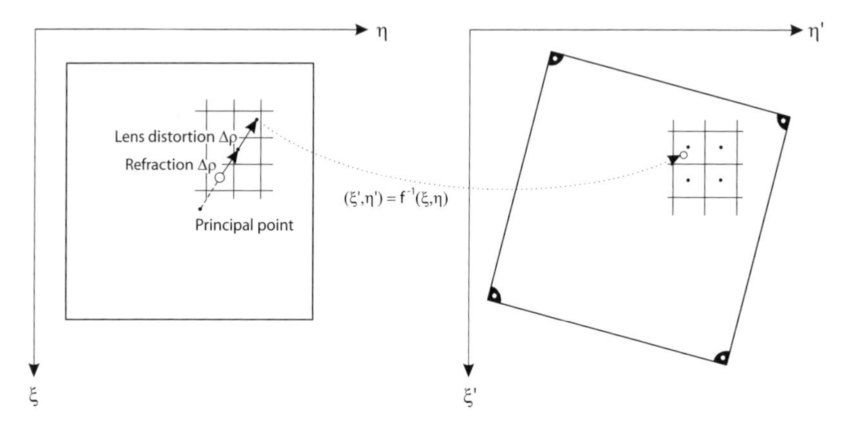

Figure 4.5-4: Digital metric image, in its position relative to the camera (left), with corrections for refraction and lens distortion; and transformed into the photograph as digitized (right)

(Figure 4.5-4, left). Corrections are then computed for the mid-point of every pixel; on the left of Figure 4.5-4, for example, corrections have been introduced for refraction (Equation (4.5-3)) and lens distortion (Section 3.1.3). These corrections are recorded in a virtual correction image with a resolution in the subpixel range. The last step is the transformation into the digital photograph, an inverse transformation which is based on the fiducial marks and which eliminates film deformation; depending on the number and arrangement of the fiducial marks, the parameters of this transformation may be chosen accordingly (Section 3.2.1.2).

The point whose position has been computed in the above three steps, is marked with a small circle on the right of Figure 4.5-4; finally, a grey level interpolation is carried out, as described in Section 2.2.3, and the interpolated grey value is transferred to the digital metric image. This position is likewise marked with a small circle in the left of Figure 4.5-4.

By choice, the raster spacing in the digital photograph is somewhat smaller than the digitizing interval in the digitized photograph. The result is a digital metric image which can be evaluated in subsequent calculations involving the collinearity equations (Section 2.1.3) and the coplanarity equations (Section 4.2.3).

4.6 Accuracy of point determination in a stereopair

When the elements of exterior orientation are known the object coordinates of a new point can be computed directly (Section 4.1). If the elements of exterior orientation are unknown the object coordinates of new points may be obtained as a by-product of "two picture resection" (Section 4.2.2). The other orientation procedures permit the computation, using these orientation elements, of new points in the stereomodel.

A stereopair from aerial survey approximates to the "normal case". Therefore, for estimating the achievable accuracy of 3D point measurements in a model the theory described in Section 2.1.7 can be applied for any photo flight, if the flight parameters are known and information about the measurement accuracy of image coordinates is given.

Leaving aside extremely large-scale aerial photographs (perhaps subject to image movement) the accuracy of object coordinates X and Y (plan accuracy) is directly proportional to the image scale and is constant in relation to the image (Equation (2.1-35)). The camera type (normal angle, wide angle and so on) has no influence on the accuracy in plan.

As may be seen form Equation (2.1-33), the height accuracy σ_Z is directly dependent on the measurement accuracy of the parallaxes, as well as being indirectly dependent on the base-height-ratio (B/Z) and on the photo scale (c/Z). σ_Z may also be regarded as either linearly or quadratically dependent in the camera-object distance Z. In both cases represented in Figure 4.6-1 the height accuracy σ_Z is linearly dependent on the camera-object distance Z. In the diagram on the left of Figure 4.6-1, both the base-height B/Z and the principal distance c are the same for both camera arrangements; in the right-hand diagram the photo scale number $m_B = Z/c$ and the base B, and consequently also the forward overlap, are the same for both. The proportionality between the height accuracy σ_Z and the object distance for different camera types holds true only for the range from normal-angle to wide-angle cameras; the relationship is weaker for super-wide-angle photographs[20].

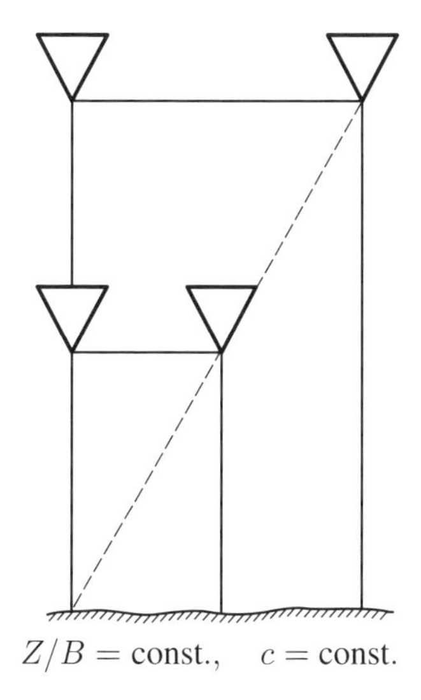

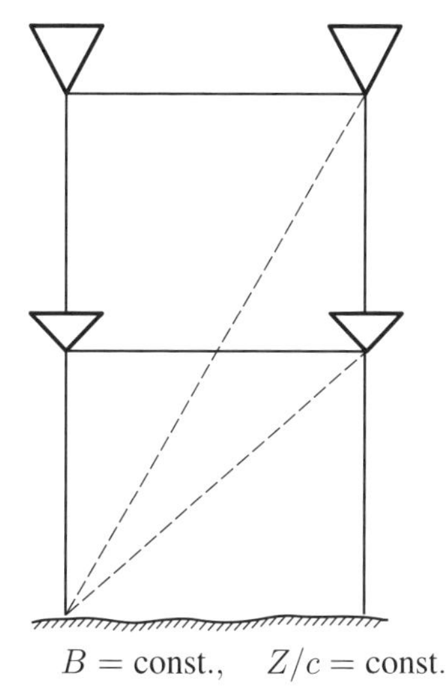

Figure 4.6-1: Height accuracy directly proportional to the camera-object distance

[20]Meier, H.K.: BuL 38, pp. 50–62, 1970. Stark, E.: BuL 44, pp. 5–14, 1976. Sievers, J., Schürer, K.: Verfeinerter Ansatz vor allem für variables Basisverhältnis und unterschiedliches Auflösungsvermögen. BuL 50, pp. 101–118, 1982.

From Equation (2.1-33) rules of thumb can easily be derived for rough estimation of accuracy. In this we limit ourselves to giving information on accuracy which is representative of the whole stereomodel. Thus we ignore readily available information (see Section B 5.2.3.1, Volume 2) about accuracies which vary within a stereomodel. To a large extent the information on accuracy is based on empirical investigations.

The conditions under which these rules of thumb apply are:

- forward overlap of 60% (possible cases deviating from this are specifically delat with; for example in the Numerical Example 2 of Section 6.7.2d)
- at least four full control points in the corners of the model (indirect georeferencing); the control coordinates must exhibit markedly better accuracy than the measured coordinates. Alternatively elements of exterior orientation from GPS and IMU (direct georeferencing) with corresponding accuracy
- image format 23 cm × 23 cm (for film-based metric cameras)
- good image quality such that with analytical plotting instruments a coordinate measurement accuracy of $\pm 6\,\mu$m at image scale can be achieved

The following values are applicable as guides in the case of targeted points or equally accurately defined points:

$$\begin{aligned} \text{Plan: } \sigma_{XY(\text{targ})} &= \pm 6\,\mu\text{m} \times \text{ image scale number } m_B \\ \text{Height: } \sigma_{XY(\text{targ})} &= \pm 0.06‰ \text{ of height above ground} \\ &\quad \text{(normal angle and wide angle)} \\ &\quad \pm 0.08‰ \text{ of height above ground} \\ &\quad \text{(super-wide angle)} \end{aligned} \tag{4.6-1}$$

If, in determining the object coordinates, one eliminates the systematic part of the image coordinate errors by means of additional parameters in the mathematical model, a further increase in accuracy of up to 50% is possible (for details see Section 5.3.5). On the other hand poor quality image material can cause distinct reduction in the numerical values of the rules of thumb (4.6-1).

These rules of thumb also apply to digital images if they have been generated by scanning aerial films with an appropriate pixel size. In practice, scanning films with 15 μm to 20 μm has turned out as acceptable trade-off between sufficient radiometric quality and good resolving power. Experience has also shown that for manual measurement if well defined points in digital images an accuracy of $1/3$ of a pixel can be achieved, i.e. 5 μm to 7 μm in the case of the above mentioned pixel sizes.

If the above-mentioned conditions do not apply, the rules of thumb will deliver unrealistic results and Equation (2.1-33) should be used only for rough estimation; for example, if:

- the accuracy of the image coordinates measurement differs significantly from 6 μm. Automatic point measurement using, for instance, lest squares matching (see Section 6.8.1.2) can reach accuracies of $1/5$ to $1/10$ of a pixel and thus much higher accuracy. On the other hand, measurements in images digitized with large pixel sizes might be less accurate. In both cases the rules could still be applied, if the results are multiplied by the ratio of the actual measurement accuracy to $6\,\mu$m.

- other photo flight parameters were used, for instance, forward overlaps significantly greater than 60%.

- digital aerial cameras were employed (such as Microsoft-Vexcel Ultracam, Intergraph DMC, Leica Geosystems ADS40). For digital cameras a generally valid rule of thumb is hard to define. There is no standardized principle for digital cameras and the geometric arrangement for image acquisition may differ significantly.

Numerical Example. Images taken with a Zeiss RMK 15/23, with a photo scale of 1 : 5000 and 60% forward overlap, were used for photogrammetric point determination. What is the accuracy of photogrammetrically derived coordinates when targeted points are involved?

Using the rules of thumb (4.6-1):

$$\begin{aligned} \sigma_{XY} &= 5000 \times 0.0006 &&= \pm 3\,\text{cm} \\ \sigma_Z &= 5000 \times 15 \times 6 \times 10^{-5} &&= \pm 4.5\,\text{cm} \end{aligned}$$

Camera arrangements can occur in practice, especially in high mountains, where the height accuracy varies quadratically with the camera-object distance. In all the cases sketched in Figure 4.6-2 the base B and the principal distance c remain the same; only the camera-object distance varies. The left-hand example represents a flight over stepped landscape using a constant base and constant height above sea level; in the right-hand diagram the bases are equal but the flying height varies.

Numerical Example. The second stereomodel on the left of Figure 4.6-2 represents pictures from a flying height above ground which is double that of the first stereomodel; the forward overlap in the second model is about 80%. As a result, the height accuracy given above (that of the first model) worsens as follows:

$$\begin{aligned} \sigma_{XY} &= \pm 3 \times 2 &&= \pm 6\,\text{cm} \\ \sigma_Z &= \pm 4.5 \times 4 &&= \pm 18\,\text{cm}(!) \end{aligned}$$

In comparison with the model with the lower flying height on the right of Figure 4.6-2, the same deterioration of accuracy also arises in the stereomodel with the greater flying height. Incidentally, an even more severe reduction in accuracy arises in that area of the model which lies outside, in the direction of the base, the area between the image centres. There is certainly a great temptation in practice to use a stereomodel with 80% forward overlap covering the entire common area—with a base which is too short.

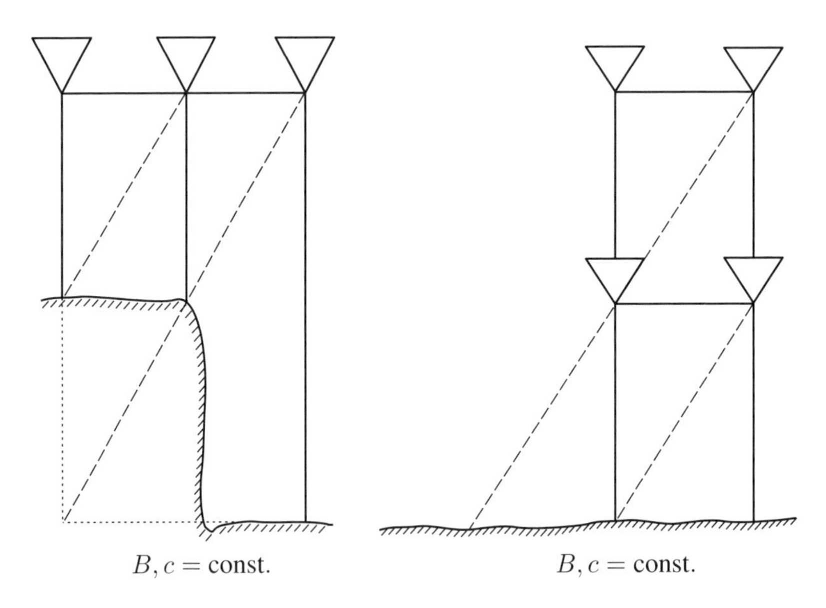

Figure 4.6-2: Height accuracy proportional to the square of the camera-object distance

Exercise 4.6-1. A photoflight has been planned over extremely mountainous country with a forward overlap of 60% and a base of 1000 m. Under these conditions the Formulae (3.7-1) show that for a 30 cm camera the flying height over the summits will be 3260 m and for a 21 cm camera it will be 2280 m. What accuracy is to be expected in the region of the summits and what in the valleys which lie 1000 m beneath the summits? (Answer: In the summit area, for the 30 cm camera, $\sigma_Z = \pm 19.6$ cm, and for the 21 cm camera $\sigma_Z = \pm 13.7$ cm. In the valleys, for the 30 cm camera, $\sigma_z = \pm 33.5$ cm, and for the 21 cm camera $\sigma_Z = \pm 28.4$ cm.)

It is worth noting that the superiority, by a factor of 1.43 ($= 19.6/13.7$), of the 21 cm camera over the 30 cm camera in the region of the summits is reduced in the valleys to 1.18 ($= 33.5/28.4$).[21]

In case of natural (non-targeted) points, the guide figures (4.6-1) for the accuracy must be increased by the uncertainty of definition. This uncertainty of definition can be related to the image as well as to the object.

In principle, the uncertainties of definition related to the image and their effects on the standard deviation of the object points may be found by repeated measurements:

[21] The suggestion for this example came from Prof. Dr. P. Waldhäusl.

$$\begin{aligned}\sigma_{XY(\text{nat})} &= \sqrt{\frac{\sum(X_i-\overline{X})^2}{n-1}} \text{ or } \sqrt{\frac{\sum(Y_i-\overline{Y})^2}{n-1}} \\ \sigma_{Z(\text{nat})} &= \sqrt{\frac{\sum(Z_i-\overline{Z})^2}{n-1}}\end{aligned} \tag{4.6-2}$$

where n ... number of measurements and $\overline{X}$, $\overline{Y}$, $\overline{Z}$ are arithmetical means

Values related to the image depend on the image quality, the film type (colour, black& white, etc.), the pixel size, etc. In a practical project one will determine the accuracy values $\sigma_{XY(\text{nat})}$ and $\sigma_{Z(\text{nat})}$ for a number of types of points using the relationships (4.6-2). In the most favourable case one obtains $\sigma_{XY(\text{nat})}$ and $\sigma_{Z(\text{nat})}$, the accuracies (4.6-1) for targeted or very precisely defined points.

The uncertainty of definition related to the object may also be determined by repeated measurements; these can be made either terrestrially or photogrammetrically using a very large photo scale. Values of the related to the ground for some types of points have been collected in Table 4.6-1.

Type of point	Plan $\sigma_{XY(O,\text{def})}$ [cm]	Height $\sigma_{Z(O,\text{def})}$ [cm]
House and fence corners	7 – 12	8 – 15
Manhole covers	4 – 6	1 – 3
Field corners	20 – 100	10 – 20
Bushes, trees	20 – 100	20 – 100

Table 4.6-1: Uncertainty of definition of natural points[22]

Starting with the object related uncertainty of definition, one must summarize by considering both contributions $\sigma_{(\text{targ})}$ and $\sigma_{(O,\text{def})}$:

$$\begin{aligned}\text{Planimetry: } \sigma_{XY(\text{nat})} &= \sqrt{\sigma^2_{XY(\text{targ})} + \sigma^2_{XY(O,\text{def})}} \\ \text{Height: } \sigma_{Z(\text{nat})} &= \sqrt{\sigma^2_{Z(\text{targ})} + \sigma^2_{Z(O,\text{def})}}\end{aligned} \tag{4.6-3}$$

Numerical Example. With what accuracy can the corners of houses and fields be measured from wide-angle photographs with about 60% forward overlap and a photo scale of 1 : 15000? (Use relevant mean values from Table 4.6-1.)

$$\begin{aligned}\text{House corners: } \sigma_{XY} &= \sqrt{(15000 \times 0.0006)^2 + 9.5^2} &&= \pm 13\,\text{cm} \\ \sigma_Z &= \sqrt{(15000 \times 15 \times 0.00006)^2 + 11.5^2} &&= \pm 18\,\text{cm} \\ \text{Field corners: } \sigma_{XY} &= \sqrt{(15000 \times 0.0006)^2 + 60^2} &&= \pm 61\,\text{cm} \\ \sigma_Z &= \sqrt{(15000 \times 15 \times 0.00006)^2} &&= \pm 20\,\text{cm}\end{aligned}$$

[22] Waldhäusl, P.: Presented Paper, Commission IV, 14. ISP Congress, Hamburg, 1980.

The standard deviation of a distance σ_S can be found directly from the coordinate error σ_{XY}:

$$\sigma_S = \sigma_{XY}\sqrt{2} \tag{4.6-4}$$

The accuracy of distances derived from photogrammetrically measured coordinates is essentially independent of the distance.

Exercise 4.6-2. Derive Formula (4.6-4) using the rules of error propagation.

In the field of close range photogrammetry one is often less interested in the absolute accuracies σ_{XY} and σ_Z and more interested in relative accuracy with respect to the largest dimension of the object S. For an object which fills the whole of the format of a standard aerial camera, therefore, the relative accuracy amounts to:

$$\frac{\sigma_{XY}}{S} = \frac{\sigma_{\xi\eta}}{23\,\text{cm}} = \frac{\pm 0.0006}{23} = \frac{1}{38000} = 2.6 \times 10^{-5} \mathrel{\hat{=}} 0.026‰ \tag{4.6-5}$$

We have, incidentally, already found this value to be the accuracy of the scale factor in absolute orientation (Table 4.4-1: $\pm 0.000029 = 0.029‰$).

Especially on account of the small format of close range cameras one obtains lower relative accuracies than with large format aerial survey cameras. For a semi-metric camera with medium format (6 cm $\times$ 6 cm, Section 3.8.8) the relative accuracy, likewise applying the value $\sigma_{\xi\eta} = \pm 6\,\mu\text{m}$, comes to:

$$\frac{\sigma_{\xi\eta}}{6\,\text{cm}} = \frac{\pm 0.0006}{6} = \frac{1}{10000} = 10^{-4} \mathrel{\hat{=}} 0.1‰ \tag{4.6-6}$$

In close range photogrammetry the photographs are frequently arranged in a less regular pattern than in aerial survey (Section 3.8.8). For this reason it is necessary to extend the simple accuracy guide (4.6-1) appropriately to suit the diverse camera arrangements of close range work. This topic is dealt with in detail in Volume 2 (Section B 4.5.2).

The empirical accuracies given by the relationships (4.6-1), (4.6-5) and (4.6-6) are chosen fairly conservatively. They can be considerably improved by means of further mathematical modelling of the photogrammetric process. This is considered in more detail in Section 5.3.5.

The rules of thumb for accuracies of photogrammetric processing given in this section and in Section 6.7 are widely accepted in photogrammetric practice. In the calling of tenders for photogrammetric work, however, tolerances enjoy a greater degree of acceptance than the standard deviation or root mean square error. Tolerances and their empirical determination are considered in more detail in Appendix 4.6-1.

Exercise 4.6-3. For purposes of road construction the heights of targeted points are to be found by photogrammetry with an accuracy of ± 5 cm using a super-wide angle camera. If the forward overlap is to be 60%, what are the photo scale and flying height? (Answer: photo scale = 1 : 7350, flying height = 625 m.)

Supplementary Exercise. What reduction in accuracy arises if, over a valley region 200 m lower, the base is not increased? (Answer: The height accuracy in the valley area is $\sigma_Z = \pm 8.7$ cm.)

Exercise 4.6-4. Is the alignment of the image rectangle of the Z/I Imaging DMC (Figure 3.7-10) well chosen for a favourable height accuracy? (Answer: No.) How would the heighting accuracy be changed if the format were to be rotated by 100 gon (90°)? (Answer: By a factor 1.7 ($= 13500/8000$).) What advantage is there in the alignment of the rectangular format chosen by Z/I Imaging? (Answer: Fewer strips of photographs for a given flying height.)

Chapter 5

Photogrammetric triangulation

We discuss first photogrammetric triangulation with aerial photographs, known as aerotriangulation or aerial triangulation. We assume metric images recorded onto film or CCD area arrays. In Section 5.5 we will discuss the digital 3-line camera.

5.1 Preliminary remarks on aerotriangulation

Even without GPS and IMU information, aerotriangulation frees photogrammetry from the constraint of determining at least three control points by ground survey methods in every stereomodel. It is possible to bridge areas without such ground survey points. In Chapter 4 we limited the discussion to methods of data capture in two overlapping photographs, but in this chapter we treat the subject of simultaneous data capture in a block of such overlapping photographs.

The results of aerotriangulation are the orientation elements of all photographs or stereomodels and the XYZ coordinates of discrete points in a global coordinate system (usually the ground coordinate system). We speak then of photogrammetric point measurement. The points may be:

- points targeted before the flight
- selected "natural" points in the photographs, usually with accompanying identification sketches
- points automatically selected in digital metric images, their image coordinates thus being known (Section 6.8.3.5 and Section B 2.4, Volume 2)

Figure 5.1-1 illustrates the principle of aerotriangulation, without reference to supporting GPS and/or IMU information (this aspect will be discussed in Section 5.4). The example shows 18 photographs in three strips. The forward overlap of photographs within a strip is about 60%, the side overlap between strips about 25%. For the orientation of this set of photographs, we have four full control points and four height control points. We require the orientation elements and the ground coordinates of new points. These new points tie models together within a strip and tie neighbouring strips together. For example, new point 7 occurs only in the photographs 12 and 13, new point 5 in the photographs 11, 12 and 13, new point 6 in the photographs 11, 12, 13, 21, 22 and 23.

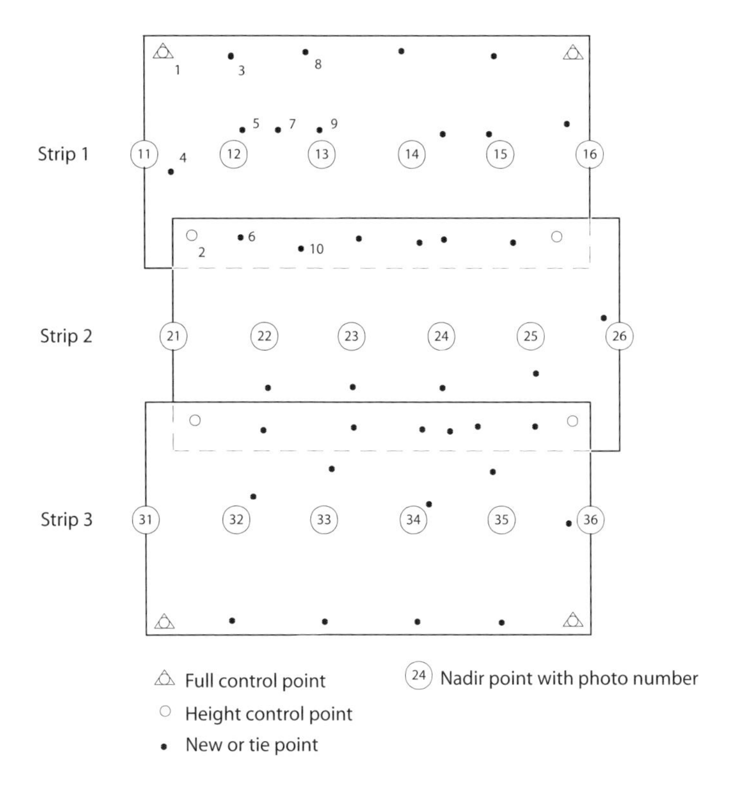

Figure 5.1-1: Set of photographs for a block adjustment

We have already introduced two techniques which can be extended to deal with blocks of photographs. The first is the numerical orientation of the two bundles of rays of a stereopair of photographs (Section 4.2.2). In this technique, the 12 elements of exterior orientation of the two photographs and the coordinates of new points in the stereomodel are computed from the known coordinates of control points. The method can be readily extended to a large number of photographs, e.g. 18 in the example of Figure 5.1-1. The method is then known as a bundle block adjustment. It is discussed in Section 5.3.

The second technique is numerical absolute orientation (Section 4.4). It is based on the assumption that a stereomodel has been formed by the numerical relative orientation of two overlapping photographs (Section 4.3). This stereomodel is then transformed into the ground coordinate system with the help of control points. The extension of the technique from a single model to a block of models implies that all models in the block—15 in the example of Figure 5.1-1—are absolutely oriented simultaneously. The method is then known as a block adjustment by independent models. It is described in the following section.

5.2 Block adjustment by independent models

We start with a block of photographs with about 60% forward overlap between photographs in the individual strips and about 25% side overlap between neighbouring strips. The adjustment of a single strip by independent models is a special case of the more general block adjustment.

The adjustment of the block by independent models begins with the model coordinates derived from the numerical relative orientations and formation of the stereomodels (Section 4.3.2). In the course of the block adjustment the individual models will be amalgamated into the single block and simultaneously transformed into the ground coordinate system. The individual stereomodels are thus the basic units of aerotriangulation by independent models.

As an introduction, we treat first the adjustment of planimetry only. This is a special case of the general block adjustment which is a problem in three dimensions. A purely planimetric adjustment is a plane problem, i.e. we are concerned only with XY coordinates.

5.2.1 Planimetric adjustment of a block

We require the XY coordinates of new points in the ground coordinate system. Given are the model coordinates xy of the relatively oriented and levelled individual models. The levelling need only be approximate, but must be good enough to ensure that the influence of the "tilted" model heights on the xy coordinates is less than the photogrammetric accuracy of the planimetric coordinates (Figure 5.2-1).

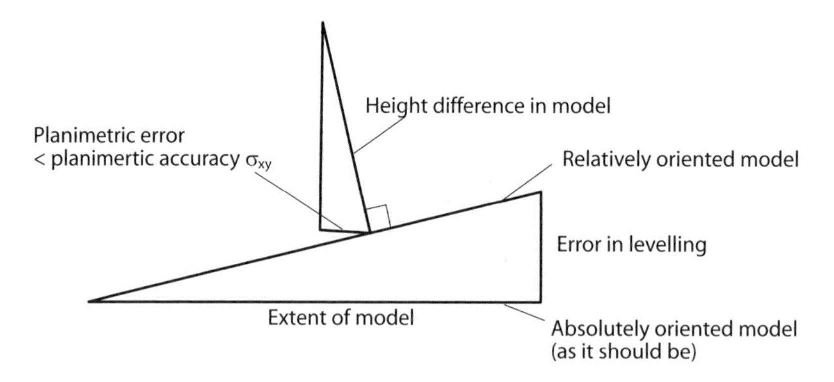

Figure 5.2-1: Effect of levelling error on planimetric coordinates

The models can be levelled, for example, by means of control points taken from relatively small-scale topographic maps. For this purpose, in addition to the actual aerotriangulation points, points in the stereomodels are chosen whose heights, or all three coordinates, can be measured approximately in a topographic map. The image coordinates of these points are also measured and used, after the relative orientation, to compute the corresponding model coordinates. The numerical absolute orientation can then be performed to level the model with these (approximate) control points.

Exercise 5.2-1. Wide-angle photographs are taken at a scale of 1 : 8000 with a 15/23 camera. A topographic map at a scale of 1 : 5000 is available, from which coordinates can be derived with an accuracy of $\pm 30\,\text{cm}$. How large may be the height differences ΔZ in a model if we use this map for the levelling of the models before a planimetric block adjustment? (Solution: $\sigma_{XY} = \pm 4.8\,\text{cm}$ (Equation (4.6-1)), $\Delta_{Z_{\text{max}}} \approx 80\,\text{m}$.)

The principle of a planimetric block adjustment is shown in Figure 5.2-2, while Table 5.2-1 contains the initial data for the block adjustment. The model coordinates are in separate, independent, local coordinate systems for each model. Each of these coordinate systems is displaced and rotated relative to the ground coordinate system and has an arbitrary scale.

As initial data for the amalgamation of the individual models into one block in the ground coordinate system we have on the one hand the tie points which exist in more than one model (points 5, 6, 7, 8 and 9) and, on the other hand, the model and ground coordinates of the field-surveyed control points (points 1, 2, 3 and 4).

The adjustment can therefore be defined as follows. The models are:

- displaced (two translations X_u, Y_u),
- rotated (rotation angle κ) and
- scaled (scale factor m)

so that:

- the tie points fit together as well as possible and
- the residual discrepancies at the control points are as small as possible.

The mathematical relation between a stereomodel (coordinates x, y) and the ground coordinate system (coordinates X, Y) can be formulated from Equations (2.1-9), (2.1-5) and (2.1-4) (plane similarity transformation):

$$\begin{pmatrix} X \\ Y \end{pmatrix} = \begin{pmatrix} X_u \\ Y_u \end{pmatrix} + m \begin{pmatrix} \cos\kappa & -\sin\kappa \\ \sin\kappa & \cos\kappa \end{pmatrix} \begin{pmatrix} x \\ y \end{pmatrix} \tag{5.2-1}$$

The non-linearities in the unknowns m and κ can be eliminated by substituting

$$\begin{aligned} m\cos\kappa &= a \\ m\sin\kappa &= b \end{aligned} \tag{5.2-2}$$

We obtain thus the linear equations

$$\begin{aligned} X &= X_u + x\,a - y\,b \\ Y &= Y_u + y\,a - x\,b \end{aligned} \tag{5.2-3}$$

The extension of the system of equations (5.2-3) to a block of models is sometimes called a chained plane similarity transformation. One possible formulation[1] of such a chained similarity transformation as a problem of adjustment of indirect observations is:

Observation equations for a control point:

$$\begin{aligned} v_x &= \underline{X}_u \qquad + x\,\underline{a} - y\,\underline{b} - X \\ v_y &= \qquad \underline{Y}_u + y\,\underline{a} + x\,\underline{b} - Y \end{aligned} \tag{5.2-4}$$

Observation equations for a tie point:

$$\begin{aligned} v_x &= \underline{X}_u \qquad + x\,\underline{a} - y\,\underline{b} - \underline{X} - 0 \\ v_y &= \qquad \underline{Y}_u + y\,\underline{a} + x\,\underline{b} - \underline{Y} - 0 \end{aligned} \tag{5.2-5}$$

The underlined terms are the unknowns. The observation equations (5.2-4) and (5.2-5) have an unusual form: although the corrections v_x and v_y derive from the inaccuracies in the model coordinates x, y, they are to be interpreted as corrections to the (known) ground coordinates X, Y (in (5.2-4)) and to the fictitious observations "0" (in (5.2-5)).

The balance of unknowns and observations for the example of Figure 5.2-2/Table 5.2-1 is:

Unknowns: $4 \times 4 = 16$ transformation elements X_u, Y_u, a, b
$5 \times 2 = 10$ tie-point coordinates X,Y
$\Longrightarrow 26$ total

Observations: $4 \times 4 \times 2 = 32$ model coordinates x,y
$\Longrightarrow 6$ redundant observations

The 32 observation equations for the 26 unknowns of the schematic example of Figure 5.2-2 are shown in Table 5.2-2.

After introducing the centroid coordinates $\bar{x}$ and $\bar{y}$, and assuming all observations have equal weight, the normal system of equations (4.1-1) shown in Table 5.2-3 is obtained. If we insert the numerical values of Table 5.2-1, we obtain the normal equations of Table 5.2-4. The introduction of coordinates referred to the centroid of each model has had the effect of reducing the submatrix of the transformation elements to a diagonal matrix.

[1] Kraus, K.: ZfV 91, pp. 123–130, 1966. Van den Hout, C.M.A.: Phia 21, pp. 171–178, 1966.

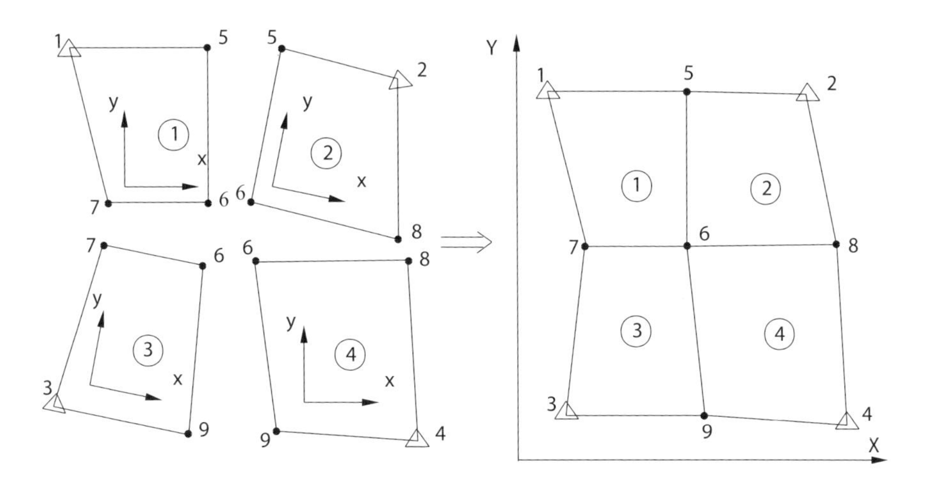

Figure 5.2-2: Planimetric block adjustment by independent models

Model 1			Model 2		
Pt.No.	x	y	Pt.No.	x	y
1	148.29	573.28	2	366.93	558.43
5	374.11	561.87	5	154.36	561.30
6	362.77	147.41	6	130.40	143.24
7	138.27	151.39	8	358.30	140.28

Model 3			Model 4		
Pt.No.	x	y	Pt.No.	x	y
3	148.59	139.40	4	359.38	135.30
6	362.10	542.71	6	140.31	578.42
7	159.40	556.85	8	359.34	549.19
9	345.67	128.76	9	141.97	149.87

Ground coordinates of control points in m		
Pt.No.	X	Y
1	4443.81	8338.54
2	7658.37	7993.67
3	4472.02	1071.18
4	8348.54	1316.60

Table 5.2-1: Initial data for the planimetric block adjustment of the block of Figure 5.2-2 (Model coordinates in [mm], ground coordinates of control points in [m])

Model		X_u^1	Y_u^1	X_u^2	Y_u^2	X_u^3	Y_u^3	X_u^4	Y_u^4	a^1	b^1	a^2	b^2	a^3	b^3	a^4	b^4	X_5	Y_5	X_6	Y_6	X_7	Y_7	X_8	Y_8	X_9	Y_9	1
Model 1	v_x	1								x_1^1	$-y_1^1$																	X_1
	v_y		1							y_1^1	x_1^1																	Y_1
	v_x	1								x_5^1	$-y_5^1$							-1										
	v_y		1							y_5^1	x_5^1								-1									
	v_x	1								x_6^1	$-y_6^1$									-1								
	v_y		1							y_6^1	x_6^1										-1							
	v_x	1								x_7^1	$-y_7^1$											-1						
	v_y		1							y_7^1	x_7^1												-1					
Model 2	v_x			1								x_5^2	$-y_5^2$					-1										
	v_y				1							y_5^2	x_5^2						-1									
	v_x			1								x_2^2	$-y_2^2$															X_2
	v_y				1							y_2^2	x_2^2															Y_2
	v_x			1								x_8^2	$-y_8^2$											-1				
	v_y				1							y_8^2	x_8^2												-1			
	v_x			1								x_6^2	$-y_6^2$							-1								
	v_y				1							y_6^2	x_6^2								-1							
Model 3	v_x					1								x_7^3	$-y_7^3$							-1						
	v_y						1							y_7^3	x_7^3								-1					
	v_x					1								x_6^3	$-y_6^3$					-1								
	v_y						1							y_6^3	x_6^3						-1							
	v_x					1								x_3^3	$-y_3^3$													X_3
	v_y						1							y_3^3	x_3^3													Y_3
	v_x					1								x_9^3	$-y_9^3$											-1		
	v_y						1							y_9^3	x_9^3												-1	
Model 4	v_x							1								x_6^4	$-y_6^4$			-1								
	v_y								1							y_6^4	x_6^4				-1							
	v_x							1								x_8^4	$-y_8^4$							-1				
	v_y								1							y_8^4	x_8^4								-1			
	v_x							1								x_9^4	$-y_9^4$									-1		
	v_y								1							y_9^4	x_9^4										-1	
	v_x							1								x_4^4	$-y_4^4$											X_4
	v_y								1							y_4^4	x_4^4											Y_4

Upper index = model number, lower index = point number

Table 5.2-2: Observation equations for Figure 5.2-2 and Table 5.2-1

X_u^i	Y_u^i	s	X_u^m	Y_u^m	s	a^i	b^i	s	a^m	b^m	X_j	Y_j	s	X_k	Y_k	
n^i											-1					$[X]$
	n^i											-1				$[Y]$
		$\ddots$									$\vdots$	$\vdots$	$\ddots$			$\vdots$
			n^m								-1		s	-1		$[X]$
				n^m								-1	s		-1	$[Y]$
					$\ddots$						$\vdots$	$\vdots$	s	$\vdots$	$\vdots$	$\vdots$
						$[(\bar{x}^i)^2 + (\bar{y}^i)^2]$					$-\bar{x}_j^i$	$-\bar{y}_j^i$				$[\bar{x}^i X + \bar{y}^i Y]$
							$[(\bar{x}^i)^2 + (\bar{y}^i)^2]$				$\bar{y}_j^i$	$-\bar{x}_j^i$				$[-\bar{y}^i X + \bar{x}^i Y]$
								$\ddots$			$\vdots$	$\vdots$	$\ddots$			$\vdots$
									$[(\bar{x}^m)^2 + (\bar{y}^m)^2]$		$-\bar{x}_j^m$	$-\bar{y}_j^m$	s	$-\bar{x}_k^m$	$-\bar{y}_k^m$	$[\bar{x}^m X + \bar{y}^m Y]$
										$[(\bar{x}^m)^2 + (\bar{y}^m)^2]$	$\bar{y}_j^m$	$-\bar{x}_j^m$	s	$\bar{y}_k^m$	$-\bar{x}_k^m$	$[-\bar{y}^m X + \bar{x}^m Y]$
											h_j					
												h_j				
							symmetrical						$\ddots$			
														h_k		
															h_k	

Upper index = model number, lower index = point number

n^i: Number of points in model i, h_j: Number of models in which the tie point P_j exists, m: number of models

Table 5.2-3: Normal equations of a planimetric block adjustment

X_u^1 Y_u^1 s X_u^4 Y_u^4 $a^1 b^1$ s a^4 b^4	X_5	Y_5	X_6	Y_6	X_7	Y_7	X_8	Y_8	X_9	Y_9	
4	−1		−1		−1						4443.81
4		−1		−1		−1					8338.54
⋱	−1		−1				−1				7658.37
		−1		−1				−1			7993.67
			−1		−1				−1		4472.02
				−1		−1				−1	1071.18
4			−1				−1		−1		8348.54
4				−1				−1		−1	8348.54
0.22575	−0.118	0.203	−0.107	−0.211	0.118	−0.207					1313.01
0.22575	−0.203	−0.118	0.211	−0.107	0.207	0.118					−1851.46
0.22365	0.098	0.210	0.122	−0.208			−0.106	−0.211			2535.99
0.22365	−0.210	0.098	0.208	0.122			0.211	−0.106			−675.32
0.21311			−0.108	0.201	0.095	0.215			−0.092	−0.213	−688.07
0.21311			−0.201	−0.108	−0.215	0.095			0.213	−0.092	792.87
0.22559			0.110	0.225			−0.109	0.196	0.108	−0.203	624.19
0.22559			−0.225	0.110			−0.196	−0.109	0.203	0.108	1962.83
	2										
		2									
			4								
				4							
symmetrical					2						
						2					
							2				
								2			
									2		
										2	

Note: the model and ground coordinates are shown in m. Although this adjustment must be made with a large number of significant digits, for reasons of space we show the coefficients here with a reduced number of significant digits.

Table 5.2-4: Normal equations for the example in Figure 5.2-2, derived from the data of Table 5.2-1

The system of normal equations (Table 5.2-3) of the planimetric adjustment of a block has a very special structure which can be expressed in matrix form as:

$$\begin{pmatrix} \mathbf{D}_1 & \mathbf{N} \\ \mathbf{N}^\top & \mathbf{D}_2 \end{pmatrix} \begin{pmatrix} \mathbf{x}_1 \\ \mathbf{x}_2 \end{pmatrix} = \begin{pmatrix} \mathbf{n}_1 \\ \mathbf{0} \end{pmatrix} \tag{5.2-6}$$

$\mathbf{x}_1$ ... unknown transformation parameters
$\mathbf{x}_2$ ... unknown tie-point coordinates
$\mathbf{D}_1, \mathbf{D}_2$ ... diagonal matrices
$\mathbf{N}^\top, \mathbf{N}$ ... sparse submatrices which give rise to correlation between the unknowns $\mathbf{x}_1$ and $\mathbf{x}_2$
$\mathbf{n}_1$ ... vector of the absolute terms of the unknowns $\mathbf{x}_1$ (the corresponding vector for the unknowns $\mathbf{x}_2$ is the null vector)

As a result of the diagonal matrices in the normal equations (5.2-6), it is particularly easy to eliminate the unknowns $\mathbf{x}_1$ or $\mathbf{x}_2$ before solving the equations by the Gaussian or the Cholesky algorithm.

a) normal equations after eliminating the transformation elements $\mathbf{x}_1$ in (5.2-6):

$$\left.\begin{aligned} \mathbf{D}_1\mathbf{x}_1 + \mathbf{N}\mathbf{x}_2 &= \mathbf{n}_1 \\ \mathbf{N}^\top\mathbf{x}_1 + \mathbf{D}_2\mathbf{x}_2 &= \mathbf{0} \end{aligned}\right\} \qquad \begin{aligned} &\mathbf{x}_1 = \mathbf{D}_1^{-1}(\mathbf{n}_1 - \mathbf{N}\mathbf{x}_2) \\ &\mathbf{N}^\top\mathbf{D}_1^{-1}\mathbf{n}_1 - \mathbf{N}^\top\mathbf{D}_1^{-1}\mathbf{N}\mathbf{x}_2 + \mathbf{D}_2\mathbf{x}_2 = \mathbf{0} \end{aligned} \tag{5.2-7}$$

$$\left(\mathbf{N}^\top\mathbf{D}_1^{-1}\mathbf{N} - \mathbf{D}_2\right)\mathbf{x}_2 = \mathbf{N}^\top\mathbf{D}_1^{-1}\mathbf{n}_1 \tag{5.2-8}$$

This method of reducing the normal equations is particularly suitable for block adjustments with a large number of models and a small number of tie points. This situation is typical in analytical photogrammetry where the operator chooses a small number of tie points.

b) normal equations after eliminating the tie-point coordinates $\mathbf{x}_2$ in (5.2-6):

$$\left.\begin{aligned} \mathbf{D}_1\mathbf{x}_1 + \mathbf{N}\mathbf{x}_2 &= \mathbf{n}_1 \\ \mathbf{N}^\top\mathbf{x}_1 + \mathbf{D}_2\mathbf{x}_2 &= \mathbf{0} \end{aligned}\right\} \qquad \begin{aligned} &\mathbf{x}_2 = -\mathbf{D}_2^{-1}(\mathbf{N}^\top\mathbf{x}_1) \\ &\mathbf{D}_1\mathbf{x}_1 - \mathbf{N}\mathbf{D}_2^{-1}\mathbf{N}^\top\mathbf{x}_1 = \mathbf{n}_1 \end{aligned} \tag{5.2-9}$$

$$\left(\mathbf{D}_1 - \mathbf{N}\mathbf{D}_2^{-1}\mathbf{N}^\top\right)\mathbf{x}_1 = \mathbf{n}_1 \tag{5.2-10}$$

This method of reducing the normal equations is particularly suitable for block adjustments with a small number of models and a large number of tie points. This situation is typical in digital photogrammetry where a large number of tie points are selected.

If new points occur only in one model (single points), we can write two observation equations (5.2-5) for each such point and compute the coordinates X, Y of the new points from Equation (5.2-8) or (5.2-9). Since these single points contribute nothing to the block adjustment by independent models however, it is better to eliminate them from the initial data and then transform them in a separate process by the transformation parameters $\mathbf{x}_1$ (Equation (5.2-7) or (5.2-10)) into the ground coordinate system.

In order to keep the computing effort and storage requirements for larger blocks within reasonable limits, computer programs[2] take advantage of the large number of zeros in the normal matrix when solving the normal equation (5.2-8) or (5.2-10). (This procedure is treated in more detail in Volume 2, Section B 5.2.2.)

Exercise 5.2-2.

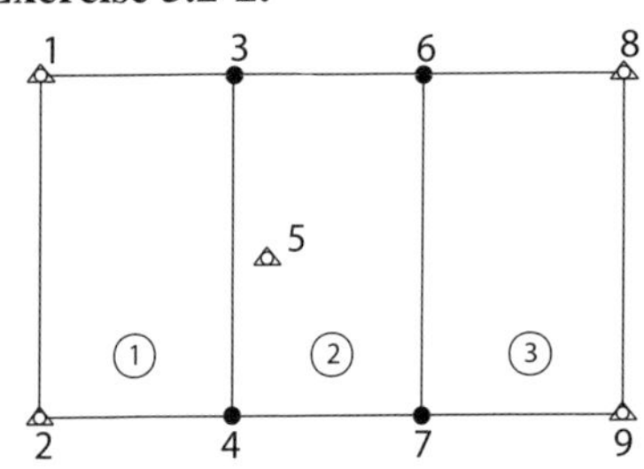

Given: The model coordinates of three models (in [mm]) and the ground coordinates (in [m]) of five control points. Required: Set up the observation equations according to Table 5.2-2 and perform the adjustment by indirect observations.

Model 1		
Pt.	x	y
1	697.46	572.52
2	698.64	463.63
3	808.18	571.81
4	808.50	475.26
Model 3		
Pt.	x	y
6	609.38	578.29
7	610.80	461.60
8	692.30	579.75
9	692.89	464.34

Model 2		
Pt.	x	y
3	686.90	542.66
4	687.32	447.69
5	759.46	494.88
6	758.68	543.16
7	760.64	440.05
Ground coordinates of control points		
Pt.	X	Y
1	1131.50	2331.50
2	1138.62	1142.22
5	3143.78	1782.21
8	3951.05	2332.99
9	3957.72	1201.05

Solution: For example $a_1 = 10.9228 \pm 0.0018$, $b_1 = -0.0542 \pm 0.0018$, $X_3 = 2341.04 \pm 0.25$ m, $Y_3 = 2317.74 \pm 0.25$ m.

5.2.2 Spatial block adjustment

In a spatial block adjustment we compute the XYZ coordinates of points in the ground coordinate system. As initial data we have the xyz model coordinates of points in the stereomodels formed in relative orientation. In addition to the model coordinates of the tie points and control points, we introduce the model coordinates of the perspective centres (Figure 5.2-3), derived from the numerical relative orientation and model formation (see solution to Exercise 4.3-3).

[2]For example: Ackermann, F., Ebner, H. and Klein, H.: BuL 38, pp. 218–224, 1970, and Ph.Eng. 39, p. 967, 1973.

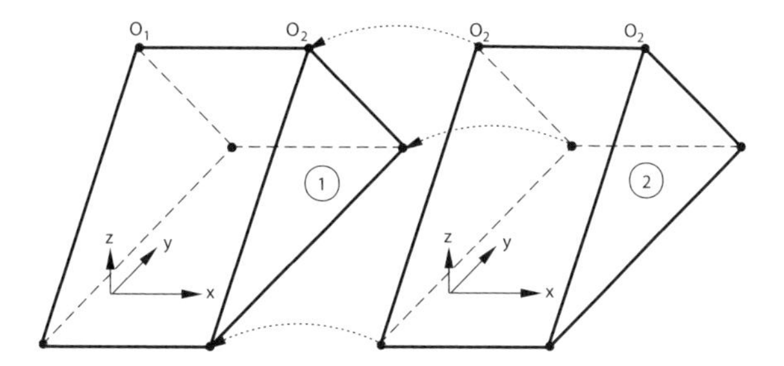

Figure 5.2-3: Model connection with perspective centres

The perspective centres stabilize the heights along the strip. A similar stabilization perpendicular to the strips is not possible, however, so that, as shown in Figure 5.2-4, chains of height control points perpendicular to the strips are required. A very good perpendicular stabilization of heights in the block could also be achieved by side overlaps between strips of about 60%.

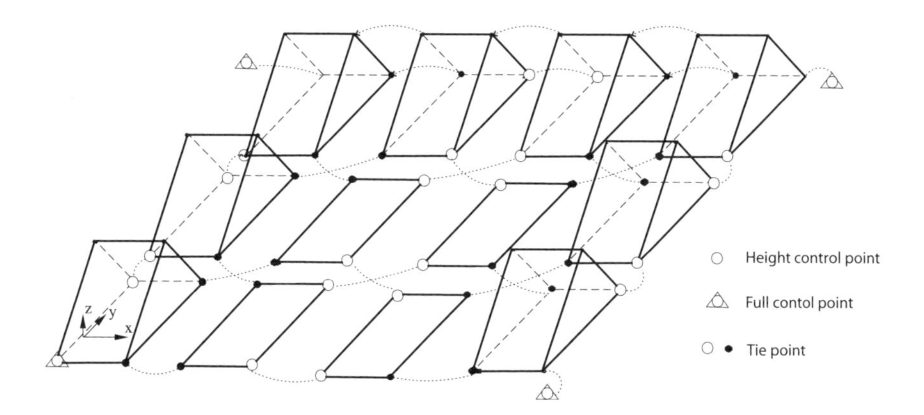

Figure 5.2-4: Spatial block adjustment by independent models

The principle of spatial block adjustment can be seen in Figures 5.2-3 and 5.2-4: the points in each model are defined in an independent, spatial coordinate system which can be transformed into the ground coordinate system by the seven elements of absolute orientation (Section 4.4). For the simultaneous absolute orientation of all models in the block, we have on the one hand the model coordinates of the tie points (including the perspective centres) and on the other the model and ground coordinates of the control points.

The principle of spatial block adjustment can therefore be defined as follows. The models are:

- displaced (three translations X_u, Y_u, Z_u),
- rotated (three rotations Ω, Φ, K) and
- scaled (scale factor m)

so that:

- the tie points (including the perspective centres) fit together as well as possible and
- the residual discrepancies at the control points are as small as possible.

The mathematical relation between the model and the ground coordinate system was defined in Equations (4.2-6) and is known as a spatial similarity transformation. The mathematical formalism of the simultaneous computation of the elements of absolute orientation of all models in the block is, by analogy with the 2D case, sometimes called a chained spatial similarity transformation.

The extension of spatial transformation (Equations (4.2-6)) to a chained spatial transformation is in principle exactly the same as the extension of a plane transformation (Equation (5.2-1)) to a chained plane transformation (Equations (5.2-4) and (5.2-5), as well as Table 5.2-2). In contrast to the plane transformation, however, a spatial transformation is non-linear. We have already seen a method of linearizing the similarity transformation of a single model in the system of Equations (4.4-5). The observation equations (4.4-6) for an adjustment by indirect observations have already been formulated. The observation equations (4.4-6) correspond to the observation equations (5.2-4) for a chained plane similarity transformation: the XYZ coordinates arising in the equations (4.4-6) correspond on the one hand to the (known) control point coordinates. On the other hand, for a new point the XYZ coordinates in equations (4.4-6) should be interpreted as (unknown) new point coordinates. Since such new points appear in several models they have a connecting function within the block of stereomodels. Just as the system of observation equations in Table 5.2-2 can be found using the observation equations (5.2-4) and (5.2-5), so the system of observation equations for the spatial block adjustment with independent models can be found from the correspondingly adapted observation equation (4.4-6). Due to space restrictions a detailed presentation will not be made.

In the system of observation equations for the spatial block adjustment with independent models, the unknowns are small additions to the approximate starting values. (In contrast, no linearization and no starting values were necessary for the system of equations for planimetric block adjustment.) Automatic generation of approximate starting values to initialize the linearization is presented in Volume 2, Section B 5.1.

5.2.3 Planimetric and height accuracy in block adjustment by independent models

The reduction in the number of field-surveyed control points—the objective of aerotriangulation—leads in general to a reduction in accuracy compared with the absolute orientation of every model on four field-surveyed points in the model corners. For the application of aerotriangulation it is therefore of prime importance to be able to estimate this reduction in accuracy.

In a spatial block adjustment, the planimetric accuracy is not affected by the accuracy of the model heights or by the layout of the height control points if the ground is relatively flat. Similarly, the height accuracy is independent of the accuracy of the model coordinates x, y and the layout of the planimetric control points. The planimetric and height accuracies are therefore treated separately. The results for planimetry are valid for planimetric as well as spatial block adjustment.

5.2.3.1 Planimetric accuracy

Since the XY coordinates of the tie points are computed from a least squares adjustment by indirect observations (Appendix 4.1-1), we can derive their accuracy—more correctly, the weight coefficients (cofactors) q_{XX} and q_{YY}—by inverting the normal equation matrix (Table 5.2-3). As a result of the symmetrical structure in X and Y, the weight coefficients q_{XX} and q_{YY} are identical; they are named q_{PP} below. We then have the accuracy $\sigma_{B,P}$ of the coordinates X and Y of a tie point after the block adjustment:

$$\sigma_{B,P} = \sqrt{q_{PP}}\,\sigma_0 = \sqrt{q_{PP}}\,\sigma_{M,P} \tag{5.2-11}$$

σ_0 is the standard error of unit weight of the adjustment, i.e. the accuracy σ_x or σ_y of a model coordinate x or y, expressed in the ground coordinate system (see the comment following Equation (5.2-5)). The quantity $\sqrt{q_{pp}}$ can therefore be regarded as a factor which, multiplied by the accuracy $\sigma_{M,P}$ of the XY coordinates in the individual models, gives the planimetric accuracy $\sigma_{B,P}$ of the block.

Figure 5.2-5[3] shows these factors $\sqrt{q_{pp}}$ for various sizes of blocks, each containing a control point in the four corners of the block. The quantity $\sigma_{\max}$ denotes the maximum planimetric error in the block, while σ_{mean} denotes the root mean square of the planimetric error of all tie points in the block, i.e.:

$$\sigma_{\text{mean}} = \sqrt{\frac{\sum_{i=1}^{n} q_{pp,i}}{n}}$$

Numerical Example. Given: Aerial photographs at a scale of 1 : 6000. Each model has 4 tie points which are as accurately defined as targeted points. The block is composed of 32 models (case d1 in Figure 5.2-5) and is adjusted by independent models. Required: $\sigma_{\max}$ of the adjusted coordinates and σ_{mean} for the entire block.

[3] These and the further diagrams of planimetric accuracy were published by Ackermann (BuL 35, pp. 114–122, 1967).

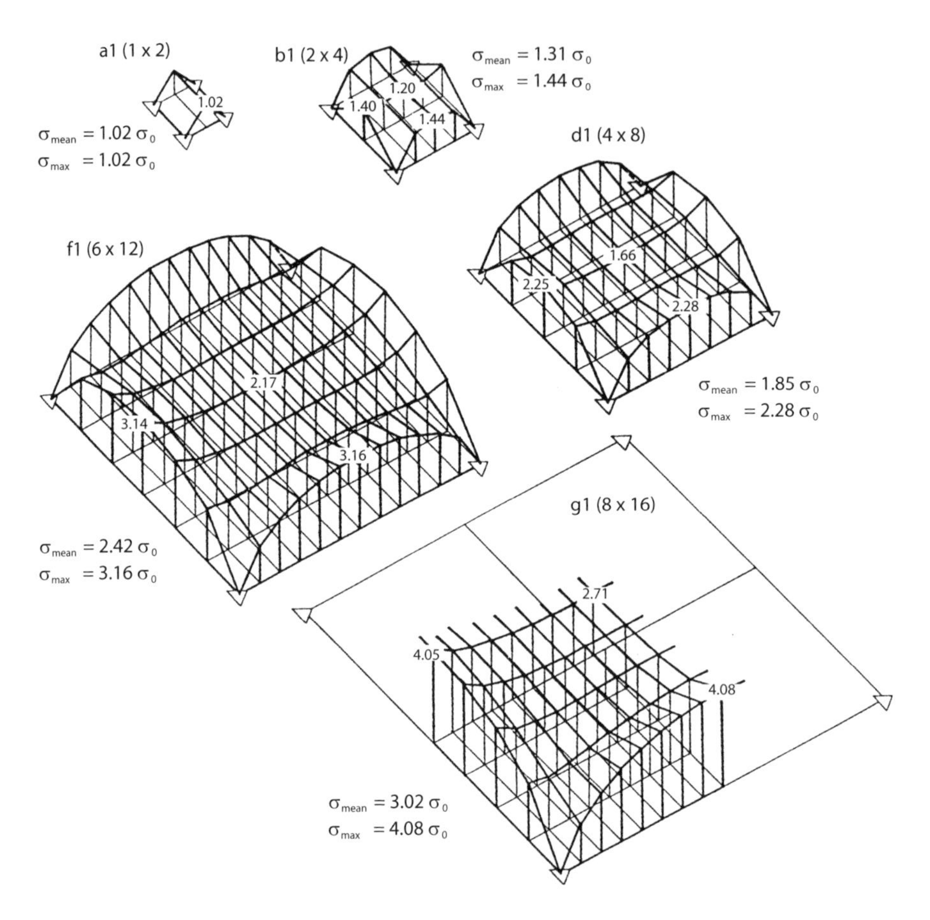

Figure 5.2-5: Planimetric accuracy with four points in the block corners, for square blocks $n_s \times n_m$ ($n_s \ldots$ number of strips, $n_m \ldots$ number of models in a strip)

Accuracy in a single model (Equation (4.6-1)):	$\sigma_{M,P}$	=	0.0006×6000
		=	$\pm 3.6\,\text{cm}$
Block accuracy (maximum error):	$\sigma_{B,P,\max}$	=	$\pm 2.28 \times 3.6$
		=	$\pm 8.2\,\text{cm}$
Block accuracy (r.m.s.):	$\sigma_{B,P,\text{mean}}$	=	$\pm 1.85 \times 3.6$
		=	$\pm 6.7\,\text{cm}$

We see from Figure 5.2-5 that:

- the accuracy falls significantly as the size of the block increases and
- the maximum error occurs in the middle of the block edges.

It is therefore obvious that if we wish to increase the overall accuracy we should pro-

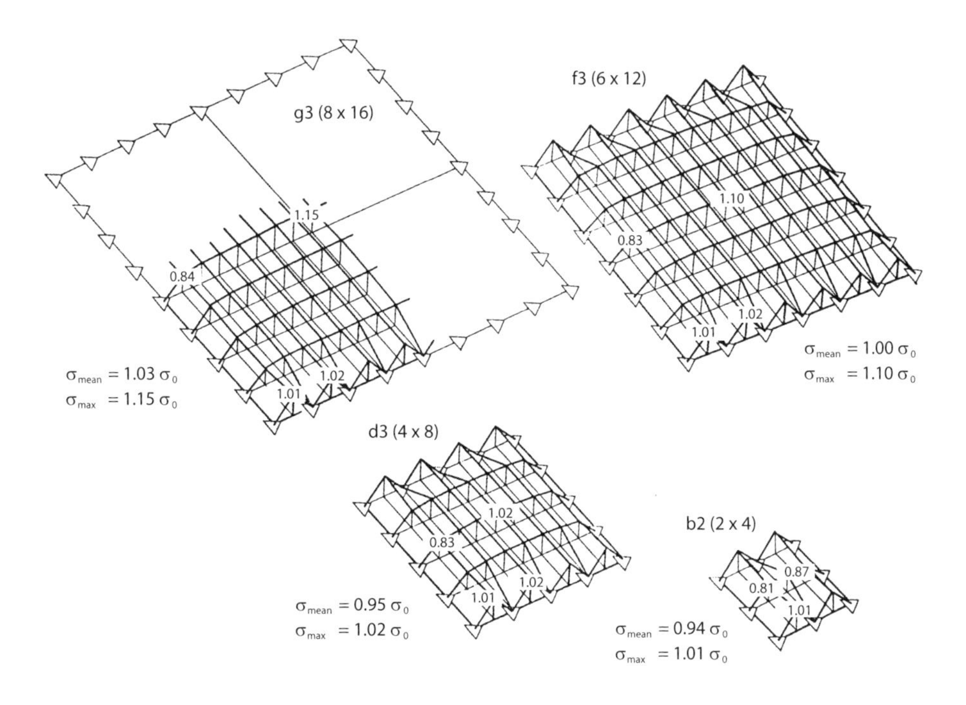

Figure 5.2-6: Planimetric accuracy with a dense pattern of control points along the block edges, for square blocks $n_s \times n_m$

vide a dense pattern of control points along the edges of the block. The success of such a strategy is shown in Figure 5.2-6.

The accuracy

- is almost independent of the size of the block and
- is close to the accuracy in a single model.

The question of further improvement in accuracy by introducing control points in the centre of the block is answered in Figure 5.2-7:

Comparison with Figure 5.2-6 shows that:

- control points inside the block bring no significant improvement in accuracy.

Some generalization of these accuracy results for square blocks has been published

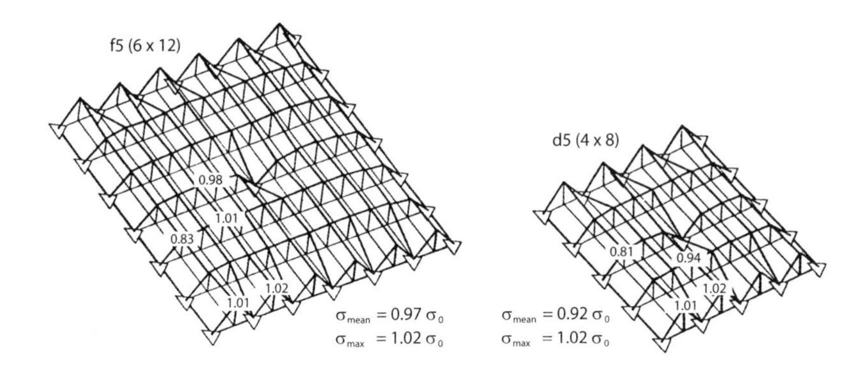

Figure 5.2-7: Planimetric accuracy with a dense pattern of control points along the edges of the block and a control point in the centre of the block, for square blocks $n_s \times n_m$

by Ackermann[4], Ebner[5] and Meissl[6] for the patterns of control points shown in Figure 5.2-8 ($P1$ corresponds to Figure 5.2-6, i.e. the number i of bridged models corresponds to two baselengths along the block edges; $P4$ corresponds to Figure 5.2-5) and summarized in Equations (5.2-12), for which six tie points per model, as opposed to four in the model corners, were assumed.

$$\begin{aligned} P1 &: \sigma_{B,P,\text{mean}} \approx (0.70 + 0.29 \log n_s)\sigma_{M,P} \\ P2 &: \sigma_{B,P,\text{mean}} \approx (0.83 + 0.02 n_s)\sigma_{M,P} \\ P3 &: \sigma_{B,P,\text{mean}} \approx (0.83 + 0.05 n_s)\sigma_{M,P} \\ P4 &: \sigma_{B,P,\text{mean}} \approx (0.47 + 0.25 n_s)\sigma_{M,P} \end{aligned} \tag{5.2-12}$$

$n_s \ldots$ number of strips in the block

Numerical Example (to verify Equations (5.2-12) for a square block with eight strips). Case 1 (block edges with dense control $\hat{=}$ g3 in Figure 5.2-6):

Equation 1 of (5.2-12): $\sigma_{B,P,\text{mean}} \approx (0.70 + 0.29 \log 8)\sigma_{M,P} \approx 0.96\sigma_{M,P}$
Figure 5.2-6: $\sigma_{B,P,\text{mean}} = 1.03\sigma_{M,P}$

Case 2 (four control points, in the corners of the block $\hat{=}$ g1 in Figure 5.2-5):

Equation 4 of (5.2-12): $\sigma_{B,P,\text{mean}} \approx (0.47 + 0.25 \times 8)\sigma_{M,P} \approx 2.47\sigma_{M,P}$
Figure 5.2-5: $\sigma_{B,P,\text{mean}} = 3.02\sigma_{M,P}$

[4] BuL 36, pp. 3–15, 1968.
[5] BuL 40, pp. 214–221, 1972.
[6] ÖZfV, pp. 61–65, 1972.

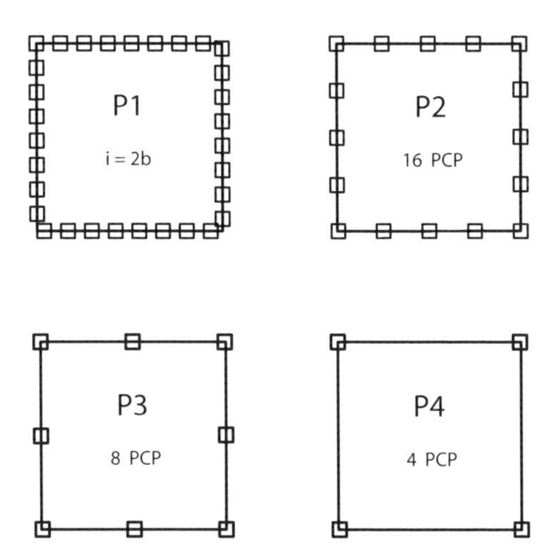

Figure 5.2-8: Pattern of planimetric control points for the Equations (5.2-12). Note that the block size is variable; for $P1$ the number of control points increases with the size of the block, for $P2-P4$ the number is constant (PCP = planimetric control point).

The difference between 0.96 and 1.03 and between 2.47 and 3.02 can be tolerated for rough estimates of block accuracy such as concern us here. The difference arises partly from the fact that Equations (5.2-12) are based upon the use of six tie points in each model while Figures 5.2-5 to 5.2-7 are based upon the use of four tie points per stereomodel.

Numerical Example (for project planning). Model control points are to be established by photogrammetry in an area of 10 km × 10 km, with a coordinate accuracy of ±10 cm. These control points will be natural points for which we assume an uncertainty of definition of ±5 cm. Eight field-surveyed control points exist in the block, four in the corners and four in the middle of the edges (P3 in Figure 5.2-8). What scale of photography must be adopted?

Trial 1, photo scale 1 : 15000:

Distance between strips for 25% side overlap (Equation (3.7-1)):	$A = 0.23 \times 15000(1 - 25/100) = 2590\,\text{m}$
Number of strips (Equation (3.7-1)):	$n_S = \lfloor 10000/2590 + 1 \rfloor = 4$
Accuracy in single model (Equation (4.6-3)):	$\sigma_{M,P} = \sqrt{9^2 + 5^2} = 10.3\,\text{cm}$
Block accuracy (Equation (5.2-12)):	$\sigma_{B,P} = (0.83 + 0.05 \times 4)10.3 = \pm 10.6\,\text{cm}$

Trial 2, photo scale 1 : 12000:

$$A = 0.23 \times 12000(1 - 25/100) = 2070\,\text{m}$$
$$n_S = \lfloor 10000/2070 + 1 \rfloor = 5$$
$$\sigma_{M,P} = \sqrt{7.2^2 + 5^2} = \pm 8.8\,\text{cm}$$
$$\sigma_{B,P} = (0.83 + 0.05 \times 5)8.8 = \pm 9.5\,\text{cm}$$

A photo scale of 1 : 12000 is therefore suitable for the task.

Exercise 5.2-3. Repeat this example of project planning with the assumption that 16 field-surveyed points are available (P2 in Figure 5.2-8). (Solution: Four strips at a photo scale of 1 : 15000 are sufficient.)

The rules for accuracy shown above are also valid, to a rough approximation, for rectangular blocks, but not for the extreme case of a single strip. Accuracy in a strip is therefore discussed separately in Section 5.2.3.4. The near-equality between block accuracy and single-model accuracy for blocks with dense control around the edges—the most important case in practice—also holds good if the edges of the block are irregular.

Up to this point, the accuracies quoted have all referred to the tie points in the corners of the models. Single points, i.e. points lying in one model only, will be less accurate. In the case d3 of Figure 5.2-6, the accuracy of single points is $1.33\sigma_{M,P}$[7], a decrease of about 33%, a value which can also be applied as a guide in blocks of other sizes and with other patterns of control. In the same publication, Ebner also studied the effect on accuracy of single points of using significantly more than four or six tie points in each model. For the same case d3 of Figure 5.2-6, 60 (!) tie points per model brought an improvement in accuracy of only about 25%. The number of tie points thus has little effect on the accuracy of the coordinates of single points after a block adjustment.

In the interest of blunder detection and blunder location, however, the statements made above need to be revised: one should use at least eight tie points in each model, two in each corner, and groups of at least two control points rather than single control points. (This aspect is treated in depth in Volume 2, Section B 7.2.2.4.)

Exercise 5.2-4. A cadastral survey is to be performed by photogrammetry in an area of 5 km × 5 km. The maximum difference between lengths derived from photogrammetrically determined coordinates and lengths measured on the ground (check lengths) is to be $\Delta s = 10\,\text{cm}$. Assume signalized points and compute the smallest photo scale together with the pattern of field-surveyed control points. Hint: A maximum error in length of $\Delta s = 10\,\text{cm}$ corresponds to a root mean-square planimetric accuracy of $\sigma_P = 10/(3\sqrt{2}) = \pm 2.4\,\text{cm}$ (division by 3: relation between maximum and root mean-square error; division by $\sqrt{2}$: Equation (4.6-4)). (Solution: Arrangement of control points as P1 in Figure 5.2-8, $1 : m_B = 1 : 4150$, 8 strips. Alternatively, arrangement of control points as P2 in Figure 5.2-8, $1 : m_B = 1 : 3960$, 9 strips. Another alternative is the arrangement of control points as P3 in Figure 5.2-8, $1 : m_B = 1 : 2790$, 12 strips.)

[7] Ebner, H.: Na. Ka. Verm. I, Heft 53, pp. 51–71, 1971.

5.2.3.2 Height accuracy

The accuracy of heights after a block adjustment can be derived from an inversion of the normal equation matrix, in the same way as for planimetric accuracy. The analogous relation to Equation (5.2-11) is then:

$$\sigma_{B,Z} = \sqrt{q_{ZZ}}\sigma_0 = \sqrt{q_{ZZ}}\sigma_{M,Z} \tag{5.2-13}$$

The height accuracy is primarily dependent on the number i of models between two chains of height control points (perpendicular to the strips). It is also good practice to improve the height accuracy along the upper and lower edges of the block by introducing height control points on these edges at intervals of $i/2$ models. The ideal pattern of height control is then as shown in Figure 5.2-9.

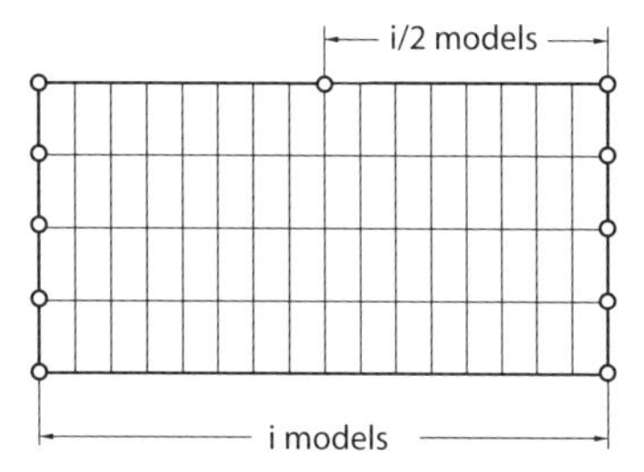

Figure 5.2-9: Ideal pattern of height control points

Figure 5.2-10[8] shows the relation between the accuracy of heights of corner points in the models and the number of models i bridged between control chains. It shows the root mean-square value for the entire block (σ_{mean}) and the maximum value σ_{max} in the most unfavourable position in the block. The corresponding equations assuming six tie points are:

$$\begin{aligned} \sigma_{B,Z,\text{mean}} &\approx (0.34 + 0.22i)\sigma_{M,Z} \\ \sigma_{B,Z,\text{max}} &\approx (0.27 + 0.31i)\sigma_{M,Z} \end{aligned} \tag{5.2-14}$$

[8]Ebner, H.: BuL 40, pp. 214–221, 1972 (Forster, B.C.: UNISURV G28, pp. 84–93, 1978, demonstrates a further generalization).

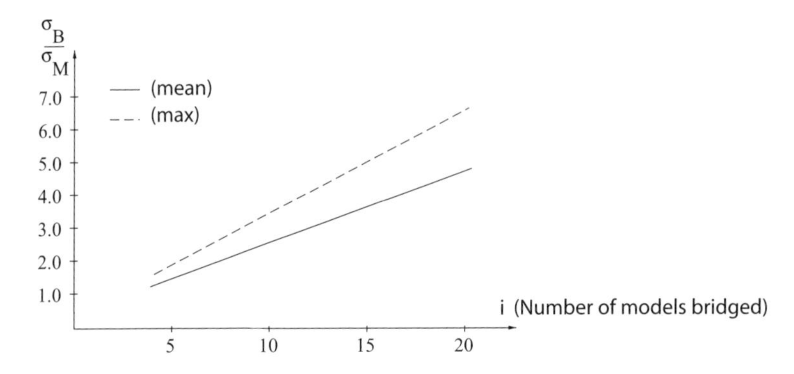

Figure 5.2-10: Height accuracy of a block adjustment (σ_B) compared with the accuracy in a single model (σ_M)

Numerical Example. Given: Aerial photographs at a scale of 1 : 6000; 15/23 camera; tie points to the accuracy of signalized points; adjustment of a 72-model block by independent models, with the control pattern shown below:

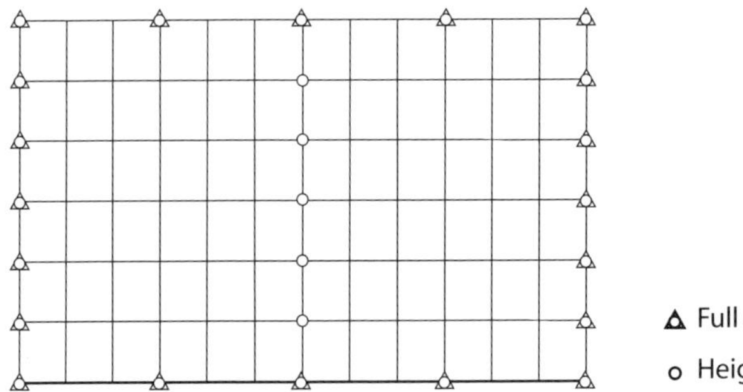

Required: σ_{max} of the adjusted heights and σ_{mean} for the height accuracy of the complete block.

Accuracy in a single model (Equation (4.6-1)):	$\sigma_{M,Z}$	$= \pm 5.4\,\text{cm}$
Block accuracy (maximum error):	$\sigma_{B,Z,\text{max}}$	$= \pm(0.27 + 0.31 \times 6)5.4 = \pm 11.5\,\text{cm}$
Block accuracy (representative value):	$\sigma_{B,Z,\text{mean}}$	$= \pm(0.34 + 0.22 \times 6)5.4 = \pm 9.0\,\text{cm}$

Exercise 5.2-5. What is the change in accuracy if four chains of height control points are used instead of three? (Solution: $\sigma_{B,Z,\text{max}} = \pm 8.2\,\text{cm}$, $\sigma_{B,Z,\text{mean}} = \pm 6.6\,\text{cm}$.)

We see that, when compared with planimetric accuracy, particularly with dense control along the edges of a block, height accuracy is significantly less favourable. An interval of about three models between chains of height control is the maximum if we wish to introduce no significant loss of accuracy compared with that of a single model—but we then affect adversely the economy of the entire process. As a compromise between accuracy and economy, an interval of at least four models between chains of control is usual. The introduction of GPS data[9] (Section 3.7.3.2) into the block adjustment frees the adjustment from this very restrictive pattern of height control (Section 5.4 and Section B 5.3.5.2.2, Volume 2).

In the interest of blunder detection and blunder location one should—as for planimetric adjustment (see the end of Section 5.2.3.1)—introduce at least eight tie points per model, two in each corner, and provide control points in groups rather than as single points (see Volume 2, Section B 7.2.2.4). In high mountain regions care should be taken that height control points do not only lie in the valleys or on mountain peaks; they should cover the full height range of the area.

5.2.3.3 Empirical planimetric and height accuracy

The theoretical accuracies of block triangulation by independent models shown above are based on the assumption of random errors in the individual models. The effects of varying accuracy within a model and the undoubted existence of correlations resulting from systematic errors are ignored.

Nevertheless, extensive empirical tests of accuracy made on field-surveyed check points[10] have confirmed the essential soundness of the theoretical results, so that we can use Equations (5.2-12) and (5.2-14) as rough approximations for project planning.

5.2.3.4 Planimetric and height accuracy of strip triangulation

The accuracy of points in a strip adjusted by independent models depends primarily on the number i of models bridged between control points. As before, we are interested in two values—a mean accuracy $\sigma_{S,\text{mean}}$ which is representative for the strip as a whole and a maximum value $\sigma_{S,\text{max}}$ of the error occurring between control points.

Since we must here also differentiate between signalized, natural and artificial points, it is convenient to relate the accuracy of strip triangulation, σ_S, to the error, σ_M, of single models introduced in Section 4.6.

The results of theoretical[11] and empirical[12] studies of accuracy of strip triangulation, which, in part, produce greatly different values, are sketched in Figure 5.2-11. The

[9]Hein, G.W., van der Vegt, H.J.W., Andersen, O., Colomina, I.: Schriftenreihe des Inst. für Photogrammetrie, Uni. Stuttgart, Heft 13, pp. 261–325, 1989. Lucas, J.R., Mader, G.L.: Journal of Surveying Engineering 115/1, pp. 78–92, 1989. Friess, P.: BuL 58, pp. 136–143, 1990.

[10]OEEPE Publ. No. 8, 1973.

[11]Ackermann, F.: DGK, C 87, 1965.

[12]Stark, E.: OEEPE, No. 8, pp. 49–82, 1973; BuL 45, pp. 183–190, 1977.

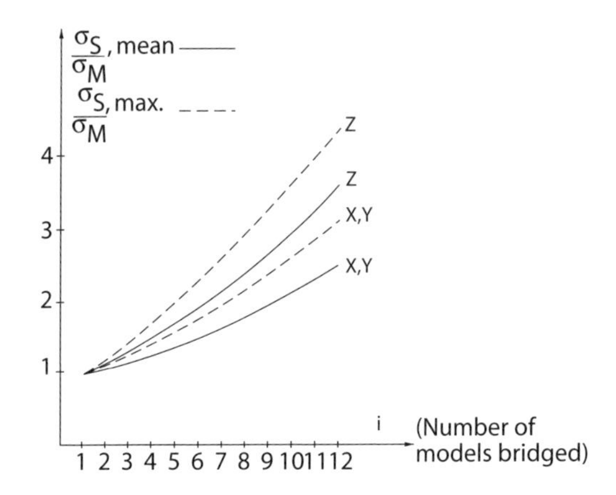

Figure 5.2-11: Accuracy of strip triangulation (σ_S) compared with the accuracy of single models (σ_M)

extreme simplification represented in this diagram has the practical advantage of simple interpretation; note, however, that differences of up to 50% from the values shown in Figure 5.2-11 can occur in individual cases.

Numerical Example (of project planning). Signalized points along a strip must be measured with a planimetric accuracy, σ_P, of ± 3 cm and a height accuracy, σ_Z, of ± 5 cm. Because of the high cost of control, as few field-surveyed control points as possible are to be used. A $15/23$ camera is to be used. The field-surveyed points are at the ends of the strip and in the middle.

Trial 1. Photo scale 1 : 4500, with the following pattern of control points.

Single-model accuracy (Equation (4.6-1)):

$$\begin{aligned} \sigma_{M,P} &= 0.0006 \times 4500 \\ &= \pm 2.7\,\text{cm} \\ \sigma_{M,Z} &= 0.00006 \times 15 \times 4500 \\ &= \pm 4.0\,\text{cm} \end{aligned}$$

Strip accuracy (Figure 5.2-11, $i = 4$):

$$\begin{aligned} \sigma_{S,P} &= \pm 2.7 \times 1.2 \\ &= \pm 3.2\,\text{cm} \\ \sigma_{S,Z} &= \pm 4.0 \times 1.5 \\ &= \pm 6.0\,\text{cm} \end{aligned}$$

Trial 2. Photo scale 1 : 3000, with the following pattern of control points.

Single-model accuracy:

$$\sigma_{M,P} = 1.8\,\text{cm}$$
$$\sigma_{M,Z} = \pm 2.7\,\text{cm}$$

Strip accuracy (Figure 5.2-11, $i = 6$):

$$\sigma_{S,P} = \pm 1.8 \times 1.4 = \pm 2.5\,\text{cm}$$
$$\sigma_{S,Z} = \pm 2.7 \times 1.9 = \pm 5.1\,\text{cm}$$

Strip accuracy in the models central between the control points:

$$\sigma_{S,P} = \pm 1.8 \times 1.7 = \pm 3.1\,\text{cm}$$
$$\sigma_{S,Z} = \pm 2.7 \times 2.2 = \pm 5.9\,\text{cm}$$

The requirements for the strip as a whole are met, while the errors in the models between the control points are only just higher than allowed.

This numerical example shows that the reduction in accuracy caused by increasing the number of bridged models is significantly less than the improvement in accuracy resulting from the increase in photo scale.

5.3 Bundle block adjustment

In a bundle adjustment of a strip or block of photographs, with at least 60% forward overlap and 20% side overlap, we compute directly the relations between image coordinates and object coordinates, without introducing model coordinates as an intermediate step. Thus, the photograph is the elementary unit in a bundle adjustment.

5.3.1 Basic principle

Figure 5.3-1 shows the basic principle of a bundle block adjustment.

The image coordinates and the associated perspective centre of a photograph define a spatial bundle of rays. The elements of exterior orientation of all bundles in a block are computed simultaneously for all photographs. The initial data consist of the image coordinates of the tie points (points existing in more than one photograph) together with the image coordinates and object coordinates of the control points.

The adjustment principle can therefore be defined as follows: the bundles of rays are:

- displaced (three translations X_0, Y_0, Z_0) and
- rotated (three rotations ω, φ, κ)

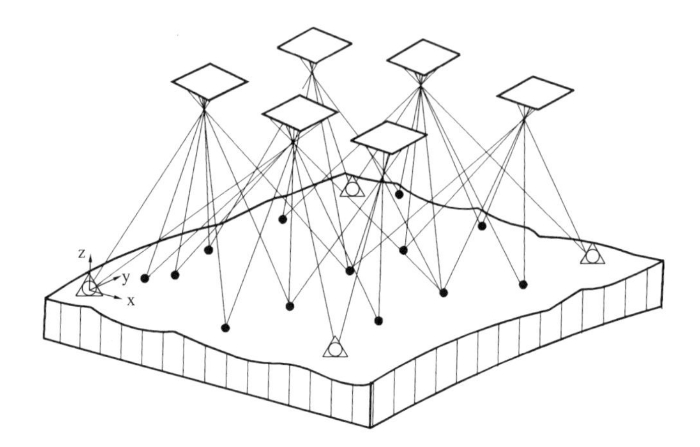

Figure 5.3-1: Principle of a bundle block adjustment

so that the rays

- intersect corresponding rays as nearly as possible at the tie points and
- pass through the control points as nearly as possible.

The mathematical relationship between the image coordinates and the global ground or object coordinate system can be found in Appendix 2.1-2. The differential quotients for the corresponding collinearity equations can be found in Appendix 2.1-3.

5.3.2 Observation and normal equations for a block of photographs

We can now use the differential quotients in Appendix 2.1-3 to write down the linearized observation equations of a least squares adjustment by indirect observations (Appendix 4.1-1), for (new) points P_i whose image coordinates have been measured in the photograph with the index j. Each measured image point yields two observation equations.

$$\begin{aligned} v_{\xi ij} = & \left(\frac{\partial \xi}{\partial X_{0j}}\right)^0 dX_{0j} + \left(\frac{\partial \xi}{\partial Y_{0j}}\right)^0 dY_{0j} + \left(\frac{\partial \xi}{\partial Z_{0j}}\right)^0 dZ_{0j} \\ & + \left(\frac{\partial \xi}{\partial \omega_j}\right)^0 d\omega_j + \left(\frac{\partial \xi}{\partial \varphi_j}\right)^0 d\varphi_j + \left(\frac{\partial \xi}{\partial \kappa_j}\right)^0 d\kappa_j \\ & + \left(\frac{\partial \xi}{\partial X_i}\right)^0 dX_i + \left(\frac{\partial \xi}{\partial Y_i}\right)^0 dY_i + \left(\frac{\partial \xi}{\partial Z_i}\right)^0 dZ_i \\ & - (\bar{\xi}_{ij} - \xi_{ij}^0) \end{aligned} \tag{5.3-1}$$

while the expression for $v_{\eta ij}$ has a similar form. The unknowns are the six elements of exterior orientation of the photograph with the index j and the three ground coordinates of point P_i. If a ray passes through a fixed point, the terms dX_i, dY_i and dZ_i disappear.

The differential quotients $(\,)^0$ are calculated from approximate values of the unknowns by the relations in Appendix 2.1-3. ξ^0_{ij} and η^0_{ij} are image coordinates computed from Equations (2.1-19) with the help of the approximations to the unknowns. $\bar{\xi}_{ij}$ and $\bar{\eta}_{ij}$ are the measured image coordinates. The necessary approximations to the unknowns can be derived in various ways. For example, for near-vertical photographs $\omega^0 = \varphi^0 = 0$; κ^0 is known from the flight plan. A block adjustment by independent models (Section 5.2) yields the coordinates $X^0Y^0Z^0$ of the perspective centres and the coordinates XYZ of the new points (this problem is taken up again in Volume 2, Section B 5.2.1).

A schematic example with four photographs will help to clarify the procedure. The balance of observations and unknowns is:

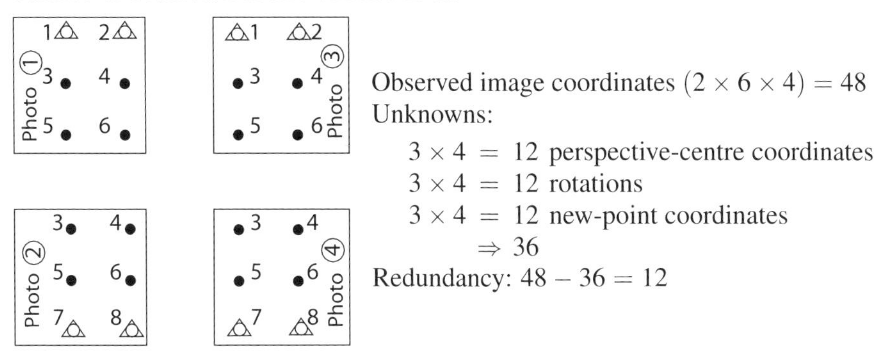

Observation equations

In place of the differential quotients $(\,)^0$ in observation equations (5.3-1), the following table uses variables a and b which are introduced in Appendix 2.1-3.

	X_{01}	Y_{01}	Z_{01}	ω_1	φ_1	κ_1	X_{02}	Y_{02}	Z_{02}	ω_2	φ_2	κ_2	
$v_{\xi 11}$	a_2^0	a_3^0	a_4^0	a_5^0	a_6^0	a_7^0							
$v_{\eta 11}$	b_2^0	b_3^0	b_4^0	b_5^0	b_6^0	b_7^0							
							Pt. 2 similarly						
$v_{\xi 31}$	a_2^0	a_3^0	a_4^0	a_5^0	a_6^0	a_7^0							
$v_{\eta 31}$	b_2^0	b_3^0	b_4^0	b_5^0	b_6^0	b_7^0							
							Pts. 4,5,6 similarly						$\dots$
$v_{\xi 12}$							a_2^0	a_3^0	a_4^0	a_5^0	a_6^0	a_7^0	
$v_{\eta 12}$							b_2^0	b_3^0	b_4^0	b_5^0	b_6^0	b_7^0	
							Pt. 2 similarly						
$v_{\xi 32}$							a_2^0	a_3^0	a_4^0	a_5^0	a_6^0	a_7^0	
$v_{\eta 32}$							b_2^0	b_3^0	b_4^0	b_5^0	b_6^0	b_7^0	
							Pts. 4,5,6 similarly						

	X_{04}	Y_{04}	Z_{04}	ω_4	φ_4	κ_4	X_3	Y_3	Z_3	$\dots$	X_6	Y_6	Z_6	l
$v_{\xi 11}$														$(\bar{\xi}_{11} - \xi_{11}^0)$
$v_{\eta 11}$														$(\bar{\eta}_{11} - \eta_{11}^0)$
$v_{\xi 31}$							a_8^0	a_9^0	a_{10}^0					$(\bar{\xi}_{31} - \xi_{31}^0)$
$v_{\eta 31}$							b_8^0	b_9^0	b_{10}^0					$(\bar{\eta}_{31} - \eta_{31}^0)$
$\dots$														
$v_{\xi 12}$														$(\bar{\xi}_{12} - \xi_{12}^0)$
$v_{\eta 12}$														$(\bar{\eta}_{12} - \eta_{12}^0)$
$v_{\xi 32}$							a_8^0	a_9^0	a_{10}^0					$(\bar{\xi}_{32} - \xi_{32}^0)$
$v_{\eta 32}$							b_8^0	b_9^0	b_{10}^0					$(\bar{\eta}_{32} - \eta_{32}^0)$

In matrix notation:

$$\mathbf{v} = \mathbf{A}\mathbf{x} - \mathbf{l} \tag{5.3-2}$$

Normal equations:

$$\mathbf{A}^\top \mathbf{A}\mathbf{x} = \mathbf{A}^\top \mathbf{l} \tag{5.3-3}$$

If we replace $\mathbf{A}^\top \mathbf{A}$ by $\mathbf{N}$ and $\mathbf{A}^\top \mathbf{l}$ by $\mathbf{n}$ the normal equations become:

$$\mathbf{N}\mathbf{x} = \mathbf{n} \tag{5.3-4}$$

5.3.3 Solution of the normal equations

The normal equations (5.3-4) of the schematic example given in Section 5.3.2 have the structure:

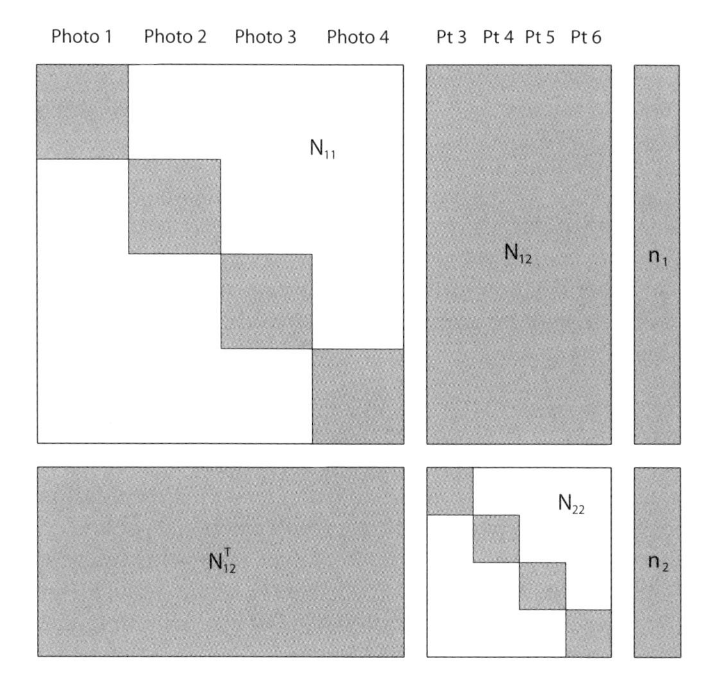

In matrix notation:

$$\begin{pmatrix} \mathbf{N}_{11} & \mathbf{N}_{12} \\ \mathbf{N}_{12}^{\mathsf{T}} & \mathbf{N}_{22} \end{pmatrix} \begin{pmatrix} \mathbf{x}_1 \\ \mathbf{x}_2 \end{pmatrix} = \begin{pmatrix} \mathbf{n}_1 \\ \mathbf{n}_2 \end{pmatrix} \tag{5.3-5}$$

The upper left matrix $\mathbf{N}_{11}$ is a hyperdiagonal matrix with submatrices each of 6×6 elements. The lower right matrix $\mathbf{N}_{22}$ is similarly a hyperdiagonal matrix with submatrices of 3×3 elements. The inversion of the two hyperdiagonal matrices is particularly simple: each submatrix can be inverted independently.

The system of normal equations reduced by the new-point coordinates $\mathbf{x}_2$ can thus be formed with no great computing effort. By analogy with Equations (5.2-9) and (5.2-10) we have from Equation (5.3-5):

$$(\mathbf{N}_{11} - \mathbf{N}_{12}\mathbf{N}_{22}^{-1}\mathbf{N}_{12}^{\mathsf{T}})\mathbf{x}_1 = \mathbf{n}_1 - \mathbf{N}_{12}\mathbf{N}_{22}^{-1}\mathbf{n}_2 \tag{5.3-6}$$

An adjustment yields corrections to the approximate values of the elements of exterior orientation of each photograph and to the approximate coordinates of the new points. If the approximations are very poor, the corrected values must be treated as new approximations. The adjustment should be repeated until there is no further significant change in the unknowns. (Further details on the solution of the normal equations will be found in Volume 2, Section B 4.4.)

Further treatment of bundle block adjustment will be found in: Wong, K.W., Elphingstone, G.M.: Ph.Eng. 39, pp. 267–274, 1973. ASPRS-Manual of Photogrammetry, 5th ed., Falls Church, 2004. Triggs, B., McLauchlan, P., Hartley, R., Fitzgibbon, A.: Bundle Adjustment—A Modern Synthesis. Vision Algorithms'99, LNCS 1883, pp. 298–372, 2000.

5.3.4 Unknowns of interior orientation and additional parameters

If amateur cameras are used instead of photogrammetric cameras, the elements of interior orientation can also be determined within the bundle block adjustment. For this purpose, the elements of interior orientation are introduced as unknowns in the observation equations (5.3-1). The solution of the normal equations (5.3-3) then yields the elements of interior orientation of the particular camera used. The larger number of unknowns in such a bundle block adjustment obviously requires a larger number of control points and tie points.

It is typical of photographs from amateur cameras that the theoretical central projection is significantly deformed by lens and film distortion. These influences can be taken into account in a bundle block adjustment by introducing correction polynomials in the observation equations, whose coefficients are determined in the adjustment. Such an adjustment is called a bundle block adjustment with additional parameters or with self-calibration. The technique is not only used for amateur-camera photographs, but more and more for photographs taken with proper photogrammetric cameras. Additional parameters provide an extremely powerful method of compensating for systematic errors. It is of no importance whether these systematic errors occur as a result of lens or film distortions or as a result of anomalies of refraction etc. This concept is further discussed in Volume 2, Sections B 3.5.6 and B 5.2.4.

Further reading: Jacobsen, K.: Phia, pp. 219–235, 1982. Kilpelä, E, Helkilä, J, Inkelä, K.: Phia, pp. 1–12, 1982.

5.3.5 Accuracy, advantages and disadvantages of bundle block adjustment

Although the mathematical model of a bundle block adjustment differs significantly from that of block adjustment by independent models—central projection with the image coordinates as observations on the one hand and spatial similarity transformation with the model coordinates as observations on the other hand—the rules of accuracy for independent-model adjustment (Section 5.2.3) can be applied, more or less, to bundle block adjustments.

The "single-model accuracy", which is required by the rules in Section 5.2.3, however, can be based on the following accuracy guides. These apply to targeted and other correspondingly accurate points in a regular block with 60% forward overlap and 20% side overlap, processed by a bundle block adjustment with additional parameters which reduce systematic errors. The values are 50% better than the guide values in Section 4.6.

In the case of analogue images, they are:

$$
\begin{array}{lll}
\text{Planimetry: } \sigma_{XY(\text{targ})} & = \pm 3\,\mu\text{m} & \text{in the photograph} \times \text{image scale } m_B \\
\text{Height: } \quad \sigma_{Z(\text{targ})} & = \pm 0.03‰ & \text{of camera distance (NA - WA)} \\
 & = \pm 0.04‰ & \text{of camera distance (SWA)}
\end{array} \tag{5.3-7}
$$

Exercise 5.3-1. Repeat the numerical examples of Section 5.2.3 under the assumption that the block triangulation has been performed by the bundle method with additional parameters rather than by independent models. (Solution: Numerical example according to Figure 5.2-5: $\sigma_{M,P} = \pm 1.8$ cm, $\sigma_{B,P,\text{mean}} = \pm 3.3$ cm; numerical example according to Figure 5.2-6: $\sigma_{M,Z} = \pm 2.7$ cm, $\sigma_{B,Z,\text{mean}} = \pm 4.5$ cm.)

Extensive details concerning the accuracy of the highly developed bundle block adjustment can be found, amongst other, in the following literature: Schwidefsky/Ackermann: Photogrammetrie. Teubner-Verlag, Stuttgart, 1976. Brown, D.C.: Ph.Eng. 43, p. 447, 1977.

At the end of this section we list the advantages and disadvantages of bundle block adjustment relative to independent-model adjustment:

Disadvantages:

- non-linear problem, for which approximations can only be established after lengthy procedures
- computer-intensive methods
- always a spatial problem, so that separate adjustments in planimetry and height are not possible

Advantages:

- most accurate method of aerotriangulation (direct relation between image and ground coordinates without the intermediate step of model formation)
- simple possibility of extending the technique to compensate for systematic errors (see Volume 2, Sections B 3.5.6 and B 5.2.4)
- simple possibility of incorporating observed elements of exterior orientation into the adjustment (see Section 5.4)
- possibility of using unconventional camera dispositions and amateur-camera photographs such as are often necessary in close range photogrammetry (Section 5.3.4)
- possibility of deriving the elements of exterior orientation to be set in particular analogue or analytical plotters. The photogrammetric measurement of control points for subsequent stereoprocessing is then unnecessary.

5.4 GPS- and IMU-assisted aerotriangulation

Aerotriangulation, as discussed in the previous paragraphs, is intended to determine the exterior orientation elements of the metric images. With GPS and IMU, which are increasingly used (Section 3.7.3.2), the exterior orientation elements can be directly determined and the images then processed using the methods of direct georeferencing (Section 4.1.1). In principle, GPS and IMU make aerotriangulation redundant. In practice, however, the following factors argue against direct georeferencing which employs only GPS and IMU information and makes no use of control points:

- GPS positioning "on the fly" is relatively demanding and cannot, at present, reach the centimetre accuracy level which is the objective of precision aerial photogrammetry.
- unavoidable cycle slips and/or multipath effects can disturb continuous GPS positioning.
- the combined calibration of the three units, GPS, IMU and imaging sensor, is a very complex exercise in which there are often considerable shortcomings (Section 3.7.3.2).
- GPS values, which are determined in a global coordinate system, must be corrected for various influences, such as the geoid, in order to correspond to the current ground (national) coordinate system in which the photogrammetric results are expected. In other words, GPS/IMU data often have datum problems with respect to the ground coordinate system.

Instead of direct georeferencing by GPS and IMU, which is known as direct sensor orientation, GPS and IMU information should be used only to support aerotriangulation. The combination of GPS and IMU data with control points for orienting image recording systems is known as integrated sensor orientation.

The following discussion concentrates on GPS-assisted aerotriangulation. The perspective centres of the metric images can, as suggested in Section 3.7.3.2 (note in particular Figure 3.7-12), be determined in the global GPS coordinate system with the aid of GPS and IMU recording. Due to the previously indicated datum problems, amongst others, the GPS coordinates X_{0j}, Y_{0j} and Z_{0j} of the perspective centres cannot be introduced as known parameters (i.e. constants) in the adjustment equations (5.3-1). (They would be better treated as "observed unknowns", see Section B 3.5.7, Volume 2.) Instead, all perspective centres fixed by GPS should be grouped into a single model in a three-dimensional coordinate system, the GPS model, and processed in a hybrid block adjustment. Here the bundles of rays from each metric image (bundle block adjustment, Section 5.3) and the GPS model just mentioned (spatial block adjustment with independent models, Section 5.2.2) are processed together. The ground coordinates of the perspective centres are regarded as unknowns in this hybrid block adjustment; the unknown perspective centres connect the ray bundles of the individual metric images and the GPS model. Each perspective centre with its three GPS coordinates provides, in addition to the adjustment equations of the bundle block adjustment, three further adjustment equations. Within the hybrid block adjustment the GPS model is subject to three translations, three rotations and a scale change. These seven unknowns principally eliminate the above mentioned datum problem between GPS data and ground coordinates.

The other errors in the GPS data, mentioned above, are largely eliminated by a division into several GPS models. In particular, a GPS model for every image strip is very suitable.

The GPS models support the perspective centres inside the photogrammetric blocks. GPS-assisted aerotriangulation therefore avoids the need for height control point chains (Section 5.2.3.2). Further details about GPS-assisted aerotriangulation and its accuracy can be found in Volume 2, Section B 5.3.5.

Where IMU data are available the process is similar. Thus, the IMU orientation angles ω_j, φ_j and κ_j should not be introduced as known parameters (constants) in the adjustment equations (5.3-1). Instead, these orientation angles should be left as unknowns in every metric image and an "orientation angle model", the IMU model, introduced into a hybrid block adjustment. This IMU model should have some unknown parameters, by analogy with the GPS model. A division into independent IMU models for individual strips is, in many cases, to be recommended.

5.5 Georeferencing of measurements made with a 3-line camera

Aerial survey flights using 3-line cameras are only made in combination with GPS and IMU (Section 3.7.2.3). GPS determines the flight path, and IMU the angular orientation, of the 3-line camera. With this GPS/IMU information, the object coordinates of new points identified in the three strips can be simply determined (Section 4.1.2). This direct solution, or direct sensor orientation, will be expanded on in the following sections. For instructive reasons, the GPS/IMU information will initially be disregarded.

In a 3-line camera (Figure 3.3-3, left), each line corresponds to a central projection with its own individual elements of exterior orientation. The six elements of exterior orientation change dynamically and are a function of time or line index. These six orientation functions are illustrated in Figure 5.5-1. Normally they are described by cubic polynomials between nodal points N_i (cubic spline interpolation). The separation of the nodes should be chosen such that the dynamically varying orientation elements are approximated to a sufficient accuracy by the successive cubic polynomials.

The abscissae t_i of the nodes N_i,which are often regularly spaced, are known; at each node the ordinates of the six orientation functions $X_0(t_i)$, $Y_0(t_i)$, $Z_0(t_i)$, $\omega(t_i)$, $\varphi(t_i)$, $\kappa(t_i)$[13] are required. Interpolation is implemented between the nodes N_i by means of cubic polynomials whose coefficients can be determined from the nodes. (Details on

[13]Figure 5.5-1 implies that the $\omega\varphi\kappa$ rotations of the sensor take place about the three coordinate axes of the aircraft. In Figure 3.7-13 this coordinate system has been designated as the body coordinate system.

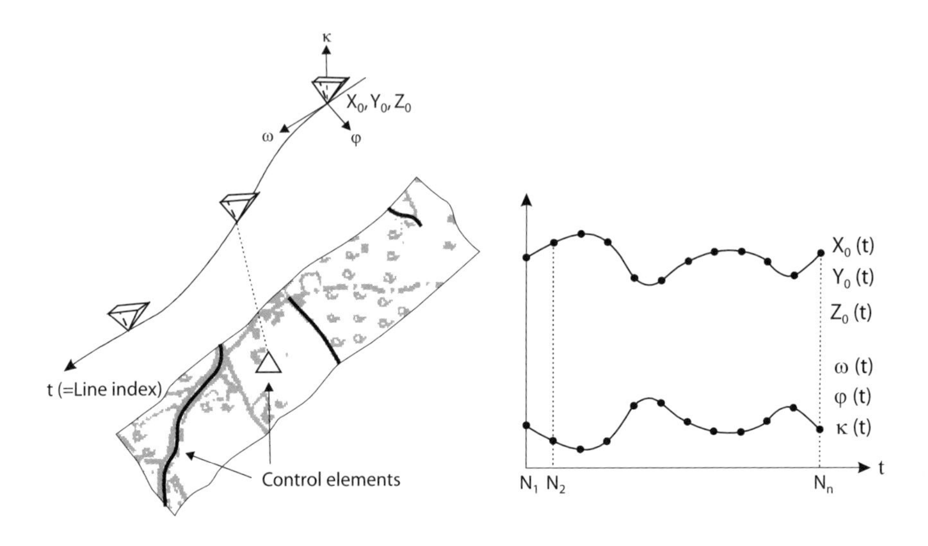

Figure 5.5-1: Flight path of a line sensor and 6 orientation functions

cubic spline interpolation can be found in Lancaster, P., Salkauskas, K.: Curve and Surface Fitting. Academic Press Ltd., 1986.)

The obvious question now arises as to how a point P with coordinates $\mathbf{x}_b = (x_b, y_b, z_b)^\top$ in the $x_b y_b z_b$ body coordinate system can be transformed into the XYZ object coordinate system. The following rotational axes are available for this exercise:

- primary rotation ω (roll) about the x_b axis until the y_b axis is parallel to the XY plane.
- secondary rotation φ (pitch) about the y_b axis (already rotated by ω) until the x_b axis also lies parallel to the XY plane.
- tertiary rotation κ (yaw/heading) about the z_b axis (already rotated by ω and φ) until the x_b and y_b axes are parallel to the X and Y coordinate axes.

Using the rotational and axial directions introduced in Figure 2.1-5, the following result is obtained by analogy with the presentation of Appendix 2.1-1:

$$\mathbf{X} = \mathbf{R}'_{\omega\varphi\kappa}\mathbf{x}_b \tag{5.5-1}$$

The rotation matrix $\mathbf{R}_{\omega\varphi\kappa}$ introduced in Section 2.1.2 needs therefore only to be transposed. (This result is not surprising given the knowledge from Section 2.1.1 about reversal of transformations involving rotation matrices.) From the numerical form of a given rotation matrix, the roll, pitch and yaw angles of an aircraft (Section 3.7.1) can be determined with the help of the transposed algebraic form of the rotation matrix (2.1-13) in the usual way.

Exercise 5.5-1. The rotation angles of a body coordinate system are to be determined from the matrix $\mathbf{R}_2$ given in the numerical example of Section 2.1.2: Solution, making use of the formulations (2.1-1-8) in Appendix 2.1-1: ω (roll angle) $= -1.0886\,\text{gon}\,(-58'47'')$, φ (pitch angle) $= 0.1499\,\text{gon}\,(8'6'')$, κ (yaw angle/heading) $= 101.3224\,\text{gon}\,(91^\circ 11'25'')$.

Further reading: Bäumker, M., Heimes, F.: OEEPE-Publication No. 43, pp. 197–212, 2002.

Before discussing the determination of the ordinates of the six orientation functions at all nodes, a particular feature of the 3-line camera (Figure 3.3-3, right) should be emphasized. If the three strips in the triplet are aligned along the same time index, then the lines arranged in vertical rows in Figure 5.5-2 have the same orientation elements; in other words: for all strips in a triplet a single set of six orientation functions $X_0(t_i)$, $Y_0(t_i)$, $Z_0(t_i)$, $\omega(t_i)$, $\varphi(t_i)$, $\kappa(t_i)$ is required.

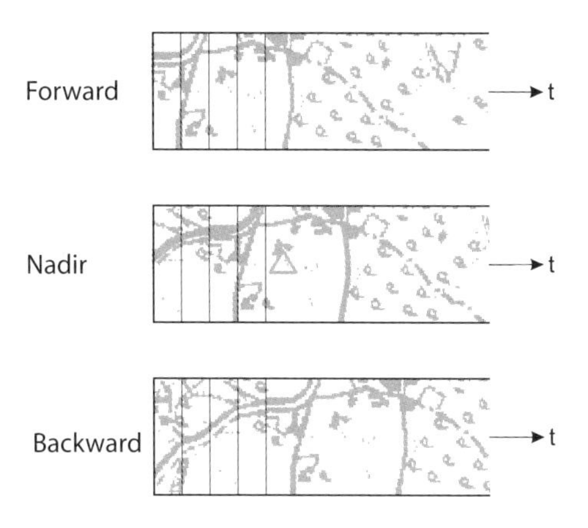

Figure 5.5-2: Image triplet from a 3-line camera showing the common line index

The basic equations for georeferencing images from a 3-line camera are the observation equations (5.3-1). However, the six exterior orientation elements X_{0j}, Y_{0j}, Z_{0j}, ω_j, φ_j, κ_j, which would be inserted here for each line triplet, are replaced by the six ordinates $X_0(t_i)$, $Y_0(t_i)$, $Z_0(t_i)$, $\omega(t_i)$, $\varphi(t_i)$, $\kappa(t_i)$ at all nodes N_i. Since the nodes N_i are considerably more widely spaced than the original lines, the number of unknowns in the adjustment is significantly reduced. (Details about this substitution can be found in Forkert, G.: Geow. Mitt. der TU Wien, vol. 41, 1994.)

Every control point identified in a triplet of strips provides $2 \times 3 = 6$ adjustment equations for determining the nodes of the six orientation functions. The ξ coordinate in the forward-looking strip corresponds to line spacing a while the negative ξ coordinate in the backward-looking strip corresponds to line separation b (Figure 4.1-1). (The η coordinates of a point identified in the three strips are only slightly different.) Every tie point determined in a triplet of strips also provides $2 \times 3 = 6$ adjustment equations for the determination of the six orientation functions. Also with the tie points the ξ coordinates are known from the a and b values. In adjustment equations based on tie points, the (unknown) ground coordinates $X_iY_iZ_i$ of a tie point also appear (see Equation (5.3-1)). A tie point in the overlapping region between two neighbouring triplets of strips provides $2 \times 2 \times 3 = 12$ adjustment equations to determine the nodes of the 6 orientation functions in each strip; the three (unknown) ground coordinates X_i, Y_i, and Z_i for such a tie point again appear in the system of equations.

Aerotriangulation, i.e. indirect georeferencing, can also be implemented with these adjustment equations. This would require many control and tie elements. Control and tie elements include control lines and tie lines in addition to control and tie points, as indicated in Figure 5.5-1, left (the relevant mathematical analysis can be found in Section B 4.3, Volume 2). The high number of control and tie elements can, however, be considerably reduced if the GPS/IMU information is included in the sensor orientation. Such a solution is very similar to GPS/IMU-assisted aerotriangulation (Section 5.4).

For this purpose, the six orientation functions are determined for all strips from the acquired GPS/IMU data. The abscissae of the nodes N_i are taken from the indirect georeferencing explained above. The ordinates of the nodes N_i can be determined by interpolation between the GPS/IMU observations at the nodes N_i. Restricting further discussion to the three orientation functions $X_0(t)$, $Y_0(t)$ and $Z_0(t)$, the result is a single GPS model for the complete area of interest. This GPS model contains three-dimensional Cartesian coordinates of all nodes N_i in a global GPS coordinate system. This GPS model can be jointly processed with the adjustment equations of the indirect georeferencing in a hybrid block adjustment, as explained in Section 5.4. In this hybrid block adjustment the ground coordinates of the nodes are treated as unknowns. The unknown nodes N_i connect the GPS model and the orientation functions $X_0(t)$, $Y_0(t)$ and $Z_0(t)$. Within the hybrid adjustment the GPS model is translated, rotated and re-scaled in order to compensate for possible datum problems and other error effects arising between the GPS data and the ground coordinate system. A division into separate GPS models for every strip can also be advantageous. In the same way as GPS models can be formed for the orientation functions $X_0(t)$, $Y_0(t)$ and $Z_0(t)$, so IMU models can be setup for the orientation functions $\omega(t)$, $\varphi(t)$ and $\kappa(t)$.

The method presented here is a combination of direct and indirect sensor orientation, i.e. an integrated sensor orientation, which has proven itself in practice.

Further reading for Section 5.5: Ries, C., Kager, H., Stadler, P.: Publikation der Deutschen Gesellschaft für Photogrammetrie und Fernerkundung, Band 11, pp. 59–66, 2002. Gervaix, F.: PFG 2002(2), pp. 85–91. Hinsken, L. et al.: ASPRS, St. Louis, USA, 2001.

5.6 Accounting for Earth curvature and distortions due to cartographic projections

Image coordinates are introduced as observations in the bundle block adjustment (Section 5.3), in the GPS- and IMU-assisted aerotriangulation (Section 5.4) and in the georeferencing of images from 3-line cameras (Section 5.5). In a prior process the image coordinates are corrected for the effects of objective lens distortion, film deformation, unflatness of the CCD detector surface, positioning errors in the detectors of CCD cameras or film scanners and refraction, as explained in Section 4.5. This section on image coordinate refinement also treats the correction for Earth curvature. The method given (Section 4.5.3) cannot be directly applied to the wide areas normally covered in aerotriangulation. Alternatives to the Earth curvature correction are therefore required.

(a) the first method, which can be applied where there are relatively minor height differences and small image tilts, makes radial corrections to the image coordinates in a similar way to refraction (see Equation (4.5-3)). This radial correction is given by the following equation (for the derivation of the formula see Section B 5.4.3, Volume 2):

$$\Delta\rho = \frac{\rho^3 h}{2Rc^2} \tag{5.6-1}$$

ρ ... radial distance from principal point
h ... flying height above ground
R ... radius of the Earth = 6370 km
c ... principal distance

The ground coordinates of the control points, and the GPS locations of the perspective centres (where available) in ground coordinates, remain unchanged. Any available IMU data relating to roll and pitch angles can also be adopted without change. However, the heading angle (yaw angle), which is also provided by the IMU, is referenced to the meridian and so a convergence of the meridian must be applied. The convergence of the meridian γ is the angle between the meridian and the North/South gridline in the ground coordinate system, frequently a Transverse Mercator coordinate system. It is given by:

$$\gamma_i = (\lambda_i - \lambda_0)\cos\varphi_i \tag{5.6-2}$$

φ_i, λ_i ... geographic coordinates of the point for which the convergence of the meridian γ_i must be calculated
λ_0 ... datum meridian in the ground coordinate system

(b) a second method fundamentally solves the problem that the object coordinate system in photogrammetry is a three-dimensional Cartesian coordinate system and the ground coordinate system is not. The control points and GPS positions of the perspective centres are given in the ground coordinate system; the photogrammetric results are expected in this coordinate system. The second method has the following steps:

- the control points and the GPS positions of the perspective centres (if available) are transformed from the ground coordinate system into a tangential coordinate system. This is tangential to the Earth ellipsoid at the centre of the area of interest.
- photogrammetric processing is done in this tangential system.
- the photogrammetric results are subsequently transformed from the tangential system into the ground coordinate system.

The IMU information requires special treatment. The discussion is here restricted to the roll and pitch angles which are referenced to the physical plumb line. The Z coordinates in the tangential Cartesian coordinate system are perpendicular to the XY plane.

Depending on the rigour of the analysis, the plumb lines are perpendicular to the spherical or ellipsoidal surface of the Earth; being specific, the tangents to the curved plum lines at the IMU's point of measurement are the datum for the IMU information. The IMU measurements must therefore be rotated by the angle between the physical plumb line and the Z axis of the tangential coordinate system. This correction, which depends on the IMU's measurement position as taken from the GPS data, can be calculated as a rotation matrix using either a spherical or ellipsoidal model of the Earth's surface and applied to the IMU data in the first calculation stage mentioned above.

Details of this rigorous method, which include accounting for the geoid, can be found in Sections B 5.3 and B 5.4, Volume 2.

Exercise 5.6-1. Table 4.5-3 shows the maximum Earth curvature in stereomodels for various image scales and camera types. Calculate the corresponding radial correction in the image using Equation (5.6-1). (Solution: for the 1st value in the 1st row: $c = 85$ mm, $h = 850$ m, $\rho = 140$ mm, $\Delta\rho = 25\,\mu$m.)

Further reading: Wang, S.: DGK, Reihe C, Nr. 263, 1980. Ressl, C.: VGI 89, Heft 2, pp. 72–82, 2001. Cramer, M.: DGK, Reihe C, Nr. 537, 2001. Heipke, C., Jacobsen, K., Wegmann, H. (Eds.): OEEPE Publication No. 43, 2002.

5.7 Triangulation in close range photogrammetry[14]

The sets of photographs in close range photogrammetry are often widely different from the examples we have discussed so far, with 60% forward overlap and 20% side overlap. Instead of the photogrammetric stereomodel we have the photogrammetric bundle of rays or directions (see Figure 3.8-11). A photograph records simultaneously a bundle of spatial directions to all object points visible in the photograph. The conversion of points initially recorded in the image plane into spatial bundles of rays occurs either by manual measurement of the image coordinates (analytical photogrammetry) or automatic target point recognition (digital photogrammetry). If measurement or recognition is monocular, object points must either be targeted or must be natural points which can easily be identified and accurately measurable in the individual photographs. For small objects with an extent of a few metres at maximum, there is also the option of projecting a pattern onto the object (see Section C 2.3, Volume 2. In Section 8.3 of this volume an interesting variant is described).

The bundles of rays of the individual photographs are tied together by common object points. Figure 5.7-1 shows such a network of photogrammetric bundles. Unmarked camera stations are normally chosen (equivalent to free stations in surveying). The camera may also be used without a tripod, so that the exterior orientation of the bundle of rays is entirely unknown. The essential criteria for the photographs are: the individual object points must appear in at least two photographs, better in at least three, and the rays must not intersect at narrow angles.

[14] A comprehensive textbook covering this topic may be found in Luhmann, T., Robson, S., Kyle, S., Harley, I.: Close Range Photogrammetry. Whittles Publishing, 2006.

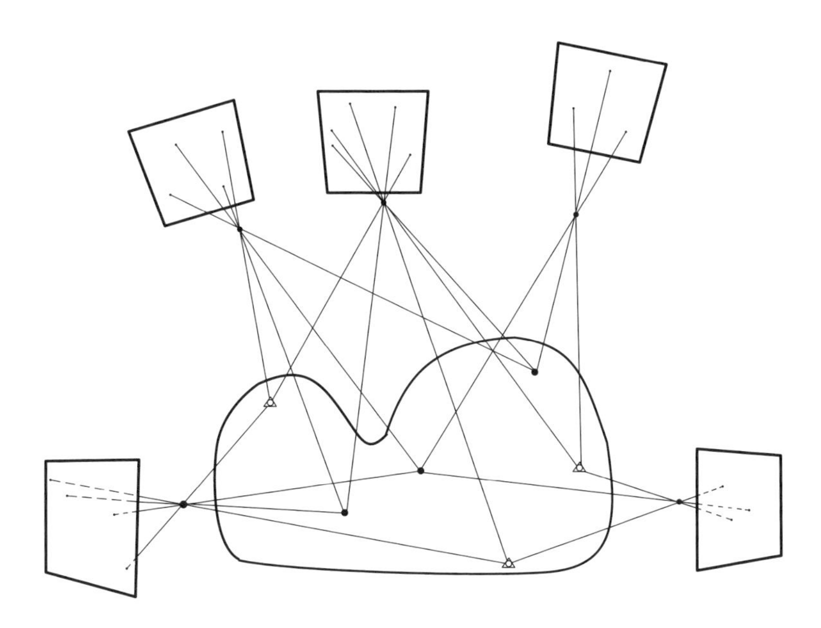

Figure 5.7-1: Network of photogrammetric bundles

A least squares adjustment of such a network of photogrammetric bundles is performed as a bundle block adjustment, described in detail in Section 5.3. The unknowns are the six elements of exterior orientation of each photograph and the three coordinates of each new point. Some control points are required on the edges of the object.

The XYZ coordinates of particular imaging stations (perspective centres), which have been determined by GPS or some other technique, can be introduced as known parameters (constants) in the adjustment equations (5.3-1). Due to the unavoidable inaccuracies in perspective centre coordinates determined in this way, it is preferable to introduce the terms as "observed unknowns" in the sense described in Section B 3.5.7 (Volume 2). In this way, perspective centres determined by GPS, or some other method, take on the role of control points.

In many cases, for example if the object is to be reconstructed in an arbitrary coordinate system, it is possible to dispense with control points and execute a free net adjustment, already mentioned in Section 4.4.1 and discussed in detail in Section B 4.6.3, Volume 2. In order to ensure that the object reconstructed in a local coordinate system has the correct scale, at least one (reasonably long) known object point separation must be included in the adjustment. The reconstructed object can be levelled by including new points along vertical lines and on horizontal planes[15].

At close ranges, a significant part is played by information on object form, such as the distribution of points on straight lines, curves, planes and surfaces in three-dimensional

[15]Some relevant publications include: Wester-Ebbinghaus, W.: IAPR 25(A3b), pp. 1109–1119, 1984, and ZfV 110, pp. 101–111, 1985. Fraser, C.: IAPR 27(B5), pp. 166–181, 1988.

space. The corresponding extension to photogrammetric triangulation is examined in Section B 3.5.5, Volume 2. Section B 4.1, Volume 2 also handles the very important and practical issue of calculating approximate values for close range bundle adjustment.

In close range work, semi-metric cameras (Section 3.8.4), amateur cameras (Section 3.8.5) and video cameras (Section 3.8.7) are often employed. Camera calibration, or at least an improvement in the calibration, is incorporated within the photogrammetric triangulation (this was discussed in Section 3.1.2 under the theme of on-the-job calibration or self-calibration). For this purpose the adjustment equations (5.3-1) are extended beyond the unknown elements of interior orientation to include additional parameters which compensate for optical distortion and other error sources in such cameras, as already discussed in Section 5.3.5.

Also at close range GPS and IMU are increasingly employed. One of the many combinations of GPS/IMU with digital cameras is the mobile mapping system presented at the end of Section 3.8.8, Figure 3.8-12. The bundle adjustment method plays a significant role in evaluating the multi-sensor data from such systems. The bundle adjustment equations (5.3-1) should be re-arranged so that the perspective centre coordinates X_{0j}, Y_{0j}, Z_{0j} do not appear as unknowns in the equations but as directly observed parameters (constants) or, better, as "observed unknowns" (Section B 3.5.7, Volume 2). If IMU data are available then the angles ω_j, φ_j, κ_j should be handled in a corresponding way. The grouping of GPS and IMU data into GPS and IMU models, as introduced in Section 5.4 for GPS and IMU-assisted aerotriangulation, is not normally required in close range work. Potential datum and calibration problems, and other error sources, can be more easily handled in terrestrial applications than in survey flights[16].

For the arbitrarily configured bundles found in close range photogrammetry, the accuracy of close range bundle adjustment can only be very approximately estimated with the previous rules of thumb (4.6-1) and (5.3-7). Here it is a condition that the camera rays intersect the object points at good angles and the entire block is surrounded by control points. The critical parameter in deciding accuracy is, as always in photogrammetry, the camera-object distance.

Numerical Example. Signalized points on a truck are to be measured using a $P31$ normal-angle metric camera ($c = 20\,\text{cm}$, see Section 3.8.3), at distances of about 5 m from the truck. Equations (5.3-7) yield the following estimates of accuracy:

$$\begin{array}{ll}\text{perpendicular to the camera axis:} & \sigma = \pm 0.003 \times 5/0.2 \quad = \pm 0.075\,\text{mm} \\ \text{along the camera axis:} & \sigma = \pm 0.00003 \times 5000 = \pm 0.15\,\text{mm}\end{array}$$

Where images fill the format with a useable extent of 100 mm, the relative accuracy (Section 4.6) is:

Perpendicular to the camera axis: $3/100000 \approx 1/33000 \mathrel{\hat{=}} 0.03‰$

Exercise 5.7-1. How would the accuracy change if a P31 wide-angle camera ($c = 10\,\text{cm}$) were used instead of the normal-angle camera? The camera-object distance

[16]Extensive further reading on mobile mapping can be found in Section 3.8.8.

could then be halved without increasing the number of photographs. On the other hand, the shorter distance (wider angle) may increase the number of dead (invisible) areas in the object. (Solution: σ (perpendicular to camera axis) $= \pm 0.075$ mm, σ (along camera axis) $= \pm 0.075$ mm.)

More comprehensive accuracy rules than the rule-of-thumb methods of Equations (4.6-1) and (5.3-7) have been given by, amongst others, Fraser[17], Grün[18] and Schlögelhofer[19]. These take into account control point layout, object shape and the number and quality of the intersections of rays through new points (Schlögelhofer's results are presented in Section B 4.5.2, Volume 2). Since networks of photogrammetric bundles vary widely from project to project, the computer programs for the adjustment must estimate the variance of each new point from the inversion of the matrix of normal equations. In this way, it is possible to check whether the intersections in the object points are too narrow, whether a sufficient number of rays intersect in a point and whether the number and layout of the control points are satisfactory (if no free network adjustment has been made). Luhmann et al.[20] also promote computer-supported measurement planning using the inverse of the normal matrix. Various orders of network optimization are defined for this concept[21].

[17] Ph.Eng. 53, pp. 487–493, 1987.

[18] IAP, Commission V, Stockholm, 1978.

[19] Geow. Mitt. der TU Wien, vol. 32, 1989

[20] Luhmann, T., Robson, S., Kyle, S., Harley, I.: Close Range Photogrammetry. Whittles Publishing, 2006.

[21] Further reading: Fraser, C.: in Atkinson (ed.): Close Range Photogrammetry and Machine Vision, pp. 256–281, Whittles Publishing, 1996. Mason, S.: Mitt. Nr. 53, Institut für Geodäsie und Photogrammetrie, ETH Zürich, 1994.

Chapter 6

Plotting instruments and stereoprocessing procedures

Stereoprocessing is central to this section, in the narrow sense of reconstruction of three-dimensional object models from corresponding points and lines in two-dimensional metric images. The determination of the elements of exterior orientation, either for the whole block (Chapter 5) or separately for individual stereomodels (Chapter 4), however, must precede the actual stereoprocessing. This section is to a large extent limited to the processing of two overlapping metric images. Automatic digital processing will certainly be considerably more robust and reliable when more than two images are dealt with simultaneously. A human operator can observe only two images in stereo simultaneously. For a human operator the process of measurement, or of matching corresponding points and lines in two overlapping photographs, can most conveniently, quickly and accurately be carried out with the help of stereoscopy. The different systems for stereoscopic observation as well as the principle of stereoscopic measurement are now considered, not only for analogue but also for digital images.

6.1 Stereoscopic observation systems

A short preamble on natural spatial vision is given before the introduction of the different stereoscopic systems used by photogrammetric operators.

6.1.1 Natural spatial vision

When we observe a point P and its surroundings, the axes of the two eyes are both directed towards P; this is known as convergence. In addition the focal length, and also the aperture, of the lenses in the eyes are changed so that a sharp image is generated on the retina; this is known as accommodation. In the plane of observation, defined by the eyebase B_A (which varies around 65 mm) and the fixed object point P, there will be two different images of the one object, one on each retina. In Figure 6.1-1 can be seen the difference in parallax p_A which relates to the difference in distance between the object point P and its neighbouring point Q. For simplification in the representation, in Figure 6.1-1 the direction of Q from the left eye is the same as that of P; their images in the left-hand retina coincide. With the head held upright, the plane of observation

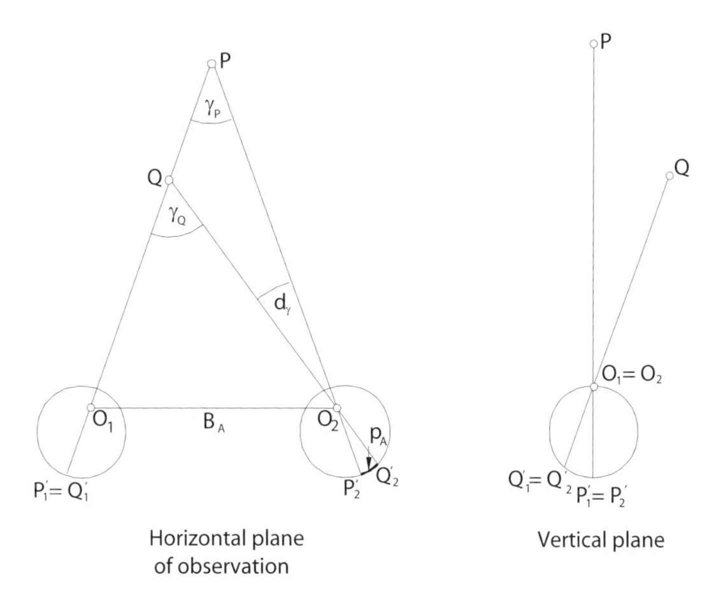

Figure 6.1-1: Natural stereoscopic vision

is approximately horizontal; for this reason we speak, in the context of natural spatial vision, of horizontal parallax and differences in horizontal parallax.

The fusion of the two retinal images, differing in horizontal parallaxes, resulting in a perception of three-dimensional space, is known as stereoscopic vision or stereoscopy. The smallest perceptible difference in angle $d\gamma = \gamma_p - \gamma_Q$ determines the resolution of stereoscopic perception or the stereoscopic acuity. This angle $d\gamma$ lies in the range of about $5''$ to $10''$. Monocular acuity, on the other hand, is of the order of about $30''$. In relation to the standard "minimum distance of distinct vision" of about 25 cm, the stereoscopic acuity lies in the interval 6 μm to 12 μm and the monocular acuity is about 40 μm[1]. Photogrammetric measuring accuracy is distinctly better (see Sections 2.1.7 and 4.6).

This improvement arises from optical or digital enlargement of the images, so that the acuity of the human eye is not the limiting factor of the man-machine processing system.

On the other hand images exhibiting such differences in horizontal parallaxes p_A can be fused into a single three-dimensional image only when the angle $d\gamma$ is smaller than about 1.3 gon (about $1°12'$). If this angle is exceeded either the foreground or the

[1] Albertz, J.: ZfV 102, pp. 490–498, 1977. In comparison with the accuracy of measuring individual image coordinates, the different monocular and stereoscopic acuities result in a higher accuracy in the measurement of ξ-parallaxes, although not in the anticipated ratio of 1 : 6 to 1 : 3. Further reading, which also contains advice on additional literature, may be found in: Schenk, T.: Digital Photogrammetry. Terra Science, 1999.

background appears to dissolve into two separate images; one can easily verify this by experimentation (but see Section 6.1.2(a)).

In contrast to the horizontal parallaxes p_A (Figure 6.1-1), which lie in a plane containing the eyebase B_A, vertical parallaxes are defined as being perpendicular to the eyebase. Homologous rays emanating from object points produce no vertical parallaxes since the rays are of necessity coplanar. In natural stereoscopic vision there are, therefore, no vertical parallaxes.

Exercise 6.1-1. For an observer lying on his side, his eyebase is vertical. Think critically about the implications for the definitions of horizontal and vertical parallaxes.

6.1.2 The observation of analogue and digital stereoscopic images

Suppose that two images have been taken with parallel axes normal to the base (the "normal case"); they will exhibit horizontal parallaxes in the direction of the base but no vertical parallaxes perpendicular to the base (Section 2.1.5). They can subsequently be presented to the two eyes. The observer interprets the horizontal parallaxes as signifying differences in depth in a virtual spatial image. One speaks also of a three-dimensional image (3D picture).

Before dealing more closely with the observation of stereoscopic images, principles should be set down as to how one arrives at stereoimages with horizontal parallaxes but without vertical parallaxes:

- "normal case" photographs taken, for example, with a stereometric camera (Section 3.8.2)
- conversion of arbitrarily oriented photographs into the "normal case" using the elements of interior and exterior orientation (the favoured method in the processing of digital metric images Section 6.8.5.2)
- projection, with oriented projectors, of arbitrarily oriented photographs onto the xy plane of a relatively oriented stereomodel (the standard method in analogue photogrammetry with optical projection Section 6.3(a))
- observation of arbitrarily oriented photographs not in a common image plane but in two separated image planes; the photographs have previously been oriented according to the elements of relative orientation (standard method in analogue photogrammetry with mechanical projection instruments, Section 6.3(b))
- observation of arbitrarily oriented photographs in a common image plane while a mechanism is running in the background using the elements of relative orientation to remove vertical parallaxes (standard method of analytical photogrammetry, Section 6.4)

(a) observation of analogue stereoscopic images

In the observation of two stereoscopic photographs lying side by side at approximately 25 cm, the minimum distance of distinct vision, an uncommon demand is made of the two eyes. With their axes almost parallel, that is with near-zero convergence, the eyes are being asked to accommodate at a short focussing distance, that is, they are being asked to overcome the unconscious reflex which normally imposes coordination of focus and convergence. After long practice and with the exercise of considerable will power, perhaps aided by the positioning of a sheet of card or paper between the two images so as to separate them visually, fusion can be achieved with the naked eye, resulting in a more or less sharp stereoscopic image.

Stereoscopic observation can, however, be made significantly easier if one places the photographs under a lens stereoscope (Figure 6.1-2). The rays entering the eyes are almost parallel and, if the photographs are a distance f below the lenses, their virtual images are at infinity and may be viewed with ease and comfort.

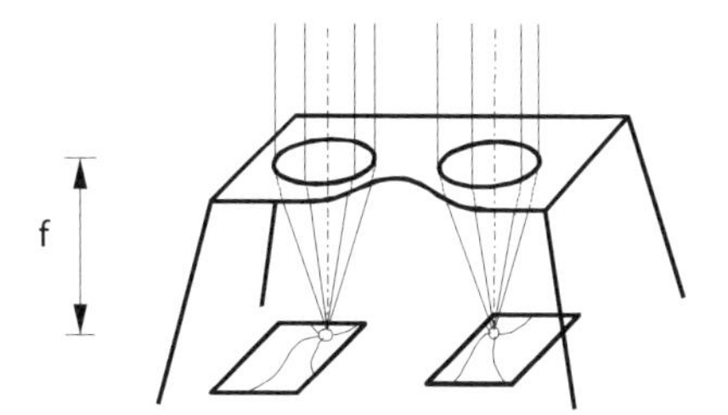

Figure 6.1-2: Lens stereoscope

Large format photographs require the use of a mirror stereoscope (Figure 6.1-3). In most such instruments optical systems are placed above the lenses L_1 and L_2 so that sections of the images can be seen magnified.

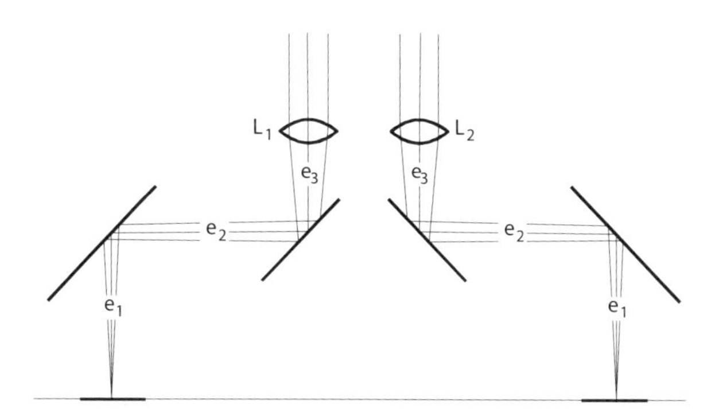

Figure 6.1-3: Mirror stereoscope ($e_1 + e_2 + e_3 = f$)

In most analogue and analytical stereoplotters (for example Figure 6.3-2b) stereoscopic observation occurs by way of lenses or mirror stereoscopes in conjunction with complicated optical systems.

Some analogue stereoplotters employ superimposed stereoscopic images (see, for example, Figure 6.3-1). Figure 6.1-4 shows superimposed stereoscopic images of a pyramid. The eyes converge and accommodate in the accustomed way at a comfortable short distance but each eye sees both of the two-dimensional images. Suppose, however, that the left-hand image has been coloured cyan and the right-hand one red and that a filter in the complementary colour has been placed in front of each eye (Section 3.2.2.1). Then each eye will see only one image in black, while the other image will disappear; the two black images will be fused in the usual manner into a three-dimensional, stereoscopic image. This anaglyph system, as it is known, is very suitable for the printing of stereoscopic images. Coloured images, however, cannot be observed in this manner.

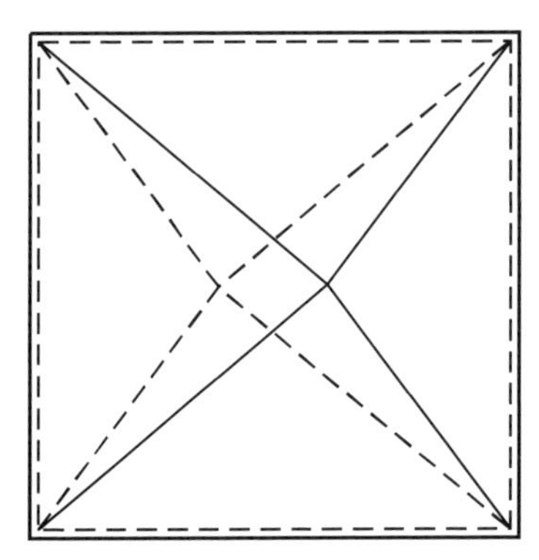

Figure 6.1-4: Superimposed stereoscopic images of a pyramid (from cameras as illustrated in Figure 6.1-8). Solid lines for the left image, broken lines for the right

(b) observation of digital stereoscopic images

Digital stereoscopic images can also be observed with either a lens stereoscope or a mirror stereoscope if the two images are presented to the eyes, either on two screens side by side or on a single, split screen. The anaglyph system mentioned above can also be applied.

Digital stereoscopic images offer other, very interesting, variations, one of which is sketched in Figure 6.1-5. The two images alternate on the same screen and are observed through spectacles incorporating LCD shutters. An infra-red system controls the synchronization of the spectacles and the screen (Figure 6.1-5); if the frequency of alternation of the images exceeds 50 Hz a continuous stereoscopic image is observed.

A second variant is shown diagrammatically in Figure 6.1-6. Under the control of a synchronizer the two images alternate in time in a bitmap memory and on a polar-

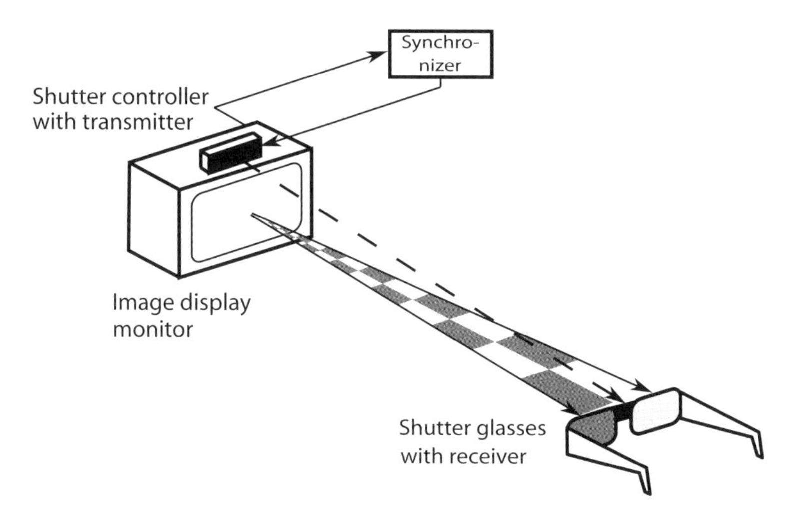

Figure 6.1-5: Shutter glasses for stereoscopy

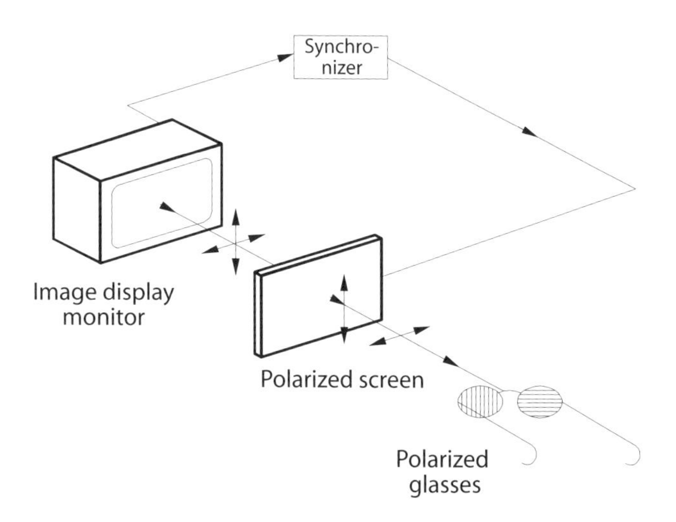

Figure 6.1-6: Polarization for stereoscopy

ized screen. Stereoscopy results with the use of correspondingly polarized spectacles (Figure 6.1-6). Polarization today also uses LCD systems.

Both these variants of stereo-observation can be used simultaneously by more than one person. As a result, difficult problems of interpretation can be discussed by a group while observing the stereomodel.

(c) additional information concerning stereoscopy

The second set of methods for the stereoscopic observation of digital images, which may be denoted passive stereoviewing, requires simple spectacles, as opposed to the first methods which may be denoted active stereoviewing. From the point of view of the operator, both variants have advantages and disadvantages, the first usually providing better image quality since there is no filter while the second variant uses light and cheap spectacles. Stereoviewing over many hours is relatively demanding whichever variant is considered, especially when the frequency of the images is low. For this reason, for detailed stereoscopic study many operators still prefer analytical stereoplotters with analogue images on film. At present great efforts are being made in research and development to produce stereoscopic images which could be viewed without spectacles, filters and so on. Such systems would offer many advantages. Interesting illusions can be produced also when observing stereoscopic images. If, for example, one interchanges the red and cyan filters in the anaglyph method, the tip of the pyramid in Figure 6.1-4 no longer appears to lie above the square base but beneath it. Instead of the orthoscopic effect the pseudoscopic effect is observed. When viewing aerial photographs pseudoscopically one gets an inversion of the relief in which mountain ridges become valleys, streams flow towards their source and so on.

Stereoscopic perception can also be disturbed. Thus the limit of 1.3 gon $(1^\circ 10')$ (Section 6.1.1) should not be exceeded in the section of the image under observation. Schwidefsky/Ackermann gives another measure for the limiting value for stereoscopic vision: the difference in scale between the sections of the left and right images being viewed should not be greater than 14%. Since in high mountains and in terrestrial images this limit is frequently exceeded, provision of individual zoom controls for each image is of great assistance in stereoplotters.

As well, the density or the colour of the two images should not be greatly different. The procedures for enhancement of digital images reviewed in Section 3.5 are also of great significance for stereoscopy. All in all, however, the human eye is very tolerant: even when one eye is presented with an RGB image (Red/Green/Blue) and the other eye with a B&W (Black and White) image a spatial impression (in colour) is obtained. The sole condition is that the intensities of the RGB image should, to a large extent, match those of the B&W image; one makes use of this when observing images from a three-line camera (Section 3.7.2.3).

Provided that the geometrical parameters of the photographed scene are of no importance but that it is the significance of individual objects that is under investigation, one frequently dispenses with a complete orientation of the images and with three-dimensional geometrical processing. Instead, for such photointerpretation one makes only a makeshift orientation as sketched in Figure 6.1-7 in which it is assumed that the photographs have been taken with the camera axis near-vertical, typically the case in air photo-interpretation. This consists simply of laying the two photographs on a plane surface for observation and aligning the principal point bases (Figure 6.1-7) whereby the vertical parallaxes are removed, admittedly only along this common straight line; with increasing distance of the image region from this favoured line, however, the vertical parallaxes increase, depending on such things as the tilts. Nonetheless, with stereometers or parallax bars, as they are called, horizontal parallaxes, and thus the ξ-parallaxes between two particular points can be measured and from these the height difference between both 3D points computed. Detailed discussion of these methods and their limitations were to be found in older textbooks.

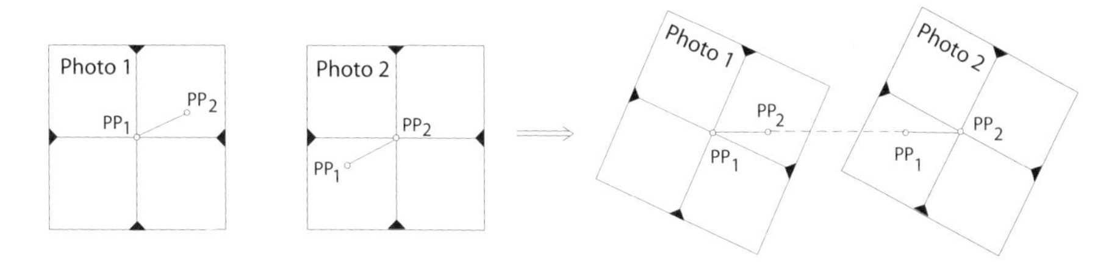

Figure 6.1-7: Orientation of near-vertical images for stereoscopy

(d) spatial impression in stereoscopic observation

The question remains to be asked concerning the extent to which the virtual spatial image arising from two stereoscopic images is distorted with respect to reality. The relationships with respect to the photograph, both wide angle and normal angle, are shown on the left of Figure 6.1-8 and on the right are sketched those relating to stereoscopic viewing. B_A is the eyebase. The virtual spatial image has come nearer to the observer by the factor H/D_A; H is the flying height and D_A the apparent distance at which the virtual model is formed.

The quantity D_A depends on the design of the stereosystem, on the observer and also on the image quality. Unexpectedly, D_A is independent of the magnification of the observing equipment and also of the separation of the two images as they lie beneath the stereoscope. One can easily satisfy oneself with respect to this physiological phenomenon by experimental magnifications and shifting of the images. For lens and mirror stereoscopes with greatly differing magnifications and different image material the apparent distance D_A is found to lie in a range[2] from 35 cm to 60 cm. In the case of digitally superimposed images, as in Figures 6.1-5 and 6.1-6, D_A is the distance between the eyebase and the screen; in this case D_A can, in principle, be arbitrarily increased from the initial value of the minimum distance of distinct vision, approximately 25 cm.

As the observed virtual 3D-image comes closer by the factor H/D_A its scale in relation to reality also decreases, to be precise, in object planes parallel to the image planes by a factor $H/(cv)$, where H is the flying height, c the principal distance of the taking camera and v the enlargement factor of the stereo-observing instrument. For digital images on the screen v is the zoom factor.

Of greatest importance is the question of the degree to which, in the observation of stereoimages, the dimensions of objects in the direction of the camera axes are exaggerated in relation to vertical distances. Using the notation of Figure 6.1-8 this vertical

[2]Miller, C.: Ph.Eng. 24, pp. 810–815, 1958, and 26, pp. 815–818, 1960. Collins, S.: Ph.Eng. 47, pp. 45–52, 1981. (For good stereo-observers D_A is small.)

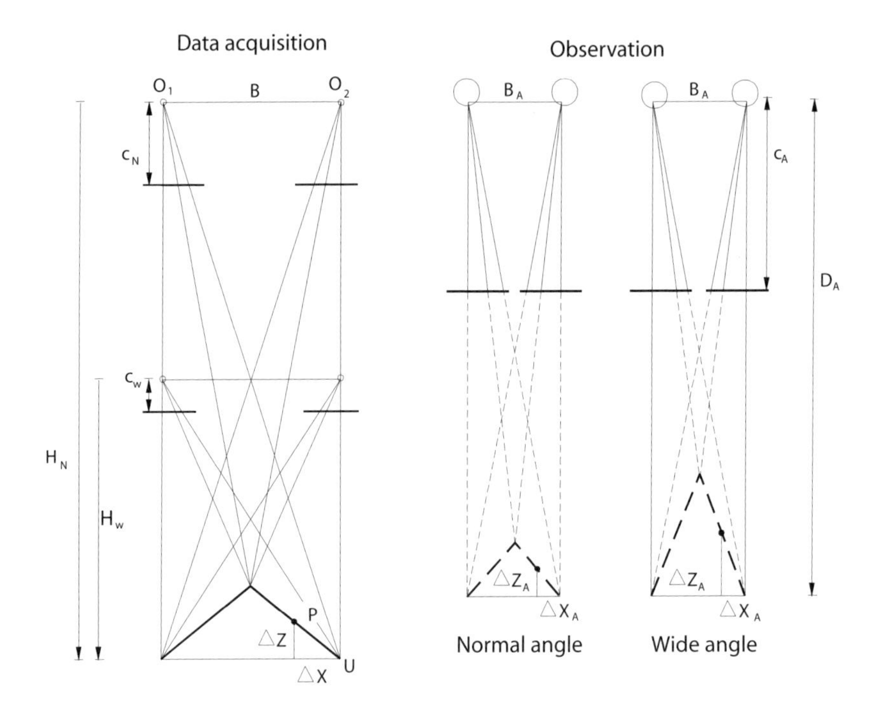

Figure 6.1-8: Taking and observing of stereoscopic images

exaggeration factor is defined as:

$$e_z = \frac{\Delta Z_A}{\Delta X_A} : \frac{\Delta Z}{\Delta X} \tag{6.1-1}$$

The values ΔX and ΔX_A can be expressed as the nth part (for Figure 6.1-8 $n = 4$) of the camera base B and of the eyebase B_A ($\approx 65\,\text{mm}$) respectively:

$$\Delta X = B/n \qquad \Delta X_A = B_A/n \tag{6.1-2}$$

The values ΔZ and ΔZ_A follow from the first relationship of the equation system (2.1-32), in which $-Z$ is substituted for the flying height H:

$$\Delta Z = H_U - H_P = \frac{cB}{p_{\xi,U}} - \frac{cB}{p_{\xi,P}} = cB\,\frac{p_{\xi,P} - p_{\xi,U}}{p_{\xi,U}\,p_{\xi,P}} \tag{6.1-3}$$

The parallax $p_{\xi,U}$ is also there in the observation of the virtual spatial image, so that the "apparent principal distance" c_A may be derived from the first relationship of the equation system (2.1-32) as follows

$$D_A = \frac{c_A B_A}{p_{\xi,U}} \Rightarrow c_A = \frac{D_A p_{\xi,U}}{B_A}$$

Then, in accordance with equation (6.1-3):

$$\Delta Z_A = c_A B_A \frac{p_{\xi,P} - p_{\xi,U}}{p_{\xi,U} p_{\xi,P}} = \frac{D_A p_{\xi,U} B_A}{B_A} \frac{p_{\xi,P} - p_{\xi,U}}{p_{\xi,U} p_{\xi,P}} = D_A \frac{p_{\xi,P} - p_{\xi,U}}{p_{\xi,P}} \qquad (6.1\text{-}4)$$

Substituting from Equation (6.1-2) to (6.1-4) in Equation (6.1-1) gives:

$$e_z = \frac{D_A p_{\xi,U}}{B_A c} = \frac{D_A}{B_A} \frac{B}{H} = \frac{B}{H} : \frac{B_A}{D_A} \qquad (6.1\text{-}5)$$

Result: The height exaggeration factor e_z is the ratio of the base/height factor B/H of the stereopair to the base/distance ratio B_A/D_A in the observation of the virtual three-dimensional image.

Table 6.1-1 gives the height exaggeration factor e_z in the case of normal angle, wide angle and super-wide angle photographs with an image format of 23 cm $\times$ 23 cm and a 60% forward overlap (base/height ratios are given in Table 3.7-1).

	Normal angle	Wide angle	Super-wide angle
c [mm]	300	150	85
B/H	1 : 3.3	1 : 1.6	1 : 0.9
e_z	1.9	3.8	6.7

Table 6.1-1: Height exaggeration factor e_z in the observation of stereoimages with 60% forward overlap when $B_A/D_A = 65 : 400 = 1 : 6.2$

A disproportionate spatial impression, an increased spatial quality, is observable when examining normal aerial photographs stereoscopically. This exaggeration of relief in the virtual spatial image plays a significant role in photo-interpretation. In photogrammetry, on the other hand, there is usually no need to worry about the phenomenon. The stereoscopic effect serves simply to hasten the matching of corresponding points in both images and to aid in accurate setting of the measuring mark on a required point in the spatial image.

Exercise 6.1-2. Make a superimposed drawing in two complementary colours of Figure 6.1-4 and observe it, both orthoscopically and pseudoscopically, through filters of complementary colours.

Exercise 6.1-3. In order to see an undistorted stereoscopic impression of images from Wild or Zeiss fixed-base stereocameras (Section 3.8.2), what camera-to-object distance should be used? (Answer: 2.5 m and 7.4 m, respectively, for camera bases of 40 cm and 120 cm.)

6.2 The principles of stereoscopic matching and measurement

A (virtual) three-dimensional measuring mark is used for the geometric analysis of the stereomodel, the three-dimensional image, presented to the operator. The operator sets this three-dimensional measuring mark on points and traces it along lines, in this manner abstracting information from the images, converting it to discrete values. The stereoplotter becomes a 3D digitizer. In this process of analysis the photogrammetric operator brings his knowledge and judgment to the choice of points and lines, interpreting their significance and capturing their meaning in encoded form. In this way human vision, recognition and recall are put to service in the processing of the stereomodel. Machine vision, which is not considered in more detail until Section 6.8, attempts to transfer this whole process to the computer.

This section concentrates on the design and operation of the three-dimensional measuring mark mentioned above, beginning with those measuring marks used with lens and mirror stereoscopes.

(a) beneath a lens or mirror stereoscope

One uses two real measuring marks which can be moved over the surface of the two-dimensional images (Figure 6.2-1). The two images fuse into a virtual three-dimensional image and, so long as no vertical parallax exists, the two measuring marks M_1 and M_2 also fuse into a single virtual three-dimensional measuring mark, known as the floating mark. If the right-hand mark moves from the position (M_2) to the correct M_2, then the virtual three-dimensional measuring mark will appear to move from the position (M) to the position M in the virtual spatial image. Shifting of the measuring marks in the two-dimensional images is perceived as a vertical displacement of the floating mark in the virtual spatial image.

After the floating mark has been set on a definite object point, the real measuring marks M_1 and M_2 each coincide with the image of that point in their respective two-dimensional photograph. With appropriate measuring systems the image coordinates can be determined in each image (Section 6.4).

Digital images may be viewed in analogue form as images on two screens or on a single split screen; in this case the cursors on the screen(s) can serve as the real measuring marks.

(b) with digitally superimposed images

The discussion is limited to digitally superimposed images that are displayed either as alternating real image sequences (Figure 6.1-5) or as alternating polarized images (Figure 6.1-6). There is only a single cursor which one sees in both images. If the cursor is not located on homologous images then the floating mark is perceived as hovering either above or below the virtual surface. In order to set the floating mark on the virtual surface one or both of the images must be shifted in the direction of the base. The cursor then becomes a 3D cursor; its apparent vertical movement relative to the virtual surface can be controlled by the operator, using a wheel on a mouse-like device. With the wheel one of the two images can be displaced in the direction of the base until the 3D cursor lies on the surface of the stereomodel (as in Figure 6.2-1). Following this stereoscopic setting on a point one knows the positions, in other words the image coordinates

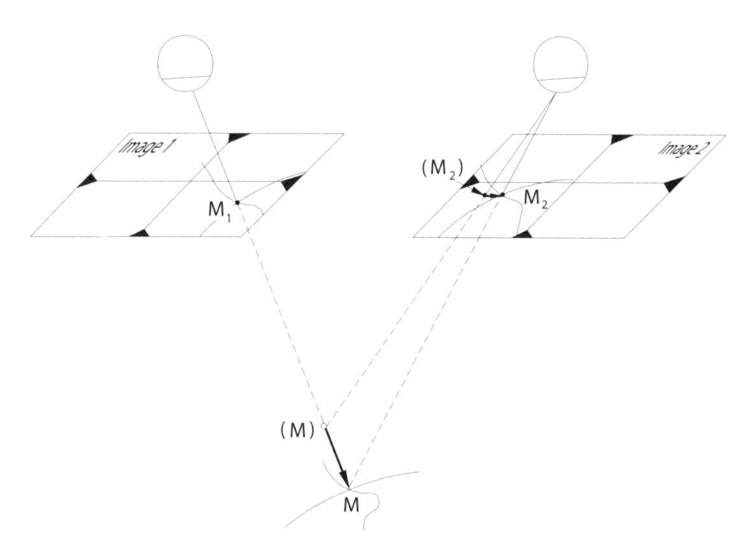

Figure 6.2-1: The principle of stereoscopic measurement with two real measuring marks

ξ and η, in both images; so both corresponding points are located and stereoscopically matched. Using the recorded image coordinates and the elements of interior and relative orientation, or in other cases the exterior orientation, the xyz coordinates are computed in real time, or in the latter case the XYZ coordinates (Sections 4.3.2 or 4.1).

6.3 Analogue stereoplotters

Between about 1910 and about 1970 probably more than a hundred different types of analogue plotters were designed and constructed. They are mentioned here for historical reasons and are of minor importance nowadays.

In an analogue stereoplotter the position and orientation of the taking cameras is reproduced at a reduced scale. The relationship is established, by means of optical and mechanical components, between the model of the object at a reduced scale and the two metric images at their original scale. Two principles of design which have been developed may be described either as optical projection or as mechanical projection.

(a) principle of optical projection

A widely used stereoplotter with optical projection was developed by the American firm Kelsh (Figure 6.3-1). Both projectors are set up for an image format of 23 cm × 23 cm. They have the same interior orientation as the camera used to take the photograph. After setting of the exterior orientation the intersection points of corresponding rays define a true model without geometric distortion.

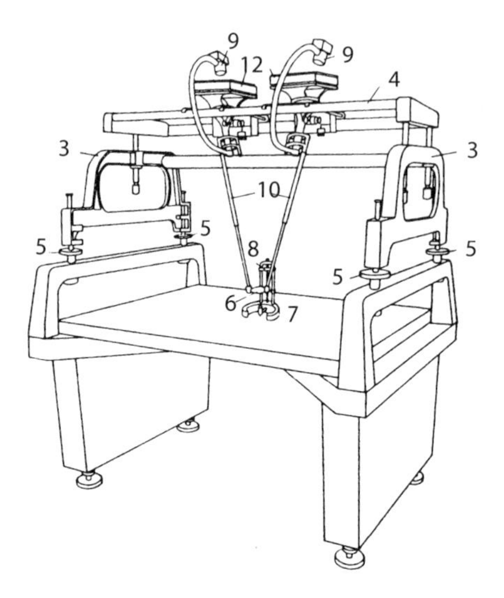

Figure 6.3-1: Kelsh-Plotter[3]

Each projector may be rotated through angles ω, φ and κ; this allows the relative orientation (independent model method) to be carried out. The rotations Ω and Φ required for absolute orientation (levelling the model) may be set using the footscrews (5). Since the model scale is determined by the ratio of the base length in the model to the base from the photoflight, the desired model scale may be set by shifts of both projectors. Note that the complete process of relative and absolute orientation may be carried out without computational aids.

Stereoscopic observation is by the anaglyph method (Section 6.1.2(a)). A light spot on the table (6) is used as the floating mark for stereoscopic setting on the model surface. The links (10) ensure that the lamps (9) project corresponding portions of the images.

(b) principle of mechanical projection

In mechanical projection each straight line joining image point, perspective centre and object point is represented by a link, known as a space rod, usually in the form of a slender rod of cylindrical cross-section which rotates about one of the perspective centres. The principle of mechanical projection is briefly explained with reference to the Wild Aviograph (Figure 6.3-2a). The object point is driven in the x, y and z directions on a three-dimensional cross slide system. The space-rod is mounted

- at the object point in a ball-joint
- at the perspective centre in stationary gymbal axes (also known as a Cardan(ic) mount)

[3]Kelsh, H.T.: Ph.Eng. 14, p. 11, 1948.

- at the image point in a ball-joint connected to the photo-carrier which may be translated in a plane

The portions of the space rod connected to the ball-joints may be lengthened or shortened along their axes, often referred to as being telescopic.

Each photo-carrier can be rotated through angles ω and φ, the axes of rotation passing through the respective perspective centre. In addition each image can be rotated in a plane about its principal point (κ rotation). The longitudinal tilt of the model Φ can be set on an adjustment screw (Φ in Figure 6.3-2a). The other inclination of the model for absolute orientation Ω is set directly with ω_1 and ω_2 on the two photo-carriers.

Corresponding to the principal distance is the distance of each perspective centre from that plane which is parallel to the tilted plane of the image and which passes through the centre of the image point ball-joint. This ball-joint, at the upper end of the space rod, moves the photo-carrier on a two-dimensional cross slide under the control of the space rod.

A measuring mark M, a light-spot illuminated by the light source L (see the left-hand part of Figure 6.3-2b), is fixed in the lower, stationary part of the tilted photo-carrier. The beam-splitter F causes the measuring mark M to be reflected into the optical system and brought into coincidence with the image point P_1.

After absolute orientation which, as has already been mentioned, can be carried out without numerical computation, the operator sees a virtual three-dimensional image not only of a section of the images but also of the measuring mark. Now he moves the x, y, z cross slides, and with them both images, until the two measuring marks coincide with the image points P_1 and P_2; that is, the floating mark has been set stereoscopically on point P and its object coordinates have been found.

An important special feature of almost all analogue stereoplotters is the Zeiss parallelogram. On the one hand it allows the perspective centres to remain in fixed positions in the instrument and, on the other hand, it avoids the difficult problem of mechanical intersection of the space rods in order to find the object point. The base is introduced by relative movement not of the perspective centres, as in the Kelsh plotter, for example (Figure 6.3-1), but of the ends of the space rods; this is achieved in the Aviograph AG1 by adjustment of b and b_Φ (Figure 6.3-2a). More information on the Zeiss parallelogram and on analogue stereoplotters is to be found in older textbooks.

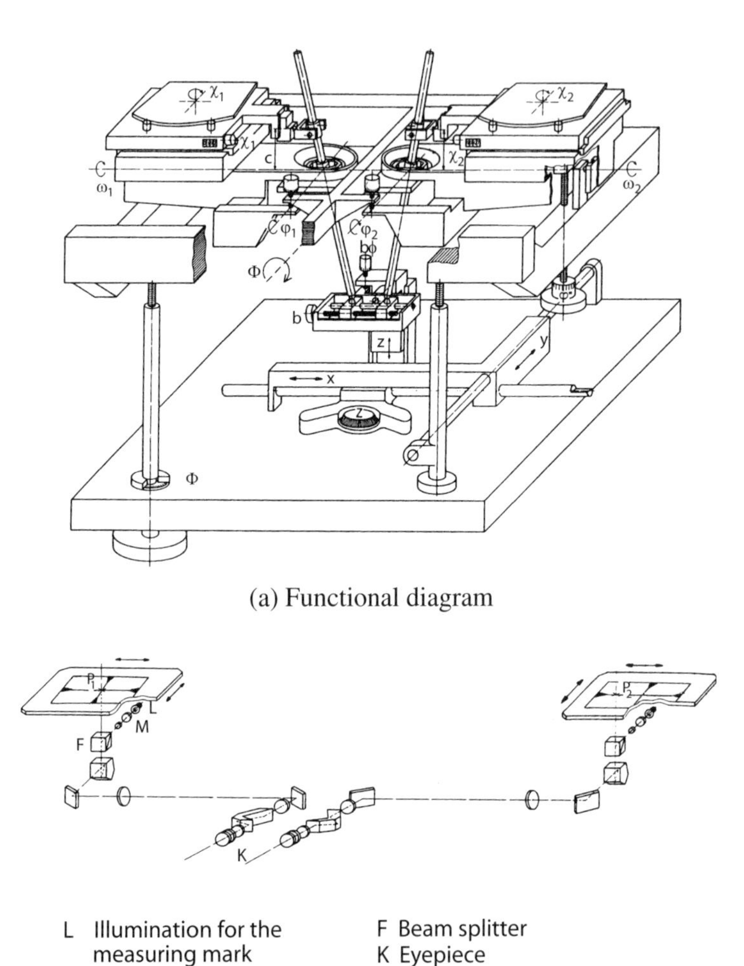

(a) Functional diagram

(b) Optical system

Figure 6.3-2: Aviograph AG1

6.4 Analytical stereoplotters

It is an interesting fact that until the end of the first decade of the 20th century virtually all photogrammetry applied analytical methods. From that time for more than half a century analytical methods were displaced by analogue procedures in plotting. This situation is rapidly changing with the advent of high-speed computers.

6.4.1 Stereocomparators

It is very simple to explain the mode of operation of a stereocomparator in terms of the cross-slide system shown in Figure 6.4-1.

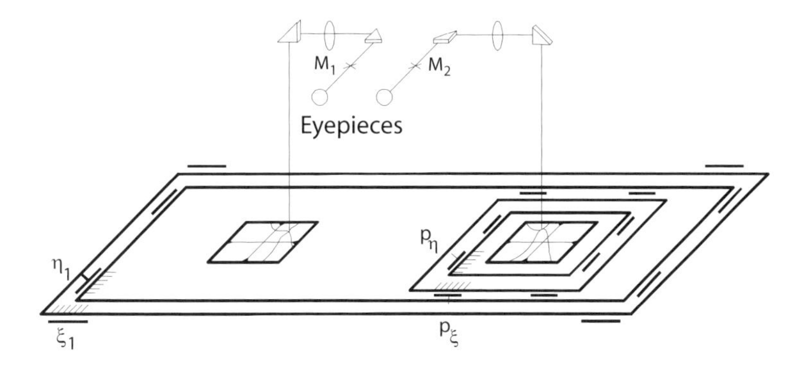

Figure 6.4-1: Diagram of a stereocomparator

The two images may be moved in coordinate directions corresponding to the perpendicular cross-slides. The two measuring marks M_1 and M_2 are fixed in the focal planes of the eyepieces. Images of parts of the photographs are also focused in these planes, so that the measuring marks may be set stereoscopically on object points (Section 6.2).

Disregarding the zero settings for the time being, such a stereoscopic setting yields the image coordinates ξ_1 and η_1 on the lower cross-slides and the parallaxes $p_\xi = \xi_1 - \xi_2$ and $p_\eta = \eta_1 - \eta_2$ on the upper cross-slides from which the image coordinates ξ_2 and η_2 may be derived. So that these uncorrected image coordinates from the stereocomparator scales may be transformed into the camera coordinate system, it is also necessary to measure the coordinates of the fiducial marks (Section 3.2.1.2).

When the image coordinates are recorded by a human operator, the XYZ object coordinates of the measured points are found by computation using analytical photogrammetry. Since in its early days photogrammetry was concerned exclusively with terrestrial photographs, usually conforming to the "normal case", the labour involved in calculation of object coordinates was kept within limits[4].

Advances in analytical photogrammetry received a very decisive stimulus with the development of electronic data-recording and electronic data processing. At this point electronic data registration should be introduced.

[4]In 1909 Eduard von Orel constructed the Stereoautograph, an instrument for "automatic" plotting from stereoscopic photographs, circumventing this computational labour (Mitteilungen des k.u.k. Militärgeographischen Institutes XXX, pp. 63–93, Wien, 1910). In England F. V. Thompson was slightly before von Orel in the design and use of the Vivian Thompson Stereoplotter (Atkinson, K.B.: Ph.Rec. 10(55), pp. 5–38, 1980. Atkinson, K.B.: Ph.Rec. 17(99), pp. 555–556, 2002). Both Thompson and Orel solved the simple equations of the "normal case" (3.8-1) by means of lineals; at one end these were fixed to the slides of the stereocomparator; at the other end they controlled a drawing device. The instruments permitted the continuous drawing of plan detail and contours. Recent publications touching on the historical development of the stereocomparator and the stereoplotter include Meier, H.-K.: ZfV 128(1), pp. 6–10, 2003, and Luhmann, T., Robson, S., Kyle, S., Harley, I.: Close Range Photogrammetry. Whittles Publishing, 2006.

6.4.2 Electronic registration of image coordinates in the monocomparator

An efficient system for the registration of image coordinates is explained on the basis of the precise comparator PK1 from the firm of Zeiss[5] (Figures 6.4-2 and 6.4-3). Two glass scales at right angles to each other are fixed with respect to the baseplate. The image to be measured lies on a photo-carriage which can be translated in two dimensions without rotation; the left-hand and the upper borders of this photo-carriage each carries a series of equidistant lines. Two linear pulse generators measure the displacements of the image. The measuring mark M is stationary. As in the instrument shown in Figure 6.4-1, it is situated in the focal plane of the optical system. The relative movements of the individual parts of the monocomparator can be studied with the aid of Figure 6.4-2 which shows, in broken and continuous lines respectively, their positions when measuring points 1 and 2.

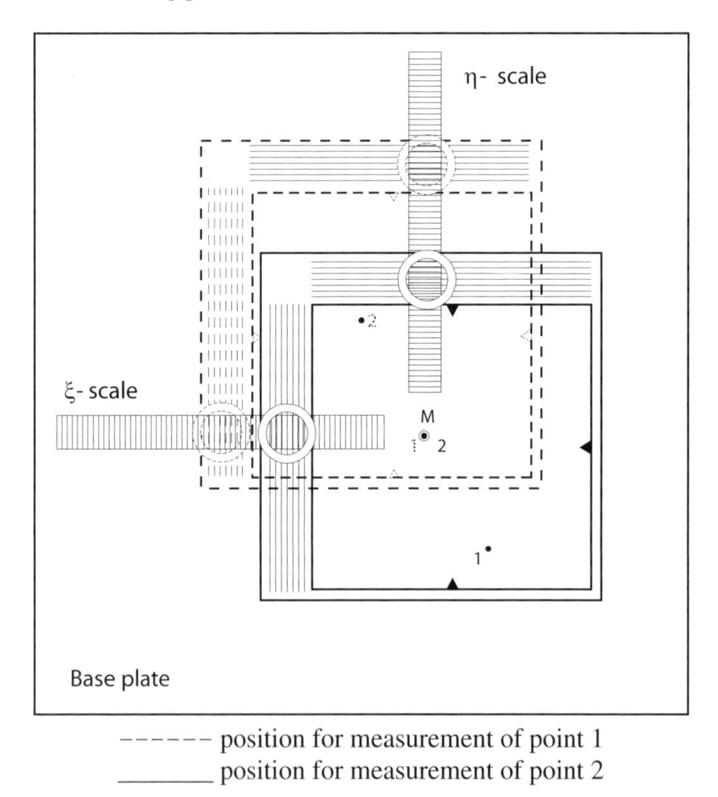

Figure 6.4-2: Principle of the Zeiss precision comparator PK1

[5]Schwebel, R.: ZfV 104, pp. 157–165, 1979.

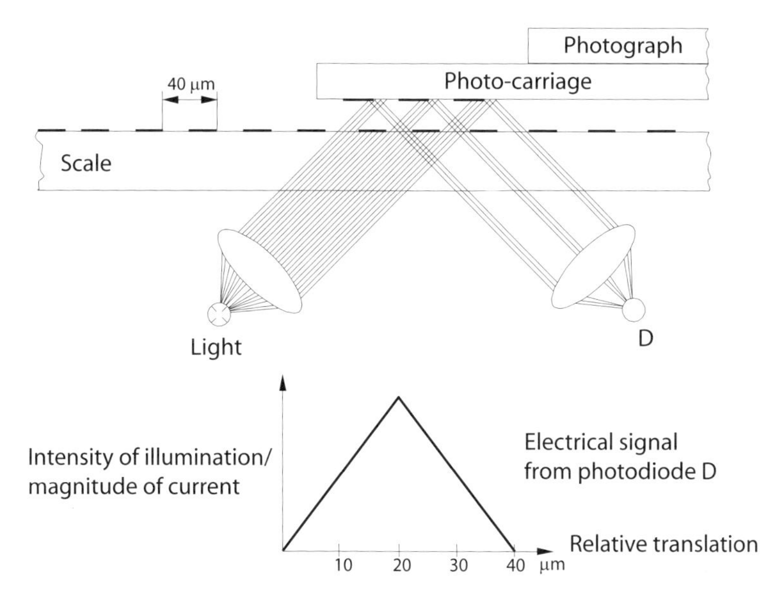

Figure 6.4-3: Linear impulse generator (Since it has no influence on the measuring principle, the parallel guidance system is not illustrated). The photograph, the photo-carrier, the light source and the photodiode move together; the scale and the measuring mark are fixed.

The linear pulse generator (Figur 6.4-3) warrants a closer examination. On the measuring scale as well as on the photo-carriage, not only the width but also the separation of the lines is 20 μm. Rays coming from the light source and falling on a line are fully absorbed; they are totally reflected from the regions between the lines on the photo-carriage. When the lines on the photo-carriage are moved 10 μm to the left from a position in which they are aligned with the graduations on the scale (Figure 6.4-3), only 25% of the emitted light falls on the photodiode D, which translates the incident light into electrical current in proportion to the intensity of illumination. If the photo-carriage is then moved to the left the current will increase. The maximum of 50% is reached when the graduations on the photo-carriage are moved 20 μm out of alignment with the graduations on the scale. A further movement to the left reduces the current, and so on.

The number of minima and maxima transmitted from the diode D to a counting system defines the image coordinates as a multiple of 20 μm intervals. Subdivision of the 20 μm intervals occurs through digitization of the electrical signal; in the case of the PK1 this is done in 20 steps which corresponds to an increment of 1 μm. The image coordinates ξ and η which are passed to the computer are obtained by measuring the extreme value and digitizing the electrical signal.

The accuracy of a comparator is very much dependent on its design, especially on the extent to which the Abbe comparator principle is fulfilled. This fundamental principle means that the distance to be measured and the measuring scale should, so far as possible, lie in one straight line.

The greater the distance d between the scale and the segment l to be measured (Figure 6.4-4), the larger the measurement error Δl attributable to out-of-squareness, such as might be caused by a defective parallel guidance system of the photo-carriage. The Abbe comparator principle is strictly adhered to in the monocomparator PK1 (Figure 6.4-2) as opposed to the stereocomparator sketched in Figure 6.4-1 in which it is not.

For this reason no great accuracy requirement is placed on the parallel guidance system for the photo-carrier in the PK1. The accuracy of the PK1 is more or less determined by the accuracy of the glass scales and the accuracy with which the measuring mark M can be held fixed in relation to the two scales. All these causes of error together result in a mean square coordinate error of only $\pm 1\,\mu$m, an error which is less than that arising for other unavoidable reasons in photogrammetry, such as irregular film distortion which can amount to about $\pm 3\,\mu$m (Section 3.2.1.1).

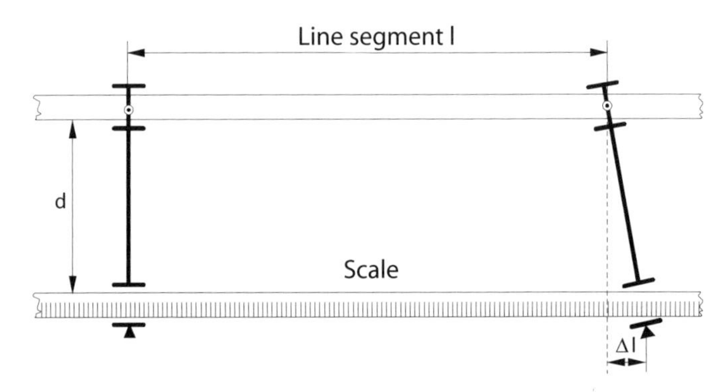

Figure 6.4-4: Measurement of segment l without regarding the Abbe comparator principle

6.4.3 Universal analytical stereoplotter

The union of a stereocomparator (Figure 6.4-1) and electronic image coordinate registration (Section 6.4.2) results in an analytical stereoprocessing instrument, with which the operator is able to register corresponding points in both images, upon which the 3D coordinates of the object point are automatically determined in the computer linked to the instrument. Each photo-carrier requires its own electronic coordinate registration system, by virtue of which such an analytical stereoplotter may be classified as an image space plotter. Image coordinates are the starting point for command of the instrument and there is no feedback from the computer to the photo-carriages or the measuring mark.

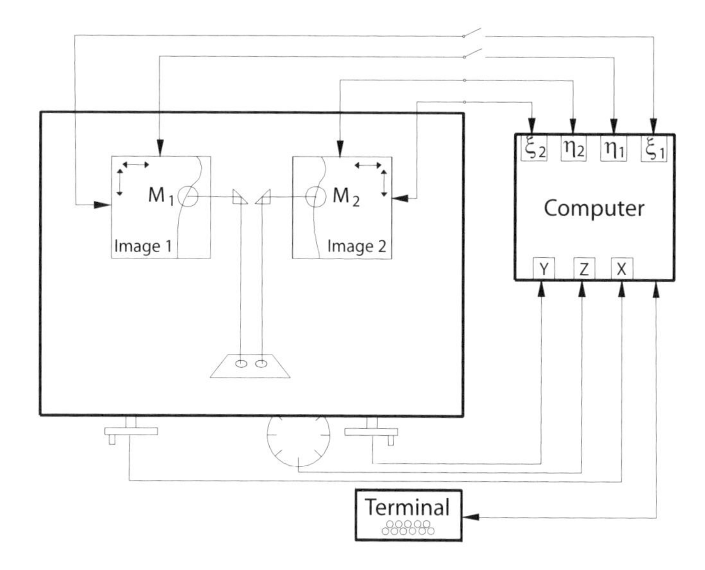

Figure 6.4-5: Universal analytical stereoplotter

The efficiency and generality of such an instrument can be increased in a fundamental way by means of a dominant incorporation of the computer in the processing, as illustrated by the schematic diagram in Figure 6.4-5. This kind of instrument is frequently classified under the rather clumsy heading of an "analytical plotter with object coordinates primary" since the starting point for command of the instrument consists of the XYZ object coordinates[6]. Using two handwheels and a footwheel the operator sets an object point in the XYZ coordinate system; alternatively, a 3D mouse may be used as described in Section 6.2. From this xyz coordinate vector and using the elements of interior and exterior orientation, the computer determines the corresponding image coordinates $\xi_1, \eta_1, \xi_2, \eta_2$, also taking account of all image refinements (Section 4.5). A prerequisite is that four servomotors should be capable of driving the two photo-carriages to the appropriate positions corresponding to image coordinates $\xi_1, \eta_1, \xi_2, \eta_2$; to accomplish this positioning operation four linear impulse generators (Figure 6.4-3) register with the computer the image coordinates corresponding to the current position of the photo-carriage. These coordinates are then compared with the required image coordinates and, corresponding to the existing differences, commands are given to the four servomotors for translation of the photo-carriers. This loop is repeated approximately 50 times per second. For all practical purposes the procedure is instantaneous; such processes, running rapidly under digital control through electronic data-processing, are known as real-time processes and programs necessary for them as real-time programs.

[6]Plans for a universal analytical stereoplotter were patented by U. Helava as early as 1957 (Phia XIV, pp. 89–96, 1957/58 and Ph.Eng. XXIV, pp. 794–797, 1958). It was another 20 years before the well-known firms extended their product range to include analytical plotters. Many references for further reading are to be found in: Petrie, G.: ISPRS Journal 45, pp. 61–89, 1990, and ITC-J, Volume 4, pp. 364–383, 1992.

The control process by which the two photographs are moved under the stationary measuring marks M_1 and M_2 (see Figure 6.4-5) is set in motion by the operator of a universal analytical plotter using rotation of the two handwheels and the footwheel, that is by the setting of the object coordinates X, Y, Z. If the object coordinates X, Y, Z describe a continuous sequence of points on the surface of the object, the operator sees the floating mark moving smoothly along this point sequence on the stereoscopically observed model. The operator must aim to turn the handwheels and the footwheel in such a way that the floating mark remains in coincidence with the surface of the stereoscopically perceived object model.

Exercise 6.4-1. Using aerial photographs ($1 : m_B = 1 : 30000$) a stereomodel is set up in an analytical plotter. After introduction of the Earth curvature correction the terrain height differs by 75 cm between the middle of the model and the corners. Check whether this discrepancy agrees with the fact of Earth curvature. (Answer: This difference is exactly in accordance with Earth curvature for this stereomodel)

6.5 Digital stereoplotting equipment

Because analogue photographs are used in analytical stereoplotters, relatively expensive components are necessary, for example drive mechanisms employing servomotors for the photo-carriages. If, on the other hand, digital photographs exist, translations of the two images can take place on the computer screen beneath stationary measuring marks. With such a digital stereo-processing instrument, therefore, the operator controls the processing by moving a "3D mouse" (Section 6.2) on the table and rotating the handwheel. The "3D mouse" defines the XYZ coordinates of a point sequence in the object model. If this digitized line runs along the object surface the operator sees the floating mark glide down this line in the stereoscopic model. Just as with an analytical stereoplotter, the task of the operator consists of moving the "3D mouse" and rotating the handwheel so that the floating mark remains in contact with the perceived stereoscopic model.

The requirement for digital displacement of both images in real time according to this principle of a digital stereoplotter is very demanding. For the observation of digital stereoimages (Section 6.1.2(b)) liquid crystals provide an image repetition frequency of at least 50 Hz for each of the two images. If two images in sequence are shifted by just one pixel, for a typical pixel size of 15 μm the speed of displacement of the image is:

$$50\,\text{Hz} \times 15\,\mu\text{m} = 0.75\,\text{mm/s} \mathrel{\hat{=}} 45\,\text{mm/min}$$

That is a relatively low speed for operator digitizing of 3D lines in the stereomodel. Therefore, for higher digitizing speeds or in the case of images with a smaller pixel size, the image repetition frequency must obviously be raised above 50 Hz.

A still greater problem than that of repetition frequency is the large image matrix of digitized photographs which, as an example, for an aerial photograph with a pixel dimension of 10 μm amounts in total to 23000×23000 pixels requiring approximately 0.5 GByte of space. The mentioned image data management can be handled as follows

(Figure 6.5-1). The screen has for example 1280 × 1024 pixels. In what is called the display memory of the graphic processor, which permits real-time access, a section of the total image is held available, perhaps four times larger than the screen. If the image excerpt on the screen approaches the edge of the segment in the display memory, the display memory is appropriately updated from the main storage of the computer, usually from a disk. For this purpose a precalculated direction of movement in the stereomodel, on the basis of the line already digitized, is very helpful. In the case when only coarse digitizing of a particular line is required (because, for example, the snake algorithm will carry out automatic refinement of the measurements, Section 6.8.7.1) a coarser image from the image pyramid (Section 3.6.1) can be held ready. As a result the region of the image from which the screen display can be updated in real-time is expanded. In many digital stereoplotters the zoom adjustment, controlled from the "3D mouse", occurs in conjunction with different levels of the image pyramid.

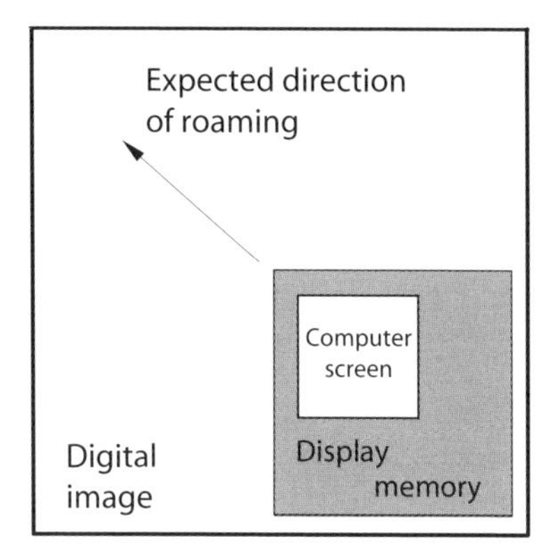

Figure 6.5-1: Different portions of an image for roaming in real-time

Further details of the roaming, so important for digital stereoplotters, may be found in Schenk, T.: Digital Photogrammetry, TerraScience, 1999, and elsewhere.

As a result, digital stereoplotters are distinguished by the fact that they do not only permit manual operation but also reach a high level of automation. This is not considered in detail until Section 6.8. First various manual procedures, with considerable computer support, are dealt with. Subsequently (Section 6.7) the accuracy of computer supported processing is discussed.

6.6 Computer-supported manual methods of analysis

Dependent on the desired results there are various ways of working. Individual object points can be selected in analytical and digital plotters by manually changing the XYZ coordinates, thus determining object coordinates (single point measurement). If one follows along a line in the model in three dimensions, such as the edge of a street or a wood, the coordinates of a dense sequence of points on this line are determined (line measurement). If the Z coordinate is left unchanged and if the floating mark is kept in contact with the surface of the stereoscopic model while changing only the X and Y coordinates, a contour line is digitized (contour measurement). When measuring lines a choice may be made between recording at constant intervals in time or at constant intervals in distance. In areas of higher curvature a greater density of points results when recording at constant time intervals, for here the operator moves more slowly. The interplay of various methods of working will be examined more closely in the following sections.

6.6.1 Recording in plan

Photogrammetric processes are always three-dimensional. The nature of many tasks is such that only two-dimensional measurement is required (measurement in plan). The third dimension, the height, is always available, but is either ignored or registered as an attribute of the plan point.

Figures 6.6-1 and 6.6-2 show typical results in plan. Figure 6.6-1 portrays photogrammetric output from an analogue instrument (Section 6.3). It is a manuscript which must be revised and completed, a process which requires about as much time as for its production on the analogue plotter. Computer support, therefore, is of great economic importance in photogrammetric plotting.

It is seen that, to a great extent in the large-scale field, computer-supported plotting delivers the final product. (Generalized representation at smaller scales, on the other hand, is carried out by cartographers, frequently by hand.) Prerequisites for creating the final product with a computer-supported system are convenient means of erasure, insertion and changing (editing) of recorded information.

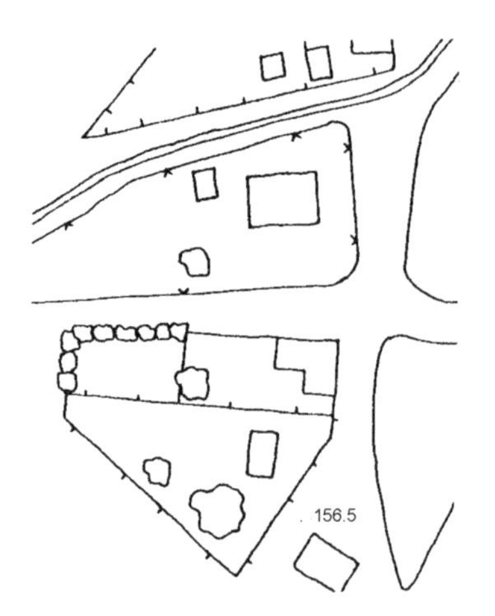

Figure 6.6-1: Photogrammetric manuscript from analogue plotting

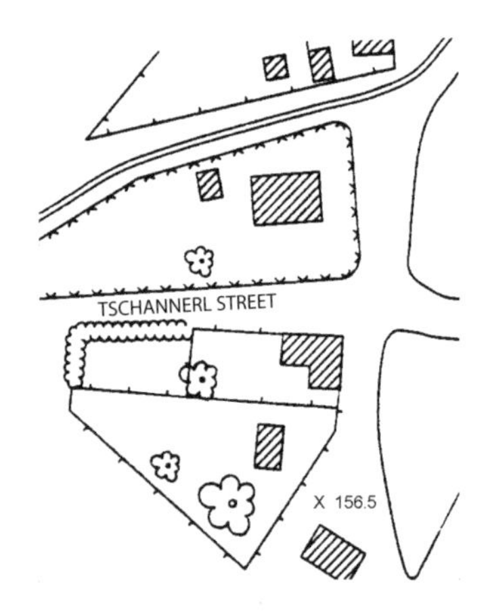

Figure 6.6-2: Graphical output from computer assisted photogrammetric plotting

Figure 6.6-2 shows a typical computer assisted photogrammetric product. Without going further into the algorithmic and technical computing side, the elements of such a plot are created as follows:

Buildings: The operator drives around the corner points of a building in turn. After storing these points one by one the operator gives the command to "square up" the connecting lines. The area defined in this way can be automatically hatched. Many computer assisted plotting systems offer the further possibility that, when the operator has driven around all but one of corner points of a building, the computer completes the final one.

Roads and railway lines: Instead of laboriously driving down long lines in plan as is necessary with an analogue plotter, the operator of a computer assisted plotter has only to set accurately on a few points; the computer then connects these by interpolation either with a polygon or with a higher order curve, as desired. A further possibility which is of special interest when plotting railway lines is that only one line is recorded in this way; the computer is instructed to add one, or more, parallel lines.

Fences and hedges: With computer assisted plotting there is also a suite of commands for various kinds of lines in the object. Thus, in the graphical output, fences, hedges and so on are directly drawn in their final cartographic form.

Symbols: As is the case for different kinds of lines, there is also a suite of commands for different kinds of object points. After the setting of the floating mark at a certain point and input of the command for the particular kind of object point, the desired symbol will be drawn at this point on the plan being output. The placement of symbols throughout the interior of a polygon can be particularly advantageous. The hatching of buildings has already been mentioned in this connection but drawing the symbols for trees, marshes and so on can also be very efficiently carried out in this way.

Spot heights: Spot heights deserve special mention. A single point is set at the desired position with the attribute that this point should be appropriately represented on the plan and should have the terrain height attached.

Text: Most computer-assisted systems also have a text generator. The operator manually chooses the position for the lettering and then enters the letters or sequence of numbers using the keyboard; these will ultimately be drawn on the plan in their final form.

Map sheet preparation: One great gain from this technology is that the map sheet can be prepared almost automatically. By that is meant the computer-controlled plotting of such items as the control points, the rectangular border, a map coordinate grid, reference to latitude and longitude, a legend or key and text. In so far as not only a plan but also elevations of the object have to be produced, a three-dimensional coordinate system is defined and in individual drawings the appropriate two-dimensional coordinate system is portrayed.

To conclude this section on plotting, field completion requires a brief mention. In field completion, details which are not recognizable in the stereomodel, or cannot be interpreted unambiguously, are gathered there and then on the spot. Field completion also includes the registration of administrative information; this means such matters as differentiation between official and private buildings, classification of paths and roads, names of streets and places, limits of political and administrative units and so on.

6.6.2 Determination of heights

When it became possible, now many decades ago, to trace contours directly in the stereomodel, it represented a great advance in topographic mapping and became the highlight of analogue stereoplotting. While the procedure can also be used with analytical and with digital stereoplotters, the direct following of contours demands understanding of topography and great skill on the part of the operator. The operator should not stare at the floating mark but should rather look ahead continuously along the contour. Directly drawn contours can be regarded as a cartographic end-product only at large scales and only for work of modest cartographic standard. Frequently it is necessary in a re-working to adjust uncertainties and to make sharp bends at the breaklines more precise[7]. Figure 6.6-3 conveys an impression of the appearance of directly traced contours before any revision.

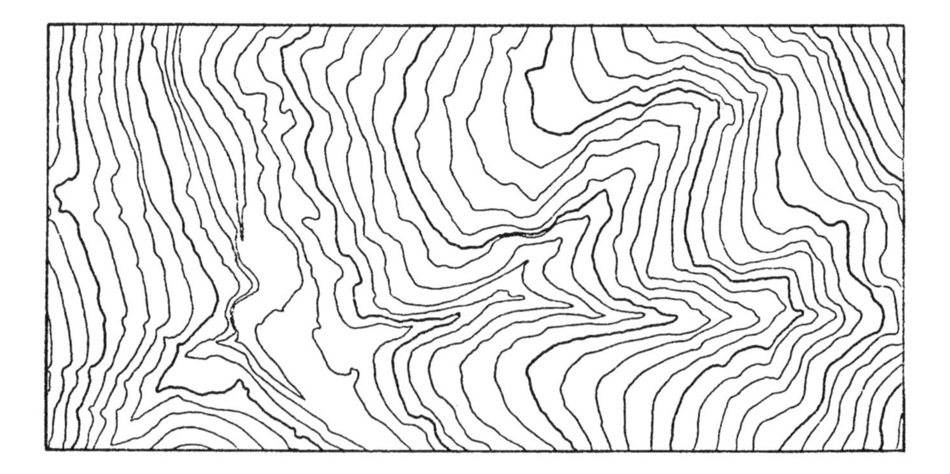

Figure 6.6-3: Directly traced contours[8]

Within forested areas in large-scale stereomodels the contour lines cannot be directly followed. One can frequently make do by measuring individual points in clearings with subsequent interpolation of the contours. Laserscanning, which is dealt with more closely in Section 8.1, offers an exceedingly interesting alternative in forested regions and when the demand for accuracy is high. In small-scale stereomodels contours can be followed directly even in forested parts. To do this one measures the tree height photogrammetrically at the edge of the forest and then, after appropriate change in

[7] Finsterwalder/Hofmann: Photogrammetrie. de Gruyter, 1968.

[8] Taken from E. Aßmus: Geow. Mitt. der TU Wien, vol. 8, 1976.

the height reading, moves the floating mark over the crowns of trees representing the forest canopy. Some operators on the other hand leave the height reading unchanged and dive, as it were, into the forest with the floating mark. In either case, contours in forest are considerably less accurate than on open land, a topic dealt with more closely in Section 6.7. Not only in forested areas but also in flat land, direct following of the contours is very risky; indeed on horizontal land contours are indeterminate.

Hence, alternative methods of photogrammetric height evaluation are to be considered which have become so competitive that, with the exception of high mountains, they have more or less displaced direct tracing of contours.

Figure 6.6-4 shows the result of a modern photogrammetric procedure whereby contours are extracted indirectly from a digital terrain model (DTM).

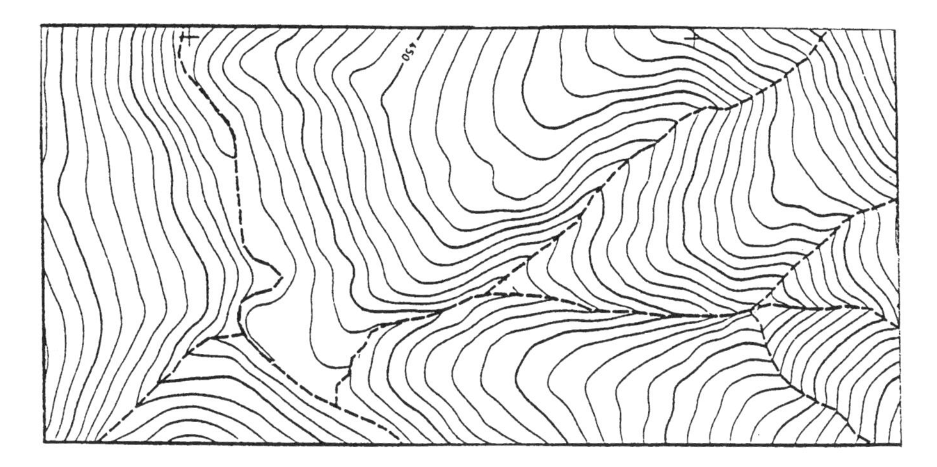

Figure 6.6-4: Contours determined indirectly from a digital terrain model (DTM), for comparison with those of Figure 6.6-3

The recording of the ground begins with the three-dimensional digitizing of the breaklines (shown as broken lines in Figure 6.6-4) marking streams and often ridges. Afterwards a static raster measurement, as it is called, takes place (Figure 6.6-5) in which the task of driving along a meandering path over the model is entrusted to the computer, the operator meanwhile having only to control the Z movement. The recording of the XYZ coordinates during this operation is usually arranged in such a way that the points lie on a square grid in the XY plane. For increased accuracy the travel along the profiles is not continuous; before the XYZ coordinates are taken the movement in XY comes to a complete stop. If satisfactory stereoscopic heighting is not possible at the pre-set XY position, because of vegetation or the presence of a house for example, that raster point is skipped or a substitute point near the defined XY point is used.

By interpolation from this data and the digitized breaklines, the next step is to produce a Digital Terrain Model (DTM), sometimes known as a DEM or a DGM (elevation and ground respectively), from which by-products, for example contour lines, can be derived. Such indirectly derived contour lines give a very good 3D impression (Figure 6.6-4); the integrated breaklines emphasize important geomorphologic elements.

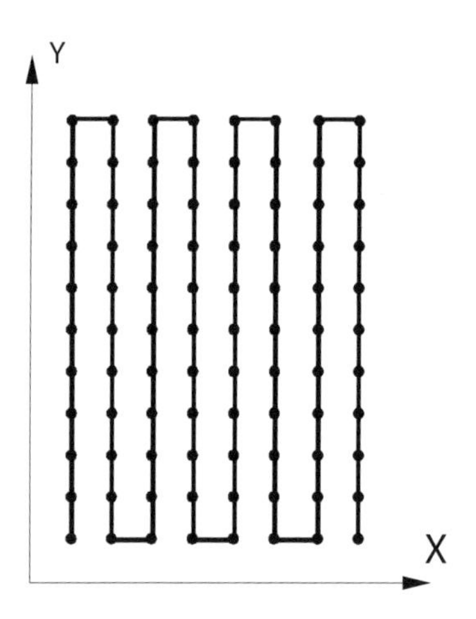

Figure 6.6-5: Static raster measurement

6.6.3 Recording of buildings

The high potential of stereophotogrammetry can be demonstrated on the basis of the recording of three-dimensional buildings. In this task the topological information, that is the connectivity between points, lines and surfaces is especially important. The topology in the case of simple buildings is captured in the computer-supported system in the form of standard models as knowledge bases. Figure 6.6-6 shows a graphical user interface with a ridge roof as a standard model. After having chosen the relevant standard model the operator has only to measure manually the corner points of the building. In this way the topological model receives its scale. Before taking a measurement on the building point in the stereoscopic model the operator clicks on the relevant building point in the topological model (Figure 6.6-6). If corner points cannot be set in the photogrammetric model they are added by the computer, using the corner points already measured and the geometric conditions implicit in the current standard model. In this work and in the case of two blended standard models (as, for example, a ridge-roofed extension to a ridged roof such as those in Figure 6.6-7) this measurement of building points and building edges is of great benefit. To this end the operator clicks in the topological standard model (Figure 6.6-6) around the middle of the ridge on which the point to be measured will lie; then he sets a point in the stereomodel on the chosen building ridge. Such a working technique is designated topology assisted recording.

To assist such a computer supported plotting system, more accurately a topology assisted system, a superposition system is required. This offers the possibility for those

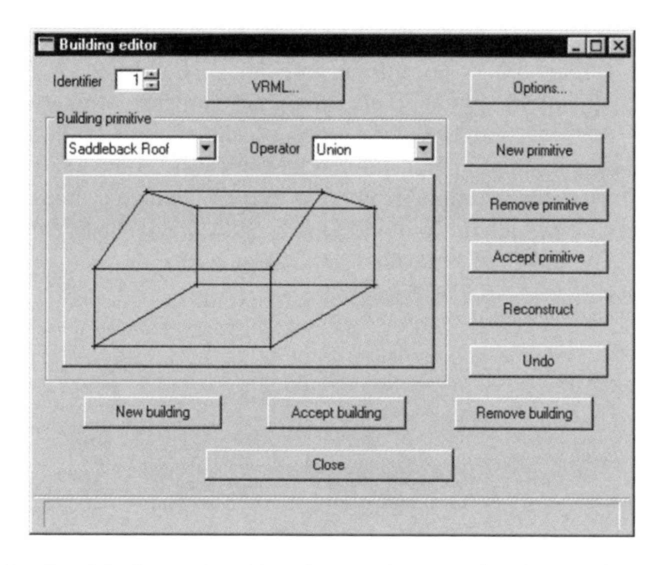

Figure 6.6-6: Graphical user interface for topology assisted recording (the standard model is known as a “primitive”)

Figure 6.6-7: Superposed building measurements in one image (Taken from: Rottensteiner, F.: Geow. Mitt., vol. 56, 2001.)

points and lines which have already been measured to be overlaid on one of the two digital photographs (Figure 6.6-7). For this purpose the XYZ object coordinates are to be transformed into the relevant photograph, using the parameters of interior and exterior orientation and taking account of the image coordinate refinement (Section 4.5). Still better is a three-dimensional superposition in the optical stereomodel; this means a superposition of the already measured features in both of the digital images. Apart from the check on accuracy which this provides, the operator is constantly kept informed of what he has already measured and what he still must do.

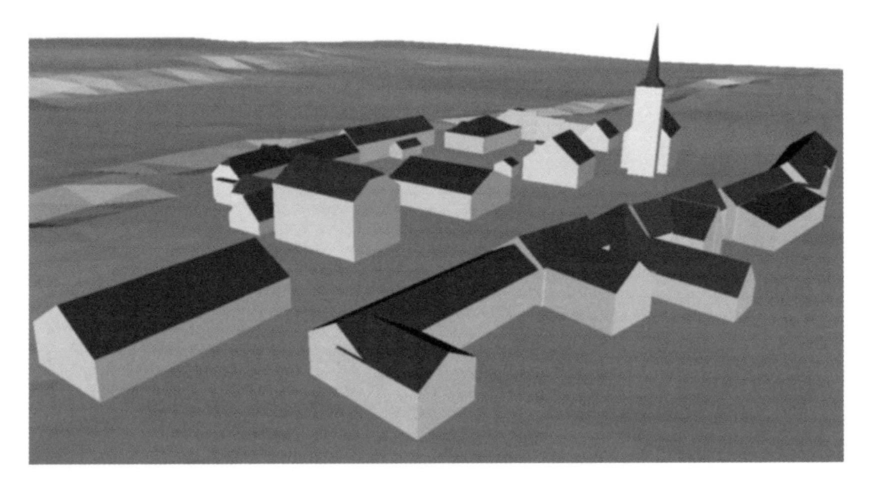

Figure 6.6-8: Oblique view of photogrammetrically processed buildings (Taken from: Rottensteiner, F.: Geow. Mitt., vol. 56, 2001.)

Figure 6.6-8 shows a rendered[9] oblique view of an assembly of buildings which coincides to a large extent with Figure 6.6-7. As a result of the topological information concerning the building surfaces it is no problem to declare, for example, that roof surfaces are dark. In Section C 1.1.3, Volume 2 there is a description of how it is possible for the surface topology to be structured in a relational database.

Incidentally, the Level of Detail (LoD) can also be efficiently regulated using the standard building model. In the above example a standard model without chimneys was used; larger extensions were, however, modelled. The final roof form in the case of a ridge roof with a ridge-roofed extension is formed by the blending of two ridge roofs as one can see for oneself in some of the blended buildings. Suites of CAD (Computer aided design) programs provide such blending functions. (The subject of more highly automated photogrammetric measurement of buildings is entered into in Section 6.8.8 in this volume.)

[9]Rendering: generation of a two-dimensional image from a three-dimensional model with the application of various lighting effects (shadows, surface reflections, etc.).

The façades of buildings cannot be acquired from aerial photographs. Façades must be captured with terrestrial photogrammetry, and evaluated either in two or in three dimensions. Figure 6.6-9 shows a typical example. The façade was plotted with an analytical stereoplotter. Computer support is especially efficient in such work. Three-dimensional geometric models of windows, ornamental features and such-like can be created, taking account not only of topology but also of proportions, and can be placed in multiple positions in the CAD model. In this way, for example, one of the windows can be accurately measured and then this window can be duplicated in positions chosen by the operator. In this kind of work the superpositioning system mentioned above is especially important for it allows the operator continually to check for any possible departures from a "standard window".

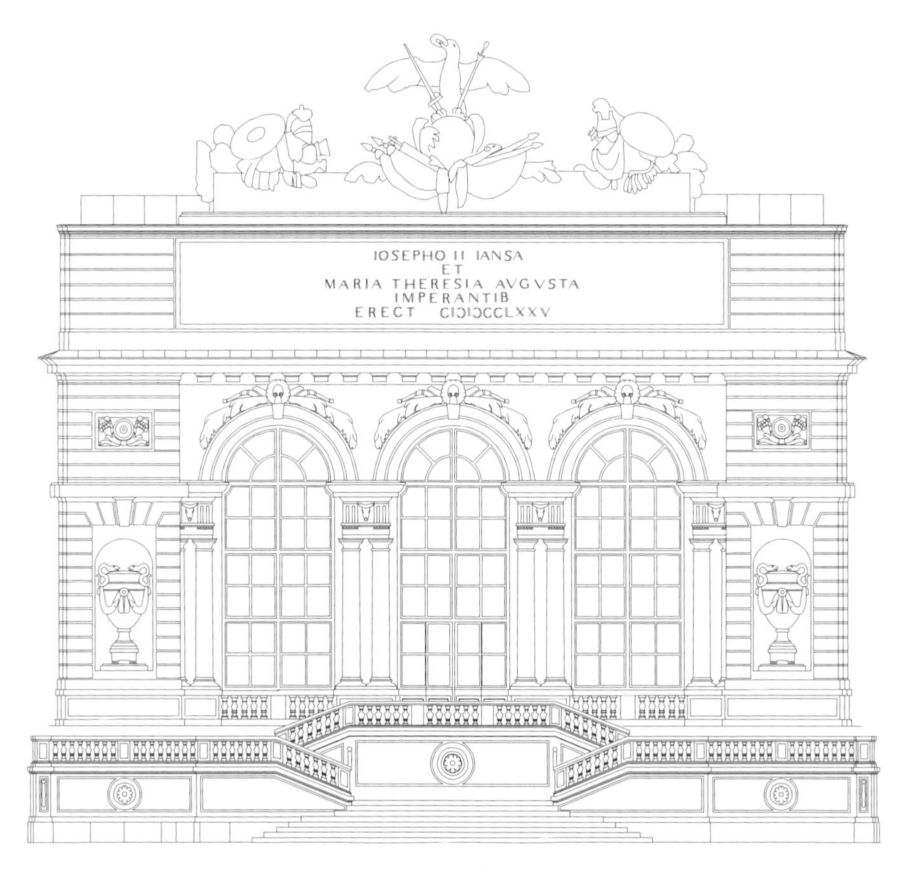

Figure 6.6-9: Façade of the Gloriette of Schloss Schönbrunn (Plotted by: J. Tschannerl, I.P.F.)

6.6.4 Transition to spatially related information systems

Photogrammetric plotters can build up relatively universal systems with which a very wide variety of products can be generated. Thus, for example, when digitizing an arbitrary three-dimensional line in real-time a running count of the plan distance can be computed and on a graphical output device a profile, showing height against distance, can be drawn, folded over and unwound. Or the computer can be commanded to drive the floating mark along geometrical figures in plan such as circles, polygons, and clothoids while the operator inputs the Z, height, coordinate. The result is a length profile over a pre-determined mathematical route traced in plan.

The developments towards universal systems outlined above at present assume less and less importance: photogrammetric stereoprocessing equipment is taking on the task only of three-dimensional digitizing. Digital object models are constructed from the resulting data. As required, various derivatives are acquired from the object models such as contours, straight and curved profiles, slopes, curvatures, oblique views, shading, and others.

In this connection several definitions are presented: digital object models are a simplification of the real world which originate through idealization and discretization and which are accessible for systematic use in electronic data processing. To the digital object models belong therefore not only the data, the three-dimensional coordinates of discrete points, but also

- encoding of the significance of points, lines and surfaces
- elements concerned with data structure
- the algorithms for conversion from the discrete points to curves and surfaces

If the digital object models are designed as the central database then the next step emerges a spatially related information system.

These definitions can be still further refined in the case of topography: digital topographic models are a simplification of the natural and cultural landscape which arise through idealization and discretization and which are accessible for systematic use in electronic data processing. The digital topographic models are the central database of a topographic information system which is a subset of superior geo-information systems.

Digital topographic models can be still further divided:

- the digital terrain model (DTM: see Figure 6.6-10) refers to the surface of the terrain, without buildings or vegetation
- the digital surface model (DSM: see Figure 6.6-10) refers to the surface with the inclusion of buildings and vegetation
- the digital building model refers to the buildings, including their roofs
- the digital street model refers to roads and street
- etc.

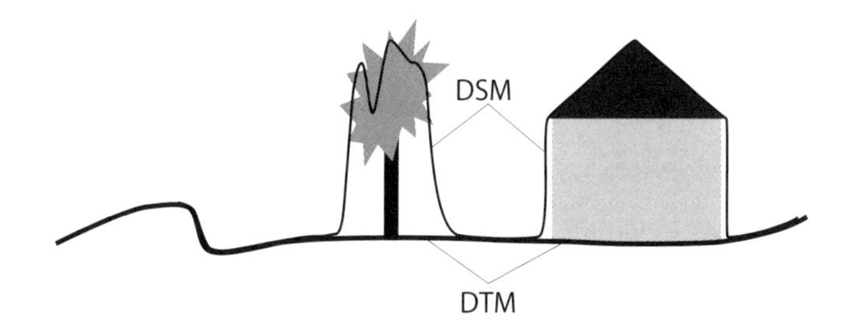

Figure 6.6-10: Terrain model (DTM) and surface model (DSM)

6.7 Operator accuracy with a computer assisted system

Measurement accuracies attained by the operator of a computer assisted photogrammetric instrument are given in this section. With technology as it stands today, there is no need to differentiate between analytical and digital equipment. Rules of thumb are given for rough estimation of accuracy, which also serve well in project planning. The structure of the section is directed towards the products discussed in Section 6.6.

6.7.1 Measurement in plan

6.7.1.1 Point measurement

With analytical and digital stereosystems, as explained in Section 6.6.1, objects are recorded, above all, through individually measuring single points which are connected by means of lines. The accuracy of point measurement was dealt with in Section 4.6 and the following numerical example should recall the statements made there.

Numerical Example. The photogrammetric plot in Figure 6.6-2 originated with a photoflight at a flying height of 1600 m using a 21 cm camera. What is the accuracy of the local path which was recorded point by point and that of the fence, likewise recorded point by point? What is the accuracy of the spot height? The photo scale comes to $1 : m_B = 1 : 7600$ (= 1600/0.21).

According to the rule of thumb (4.6-1) the accuracies for very well defined points are as follows:

$$\text{Plan:} \quad \sigma_{XY} = \pm 4.6\,\text{cm} \ (= 7600 \times 0.0006)$$

$$\text{Height:} \quad \sigma_Z = \pm 9.6\,\text{cm} \ (= 160000 \times 0.00006)$$

For natural points the uncertainty of definition in plan and in height, $\sigma_{XY(0,\text{def})}$ and $\sigma_{Z(0,\text{def})}$, according to the relationships (4.6-3), must be taken into account. We make the following assumptions: for the local path $\sigma_{XY(0,\text{def})} = \sigma_{Z(0,\text{def})} = \pm 1$ cm and for the fence $\sigma_{XY(0,\text{def})} = \sigma_{Z(0,\text{def})} = \pm 8$ cm, while for the spot height $\sigma_{Z(0,\text{def})} = \pm 2$ cm. Using these values the relationships (4.6-3) yield the following accuracies for the plan drawing of Figure 6.6-2:

$$\begin{array}{llll} \text{Local path:} & \sigma_{XY} = \pm 4.7\,\text{cm} & \sigma_Z = \pm 9.7\,\text{cm} \\ \text{Fence:} & \sigma_{XY} = \pm 9.2\,\text{cm} & \sigma_Z = \pm 12.5\,\text{cm} \\ \text{Spot height:} & & \sigma_Z = \pm 9.8\,\text{cm} \end{array}$$

Note: the point density along the lines must be chosen such that no significant error of interpolation will be introduced. The object-specific uncertainties of definition $\sigma_{XY(0,\text{def})}$ and $\sigma_{Z(0,\text{def})}$ should be evaluated in the photogrammetric instrument, preferably with large-scale images, by means of repeated measurements.

Exercise 6.7-1. The accuracy needs to be improved, above all in height. For a practical project, for example with respect to the local path, one expects accuracy of $\sigma_{XY} = \pm 2\,\text{cm}$ and $\sigma_Z = \pm 2.5\,\text{cm}$. How can this accuracy be achieved? (Answer: Use of a 15 cm camera (wide-angle) instead of a 21 cm camera. Making full use of the accuracy potential of bundle block adjustment with additional parameters (Section 5.3.4) and reducing the flying height.)

For a first trial:
Flying height 800 m, implying $1 : m_B = 1 : 5300$.

$$\begin{array}{lll} \text{For very accurately defined points:} & \sigma_{XY} = \pm 1.6\,\text{cm} & \sigma_Z = \pm 2.4\,\text{cm} \\ \text{For the local path:} & \sigma_{XY} = \pm 1.9\,\text{cm} & \sigma_Z = \pm 2.6\,\text{cm} \end{array}$$

The requirement for plan accuracy is met. In height the value is slightly excessive; this can be remedied by a slight reduction in flying height.

6.7.1.2 Processing of lines

Continuous guiding of the floating mark along a line is afflicted with considerably greater inaccuracies than in single point setting. By empirical investigation as early as 1954 Heißler[10] found an accuracy of about

$$\sigma_L = \pm 45\,\mu\text{m at photo scale} \tag{6.7-1}$$

Considering a drawing accuracy specification in final cartography of ±0.2 mm, an enlargement ratio from image to graphical output no greater than 1 : 4.5 may be chosen. Such an enlargement ratio is perfectly normal when mapping at scales from 1 : 1000 to 1 : 2000. It is not possible in small scale mapping, however, to use such high enlargements from image to plot, which would imply the use of very small-scale photographs; this is determined not by limits on graphical accuracy but by limits on interpretability of the content of images at very small scales. Thus, for example, the operator who must portray lines in plan in a topographical map at 1 : 50000 can only just make them out in photographs at about 1 : 60000. The optimal choice of photo scale in view of

[10]BuL 22, pp. 67–79, 1954.

- accuracy in large scale mapping and
- interpretability in small-scale mapping

is expressed in the following rule of thumb (6.7-2) or in Table 6.7-1 (the parameter k varies between 200 and 300 depending on the quality of the aerial photographs and on the demands on the end-product):

$$m_B = k\sqrt{m_M} \tag{6.7-2}$$

$1 : m_M$	m_B
1 : 1000	6300 – 9500
1 : 5000	14000 – 21000
1 : 10000	20000 – 30000
1 : 25000	32000 – 47000
1 : 50000	45000 – 67000

Table 6.7-1: Relationship between map scale m_M and photo scale m_B

The draughting accuracy of ± 0.2 mm plays a central role in the considerations of Heißler, mentioned above. With the digital graphical output of the results produced using analytical or digital instruments, the relationship between draughting accuracy and map scale is largely lost. The operator can choose, almost, any arbitrary scale for the graphical representation. Just as tricky are the zoom possibilities of the graphical systems which permit arbitrary enlargement of the content and give the user the impression, because of the fine lines even when the enlargement is very high, of superior accuracy of the data set. These risks can be removed if, at the same time as the graphical representation is magnified, the lines are intensified. The map scale numbers m_M given in Table 6.7-1 and in the Formula (6.7-2) refer to drawings in which one expects an accuracy of ± 0.2 mm. Graphical representations with considerably worse accuracy than this do not warrant description as photogrammetric products.

6.7.2 Height determination

6.7.2.1 Directly drawn contours

The conventional method of contouring in analytical and digital instruments is by direct tracing in the model (Figure 6.6-3 shows an example). The accuracy of such contours is dependent on the slope α of the land as first expressed by Koppe:

$$\sigma_H = \sigma_Z + \sigma_C \tan \alpha \tag{6.7-3}$$

σ_Z = 0.2‰ of the distance from camera to ground, which is the accuracy of continuously traced lines from photographs with 60% forward overlap. Compared with the accuracy of Equation (4.6-1) this is considerably worse since this is a dynamic event unlike the measurement of an individual point. In areas of forest this value is too optimistic.

σ_C = 100 μm (at photo scale) which is the plan accuracy of directly drawn contours. This is distinctly worse than the plan accuracy of Formula (6.7-1) since the contours in the stereomodel are not marked in the photographs but have to be sought.

For a statement of the accuracy—with reference to the object—of directly drawn contours, in the case of aerial photographs in open terrain, Equation (6.7-3) is modified as follows:

$$\sigma_H = \pm\left(0.2‰ \quad \text{of} \quad h + \frac{0.10}{c} h \tan\alpha\right) \tag{6.7-4}$$

h ... flying height above ground;
c ... principal distance (in [mm]);
$\tan\alpha$ ... slope of the ground

The dimensions for σ_H are the same as those chosen for h.

Numerical Example. The contours of Figure 6.6-3 originated with a photoflight using a 15 cm camera and a photo scale of 1 : 30000. Equation (6.6-3) gives the following accuracy for the contours (flying height = 4500 m):

$$\sigma_H[\text{m}] = \pm(4500 \times 0.0002 + (0.10/150)4500 \tan\alpha)$$
$$\sigma_H[\text{m}] = \pm(0.9 + 3 \tan\alpha)$$

In flat land, therefore, the height accuracy comes to ± 0.9 m; on a 10% slope to ± 1.2 m; and on a 25% slope to ± 1.7 m; and so on. In forested areas these figures are increased by a further ± 2.0 m.

6.7.2.2 Relationship between contour interval and heighting accuracy

In topographic mapping it is usual to choose a contour interval which, in metres, is one thousandth of the map scale number; thus for a scale of 1 : 5000 a contour interval of 5 m.[11] In flat country the interval should be reduced. In order that the character of the contours should not be negatively influenced, the mean square error of the contours should not exceed 1/4 to 1/8 of the contour interval, according to the quality demanded for the end product.

[11] Imhof favoured distinctly smaller contour intervals (Hake, G., Grünreich, P., Meng, L.: Kartographie. 8th ed., Walter de Gruyter, 2002).

Numerical Example (Continuation of the numerical example of Section 6.7.2.1). The contour lines of Figure 6.6-3 were produced for an orthophoto at 1 : 10000 with a contour interval of 10 m, which is in accordance with the above rule of thumb. Depending on the quality stipulated for the end product, the accuracy of the contours should therefore lie between ± 1.25 m and ± 2.5 m. Since the maximum slope of the terrain is 25%, application of Formula (6.7-4) results in a heighting accuracy of ± 1.7 m—as is already reported above. The height accuracy of the contours is therefore too poor to meet very high quality map requirements; for medium quality it is satisfactory. This statement is valid for open country. In forested areas, with 25% slope, the height accuracy deteriorates to about ± 2.6 m ($= \sqrt{1.7^2 + 2.0^2}$), so that, even for very modest quality requirements, the accuracy of the contours is too poor. The contour lines which touch each other in Figure 6.6-3, bear eloquent witness to the poor height accuracy. Such contancting lines may appear, or even cross one another, especially in forested areas.

Exercise 6.7-2. Re-consider the numerical example of Sections 6.7.2.1 and 6.7.2.2 under the assumption that, instead of a 15 cm camera, a 21 cm camera is used. For reasons considered more closely in Section 7.3.2, the 15 cm camera is not popular for orthophoto work. (Answer: The vertical accuracy of contours for a 25% slope is, in open country, $\sigma_H = \pm 2.1$ m and, in forested country, $\sigma_H = \pm 2.9$ m. This means that, perhaps, instead of a 10 m vertical interval a 12.5 m interval should be selected.)

6.7.2.3 Contours obtained indirectly from a DTM

Contours obtained indirectly via a digital terrain model (DTM) have the accuracy of the DTM. This amounts to:

$$\sigma_H = \pm \left(0.15‰ \quad \text{of} \quad h + \frac{0.15}{c} h \tan \alpha \right) \tag{6.7-5}$$

h ... flying height above ground;
c ... principal distance ([mm]);
$\tan \alpha$ ... slope of the ground

The dimensions for σ_H are the same as those chosen for h.

The constant term in the expression (6.7-5), 0.15‰ of the flying height, corresponds approximately to the vertical accuracy of static raster measurement (Figure 6.6-5). In forested country another value has to be added, as in Formula (6.7-3).

Numerical Example. The contours in Figure 6.6-4, derived indirectly from a DTM, have the following accuracy (the parameters for the photoflight are those of Section 6.7.2.1:

$$\sigma_H[\text{m}] = \pm(4500 \times 0.00015 + 0.15 \times 4500/150 \tan \alpha)$$
$$\sigma_H[\text{m}] = \pm(0.7 + 4.5 \tan \alpha)$$

In flat, open country, therefore, the vertical accuracy of the contours comes to ± 0.7 m, in land with a 10% slope to ± 1.1 m, and with a 25% slope it is ± 1.8 m.

The choice of spacing for the static raster measurements, the choice of vertical interval in relation to the grid spacing of the DTM and also the geomorphologic quality of the interpolated contours should be closely considered.

6.7.2.4 Measurement of buildings

The recording of buildings, introduced in Section 6.6.3, proceeds through single point measurement in stereoplotters, mainly digital stereoplotters. Hence, the accuracy corresponds to that of single point measurement as reviewed in Section 4.6.

Numerical Example. For the buildings of Figure 6.6-8 image material was available at a scale of 1 : 4500, taken with a 15 cm camera. On the basis of experience in this and similar projects, the following may be taken from Table 4.6-1 as the appropriate values for uncertainty of definition: $\sigma_{XY(0,\mathrm{def})} = \pm 7\,\mathrm{cm}$, $\sigma_{Z(0,\mathrm{def})} = \pm 8\,\mathrm{cm}$. Assuming these values, the following accuracies for stereoscopic recording of buildings are derived:

$$\sigma_{XY} = \pm 7.5\,\mathrm{cm} \quad \sigma_H = \pm 9.0\,\mathrm{cm}$$

The accuracy of points on buildings measured from large-scale images will therefore be dominated by uncertainties of definition.

Numerical Example. For the recording of the façade of Figure 6.6-9 image material was available at a scale of 1 : 290, taken with the 45 mm camera Wild P31 (Table 3.8-2). The distance from camera to object was 13 m (= 290 × 0.045). Because of visibility problems increasing with enlargement of the base, the full format of the P31 in the base direction was not exploited. Two stereopairs were chosen each with a base of 7 m, giving a photo-base of 24 mm (= 7000/290). As is general at close range, the accuracy should be estimated not from Equation (4.6-1) for very well defined points, but from the relationship (2.1-35) in which one normally takes $\pm 6\,\mu\mathrm{m}$ as the measuring accuracy at image scale.

Accuracy in the façade,
at right angles to the camera axis, $= \pm 1.7\,\mathrm{mm} (= 290 \times 0.006)$

Accuracy perpendicular to the façade,
in the direction of the camera axis, $= \pm 3.2\,\mathrm{mm} (= 1.7 \times 13/7)$

Additionally, using Equation (4.6-3), an object-related uncertainty of definition should be taken into account; for a point on a façade this is about ± 5 mm.

Exercise 6.7-3. With the same positions of the camera, but using a 100 mm P31 camera instead of the 45 mm camera, how would the accuracies of this numerical example change? (Answer: Accuracy in the façade = ± 0.8 mm; accuracy perpendicular to the façade = ± 1.4 mm. This increased accuracy comes at the cost of losing the upper portions of the façade—when using the “landscape” orientation of the camera they do not fall within the image.)

6.7.3 Checking of the results

Photogrammetric processing should be validated at certain intervals by means of check measurements which, for example, are carried out with a photo scale distinctly larger than the original or with GPS measurements on the ground. Such comparative measurements expose all error sources, beginning with potential errors in calibration of the camera, through possible inadequacies in the operator, up to errors in the photogrammetric instrument and shortcomings in the software. This expensive procedure can, to a large extent, be replaced by checking of the model joins. In general errors are at their largest at the margins of the models, so that the discrepancies in measurements between two different stereomodels allow a good quality control. In this procedure errors in relative and absolute orientation are also revealed. Errors arising in the overlap zones of the stereomodels also give a very reliable basis for consideration of the oft-mentioned uncertainties of definition for object points and lines.

For this section and the preceding Section 6.7.2, Appendix 4.6-1, dealing with empirical determination of standard deviations and tolerances, can be of interest as well.

6.8 Automatic and semi-automatic processing methods

Digital photographs open up the possibility of automatic processing. In place of automatic processing methods, names such as machine processing or procedures based on computer vision are also common. Some photogrammetric problems can be solved fully automatically, others only partly automatically. Partly automated procedures, which are treated in Sections 6.8.7 and 6.8.8, are known as semi-automatic procedures.

The topic of automatic processing commences here with the correlation algorithms. They are central to many automatic procedures. Correlation processes take the place of humans at photogrammetric instruments. Instead of human sight and action, machine sight and action take over.

6.8.1 Correlation, or image matching, algorithms

Correlation algorithms solve the task of finding corresponding patches in two images, usually images taken from different camera positions. In this connection one speaks of image matching and also of the maximum similarity or best agreement of the two image patches. One of the two images can even be a geometric figure, in which case the term pattern recognition is very appropriate. In the following passages a well-known correlation algorithm is explained on the basis of such pattern recognition. The geometric figure might be a cross which may exist in the form of an artificially generated digital image (Figure 6.8-1, left). This image is known as the reference image or the reference matrix. One speaks also of a pattern matrix or a template. The second image may be known as the search image.

The position of the cross, that is of the reference image, is sought in the second image. The terms search window or search matrix are used. In Figure 6.8-1 a 5×5 template is compared with a 12×12 search window; for the sake of simplicity the grey levels are assumed to lie only between 1 and 9.

The cross obviously lies in the search image in position $i = 8$ and $j = 7$. The automatic search for this position is made more difficult because

- on the one hand the search image has faded out, that is the grey levels contain accidental defects, and
- on the other hand, on account of the finite size of the detectors with which the search window was created, the grey levels are smeared, that is mixed pixels appear at the edges.

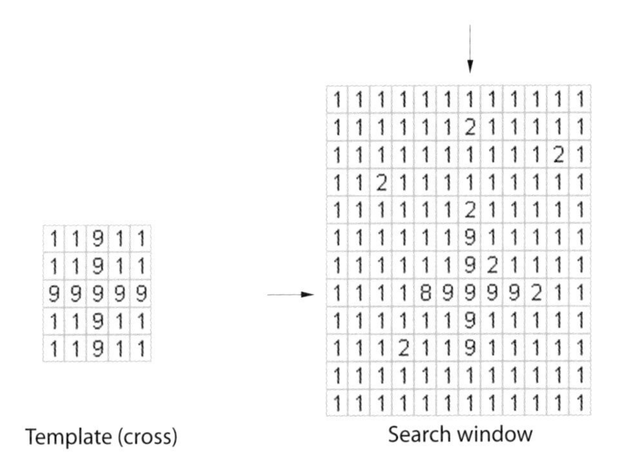

Figure 6.8-1: Reference matrix (template) and search matrix (search window)

For the sake of simplicity the solution of this correlation task is demonstrated, not in two dimensions but with reference to a one-dimensional example (Figure 6.8-2).

6.8.1.1 Correlation coefficient as a measure of similarity

The desired position of the template in the search window is found by calculating correlations. A measure for the correlation or similarity is the correlation coefficient r. This is computed as follows from the standard deviations σ_r and σ_s of the grey levels g_r and g_s in both images and also the covariance σ_{rs} between the grey levels of the two images:

$$r = \frac{\sigma_{rs}}{\sigma_r \sigma_s} = \frac{\sum(g_r - \overline{g_r})(g_s - \overline{g_s})}{\sqrt{\sum(g_r - \overline{g_r})^2 \sum(g_s - \overline{g_s})^2}} \tag{6.8-1}$$

in which $\overline{g_r}$ and $\overline{g_s}$ are the arithmetic means of the grey levels of the template and those of the corresponding section of the search window respectively.

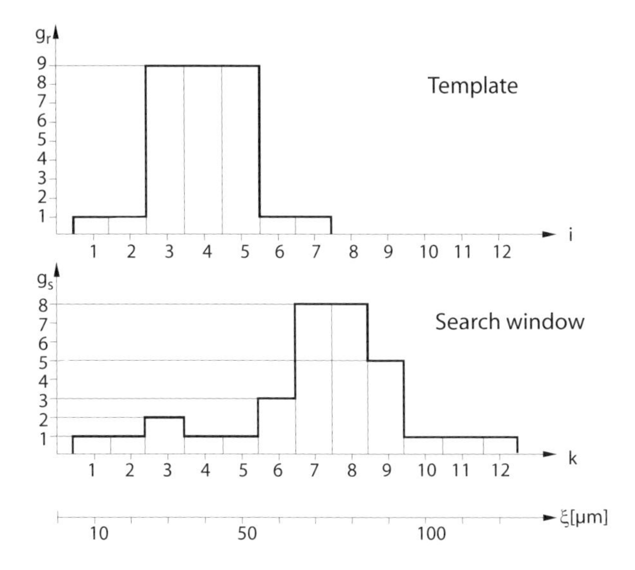

Figure 6.8-2: Template and search window of a one-dimensional pattern

The correlation coefficient r is evaluated for all possible positions of the template in the search window. The position with the largest correlation coefficient r is the position sought.

Numerical Example (with reference to Figure 6.8-2). The correlation coefficient for the 1st position, computed with $i = 1, \ldots, 7$ (or $k = 1, \ldots, 7$), is:

$$r_1 = \frac{\sum(g_r - 4.43)(g_s - 2.43)}{\sqrt{\sum(g_r - 4.43)^2 \sum(g_s - 2.43)^2}} = \frac{-26.3}{10.5 \times 6.3} = -0.40$$

The correlation coefficients for all possible positions are collected in the following Table 6.8-1 in which the matching ξ coordinates (the pixel size $\Delta\xi$ = 10 μm) are also given:

Position	1st	2nd	3rd	4th	5th	6th
k	1 – 7	2 – 8	3 – 9	4 – 10	5 – 11	6 – 12
r	−0.40	−0.51	0.00	0.73	**0.92**	0.24
ξ [μm]	40	50	60	70	80	90

Table 6.8-1: Correlation coefficients for the example outlined in Figure 6.8-2

The 5th position is that sought; thus ξ = 80 μm. The neighbouring correlation coefficients ($r_4 = 0.73$ and $r_6 = 0.24$) mean that the optimal position must lie somewhat before $\xi = 80\,\mu$m. Thus a fit in the subpixel region is sought in a second step.

6.8.1.2 Correlation in the subpixel region

The following will provide a fit in the subpixel region

- formation of a continuous correlation function in the neighbourhood of the largest correlation coefficients and
- determination of the maximum of this correlation function by setting the first derivative to zero

Numerical Example (continued). The following second order polynomial is found from the last three ξ coordinates:

$$r(\xi) = 24.96 + 0.67150\xi - 0.00435\xi^2$$

The first derivative is:

$$r'(\xi) = 0.67150 - 0.00870\xi$$

Setting the first derivative to zero results in:

$$\xi = 77.2\,\mu\text{m}$$

Further details of this method, in particular in the subpixel region with two-dimensional images, are to be found in Section B 6.1.2.2, Volume 2.

A very widely used method of correlation in the subpixel region employs a least squares estimation method (Appendix 4.1-1) and is known as least squares matching (LSM). In LSM one starts from an approximate fit which can be first found using the correlation procedure described in Section 6.8.1.1. If, after the approximate fit, the template and the search window are referred to the same coordinate system, in our case that of ξ coordinates, then the required translation of the template with respect to the search window is only small, usually less than a pixel (Figure 6.8-3). Looking at the template and the search image together in Figure 6.8-3 one can conclude that the template should be moved somewhat to the left for an optimal fit, that is $\xi < 80\,\mu$m.

Let this small translation be b. The positions of the two grey level ensembles g_r and g_s are therefore related as follows:

$$g_s(\xi) = g_r(\xi + b) \tag{6.8-2}$$

Let the grey level g_s contain a random component v. The right-hand side of Equation (6.8-2) must be linearized; since b is small, this results in the following equation:

$$v = \frac{\partial g_r(\xi)}{\partial \xi} b - (g_s(\xi) - g_r(\xi)) \tag{6.8-3}$$

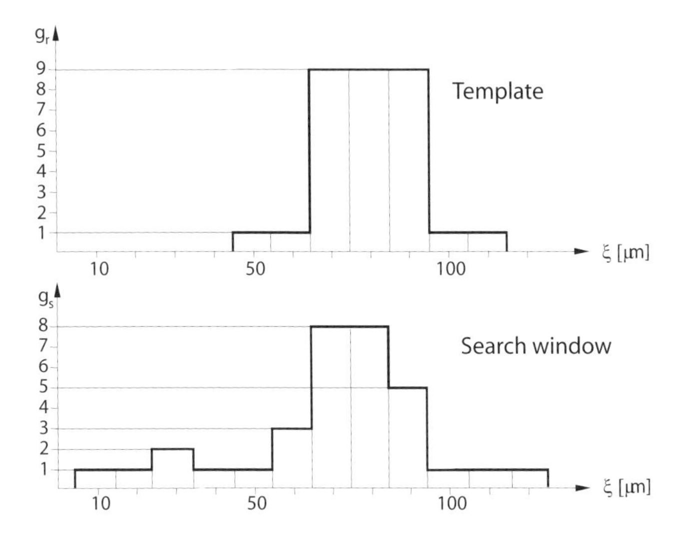

Figure 6.8-3: Template and search window after an approximate fit

$g_r(\xi), g_s(\xi)$ = corresponding grey levels in the two images, the number of such pairs of grey levels being determined by the size of the template

$(\partial g_r(\xi)/\partial\xi)$ = the slope of the grey level profile of the template at the particular position; that is, $(\partial g_r(\xi)/\partial\xi) = \Delta g/\Delta\xi$. For the first pixel Δg is taken as $\Delta g = g_2 - g_1$ and for the last pixel, the nth pixel, $\Delta g = g_n - g_{n-1}$; for those pixels lying in between $(1 < i < n)$ Δg is expediently chosen as $\Delta g = g_{i+1} - g_{i-1}$; for these interior pixels $\Delta\xi$ is twice the pixel dimension.

The procedure follows the formal system of adjustment by the method of indirect observations (Appendix 4.1-1)

$$\mathbf{v} = \mathbf{A}^\top\hat{\mathbf{v}} - \mathbf{l}, \qquad \hat{\mathbf{v}} = (\mathbf{A}^\top\mathbf{A})^{-1}\mathbf{A}^\top\mathbf{l} \tag{6.8-4}$$

The accuracy of the position unknown b comes out of the computation (Appendix 4.1-1) after the standard deviation of unit weight σ_0 has first been estimated from the elements of the vector $\mathbf{v}$. The standard deviation σ_b of the unknown b comes from the following:

$$\mathbf{N} = (\mathbf{A}^\top\mathbf{A}), \qquad \mathbf{Q} = \mathbf{N}^{-1}, \qquad \sigma_b = \sqrt{q_{bb}}\sigma_0 \tag{6.8-5}$$

The possibility of such error computation is a great advantage of the LSM over other correlation methods.

For our example the least squares equations are as follows (6.8-3):

$$\mathbf{v} = \begin{pmatrix} 0 \\ 0.4 \\ 0.4 \\ 0 \\ -0.4 \\ -0.4 \\ 0 \end{pmatrix} b - \begin{pmatrix} 0 \\ 2 \\ -1 \\ -1 \\ -4 \\ 0 \\ 0 \end{pmatrix}$$

From the normal equation system $0.64b = 2$, the required translation is found to be $b = 3.1$; hence the final position of the template in the search window has the following ξ coordinate: $\xi = 80 - 3.1 = 76.9\,\mu\text{m}$.

The standard deviation of unit weight σ_0 may be obtained from the values

$$\mathbf{v} = (0, -0.76, 2.24, 1.00, 2.76, -1.24, 0)^{\top}$$

$\Rightarrow \sigma_0 = \sqrt{15.75/(7-1)} = \pm 1.62$ greylevels. The standard deviation of the ξ coordinate, that is of the result of the correlation in the subpixel region, amounts to (Equation (6.8-5)):

$$\sigma_\xi = \sigma_b = \sqrt{1/0.64}\,\sigma_0 = \pm 1.25 \frac{\mu\text{m}}{\text{greylevels}}\,\sigma_0 = \pm 2.0\,\mu\text{m}$$

In this simulated example the accuracy of positioning in the subpixel region is almost an order of magnitude better than the pixel size, which in our example is $10\,\mu\text{m}$. This high accuracy has been confirmed many times in comprehensive correlations with real image data.

The reference image and the search image frequently differ from each other not only to a small extent in positioning but also in grey levels. Following the example of Section 3.5.1, a parameter c (contrast adjustment) and a parameter d (brightness adjustment) may be introduced into the grey level g_r. Equation (6.8-2) is extended by the inclusion of these two parameters:

$$g_s(\xi) = c\,g_r(\xi + b) + d \tag{6.8-6}$$

These equations can be linearized and re-arranged into correction equations for an adjustment by the method of indirect observations. The unknowns are the three parameters b, c and d. The simultaneous determination of position in the subpixel region and of the adjustments for contrast and brightness is not always to be recommended as it can seriously impair the convergence, especially with a small correlation window and very faded grey levels. In these cases the adjustments for contrast and brightness should be performed before the LSM and a subsequent correlation in the subpixel region be performed using Equation (6.8-3).

Exercise 6.8-1. Linearize Equation (6.8-6) and repeat the numerical example. Hint: The unknown b should be replaced by the unknown $\bar{b} = bc$. (Answer: $c = 0.69$, $d = 0.81$, $\bar{b} = 3.1\,\mu\text{m}$, $b = 4.5\,\mu\text{m}$.)

The formulation for LSM correlation in the subpixel region can very simply be extended for the two-dimensional correlation case. In this case the process is limited to the two translations b_ξ and b_η. Adjustments for contrast and brightness for the two image regions to be correlated are, therefore, abandoned. Following the example of Equation (6.8-3), the correction equations for the two unknowns b_ξ and b_η become:

$$v = \left(\frac{\partial g_r}{\partial \xi}\right)_{(\xi,\eta)} b_\xi + \left(\frac{\partial g_r}{\partial \eta}\right)_{(\xi,\eta)} b_\eta - (g_s(\xi,\eta) - g_r(\xi,\eta)) \qquad (6.8\text{-}7)$$

From this the following normal equation matrix for two-dimensional correlation is derived:

$$\mathbf{N} = \begin{pmatrix} \sum(\partial g_r/\partial\xi)^2 & \sum(\partial g_r/\partial\xi)(\partial g_r/\partial\eta) \\ \sum(\partial g_r/\partial\xi)(\partial g_r/\partial\eta) & \sum(\partial g_r/\partial\eta)^2 \end{pmatrix} \qquad (6.8\text{-}8)$$

In this matrix the summations are over all pixels in the particular correlation window. Equation (6.8-8) is the basis for what is known as an interest operator (Section 6.8.1.3) and for feature-based matching which is dealt with in Section 6.8.1.4. Before that, some general observations should be made concerning correlation:

(a) the accuracy of the positioning depends decisively on step changes in grey levels in the images. Large changes generate very large elements on the principal diagonal of the matrix **N** of the normal equations (Equations (6.8-4), (6.8-5) and (6.8-8)); large elements on the principal diagonal of the normal equation matrix lead to small weight coefficients (Equation (6.8-5)) and therefore also to small standard deviations of positioning.

(b) enlargement of the reference matrix raises the accuracy if it means that additional step changes in grey level are thereby introduced. This increase in accuracy can also be estimated from the inverse of the normal equation matrix.

(c) independent of its individual position in the digital image, a reference image always results in the same inverse $\mathbf{N}^{-1}$ of the normal equation matrix (Equation (6.8-8)); therefore, in order to recognize the same pattern in different places in the image, one has to evaluate $\mathbf{N}^{-1}$ only once. This characteristic is important in the processing of réseau photographs.

(d) the correction equations (6.8-7) can be weighted differently; for example the central picture element can be given a higher weight. This brings to mind the higher sensitivity of the fovea in the human eye.

(e) in the basic equation (6.8-2) aside from a translation b, a scale factor in the ξ coordinate of the reference image can also be introduced. Such an extension is necessary, for example, when template and search window have different pixel sizes. In the two-dimensional case the parameters of a similarity transformation or an affine transformation (Section 2.1.1) can be introduced.

(f) in a more complex implementation the LSM usually requires a number of iterations. Between individual iterations a resampling (Section 2.2.3) of the $\xi\eta$ raster of one of the two images is carried out.

Exercise 6.8-2. How can Equation (6.8-1) be re-arranged in order to economize on computing time? Repeat the example of Table 6.8-1 only with use of the covariance σ_{rs} as the measure of similarity. Is it adequate to enlist the covariance as similarity measure? (Answer: Equation B (6.1-1), Volume 2. In principle the covariance σ_{rs} can be used as a measure of similarity. All the same, for the introduction of threshold values which in practice are important, the correlation coefficient, which varies between zero and one, is significantly better suited.)

Exercise 6.8-3. Repeat the whole of the exercise of this section under the assumption that the reference image comprises nine pixels which have the values $g_1 = g_2 = g_8 = g_9 = 1, g_3 = g_7 = 4, g_4 = g_5 = g_6 = 9$. (Result: $\xi = 76.8 \pm 2.4\,\mu$m). Why does a reduction in accuracy now occur? (Answer: The sharpness of the grey level gradations has become less pronounced.)

Exercise 6.8-4. Repeat the whole of the exercise of this section under the assumption that the pixels are not $10\,\mu$m but $5\,\mu$m in size. This results in a considerable rise in accuracy. (Result: $\hat{\sigma}_b = \pm 1.06\,\mu$m.)

Exercise 6.8-5. Reduction in pixel size usually results in an increase in the noise component in the digitized image. As a result the increased noise component is carried into the computation in which some of the grey levels in the search window are arbitrarily changed to $g_4 = 2, g_7 = 9, g_{11} = 3$ in the $10\,\mu$m image. How accurate is the positioning of the template now with the $5\,\mu$m pixels in the search window? (Result: $\hat{\sigma}_b = \pm 1.11\,\mu$m.)

Further reading: Ackermann, F.: Schriftenr. d. Inst. f. Photogr. d. Uni Stuttgart, Heft 9, pp. 231–243, 1984. Rosenholm, D.: The Photogrammetric Record, 12(70), pp. 493–512, 1987. Trinder, J.: IAPR 27(B3), pp. 784–792, Kyoto, 1988. Luhmann, T., Robson, S., Kyle, S., Harley, I.: Close Range Photogrammetry. Whittles Publishing, 2006. Schenk, T.: Digital Photogrammetry, TerraScience, 1999. See also Sections B 6.1.2 and C 2.2.1 in Volume 2.

6.8.1.3 Interest operators

Since the quality of the correlation depends strongly on the texture of both of the corresponding image patches it is to be recommended in many cases, before the actual correlation, to choose image positions especially suitable for correlation. Interest operators, as they are called, are suitable for this purpose. In this and the following section we assume two images overlapping each other. The reference matrix is thus a (smaller) excerpt from one of the two images.

A very well known interest operator uses the accuracy measure for the two translations b_ξ and b_η of the LSM from Equations (6.8-7) and (6.8-8). For this purpose one forms the grey level differences $(\partial g/\partial \xi) := g(\xi,\eta)_{i+1,j} - g(\xi,\eta)_{i-1,j}$ and $(\partial g/\partial \eta) := g(\xi,\eta)_{i,j+1} - g(\xi,\eta)_{i,j-1}$ and establishes, in the neighbourhood of one of the pixels, for example for a 7×7 window, the matrix $\mathbf{N}$ of the normal equations (Equation (6.8-8)). The inverse $\mathbf{Q}$ of the normal equation matrix is directly usable for the analysis of

Figure 6.8-4: Marked pixels with a small value of $\mathrm{tr}\,\mathbf{Q}$

accuracy. The trace of the matrix $\mathbf{Q}$, denoted by "$\mathrm{tr}\,\mathbf{Q}$", should be as small as possible. The central pixels of such a window in which $\mathrm{tr}\,\mathbf{Q}$ is less than a specified threshold value are marked in Figure 6.8-4.

Some markings lie on straight-line grey level edges, where an accurate correlation is possible only in a direction perpendicular to the edge; in the direction along the edge no accurate correlation is possible. On this ground the criterion for this interest operator is extended insofar as not only a small value for $\mathrm{tr}\,\mathbf{Q}$ is sought but also a good distribution of edges in all directions. This second criterion can also be tested using the matrix $\mathbf{N}$ of the normal equations. Details can be found in Section B 6.4.1.1, Volume 2.

Interest operators can, for example, be applied in both images of a stereopair, independently of each other. If the criteria for the interest operators are met in the corresponding places in both images, it is guaranteed that an accurate correlation of the two image patches will be possible. The inclusion of interest operators makes automated processing markedly superior to operator processing. Nevertheless, the good operator would still be inclined, even with poor texture in both images, to measure stereoscopically.

Förstner was the first to report the interest operator outlined above which analyses the inverse of the normal equation matrix (6.8-8). In the literature it is known as the Förstner operator. Another interest operator comes from Moravec. For this interest operator the variance of the grey level differences of neighbouring pixels in a window is computed, and compared with a threshold value, along both the rows and the columns of the image as well as along both diagonals. Large variances offer good prospects for correlation.

6.8.1.4 Feature based matching

An interest operator extracts features in a digital image; they are more or less the intersection points of grey level edges. These points are positioned with single pixel accuracy in the image matrix, in accordance with the Förstner operator introduced in Section 6.8.1.3. But, as a secondary product, the values for the chosen interest criterion also come up for the neighbouring pixels. As a result the possibility arises of finding, from all interest values, a continuous function over a surrounding area and of setting its first derivative to zero. In this way the position of a feature point in the subpixel region is found (an example of such a positioning in the subpixel region is to be found at the beginning of Section 6.8.1.2).

In the processing of a stereopair of images the extraction of features in the subpixel region is carried out in both images independently of each other. Subsequently, corresponding features in the two images are assigned to each other; this is yet to be spoken of (see, for example, Section 6.8.3.2). If two features are found to correspond the task of correlation is solved as well. Correlation through extracted features is known as feature based matching as opposed to area based matching. LSM, for example, is an area based matching correlation since within a region, or an area, the pixels from the reference image and the search image are directly connected with each other.

Some points to note: The accuracy of feature based matching, lying somewhere around $1/4$ pixel, is distinctly worse than that of LSM. Feature based correlation is, as a rule, more robust than LSM. A substantial disadvantage of LSM is that the approximate positioning required before the LSM step must be relatively accurate. Feature based correlation is rotation invariant, that is, the two images can be rotated relative to each other. Two-dimensional correlation using correlation coefficients (Section 6.8.1.1) cannot be applied to images rotated with respect to one another.

Literature: Förstner, W., Gülch, E.: Proc. Intercommission Workshop on "Fast Processing of Photogrammetric Data", pp. 281–305, Interlaken, 1987. Baltsavias, E.: Dissertation, ETH Zürich, Institut für Geodäsie und Photogrammetrie, Nr. 49, 1991.

6.8.1.5 Simultaneous correlation of more than two images

The human being can correlate only two images. More than two images can be simultaneously correlated using machine vision. For the simultaneous correlation of the three images from a three line camera, or three line scanner (Figure 5.5-2), an extension of correlation from two to at least three images is necessary. In Figure 6.8-5 three image excerpts are sketched, as they relate to each other after a coarse fitting. It is very clear that a fitting in the subpixel regions will bring about an improvement. Equation (6.8-7) can be used for this kind of subpixel fitting. Thus any possible simultaneous contrast and brightness fitting is abandoned; it suffices to use only translations b_ξ and b_η while transformation with additional parameters (as is given in Section 6.8.1.2) is discarded.

It is possible to fit image 2 with image 1 and image 3 with image 1 more accurately using Equations (6.8-7). If image 1 is shifted during these two fittings, two different positions for image 1 will have been found. If, during both the first positioning and the second, image 1 is held fixed, this shortcoming can be remedied: image 2 is shifted in the first case and image 3 is shifted in the second case. Up to this point the remaining possible pairing, image 3 with image 2, has been disregarded. Separate pairwise

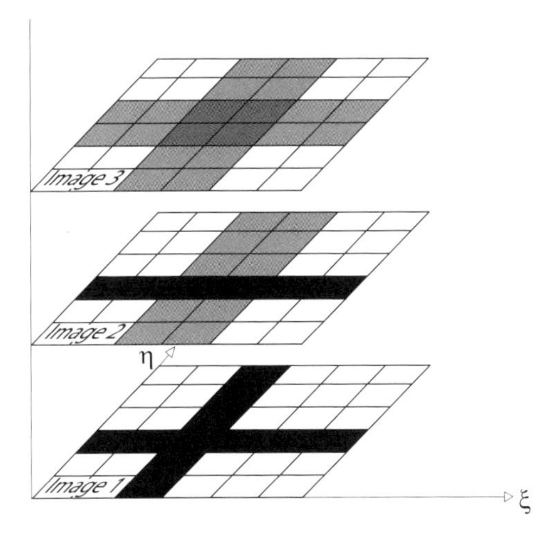

Figure 6.8-5: Multi-image correlation

fittings thus have the disadvantage that not all possible pairings, in our case three, are simultaneously looked at and that multiple positions are possible.

A possible variant is that one of the three images should be designated as master; we choose image 1. A subpixel fit of image 2 to image 1 can be carried out using the following equations, which result from relevant substitutions in Equations (6.8-7):

$$v = (\partial g_2/\partial\xi)b_{\xi,21} + (\partial g_2/\partial\eta)b_{\eta,21} - (g_1(\xi,\eta) - g_2(\xi,\eta)) \tag{6.8-9}$$

Similarly, for the subpixel fit of image 3 to the (master) image 1, the following equations are derived:

$$v = (\partial g_3/\partial\xi)b_{\xi,31} + (\partial g_3/\partial\eta)b_{\eta,31} - (g_1(\xi,\eta) - g_3(\xi,\eta)) \tag{6.8-10}$$

The formation of the normal equations from Equations (6.8-9) together with Equations (6.8-10) leads to a system of normal equations in the four unknowns, which decompose into two smaller systems of normal equations, in one case in the unknowns $b_{\xi,21}$ and $b_{\eta,21}$ and in the other case in the unknowns $b_{\xi,31}$ and $b_{\eta,31}$. While these are indeed joint equations involving the grey levels of the three images, the grey level differences between image 3 and image 2 are not, thus far, minimized.

Additional equations are, therefore, sought of which the absolute term contains the grey level difference $(g_2(\xi,\eta) - g_3(\xi,\eta))$ and in which no further positioning unknowns appear. The four positioning unknowns in the Equations (6.8-9) and (6.8-10) actually define the geometric problem clearly because $b_{\xi,32}$ appears in the difference $(b_{\xi,21} - b_{\xi,31})$ and $b_{\eta,32}$ appears in the difference $(b_{\eta,21} - b_{\eta,31})$. The extra equation which is being sought comes from the difference of the two Equations (6.8-9) and (6.8-10):

$$\begin{aligned} v = & \frac{\partial g_2}{\partial \xi} b_{\xi,21} + \frac{\partial g_2}{\partial \eta} b_{\eta,21} \\ & - \frac{\partial g_3}{\partial \xi} b_{\xi,31} - \frac{\partial g_3}{\partial \eta} b_{\eta,31} - (g_3(\xi,\eta) - g_2(\xi,\eta)) \end{aligned} \tag{6.8-11}$$

The normal equations formed from Equations (6.8-9), (6.8-10) and (6.8-11) lead to a simultaneous correlation of three images. In the case of four or more images corresponding equations are added. The LSM algorithm outlined can be called multiple-patch matching.

Exercise 6.8-6. Consider the correction equations for a four image correlation. (Answer: There are six unknown translations; the absolute terms read as follows: $(g_1 - g_2)$, $(g_1 - g_3)$, $(g_1 - g_4)$, $(g_3 - g_2)$, $(g_4 - g_2)$ and $(g_4 - g_3)$.)

A more general method, using a fictitious (master) image was given in Krupnik, A. (PE&RS 62, pp. 1151–1155, 1996), in which references to further literature can be found.

6.8.2 Automated interior orientation

In the case of photographs taken with a digital metric camera, the principal distance and the position of the principal point are known. The interior orientation of digital metric cameras is directly available for use.

If, on the other hand, the photographs were taken with a film-based camera and the digital metric image originates in a subsequent digitization (Section 3.4), then the processing begins with the location of the individual fiducial marks. The same kind of task arises when the images are taken with réseau cameras (Section 3.8.4) or when digitizing of a réseau plate is utilized (Figure 3.4-3).

What follows concentrates on the automated location of the fiducial marks. A reference matrix is produced for the particular fiducial mark. This is therefore a typical exercise in pattern recognition.

In order to minimize the time and cost of the search process:

- image pyramids are produced from the photographs (Section 3.6.1) and
- relatively small search regions are defined in the vicinity of the standard positions of the fiducial marks (as, for example in Figure 3.7-7; it is required that when the image is laid in the scanner it should be rotated and shifted as little as possible from a standard position).

For the different levels of pyramid one requires different reference templates to match different extracts from the image of the fiducial marks. Figure 6.8-6 shows these extracts for a typical fiducial mark. The pixel size is reduced in accordance with the

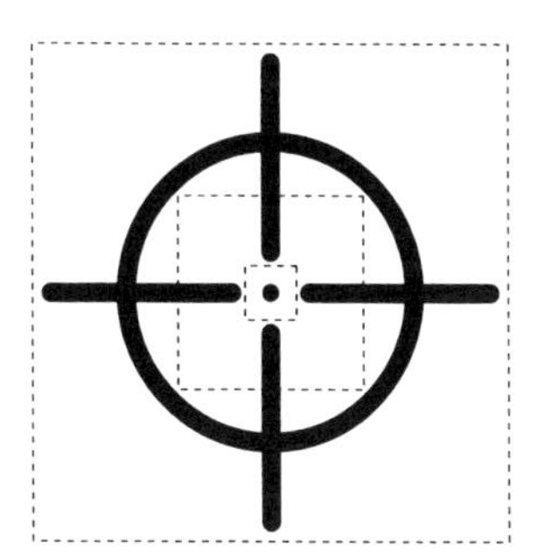

Figure 6.8-6: Negative image of a fiducial mark with three different extract sizes

reduction in size of the extract. In the step using the coarsest pyramid the largest extract, with a wealth of detail, is required for recognition. At the level of this pyramid the correlation is performed with only the correlation coefficient as a measure of similarity (Section 6.8.1.1). At the next pyramid level the result from the coarsest level is invoked as an approximate position so that a relatively small search region can be chosen. Advancing in this manner is continued until the finest pyramid level. A subpixel correlation (Section 6.8.1.2) is always carried out at the level of the finest pyramid. In practice, at the finest level, the achievable accuracy lies between $1/10$ and $1/5$ pixel.

Accurate location of the fiducial mark in the digitized metric image is one task; the other task is to establish which particular fiducial mark has been located. This is not a trivial task since the image may have been placed in the scanner right way round, upside down, rotated through 90 or whatever. It would be best if each individual fiducial had its own unambiguous identification markings, for example in the form of a pattern of lines. Another possibility which includes an additional, asymmetric fiducial mark may be mentioned with reference to Figure 3.7-7.

Section B 6.1, Volume 2 contains information on those forms of fiducial marks which are favourable for automatic measurement, on efficient computation for correlation coefficients, on the use of binary images instead of grey value images, etc. Examples of further literature on these subjects: Heipke, C.: ISPRS-Journal 52, pp. 1–19, 1997. Schenk, T.: Digital Photogrammetry. TerraScience, 1999.

6.8.3 Automated relative orientation and automated determination of tie points

The relative orientation of two overlapping metric images is a central task both in photogrammetry and in computer vision. The mathematical bases of relative orientation were set out in Section 4.3. It remains unsaid, however, how the image coordinates of corresponding points in both digital images can be found. In recent years the automation of this measuring process has been developed successfully. The strategies that have been implemented for its solution in software packages depend above all on the configuration of the camera stations and the orientations of the images as well as on the form of the object.

6.8.3.1 Near-vertical photographs with 60% forward overlap taken over land with small height differences

Near-vertical photographs with 60% forward overlap are discussed first. Figure 6.8-7 shows such a stereopair. For correlation, reference matrices are chosen in the first image and search regions are defined in the second image. In order to get a good distribution of orientation points the system chooses the reference and search matrices in the standard positions, some times called the Gruber points (Section 4.3.4.2). This choice takes place at the coarsest pyramid level. In order to increase redundancy in the correlation, relatively large reference matrices are chosen. In highly distorted photographs, however, problems can arise with reference matrices that are too large; this can happen when there are large height differences in relation to the camera-object distance. The correlations in this first step can be carried out using correlation coefficients (Section 6.8.1.1); an estimate with subpixel accuracy is not necessary.

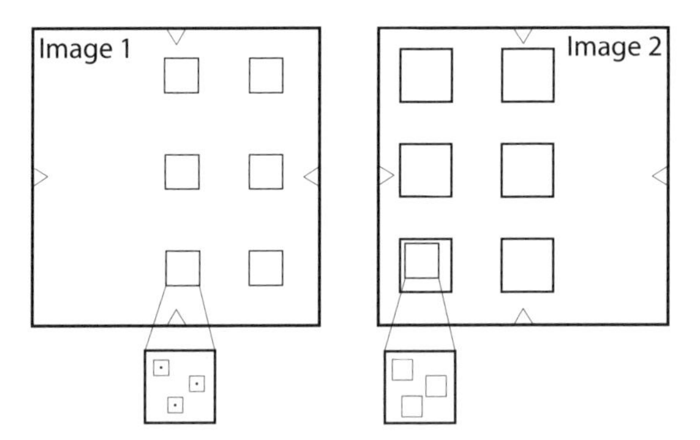

Figure 6.8-7: Reference matrices in the first image and search matrices in the second image for the coarsest pyramid level; additional reference matrices, containing interest points, and corresponding search matrices for a finer pyramid level in the neighbourhood of standard point number 5.

The corresponding correlation windows found in this way are very good approximations to be used for correlations at a finer pyramid level. As indicated in Figure 6.8-7 for standard position number 5 in the first image, a number of small reference matrices can be selected inside such a window. Feature points, which may be found using an interest operator (Section 6.8.1.3) and around which the reference matrices are extended, can be of great service in making this choice. (Incidentally, the choice of the reference matrices in the first image at the coarsest pyramid stage can also be supported with interest operators.) As illustrated in Figure 6.8-7, the search window in the second image can likewise be chosen to be relatively small, but clearly larger than the reference matrix.

It is sufficient for the relative orientation to run through three or four pyramid levels. Only at the level of the finest pyramid should subpixel estimation be performed, for example using LSM (Section 6.8.1.2), after matching using correlation coefficients. The result is six homologous locations in each of which, in general, a number of homologous points lie. As a result of the high number of orientation points provided by the automated procedure, the accuracy of the orientation elements is as a rule distinctly better—on average by a factor between two and three—than the accuracies given in Table 4.3-1 when using six orientation points.

6.8.3.2 Near-vertical photographs with 60% forward overlap taken over land with large height differences

With big height differences in relation to the camera-object distance the strategy outlined in Section 6.8.3.1 can fail, especially when there are large areas which are not visible in one or the other image. A first countermeasure consists of choosing search regions in the second image which are perceptibly larger, above all in the direction of the base. On account of the severe difference in scale which stems from the large height difference, correlation coefficient and LSM methods can also fail. For this reason feature based correlation (Section 6.8.1.4), which is independent of scale and rotation, is used; that is, in each image feature points are extracted with an interest operator and corresponding image points are subsequently sought. At the level of the coarsest image pyramid the properties of near-vertical photographs, which display only small differences in η coordinates between left and right images, are exploited in support of these correspondence analyses. The results from the coarser pyramids are good approximations, in both coordinate directions, for the finer pyramid levels. Relatively many errors are made in the process of finding matching points. The computation of the relative orientation must, therefore, be combined with gross error analysis. Data snooping and robust estimation, which are dealt with in detail in Section B 7.2, Volume 2, are suitable for this purpose.

Literature: Heipke, C.: ISPRS-J 52(1), pp. 1–19, 1997, which contains an extensive literature list, is relevant to Sections 6.8.3.1 and 6.8.3.2.

6.8.3.3 Arbitrary configurations of photographs and objects with very complex forms

In close range work, especially, many of the photographic configurations used (multi-image photogrammetry) would exemplify the above heading. Automated relative orientation in such a photogrammetric network is very difficult. First, using interest operators, feature points are extracted, independently of each other, in both images. The so-called epipolar geometry, which is not considered more closely until Section 6.8.5, plays a central role in the subsequent process of allocating corresponding feature points to each other. To be sure, epipolar geometry requires approximate values for the orientation elements; it may be possible to recover these by directly observing and identifying at least five homologous points and subsequently computing relative orientation using the methods of Sections 4.3.2 or 4.3.3. In situations where the orientation must be carried out without human intervention, the analysis of correspondence may be placed on a good foundation by the introduction of object recognition. Here, for example, feature points extracted using an interest operator are compared with each other and topologically similar patterns are sought in the first and second images.

Literature: Haralick, R., Shapiro, L.: Computer and Robot Vision. Addison-Wesley, Reading, USA, 1992. Van Gool, L., Tuytelaars, T., Ferrari, V., Strecha, C., Vanden Wyngaerd, J., Vergauwen, M.: IAPRS 34(3A), pp. 3–14, Graz, 2002.

6.8.3.4 Line-based (edge-based) relative orientation

Prompted by the fact that human stereoscopic vision uses more lines and fewer points, Schenk[12] proposed a method for automated relative orientation which uses extracted edges for the correlation. To this end, edges are first extracted independently in both images (Figure 6.8-21 shows an example in which relevant edges still have to be selected and extracted). The form of the edges, for example the direction of the tangent as a function of the edge length, is then analysed and similarities in the lines of both images are selected. When corresponding edges have been found either of two procedures can be followed:

- either, distinctive points along corresponding edges, for example corners, can be found and the relative orientation can be computed using these points following the algorithm of Section 4.3
- or, a method based on associated lines can be used for the relative orientation.

Formulae for relative orientation based on straight lines can be found, for example, in the work of Schwermann and Luhmann[13].

For relative orientation using only straight lines, it is necessary to observe three images together. An explanation: a straight line in the first image, together with the perspective centre of that image, defines a plane; the corresponding straight line in the second image, together with the perspective centre of the second image, defines a second plane, which intersects the plane from the first image in object space. This is a necessary, but not a sufficient constraint for relative orientation, since all such pairs will intersect in space, independent of the relative orientation of the two images. Therefore, no solution to relative orientation is possible, even with many pairs of corresponding straight lines in two images. A third image with a straight line corresponding to the lines in the first two images provides additional lines of intersection in object space; for correct relative orientation all of these lines must coincide. This can be used for a combined relative orientation of an image triple.

In the future line photogrammetry, merely touched upon in this section, will gain in importance because, in the extraction of features, lines are more suitable than points.

[12] Schenk, T.: Digital Photogrammetry. TerraScience, 1999.

[13] Schwermann, R.: Veröffentlichung des Geod. Inst. der RWTH Aachen, Nr. 52, 1995. Luhmann, T., Robson, S., Kyle, S., Harley, I.: Close Range Photogrammetry. Whittles Publishing, 2006.

6.8.3.5 Tie points for automated aerotriangulation

Up to this point in Section 6.8.3 only the automated establishment of corresponding points for relative orientation has been dealt with. The determination of tie points for a bundle block adjustment (Section 5.3) proceeds in a similar fashion. The positions of the reference matrices and the definitions of the search matrices for a block of near-vertical photographs with 60% forward and 20% side overlaps are sketched in Figure 6.8-8; it is an extension of the pattern of Figure 6.8-7 from two metric images to many.

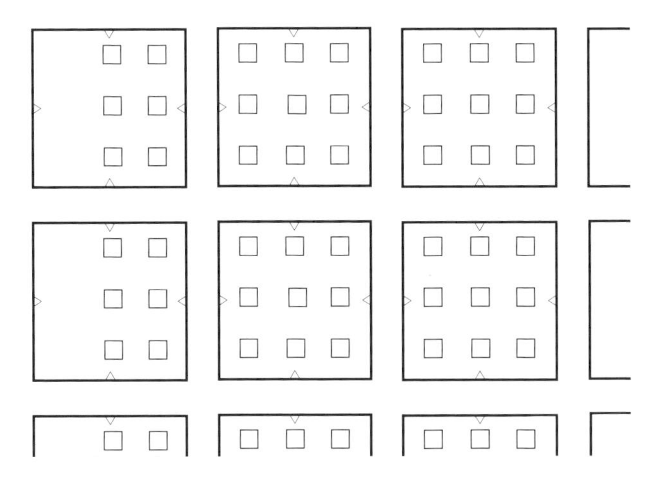

Figure 6.8-8: Upper-left portion of a block of photographs with 60% forward and 20% side overlaps showing reference matrices and search matrices

Using the strategies and algorithms given in Sections 6.8.3.1 and 6.8.3.2 one could establish the tie points pairwise, including the tie points between the strips. Of course in this way one would get only points that tied together the two images concerned. The accuracy of a bundle block adjustment is, however, increased in a fundamental way if it is possible to find unambiguous points lying in the overlap zones of all affected images. As can be gathered from Figure 6.8-8, that point which lies in the lower row of the second image of the first strip connects six images together. To determine a tie point in this area, therefore, the six-image correlation is to be applied (Section 6.8.1.5).

Literature: Fritsch, D., Tsingas, V., Schneider, W.: ZPF 62, pp. 214–223, 1994. Schenk, T.: ISPRS-J 52, pp. 110–121, 1997.

6.8.4 Automated location of control points

For indirect georeferencing, which also implies indirect sensor orientation, as well as for verification either of direct georeferencing or of direct sensor orientation, control points are required in the object coordinate system; generally speaking these are given in the national coordinate system (see the introduction of Section 4.1.1 and that of Section 5.4). Location of control points in digital photographs is required at different stages of the photogrammetric process; in a sequence based on their importance in practice these are:

- in metric photographs to which a bundle block adjustment (Section 5.3) is to be applied
- in each of two metric photographs constituting a stereomodel which is to be included in a block adjustment using the method of independent models (Section 5.2)
- in a single metric photograph which has to be oriented using three-dimensional resection (Section 4.2.1), for example in the production of orthophotos (Section 7.3)
- in each of two metric photographs for their combined, single-stage orientation (Section 4.2.2)
- in both of the two metric photographs constituting a stereomodel for which absolute orientation is required (Section 4.4)

Finding the control points in the images depends overwhelmingly on whether targeted or natural control points are available. The location of targeted (signalized) points is a pattern recognition task, comparable to the automated location of fiducial marks (Section 6.8.2). Undeniably, the location of targeted control points is considerably more difficult than the location of fiducial marks for the following reasons:

- in the reference pattern, generally speaking, there is no supplementary geometrical information, such as plays an important role in the recognition of targets.
- as a rule, the background to the target is not uniform; there is disturbing information present.
- getting approximate values for the search window is not easily possible.
- the targets are usually portrayed at different sizes in the image.

For aerotriangulation, which is the standard procedure for the determination of the orientation elements of metric photographs, only a small number of control points are necessary (Sections 5.2.3 and 5.3.5); for this reason the approximate positioning and identification are frequently undertaken by the operator and only the fine measurement is entrusted to the computer.

Further details are to be found in Section B 6.2, Volume 2 and in the literature: for example Gülch, E.: IAPR 31(B3), pp. 279–284, 1996. Rottensteiner, F., Prinz, R.: VGI 84(2), pp. 189–195, 1996.

Targeted control points are to be found almost exclusively in close range photogrammetry. In aerotriangulation the control points are usually natural points, such as corners of buildings, corners of the boundaries of agricultural land, and so on, but also larger ensembles of topographic elements. The recognition process for such control points is exceedingly difficult, especially in obtaining suitable reference matrices (further details and some sources of literature are to be found in Section B 6.3, Volume 2). Höhle[14] extracted such a reference image from the topographic database of a geographic information system (GIS) in conjunction with an obsolete, digital orthophoto. It goes without saying that relatively many erroneous identifications resulted, but the high number of "GIS control points" in general permitted very reliable error removal.

6.8.5 Inclusion of epipolar geometry in the correlation

With the aid of epipolar lines, a two-dimensional correlation for finding corresponding points in two relatively oriented images can be reduced to a one-dimensional correlation problem. Such a one-dimensional correlation is illustrated in the stereopair of Figure 6.8-9. Points are selected in the left-hand image, for example with an interest operator (Section 6.8.1.3). The corresponding point in the right-hand image can lie only on the particular corresponding epipolar line. The reference matrix of the left-hand image is therefore shifted along the epipolar band in the right-hand image until the optimal correlation is reached. The saving in computer time arising from the one-dimensional correlation is of great significance in practical photogrammetry. Not only is there a great saving in computer time connected with the use of epipolar lines, however, but the reliability of the correlation is also much improved.

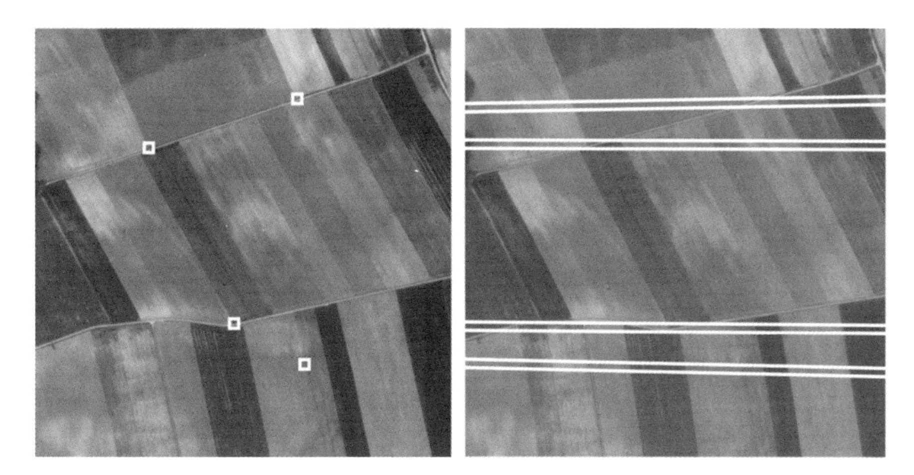

Figure 6.8-9: Stereopair with four points in the left image and corresponding epipolar bands in the right image

[14]Höhle, J., Potuckova, M.: PFG 6/2001, pp. 397–404, see also Morgado, A., Dowman, I.: ISPRS-J 52, pp. 169–182, 1997.

6.8.5.1 Epipolar geometry after relative orientation using rotations only

Epipolar geometry is illustrated in Figure 6.8-10. An epipolar line is the line of intersection of the plane of the photograph and an epipolar plane or basal plane. A basal plane is a plane containing the two perspective centres O_1 and O_2 and a relevant object point P. All epipolar rays in a photograph intersect in the epipole K. The epipole is the point of intersection of the base O_1O_2 and the plane of the photograph. An epipole in one photograph is therefore the image of the other perspective centre. Thus a stereopair has two epipoles, the points K_1 and K_2 in Figure 6.8-10. Corresponding image points lie on the two epipoles. In the "normal case" of photogrammetry, the epipoles are imaginary points, coinciding in a single point at infinity in the direction of the base. For this reason it is probably preferable to use the term basal plane.

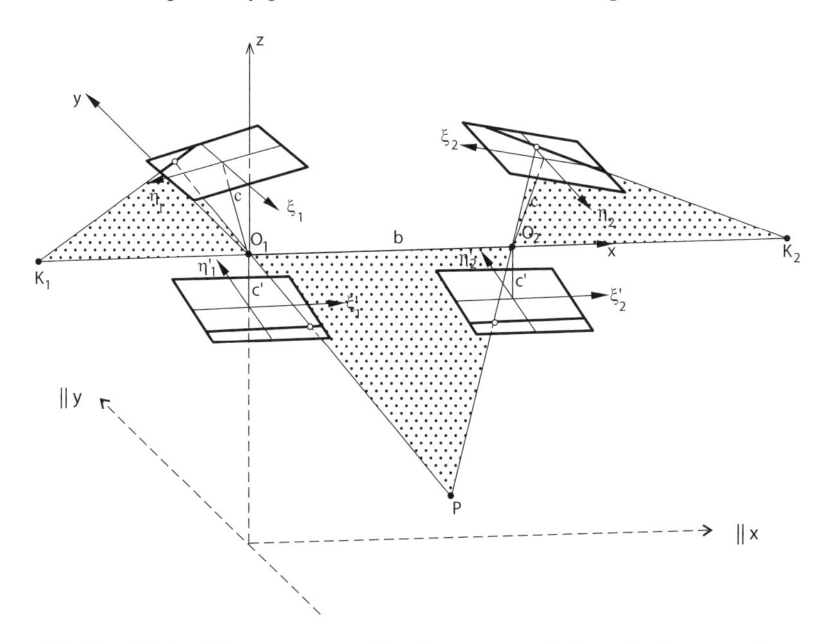

Figure 6.8-10: Original images, normalized images and the epipolar rays for two corresponding image points, radiating from the epipoles K_1 and K_2

When the relevant epipolar rays have been found in both images, a one-dimensional correlation can be carried out along these two epipolar rays. (The difficulties which arise from the diagonal course of the epipolar rays in the matrices of the original images should not be entered into here.) The obvious points from which to determine the epipolar rays are the epipoles K_1 and K_2 (see Figure 6.8-10). The image coordinates of K_1, for example, are obtained by substituting the model coordinates $(b, 0, 0)$ of the perspective centre O_2 in the collinearity equations (2.1-19) for the first image (see Equations (6.8-12)):

$$\xi_{K_1} = -c\frac{r_{11}b + r_{21}0 + r_{31}0}{r_{13}b + r_{23}0 + r_{33}0} \qquad \eta_{K_1} = -c\frac{r_{12}b + r_{22}0 + r_{32}0}{r_{13}b + r_{23}0 + r_{33}0}$$

Or, for the general case:

$$\xi_K = -c\frac{r_{11}}{r_{13}} \qquad \eta_K = -c\frac{r_{12}}{r_{13}} \tag{6.8-12}$$

Exercise 6.8-7. Derive the coordinates of the epipoles of both of the photographs of Section 4.3.1 using the orientation angles from Section 4.3.2. Answer: The rotation matrices $\mathbf{R}_1$ and $\mathbf{R}_2$ must first be evaluated from the orientation angles:

$$\mathbf{R}_1 = \begin{pmatrix} 0.999615 & -0.026817 & -0.007148 \\ 0.026818 & 0.999640 & 0.000000 \\ 0.007145 & -0.000192 & 0.999974 \end{pmatrix}$$

$$\mathbf{R}_2 = \begin{pmatrix} 0.999912 & 0.013170 & -0.001502 \\ -0.013200 & 0.999676 & -0.021780 \\ 0.001215 & 0.021798 & 0.999762 \end{pmatrix}$$

The image coordinates of the epipoles, in [mm], are hence derived as: $\xi_{K_1} = 21351$, $\eta_{K_1} = -573$, $\xi_{K_2} = 101613$, $\eta_{K_2} = 1338$.

6.8.5.2 Epipolar geometry in normalized images

The epipolar geometry of normalized images is especially simple. Normalized images are indicated in Figure 6.8-10 beneath the perspective centres. They correspond to the "normal case" of stereoprocessing (Section 2.1.5). In Section 2.1.5 the fact was recorded that in such an image pair no η parallaxes appear but only ξ parallaxes. This statement can also be put in terms of epipolar geometry: in the "normal case" the epipoles are infinitely distant so that in every instance two corresponding epipolar rays in the two images exhibit the same η coordinate (Figure 6.8-10).

In human vision (Section 6.1.1), also, there are no η parallaxes (vertical parallaxes) but only ξ parallaxes (horizontal parallaxes). With such an image pair a person correlates the two images received on the retinas and from this derives the succession in depth of visible objects. Hence, in machine vision, correlation of normalized images emulates natural human stereoscopic vision. The ξ parallaxes in normalized images constitute what, in computer vision, is referred to as a disparity map.

Normalized images are obtained in photogrammetry if the photographs are taken as "normal case" images, for example with a stereometric camera (Section 3.8.2). But normalized images can also be produced from arbitrarily configured photography. It is a precondition that the elements of orientation are known for both images. It is supposed, first of all, that relative orientation by means of rotations only has previously been carried out (independent relative orientation, see Section 4.3). The connections are represented in Figure 6.8-10. ξ_1,η_1 and ξ_2,η_2 are the image coordinates of the original metric image, ξ_1',η_1' and ξ_2',η_2' those of the equivalent normalized images. The principal distance of the normalized images is c'. The origin of the local object coordinate system, or model coordinate system, is set at O_1. The x axis is taken in the direction of the perspective centre O_2. The z axis becomes, as it were, the camera axis of the first normalized image. The y axis is perpendicular to the xz plane.

The mathematical relationship between the image coordinates ξ and η of one of the original metric images and the image coordinates ξ' and η' of the equivalent normalized image follows from the collinearity equations (2.1-19), in which

- the specific position of the object coordinate system is noted
- the (negative) principal distance c' is substituted in place of $(Z - Z_O)$
- ξ_O and η_O are set to zero

$$
\begin{aligned}
\xi &= -c\frac{r_{11}\xi' + r_{21}\eta' - r_{31}c'}{r_{13}\xi' + r_{23}\eta' - r_{33}c'} \\
\eta &= -c\frac{r_{12}\xi' + r_{22}\eta' - r_{13}c'}{r_{13}\xi' + r_{23}\eta' - r_{33}c'}
\end{aligned}
\qquad (6.8\text{-}13)
$$

Equations (6.8-13) may be re-arranged to give expressions for ξ' and η', corresponding to the Equations (2.1-20):

$$
\begin{aligned}
\xi' &= -c'\frac{r_{11}\xi + r_{12}\eta - r_{13}c}{r_{31}\xi + r_{32}\eta - r_{33}c} \\
\eta' &= -c'\frac{r_{21}\xi + r_{22}\eta - r_{23}c}{r_{31}\xi + r_{32}\eta - r_{33}c}
\end{aligned}
\qquad (6.8\text{-}14)
$$

r_{ik} ... are elements of the three-dimensional rotation matrix $\mathbf{R}$ derived from the angles of relative orientation performed using rotations only; that is, using $\omega_1 = 0, \varphi_1, \kappa_1$ for the first image and $\omega_2, \varphi_2, \kappa_2$ for the second image. (The solution for Exercise 6.8-7 contains an example.)

The conditions for the transformation of the original digital metric images into normalized digital images are set out in these equations. A new image matrix is defined in the normalized image (Figure 6.8-11). In order that no pixels will be lost from the original image, the principal distance c' is chosen to be somewhat larger than the original principal distance c. Taking this into consideration we are starting from the assumption that the pixel sizes in the original and in the normalized images are chosen to be the same. For every one of the pixels in the normalized image, with coordinates ξ', η', c', the required position in the original image can be found (Equations (6.8-13)). There a grey level interpolation over the neighbouring pixels can be carried out, as explained in Section 2.2.3. The result of this resampling are two normalized digital images, and therefore a normalized digital stereopair.

With a normalized digital stereopair the complete stereomodel can be processed using one-dimensional correlation. Corresponding image points in the two normalized images lie on two lines in the images with the same η coordinates ($\eta_1 = \eta_2$). Reference matrices, or more accurately reference vectors, are best chosen in one image in the

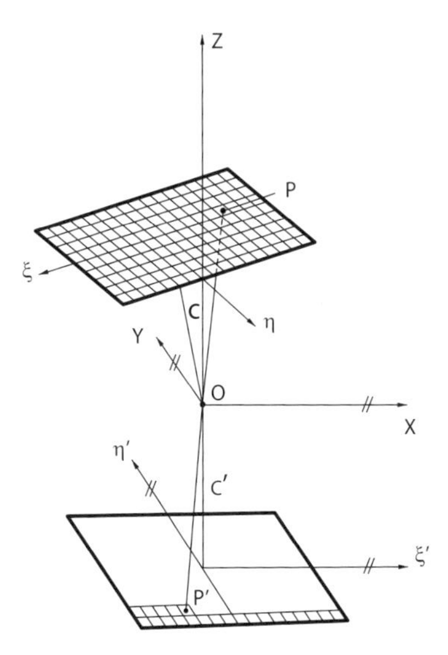

Figure 6.8-11: Original digital image and digital normalized image

neighbourhood of intersection points of grey level edges with the η coordinate lines. The larger search matrices, or more accurately search vectors, should be chosen in the other image, shifted by the base length (at image scale) along the η coordinate. In cases where the orientation elements still contain significant errors the search should be made along bands of lesser or greater width lying along the epipolar lines instead of along the epipolar lines. A small rectangle is used as the reference matrix; the search matrix is also defined as rectangle, but one that is somewhat wider and longer.

6.8.5.3 Epipolar geometry in original, tilted metric photographs

The transformation of tilted photographs into normalized images is relatively costly. When correlation is necessary for only a few points, the correlation should be carried out in the original, tilted photographs along the epipolar rays. For this purpose a method must just be given for finding the epipolar rays in tilted metric photographs.

The first important step, determining the epipolar points K_1 and K_2, has already been explained in Section 6.8.5.1. Then a point P_1 is arbitrarily chosen in the first original image. The points P_1 and K_1 define the epipolar ray in the original image (Figure 6.8-10). Subsequently the point $P_1(\xi_1,\eta_1)$ is transformed into the first normalized image using Equations (6.8-14). The resulting point is $P_1'(\xi_1',\eta_1')$. This point can be carried over directly into the second normalized image as $\overline{P}_1'$:$\bar{\xi}_2' = \xi_1'; \bar{\eta}_2' = \eta_1'$. A constant could be added to the ξ coordinate correspondingly approximately to the base in the photograph. Finally the point $\overline{P}_2'(\bar{\xi}_2', \bar{\eta}_2', c')$ is transformed into the second original image using Equations (6.8-13). The points P_2 and K_2 in the second original photograph fix the epipolar ray in that photograph; it corresponds to the epipolar ray through the points P_1 and K_1 in the first original photograph.

Exercise 6.8-8. Taking the image coordinates ξ_1 and η_1 in the first original image as starting point, find the epipolar rays for the eight points of the numerical example in Section 4.3.1. The rotation matrices of Exercise 6.8-7 should be used for this. (Check: Ignoring accidental errors, the points with the image coordinates ξ_2 and η_2 must lie on the epipolar rays in the second original image.)

6.8.5.4 Derivation of normalized images using the elements of exterior orientation

In Figure 6.8-10 it is assumed that the correlation has been preceded by a relative orientation using rotations only. In many cases, for example with direct orientation of the sensor using GPS and IMU (Section 4.1), this assumption does not apply. Either the formulae used above are re-arranged in terms of the elements of exterior orientation or, using a suitable transformation, the relationships of Figure 6.8-10 are established. Here, the second route is taken.

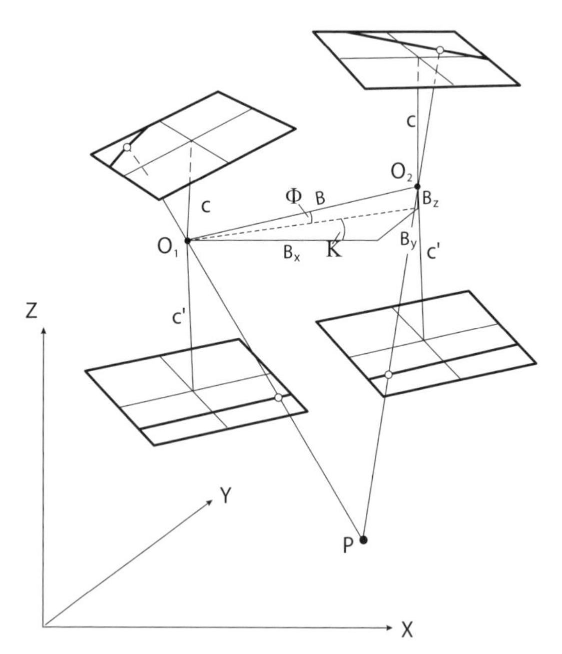

Figure 6.8-12: Normalized images using the elements of exterior orientation

Two metric images, after determination of the elements of exterior orientation, are illustrated in Figure 6.8-12. The camera axes (rays from the principal points through the perspective centres) of both normalized images should be normal to the base B and parallel to each other; the principal distances are c'; these requirements determine the angles K_1 and Φ_2. The Ω direction of the camera axes (the parallel camera axes of the normalized images) can be set more or less arbitrarily. A good choice is the mean of the angles ω_1 and ω_2 from the exterior orientation:

$$\Omega = \frac{\omega_1 + \omega_2}{2} \tag{6.8-15}$$

In the transition from the original images to the normalized images, the angles K and Φ are to be taken into account too; they are derived from the base components (Figure 6.8-12):

$$\mathrm{K} = \arctan\frac{B_y}{B_x} \tag{6.8-16}$$

$$\Phi = \arctan\frac{B_z}{\sqrt{B_x^2 + B_y^2}} \tag{6.8-17}$$

A 3×3 matrix representing a single rotation can be formed for each of the three angles Ω, Φ and K (Appendix 2.1-1). The three rotation matrices $\mathbf{R}_\Omega$, $\mathbf{R}_\Phi$ and $\mathbf{R}_\mathrm{K}$ must be combined with the known rotation matrices from the exterior orientation; these are, for the first photograph, the rotation matrix $\mathbf{R}_1(\omega_1, \varphi_1, \kappa_1)$ and, for the second, $\mathbf{R}_2(\omega_2, \varphi_2, \kappa_2)$. In so doing, the correct order of the matrices must be observed. The object coordinates $\mathbf{X}$ and the image coordinates $\boldsymbol{\xi}_1$ from the original image 1 are related as follows through the rotation matrix $\mathbf{R}_1$ of the exterior orientation (from Equation (2.1-2-5) of Appendix 2.1-2; in the present instance the translations and the scale factor of Equation (2.1-2-5) have no significance): $\mathbf{X} = \mathbf{R}_1\boldsymbol{\xi}_1$. The image coordinates $\boldsymbol{\xi}'_1$ in the normalized image 1 are found from the $\mathbf{X}$ coordinates after rotations through angles K (primary), Φ (secondary) and Ω (tertiary); taking note of the correct order of multiplication of the matrices (Appendix 2.1-1) the following result is obtained: $\boldsymbol{\xi}'_1 = \mathbf{R}_\mathrm{K}\mathbf{R}_\Phi\mathbf{R}_\Omega\mathbf{X}$. After substitution of $\mathbf{X} = \mathbf{R}_1\boldsymbol{\xi}_1$, the following relationship between the coordinates of original image 1 and those of the normalized image 1 is obtained (the corresponding relationship for image 2 is added):

$$\boldsymbol{\xi}'_1 = \mathbf{R}_\mathrm{K}\mathbf{R}_\Phi\mathbf{R}_\Omega\mathbf{R}_1\boldsymbol{\xi}_1 = \overline{\mathbf{R}}_1\boldsymbol{\xi}_1 \qquad \boldsymbol{\xi}'_2 = \mathbf{R}_\mathrm{K}\mathbf{R}_\Phi\mathbf{R}_\Omega\mathbf{R}_2\boldsymbol{\xi}_2 = \overline{\mathbf{R}}_2\boldsymbol{\xi}_2 \tag{6.8-18}$$

The relationship between the $\xi\eta$ system of the original image and the $\xi'\eta'$ system of the normalized image is given in detail in Equations (6.8-13) and (6.8-14). The elements r_{ik} appearing in these equations are to be replaced by the elements $r_{1,ik}$ from the rotation matrix $\overline{\mathbf{R}}_1$ for image 1 and by the elements $r_{2,ik}$ from the rotation matrix $\overline{\mathbf{R}}_2$ for image 2. As is evident from Equations (6.8-18) the rotation matrices $\overline{\mathbf{R}}_1$ and $\overline{\mathbf{R}}_2$ arise from multiplication of the relevant four rotation matrices.

Exercise 6.8-9. If the angles Φ and K of Figure 6.8-12 are chosen as $\Phi = \arctan(B_z/B_x)$ and $\mathrm{K} = \arctan(B_y/\sqrt{B_x^2 + B_z^2})$, what is the relationship between the original image and the normalized image? (Answer: For image 1, for example, $\bar{\boldsymbol{\xi}}_1 = \mathbf{R}_\Phi\mathbf{R}_\mathrm{K}\mathbf{R}_\Omega\mathbf{R}_1\boldsymbol{\xi}_1$.)

6.8.5.5 Epipolar geometry in images which have been oriented relatively using projective geometry

In Section 4.3.3 acquaintance was made with an alternative method of relative orientation with its origin in projective geometry. It is suitable, above all, for photographs of which the interior orientation is unknown. With at least eight corresponding points in both original images the eight unknowns f_{ik} of this relative orientation can be found. For one-dimensional correlation in such non-metric images the corresponding epipolar rays are necessary. They have already been found: in connection with Equation (4.3-21) it was realized that, for a point in one image, there is a line in the other image on which its corresponding point must lie. Starting from Equation (4.3-22), the equation of the epipolar ray in the first image corresponding to a point with image coordinates ξ_2, η_2 in the second image is found to be:

$$a\xi_1 + b\eta_1 + c = 0 \qquad (6.8\text{-}19a)$$

where[15]

$$\begin{pmatrix} a \\ b \\ c \end{pmatrix} = \begin{pmatrix} \xi_2 f_{11} + \eta_2 f_{12} + f_{13} \\ \xi_2 f_{21} + \eta_2 f_{22} + f_{23} \\ \xi_2 f_{31} + \eta_2 f_{32} + 1 \end{pmatrix} \text{ from (4.3-23) hence } = \mathbf{F} \begin{pmatrix} \xi_2 \\ \eta_2 \\ 1 \end{pmatrix} \qquad (6.8\text{-}19b)$$

In a similar way, using $\mathbf{F}^\top$, the transpose of $\mathbf{F}$, the epipolar ray in the second image can be found corresponding to the point ξ_1, η_1 in the first image.

An insight: the epipolar rays can be defined directly using the $\mathbf{F}$ matrix, without recourse to the epipole. This insight can also be applied to the task of correlation in metric images. In Equation (4.3-20) it is stated how the fundamental matrix may be arrived at in terms of the elements of interior orientation (matrices $\mathbf{C}_1$ and $\mathbf{C}_2$, Equations (4.3-19)) and the elements of relative orientation (matrices $\mathbf{R}_1$ and $\mathbf{R}_2$, Equations (4.3-9)). As a result the epipolar rays, which are important for correlation in metric images, may be defined by means of the Equations (6.8-19). Although the epipoles are not needed for this, relationships should be stated from which the coordinates of the epipoles may be found: ξ_{K_1}, η_{K_1} of K_1 in the first image and ξ_{K_2}, η_{K_2} of K_2 in the second image:

$$\mathbf{F}^\top \begin{pmatrix} \xi_{K_1} \\ \eta_{K_1} \\ 1 \end{pmatrix} = \mathbf{0} \qquad \mathbf{F} \begin{pmatrix} \xi_{K_2} \\ \eta_{K_2} \\ 1 \end{pmatrix} = \mathbf{0} \qquad (6.8\text{-}20)$$

Finding the coordinates of three-dimensional object models using the fundamental matrix can be understood from the relevant literature already given at the end of Section 4.3.3.

Exercise 6.8-10. In Exercise 4.3-6 the **F** matrix was derived from 8 corresponding points in two photographs. Using that **F** matrix and the relationships (6.8-20) find the epipoles. (Answer: $\xi_{K_1} = 6406\,\text{mm}$, $\eta_{K_1} = -206\,\text{mm}$, $\xi_{K_2} = 8456\,\text{mm}$, $\eta_{K_2} = 62\,\text{mm}$.) Compare these results with those of Exercise 6.8-7. Comment: The large difference between the two results stems, among other things, from the fact that the

[15] This relationship fails if either or both of ξ_2 and η_2 are infinitely large, that is if the image point lies at infinity. This insight has already been gained at the end of Section 4.3.3. It was established there that, in the exact "normal case", relative orientation is not possible using Equation (4.3-22).

elements of interior orientation were used in one case and not in the other. As well, the tilts of the photographs are very small so that in each case the coordinates of the epipole are very unreliable.

6.8.5.6 Epipolar geometry in three images

Epipolar geometry can also be used efficiently for identification of points. Assume that on a surface, for example the body of an automobile, many small circular object points are available (for example a grid of projected points, Section C 2.3.1.2, Volume 2). Photographs of the surface should be taken in such a way that every object point is captured on at least three images. These points can be successfully found in a particular photograph with a circular reference matrix, as explained in Sections 6.8.1.1 and 6.8.1.2. The weighted centroid method has proved itself for such a task: within a window inside which the circular disk lies, the coordinates ξ_M and η_M of the centre of the circle are calculated as the weighted arithmetic mean of the ξ and η pixel coordinates using the grey levels g_{ij} as weights:

$$\xi_M = \frac{\sum \xi_{ij} g_{ij}}{\sum g_{ij}} \qquad \eta_M = \frac{\sum \eta_{ij} g_{ij}}{\sum g_{ij}} \tag{6.8-21}$$

The summations in Equations (6.8-21) run over all pixels within the chosen window. In this calculation those pixels are removed for which the grey value lies below a suitably chosen limit (for example all pixels with a grey level smaller than 10, if the circles are light coloured against a dark background). The problem of positioning the window within the neighbourhood of a particular circular area still remains open. One solution consists of moving the window one pixel at a time left or right and up or down until the sum of the grey values g_{ij} reaches a maximum. This solution leads to success, though, only when the size of the window is chosen to be less than the separation of neighbouring circular areas. If the circular areas are not too small and are against a dark background the standard deviation of this method lies around ± 0.05 pixel, thus a very high accuracy, comparable to that of LSM (Section 6.8.1.2).

In the following paragraph it is assumed that the circular points in all three images have been found. Epipolar geometry offers an interesting solution to the problem of identifying corresponding points, illustrated in Figure 6.8-13.

A point P is chosen in image 1; the corresponding point is sought in image 2. The epipolar ray $K_1 \to 2$, in image 2, cuts down the set of points to a one-dimensional group; when there are many points, no unambiguous solution is reached, however; three points which come into consideration are marked in Figure 6.8-13. Certainty can be restored with the help of image 3. For this purpose epipolar rays $K_i 2 \to 3$ are established in image 3 corresponding to all candidate points in image 2; in the example of Figure 6.8-13 there are three such candidate points. Finally the epipolar ray $K_1 \to 3$ is found in image 3, corresponding to the point P chosen in image 1. The required point P in image 3 lies on an intersection of the epipolar rays $K_i 2 \to 3$ with the epipolar ray $K_1 \to 3$. From this the corresponding point is also found in image 2. Should the points lie very close to each other or if the orientation elements are not very accurately

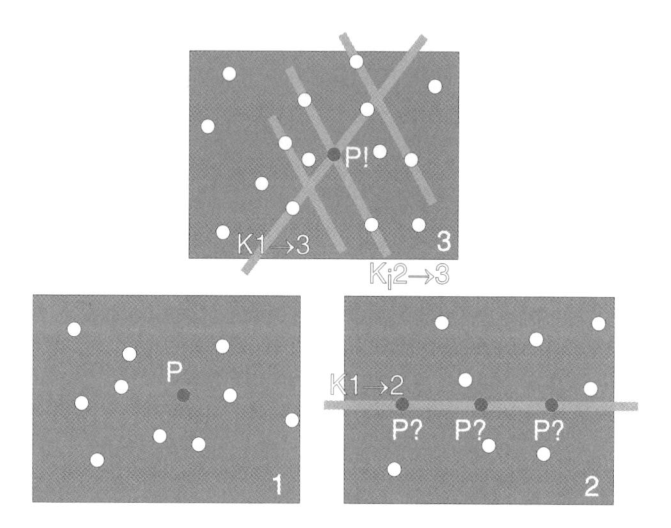

Figure 6.8-13: Epipolar geometry for the identification of the point P in three images

known the inclusion of a fourth photograph in the epipolar geometry system can be very valuable.

Exercise 6.8-11. How does the identification method explained in this section perform if the three camera stations are arranged along a straight line? (Answer: Since the epipoles are the images of the perspective centres of the other images, in this case two epipoles coincide. Therefore identification of corresponding points is not possible with this method.)

Literature: Dold, J., Maas, H.-G.: IAPR 30(5), pp. 65–70, Melbourne, 1994.

6.8.6 Automated recording of surfaces

In cases where there is an artificial pattern of points on a surface, for example points marked by circles as was discussed in Section 6.8.5.6, the surface can be recorded fully automatically. If the surface exhibits no such texture, another strategy must be followed. The recording of the surface of the Earth is central to the following discussion, although the same solution can also be applied to many kinds of tasks in close range photogrammetry.

Characteristic of this solution is that, instead of finding homologous areas or points of one image directly in the other image and then performing object reconstruction using three-dimensional intersection, the establishment of the relationships among the images and the reconstruction of the object are completed in a single, unified process.

Figure 6.8-14 illustrates the simplest procedure of this kind, known as the VLL relation (vertical line locus) which starts after the exterior orientation of the two photographs has been completed. The XY coordinates are set in advance, preferably over a regular grid (Figure 6.6-5). Beginning with one of the grid points a series of equally spaced Z coordinates is defined along the vertical line through that point. The strategy begins with points at particular Z levels being transformed into both photographs using the collinearity equations (2.1-19). In this way candidates for homologous windows are found in both photographs. Taking excerpts of equal size in both images, a measure of similarity, for example the correlation coefficient (6.8-1), is computed for each window pair. The required object point and its XYZ coordinates have been found when the maximum value of this similarity measure is reached.

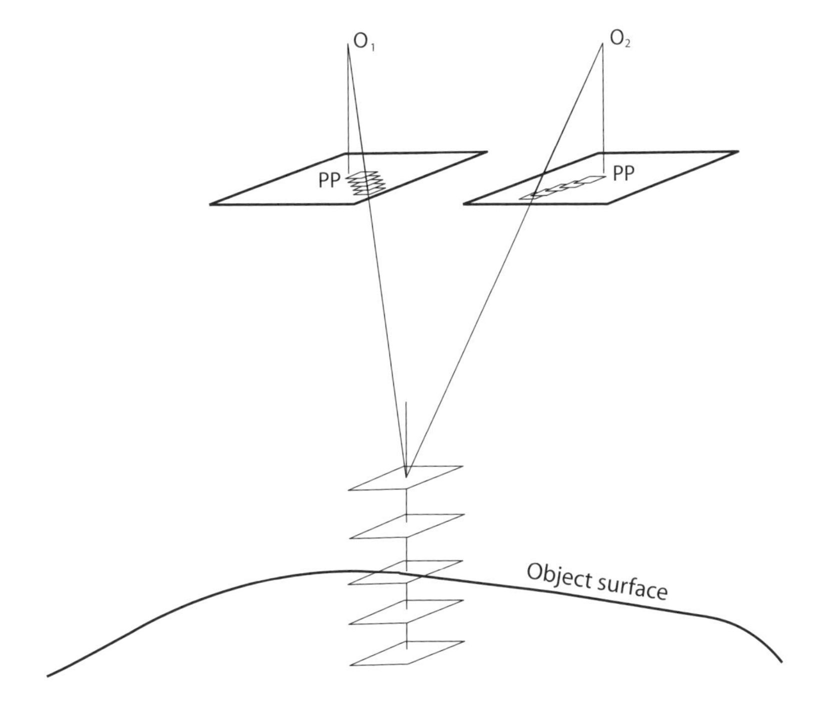

Figure 6.8-14: VLL procedure

This raster-driven correlation method gives only a point cloud and no break-lines on the ground. These have to be obtained by an operator and, together with the ground points obtained by correlation, brought together in a digital terrain model (DTM). Contour lines derived in this way have the appearance portrayed in Figure 6.6-4.

The VLL procedure delivers satisfactory results only under the following conditions (Section C 2.2.1, Volume 2):

- the image regions to be correlated must exhibit a good natural texture.
- the ground must be open (without buildings and not forested).

- the slope of the terrain may not be great (the maximum rotation of the corresponding area of the image is 30° and the maximum scale difference of the corresponding excerpts of the images is 25%).

The VLL procedure can be refined in various respects:

- from the different correlation coefficients r and the corresponding Z values, a continuous function $r(Z)$ can be found; if the first derivative of this function is set to zero the desired Z value is found with sub-pixel accuracy.
- the predetermined XY raster points can lead to pairs of windows in the photographs in which no suitable correlation is possible (too great a slope of the terrain or too much vegetation); at other XY raster points where a perfectly satisfactory correlation can be made the measured point may lie above the ground, as for example on a house roof; at such points still more "intelligence" must be built into the processing so that the correlator reacts in a similar way to a human operator (see the end of Section 6.6.2) and bypasses the raster point or chooses a substitute point nearby; the elimination of points on roofs can also take place in post-processing (Section 8.1.2.2).
- in place of the horizontal square element in the object (Figure 6.8-14) an inclined object element can be used; the slope can be estimated either from additional geometrical parameters in the LSM (Section 6.8.1.2(e)) or from the XYZ coordinates of neighbouring points already found through correlation.
- instead of isolated elements at each raster point the surface can also be approximated by continuously connected facets and in a computational procedure the correlations and all parameters of the facet structure be derived.

The object sends out not only geometric information but also additional radiometric information. In the case of digital photography we must consider not only geometric rays emitted from an object point, but also their electromagnetic properties which produced the grey, or colour, values in the photographs. In the reconstruction of the object surface the homologous geometrical rays must intersect in the object point and the homologous grey or colour values must produce identical radiation values in object space. To this end radiation models which take the line of sight into account should, above all, be introduced into the processing.

Literature: Grün, A., Baltsavias, E.: PE&RS 54, pp. 633–641, 1988. Heipke, C.: DGK, Reihe C, Heft 366, München, 1990 and PE&RS 58, pp. 317–323, 1992. Helava, U.: PE&RS 54, pp. 711–714, 1988. Wrobel, B.: IAPR 27(B3), pp. 806–821, Kyoto, 1988, and BuL 55, pp. 93–101, 1987. Zheng, Y.-J.: PE&RS 59, pp. 489–498, 1993. Nevatia, R.: IAPR 31(B3), pp. 567–574, 1996. See also Section C 2.2.4, Volume 2.

6.8.7 Semi-automated processing for plan

Fully automatic analysis of a scene captured on two or more photographs places the highest of demands on recognition and modelling processes. In computer vision such complex image analyses are performed for many problems. These processing methods from computer vision can frequently be transferred to close range photogrammetry. For complete processing in plan from aerial photographs, on the other hand, there are still few reliable and accurate solutions. That is the reason why photogrammetric processing in plan is solved semi-automatically, the recognition part of the task rendered by man (Section 6.6.1) and the measurement and modelling part automated. This section is limited to determination of the boundaries of different kinds of land use, to the transport network and so on; the independent Section 6.8.8 is dedicated to the semi-automated processing of buildings.

6.8.7.1 Active contours (snakes)

Semi-automated processing in plan can be put into practice using computed curves known as snakes. They are explained with reference to Figure 6.8-15: in the stereomodel the operator selects the forest boundary and sets the floating mark on a few points (Section 6.5). Then he initiates an automatic measuring process which, starting out from a coarse polygonal line, lays a curve along the forest boundary. This adaptation can take place in both photographs independently of each other. Intersection with epipolar rays produces the $\xi\eta$ coordinates of corresponding points in both images so that the XYZ coordinates of points along the three-dimensional curve in the object model can be derived[16]. Other strategies are conceivable which, for example, carry out adaptation to the image curves in both photographs simultaneously, while also taking the 3D spatial curve into account. (Details are to be found in the literature given in this section.)

Adaptation of the coarse polygonal line to a curved grey level edge in a photograph still needs closer consideration. The curved edge may be modelled for example by a spline function $\mathrm{s}(t)$; s is vector valued,i.e. there is a spline function for each of the two image coordinates $\xi(t)$ and $\eta(t)$. The parameter t can be taken as curve length from the starting point.

The spline function $\mathrm{s}(t)$ should, in our example, be smoothed as much as possible. This geometric constraint is arrived at by minimizing the first and second derivatives, that is by minimizing $\mathrm{s}'(t)$ and $\mathrm{s}''(t)$. The appropriate corrections for the corresponding observation equations in the sense of a least squares estimation[17] are denoted by $\mathbf{v}_{\mathrm{s}'}$ and $\mathbf{v}_{\mathrm{s}''}$.

[16] Should the particular epipolar ray make only a very shallow intersection, or even meet the curve tangentially, the determination of the XYZ object coordinates becomes very inaccurate or, in some cases, indeterminate. Details of the problem are to be found in the third item in the comment list of Section 6.8.8.

[17] Instead of the least squares solution used in this section the optimization of snakes can be arrived at by calculus of variations, by dynamic programming, by level-set formulation (see, for example, Osher, St., Paragios, N.: Geometric Level-Set Methods in Imaging, Vision and Graphics. Springer, 2003) or by simulated annealing (van Laarhoven, P.J.M.: Simulated annealing—Theory and Applications. Reidel, 1992).

Figure 6.8-15: Three operator-placed starting points (left); interim result (middle); and final position (right) of the curve as forest boundary

The corrections $\mathbf{v}_{s'}$ and $\mathbf{v}_{s''}$ are internal geometric constraints on the desired outline. An external constraint is that the boundary line should run as close as possible to the digitized polygon points. The corrections connected with this in a least squares estimation are denoted by $\mathbf{v}_{\text{ext}}$. (Details of the mathematical formulation are to be found in Sections B 3.5.4, Volume 2.)

The third group of corrections $\mathbf{v}_{\text{pho}}$ relates the curve to the image content. The LSM scheme (Section 6.8.1.2) can be invoked. A reference matrix, in this case a reference matrix for a forest boundary, is defined along the approximate curve. To the left of the approximate curve a dark grey level is used and to the right of it a light grey level. Both grey values will be taken from the mid-range of values in the given photograph. A two-dimensional LSM formulation (6.8-7), which allows not only two translations b_ξ and b_η but also permits a deformation of the reference image according to the spline function chosen above, will improve the curve in a combined adjustment. (With longer curves the reference image is broken into shorter pieces.)

Applying appropriately chosen weights $\mathbf{P}$, the combined adjustment minimizes the following function:

$$\begin{aligned}\mathbf{v}^\top\mathbf{P}\mathbf{v} &= (\mathbf{v}_{s'}^\top\mathbf{P}_{s'}\mathbf{v}_{s'} + \mathbf{v}_{s''}^\top\mathbf{P}_{s''}\mathbf{v}_{s''}) + \mathbf{v}_{\text{ext}}^\top\mathbf{P}_{\text{ext}}\mathbf{v}_{\text{ext}} + \mathbf{v}_{\text{pho}}^\top\mathbf{P}_{\text{pho}}\mathbf{v}_{\text{pho}} \\ &= \mathbf{E}_{\text{int}} + \mathbf{E}_{\text{ext}} + \mathbf{E}_{\text{pho}}\end{aligned} \tag{6.8-22}$$

In the terminology of snakes, the function to be minimized is known as an energy function with the terms $\mathbf{E}_{\text{int}}$ for the internal energy, $\mathbf{E}_{\text{ext}}$ for the external energy and $\mathbf{E}_{\text{pho}}$ for the photometric energy.

The snakes strategy introduced here for semi-automatic photogrammetric processing follows the example of the publication by Grün and Li, which also contains instances of different applications and an extension to the simultaneous determination of curves in several overlapping photographs.

This section should conclude with further indication of the flexibility of the snakes formulation by mentioning some of the different ways in which the $\mathbf{E}_{\text{pho}}$ terms may be framed. In the above formulation, different grey levels left and right of the provisional curve were used. Frequently, however, differences in grey levels along a forest border are not very meaningful, so that texture ought to be invoked in the place of grey levels. The interior of a forest is highly textured whereas outside the forest there is usually little texture. From the photographs in question a representative measure of texture is made to the left of the provisional curve and likewise a representative measure of texture is made to the right. Texture measures can be determined with the help of the Förstner operator or the Moravec operator (Section 6.8.1.3).

Additional Literature: Kass, M., Witkin, A., Terzopoulos, D.: International Journal of Computer Vision 1(4), pp. 321–331, 1988. Kerschner, M.: IAPR 32(3/1), pp. 244–249, Ohio, 1998.

6.8.7.2 Sequential processing

After the operator has chosen the particular kind of line to be used in plan and has digitized the starting points, the snake solution runs through smoothly in one go, so long as the operator does not intervene because he considers that the lines in the photographs look unsatisfactory. In the present section this closed solution described above is compared with another method which consists of a number of sequential steps.

The key phases of this semi-automatic, three-dimensional processing method are as follows:

- in each of the two photographs, independently of each other, the lines are extracted.
- then corresponding lines are matched to each other; here epipolar geometry provides very valuable assistance.
- the matching of corresponding lines is supported by means of human stereo-observation. This gives the operator an opportunity to prescribe the meaning of individual lines from a catalogue of object types.
- the image coordinates of homologous points come from epipolar geometry. If the particular epipolar line just grazes, or barely cuts, the particular plan line, the image coordinates are very unreliable; the best remedy is the inclusion of a third metric image.
- with the image coordinates of the homologous points and the known elements of exterior orientation of both, or in some cases more, metric images the XYZ coordinates of a succession of points in the 3D object model follow.

Operator involvement in the matching of corresponding lines can be very costly; by accepting a certain loss of quality, this cost can be reduced when choosing the following work programmes:

- capturing and modelling of the object surface, that is for example a terrain model from aerial photographs (Section 6.8.6)
- production of a digital orthophoto from one of the two metric images (Section 7.3)
- line extraction in the two-dimensional digital orthophoto; the result is the XY coordinates for the succession of points in the two-dimensional object model
- in cases where a three-dimensional object model is called for, the Z coordinates, heights corresponding to the XY coordinates of the DTM points, can be extracted by means of interpolation.

Line extraction requires still closer examination. In the section "Filtering in the spatial domain" (Section 3.5.2.1) readers have already become acquainted with the basic ideas of edge extraction which, in the case of processing in plan, is indicated as better than line extraction[18]. A typical line-extracting operator is the Laplace operator (3.5-14). Figure 6.8-16 is the result of a convolution of the right-hand matrix of Figure 6.8-1 with the Laplace operator. The lines in the original image come out as zero-crossings in the convoluted image. Through linear interpolation between the negative and positive pixel values along the rows and along the columns, the lines are found with subpixel accuracy. (The achievable high accuracy can be diminished, however, because of the shift in position connected with compression of the photographs, Section 3.6.2). In Figure 6.8-16 the outline of a cross is obtained.

But isolated zero lines also occur; they are caused, for example, by noise in the original grey values. In the original Figure 6.8-1 the grey value differences at the isolated points were around 1; in Figure 6.8-16, which is formed through convolution with the Laplace operator, they are around 5. This statement can be generalized: the Laplace operator and comparable operators, which are based on second differences, reinforce the grey value differences between neighbouring pixels (in our example the maximum grey value difference in the original image is 8 ($= 9 - 1$) and in the convoluted image it is 32 ($= 8 + 24$).

[18] In the image-processing literature there is frequently a differentiation between edges and lines. At an edge an abrupt transition occurs from one grey level to another grey level. Along a grey level profile perpendicular to the edge, the position of the edge is at a maximum of the first derivative or at a zero of the second derivative of the profile function. At a line, coming from either side, there is a relatively more abrupt transition from one grey level value (possibly, indeed, from two different grey value levels) to one lying in the middle of a common grey value level. The band of pixels with the common grey value level, the actual line, is very narrow. If one imagines a grey value profile perpendicular to the line, the position of the line is at a zero of the first derivative of the profile function. An operator which extracts edges gives the envelope of a line. The thinner the line the more merged together the envelope. The edge extraction operator can thus be used in a similar manner for extraction of lines.

0	0	0	0	1	-4	1	0	0	1
0	1	0	0	0	1	0	0	1	-4
1	-4	1	0	0	1	0	0	0	1
0	1	0	0	1	4	1	0	0	0
0	0	0	0	8	-23	9	0	0	0
0	0	0	7	16	-15	12	9	1	0
0	0	7	-20	-17	0	-15	-23	4	1
0	0	1	7	16	-16	16	8	1	0
0	1	-4	1	8	-24	8	0	0	0
0	0	1	0	0	8	0	0	0	0

Figure 6.8-16: Convolution of the right-hand matrix of Figure 6.8-1 with the Laplace operator (3.5-14) (the convoluted matrix is smaller than the input matrix by two columns and two rows)

The troublesome consequences of noise in line extraction with the Laplace operator can as a result be removed to a large extent if the original image is first subjected to a low-pass filter (for example a Gauss filter (Sections 3.5.2.1 and 3.5.2.2)) before the Laplace operator.

Both of these processes are combined in the *LoG* operator, the Laplacian of Gaussian. The *LoG* operator filters out the noise (a low-pass filter) and simultaneously provides an image in which the zero crossings represent the lines. This operator is obtained when, in imitation of the differentiation of the Laplace operator (Section 3.5.2.1), the Gaussian bell curve, with which the elements of the Gauss filter are fixed, is twice differentiated and the second derivatives in both coordinate directions—the total differential—are combined in one function.

If one chooses the determining parameter σ of the Gaussian low-pass filtering to be the same in both coordinate directions, that is $\sigma = \sigma_\xi = \sigma_\eta$, and assumes independence between the two coordinate directions, that is $\sigma_{\xi\eta} = 0$, the two-dimensional normal distribution is obtained (Figure 6.8-17)[19]:

$$W(\xi,\eta) = W(r) = \frac{1}{2\pi\sigma^2} e^{-\frac{\xi^2+\eta^2}{2\sigma^2}} \qquad r^2 = \xi^2 + \eta^2 \tag{6.8-23}$$

Differentiating with respect to ξ and with respect to η:

$$\left(\frac{\partial W}{\partial \xi}\right)(\xi,\eta) = -\frac{\xi}{2\pi\sigma^4} e^{-\frac{\xi^2+\eta^2}{2\sigma^2}} \qquad \left(\frac{\partial W}{\partial \eta}\right)(\xi,\eta) = -\frac{\eta}{2\pi\sigma^4} e^{-\frac{\xi^2+\eta^2}{2\sigma^2}} \tag{6.8-24a}$$

Differentiating again with respect to ξ and with respect to η:

$$\left(\frac{\partial^2 W}{\partial \xi^2}\right)(\xi,\eta) = -\frac{\xi^2-\sigma^2}{2\pi\sigma^6} e^{-\frac{\xi^2+\eta^2}{2\sigma^2}} \qquad \left(\frac{\partial^2 W}{\partial \eta^2}\right)(\xi,\eta) = -\frac{\eta^2-\sigma^2}{2\pi\sigma^6} e^{-\frac{\xi^2+\eta^2}{2\sigma^2}} \tag{6.8-24b}$$

[19]For example: Mikhail, E.: Observation and Least Squares. IEP-A Dun-Donnelley Publisher, New York, 1976.

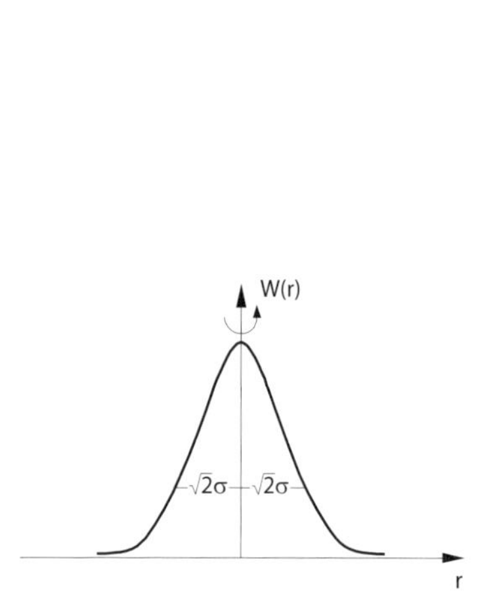

Figure 6.8-17: Gaussian bell curve for the two-dimensional normal distribution

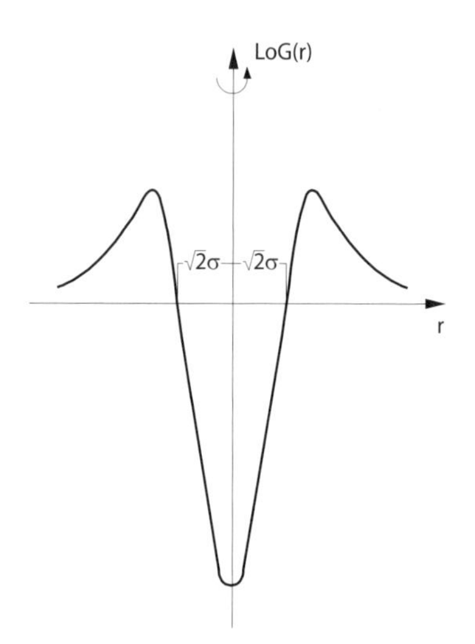

Figure 6.8-18: Two-dimensional *LoG* function

As in the derivation (3.5-13) of the Laplace operator from the second differences (3.5-12), here the second derivatives are added and are designated the *LoG* function:

$$LoG := (\nabla^2 W(\xi,\eta)) = \frac{\xi^2+\eta^2-2\sigma^2}{2\pi\sigma^6} e^{-\frac{\xi^2+\eta^2}{2\sigma^2}} \tag{6.8-25}$$

The *LoG* function has the shape of an inverted sombrero (Figure 6.8-18). The one and only parameter that is free to be chosen is σ. A small σ results in many zero points. A large σ suppresses the pseudo-lines, which are caused above all by the noise; it is more or less the case that only the relevant lines in the original image are extracted. The *LoG* convolution operator can be derived from the *LoG* function (6.8-25). For $\sigma = \sqrt{2}$, for example, the following convolution matrix is obtained; it is limited to a 9×9 matrix; the values lying outside this 9×9 matrix, which are small, are not shown:

$$LoG = 10^{-3} \times \begin{pmatrix} 0 & 1 & 2 & 4 & 4 & 4 & 2 & 1 & 0 \\ 1 & 3 & 7 & 10 & 10 & 10 & 7 & 3 & 1 \\ 2 & 7 & 11 & 6 & 0 & 6 & 11 & 7 & 2 \\ 4 & 10 & 6 & -24 & -46 & -24 & 6 & 10 & 4 \\ 4 & 10 & 0 & -46 & -80 & -46 & 0 & 10 & 4 \\ 4 & 10 & 6 & -24 & -46 & -24 & 6 & 10 & 4 \\ 2 & 7 & 11 & 6 & 0 & 6 & 11 & 7 & 2 \\ 1 & 3 & 7 & 10 & 10 & 10 & 7 & 3 & 1 \\ 0 & 1 & 2 & 4 & 4 & 4 & 2 & 1 & 0 \end{pmatrix} \tag{6.8-26}$$

The advantage of the *LoG* operator over the Laplace operator for the extraction of lines is demonstrated with the aid of a practical example. Figure 6.8-19 shows the original

Figure 6.8-19: Original image

Figure 6.8-20: With lines extracted using the Laplace operator

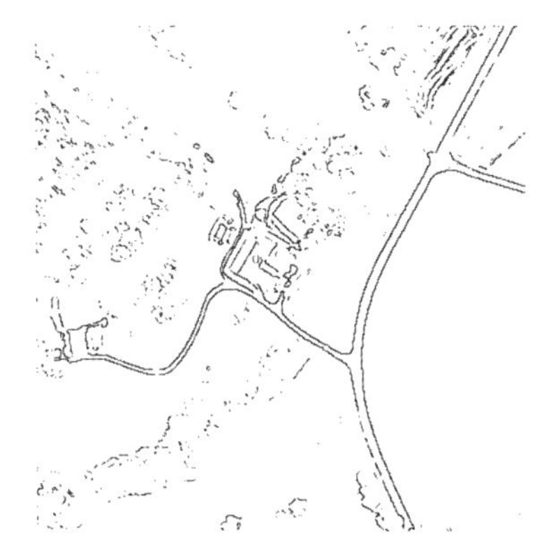

Figure 6.8-21: With lines extracted using the LoG operator

image. Figure 6.8-20 reproduces the image after convolution with the Laplace operator (3.5-14); Figure 6.8-21 is the result of a convolution with the LoG operator (6.8-26). The advantage of the LoG operator is clearly recognized; the many pseudo-lines are suppressed and the relevant lines are clearly emphasized.

Before leaving this section the problem of the shift in position of lines extracted with the LoG operator should be touched on. Since the LoG operator is an edge-extracting operator, the sum of its elements should be zero (Section 3.5.2.1). Checking this for the matrix (6.8-26) though gives a sum of -0.008. This defect leads to a shift of the extracted lines that must be taken into account when high accuracy is demanded. The shift is dependent both on the deficiency in the edge operator and on the contrast at the particular edge (Section C 2.2.2,Volume 2).

Incidentally, similar shifts also appear on application of the Förstner operator to edge extraction (Section C 2.2.2,Volume 2). The Förstner operator was introduced in Section 6.8.1.3 as the interest operator for finding well correlated image points. With this operator points in the image are found at which a high contrast is present and a good distribution of grey level edges in all directions.

For the extraction of lines the Förstner operator must be modified; patches with high contrast are indeed sought but they should also have a very limited distribution of edges in all directions, in the ideal case only a line. The Förstner operator modified in this way delivers points and the directions of the lines at those points. From such a sequence of points and associated tangents a complex curve may be produced.

Exercise 6.8-12. Develop the *LoG* operator for $\sigma = 2$. Result: The central element is -0.0199; the element which is distant two pixels from the centre is -0.0037; the element a distance of $2\sqrt{2}$ from the central point is zero; and so on.

Exercise 6.8-13. Establish the one-dimensional *LoG* operator. Take, as starting point, the one-dimensional normal distribution, $W(\xi) = 1/(\sigma\sqrt{2\pi})\exp(-\xi^2/(2\sigma^2))$. σ is the abscissa of the inflection point in the Gaussian bell curve and also the abscissa of the zero point in the *LoG* operator. Result: $LoG = (\xi^2 - \sigma^2)/(\sigma^5\sqrt{2\pi})\exp(-\xi^2/(2\sigma^2))$.

Literature: Schenk, T., Li, J.-C., Toth, C.: PE&RS 57, pp. 1057–1064, 1991. Schenk, T.: Digital Photogrammetry. TerraScience, 1999. Couloigner, I., Ranchin, T.: PE&RS 66, pp. 867–874, 2000. Baumgartner, A., Hinz, S., Wiedemann, C.: IAPR 34(3B), pp. 28–31, 2002.

6.8.8 Semi-automatic measurement of buildings

The semi-automatic measurement of buildings usually begins with the operator making topology-supported measurement, as discussed in Section 6.6.3. The corner points of the buildings are only roughly set; the fine measurement, which is the subject of the following passages, is taken over by the computer.

The starting situation for the automatic fine measurement is portrayed in Figure 6.8-22. The buildings have already been reconstructed with moderate accuracy. Using the known elements of interior and exterior orientation, these relatively inaccurate building edges are transformed into the relevant metric images. Figure 6.8-22 is limited to the roof ridge in two photographs.

Before this transformation as illustrated in Figure 6.8-22, line extraction was carried out in the photographs involved (Section 6.8.7.2). The points used in extracting the lines are illustrated in Figure 6.8-23, as well as the transformed straight lines of the roof ridge, accentuated in Figure 6.8-22. In both of the photographs an ε frame (i.e. a tolerance band) has been laid around the transformed (approximate) roof ridge. Subsequently all points used in extracting the lines inside the particular ε frame are chosen and an improvement of the straight-line roof ridge is brought about in the object model.

Two linearized observation equations can be set up for each point of the extracted line in each of the two metric photographs (in image 1 they are the points 1, 2, 3, 4 and, in image 2, the points 5, 6, 7, 8, 9). These equations correspond to Equations (4.1-3) for

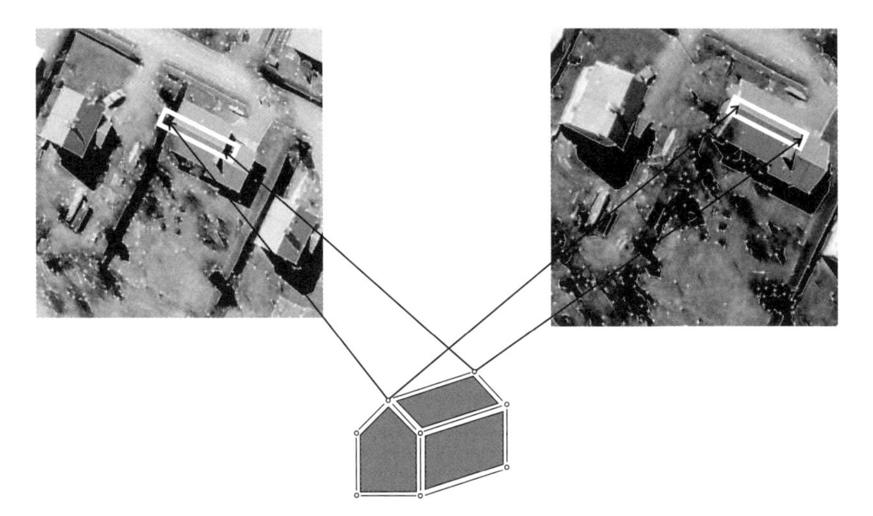

Figure 6.8-22: The approximate building model, transformed into and inserted in, two photographs from which line extraction is planned. The area surrounding the roof ridge has been highlighted (after Rottensteiner, F.: Geow. Mitt. der TU Wien, vol. 56, 2001).

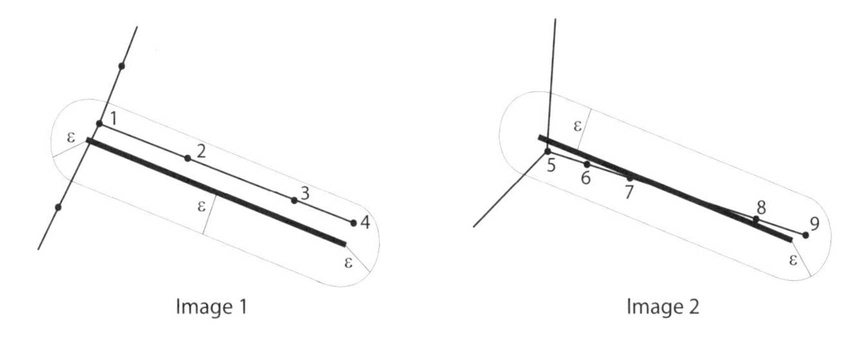

Figure 6.8-23: The approximated roof ridge in both metric images and the image points found during line extraction inside an ε frame

three-dimensional intersection but, since there are no homologous points as in three-dimensional intersection, for each image point there are three unknown object coordinates X, Y, Z. In the example of Figure 6.8-23 this means that, as against 18 $(= 9 \times 2)$ observation equations, there are 27 $(= 9 \times 3)$ unknown XYZ coordinates.

In order to make the adjustment equations soluble and to achieve straightness in the roof ridge, additional condition equations must be introduced into the least squares estimation. In order that three points with indices i, j, k should lie on one straight line, the following two conditions should be met:

$$\text{In the } XZ \text{ coordinate plane: } \frac{X_k - X_i}{Z_k - Z_i} = \frac{X_j - X_i}{Z_j - Z_i} \tag{6.8-27}$$

$$\text{In the } YZ \text{ coordinate plane: } \frac{Y_k - Y_i}{Z_k - Z_i} = \frac{Y_j - Y_i}{Z_j - Z_i} \tag{6.8-28}$$

A third, corresponding, equation in the XY coordinate plane could indeed be formulated but it is redundant, for it could be obtained by division of Equations (6.8-27) and (6.8-28). The next point, and each further point, provides two condition equations of the form of (6.8-27) and (6.8-28). Since 9 points occur in this example, there are in total 14 $(= (9-2) \times 2)$ condition equations, so that for the determination of the 27 unknown XYZ coordinates mentioned above a total of 32 $(= 18 + 14)$ equations is available. A discussion of this type of estimation problem, adjustment with indirect observations and condition equations, can be found in textbooks[20] on least squares estimation.

This adjustment results in a set of points, $1-9$, with their XYZ coordinates, all lying along the straight line of the 3D roof ridge. The observed image coordinates ξ and η in the photographs receive "corrections", the sum of the squares of which has been minimized.

This section on the refinement of the solution for the roof ridge should conclude with some comments:

- the conditions corresponding to the Equations (6.8-27) and (6.8-28) should be established for those pairs of coordinate planes in which the coordinate differences are greatest.
- in order to avoid numerical problems in the least squares estimation, the separations of the points along the extracted lines should not be very small. In every condition equation the first and last points of a line should be included; points to be replaced should always be confined to interior points.
- if the roof ridge runs parallel to the camera base the bundles of rays from both photographs lie in the same plane; therefore the intersection conditions for points along the 3D line in the object will be unsatisfactory. If the roof ridge portrayed in the photographs runs along the epipolar rays in both images, a similar critical situation can also arise. The best remedy is the inclusion of a third image. The use of more than two photographs also has an additional advantage; in cases where part of a building is hidden in one photograph that problem, too, can be overcome.
- likewise, with robust estimation and data-snooping (Sections B 7.2.1.4 and B 7.2.1.5, Volume 2) wrongly matched image points can be reliably eliminated.
- analysis of the "corrections" can also lead to the realization that the roof ridge is not a straight line but should be approximated by a 3D polygonal path or a 3D

[20] E.g. Mikhail, E.: Observations and Least Squares. IEP-A Dun-Donnelley Publisher, New York, 1976.

curve. In this case, the 3D straight line should be replaced with a 3D polygonal path or curve.

- the use of image pyramids from the photographs (Section 3.6.1), which can be generated with low-pass filtering, is recommended for the automatic refinement process. The coarsest pyramid is used to start with; extraction at this level finds only prominent lines, so there are no disturbing points from lines of minor importance. When passing to finer levels the ε frame is progressively reduced; to a large extent, errors in identifying corresponding points are excluded in this way.
- other a priori knowledge besides straightness of lines is brought into the refinement, for example the horizontality of roof ridges in new buildings. In this case the condition equation
$$Z_j - Z_i = 0 \qquad (6.8\text{-}29)$$
is additionally introduced in the least squares estimation. The second and further points each give one such condition equation. Thus for our example from Figure 6.8-23, eight further condition equations (6.8-29) appear; this means that 40 equations are available for the determination of the 27 unknowns.
- instead of the XYZ point sequence along the adjusted 3D straight line, the parameters of the straight line can be used directly in the automatic refinement. Such an expression has few unknowns; its formulation is didactically more demanding.
- automatic refinement should not be applied for each 3D line in isolation, but simultaneously for all edges of a whole building (Figure 6.6-6 shows an example) or at least for a part of a building which is indicated as a building primitive[21]. The advantages of a simultaneous treatment are obvious; especially that extensive a priori knowledge can be elegantly introduced in the form of numerous kinds of conditions (building walls perpendicular to each other, roof planes with equal inclination, and so on). The example shown in Figures 6.6-7 and 6.6-8 resulted from this kind of complex automatic refinement.

Additional Literature on semi-automatic processing of buildings: Gülch, E., Müller, H.: in Baltsavias, E., Grün, A., van Gool, L. (Eds.): Automatic extraction of man-made objects from aerial and space images (III), pp. 103–114, Balkema Publishers, Lisse, 2001. Grün, A., Wang, X.: ibid., pp. 93–101, 2001. Zhou, G., Li, D.: PE&RS 67, pp. 107–116, 2001.

6.8.9 Accuracy and reliability of results obtained by automated or semi-automated means

There is as yet no long tradition of semi-automated and automated measurement procedures in photogrammetry. It is not surprising, therefore, that there are no rules of thumb for their accuracy.

[21] Rottensteiner, F.: Geow. Mitt., vol. 56, 2001.

Certainly semi-automated and automated measurement methods are displacing human operators more and more. These modern methods of processing, however, will be accepted only if they achieve at least the same quality of the work as for that carried out by human operators, which was reported in Section 6.7.

For project planning, therefore, the rules of thumb given in Section 6.7 can be used, independently of whether the work is done semi-automatically, automatically or by a human operator. But it is necessary to see to it that the prerequisites for good semi-automated and automated measurement procedures are present, for example relatively high resolution in digitizing of the photographs, correlation algorithms which give sub-pixel results and so on.

From interim published accuracies achieved in specialist projects it can be concluded that semi-automated and automated procedures have a higher potential accuracy than that of work done by human operators. When it comes to single point measurement, it may indeed be the case that the work of human operators is more accurate; the much higher number of points and lines which are amassed in the automated procedures compensate for the accuracy deficit in single-point measurement. Usually over-compensation is made.

The reliability of results obtained automatically can be guaranteed only for simple tasks. If the nature of the task is complex, as in capturing topographic information from aerial photographs, a visual check should always be planned, such as by superimposing the results on the photographs.

6.8.10 Special features of the three-line camera

The use of images from a three-line camera, or three-line scanner (Section 3.7.2.3), involves some special features which will be touched on in this section. Since a three-line camera is always flown with GPS and IMU support, direct georeferencing (Section 4.1.2) can be performed. It remains an unanswered question though, how automatic localization of corresponding points in the three image strips takes place.

Good conditions for automatic localization occur because the images are first normalized using information from the GPS/IMU, in order to remove large distortions arising because of the dynamics of the flight. This step is comparable with the production of normalized images from arbitrarily configured images from conventional frame cameras (Section 6.8.5.2). The mathematical relationships need not be repeated here. The procedure corresponds to an indirect transformation of the images from the three-line camera into an adopted horizontal plane.

Corresponding points in images from a normalized strip have almost identical η coordinates. On the contrary the ξ coordinates of corresponding points differ very greatly. On the one hand the differences come from the way in which the rows a and b in the camera are staggered (Figure 3.3-3, right) and on the other hand from the height differences of the ground. Normalized strip images have characteristics comparable to those of normalized images from stereophotogrammetry with conventional metric cameras

(Section 6.8.5.2): correlation can proceed along narrow bands which exhibit the same η coordinates.

Correlation to establish corresponding points in the normalized strip image-triples for automated and semi-automated procedures (Section 6.8) can be carried out with three-image correlation (Section 6.8.1.5). This three-image correlation is a significant advantage of the three-line camera in comparison with the conventional aerial photographs with 60% forward overlap where, over large areas, correlation can be performed only with two-image correlation.

After the normalization of the image-triple using the GPS/IMU information, it is unusual for the processing for height, plan and buildings to begin immediately; normally the GPS/IMU orientation will first be improved. For this purpose many tie points on the upper and the lower edges and in the middle of the image-triple are automatically located and an integrated sensor orientation (Section 5.5) carried out.

After this improvement of the orientation there are no longer any residual η parallaxes between the images of the strip-triple. The images can be observed in digital workstations (Section 6.5) and the heighting, measurement in plan and recording of the buildings can be carried out by an operator, with computer assistance (Section 6.6). The operator has the possibility of choosing any particular pair of strips and observing their image-triples stereoscopically.

This choice between normal angle, wide angle and intermediate angle (Section 3.7.2.3) depends on the desired accuracy and the visibility on the ground. Work-stations that are set up for the processing of images from a three-line camera allow a rapid switchover between particular pairs of images of a triple. Such workstations take advantage of the fact that in flight direction there are no stereomodel boundaries.

Literature additional to that given at the end of Section 5.5: Grün, A., Li, Z.: PFG 2003(2), pp. 85–98. Sandau, R. et al.: IAPRS 33(B1), pp. 258–265, Amsterdam, 2000.

Chapter 7

Orthophotos and single image analysis

As is well known, a single image is not sufficient for reconstructing a spatial object. However, methods have been developed which only use one image. Such single image analyses are conditional on knowing the geometric shape of an object; for example that the situation is dealing with an object plane or that a digital surface model is available.

Single image analysis frequently takes place in two stages. In the first stage an analogue or digital photograph, which is normally distorted, is converted into a geometrically correct photograph or orthophoto. In a second stage this orthophoto is evaluated by analogue, analytical or digital means.

In many cases, by adding graphical elements, such orthophotos are converted to orthophoto maps, in particular those based on aerial photographs, and known simply as photomaps. In undeveloped regions, such photomaps are significantly more advantageous than line maps. Archaeologists, soil scientists, foresters, agriculturalists, geographers, geologists, planners and ecologists often do not find, in a line map, the details import to them. Maps which provide the full content of analogue or digital photographs are a better solution in this case. Even in developed areas, however, there is a demand for the image content in the photos. Orthophoto maps are also attractive because they are significantly faster and cheaper to produce than line maps.

Increasingly, digital orthophotos are employed as data in topographical information systems. To a considerable extent, orthophotos bring the real world into a Geographic Information System (GIS), in which the topographic information system is a subsystem. The digital orthophoto is therefore an excellent orientation aid for the GIS user. (This opinion was expressed in Section 6.6.3 for photogrammetric vector analysis.) A GIS user coming from one of the related disciplines mentioned above can personally undertake photo interpretation on the basis of the digital orthophotos and digitize the results in vector mode. In this way these disciplines can build up their own thematic data base.

Digital orthophotos play a large role in the visualization of both developed and undeveloped regions. For this purpose the orthophoto pixels are projected onto the digital terrain model and viewed obliquely. These three-dimensional visualizations can also be presented as animations (Section 7.6).

However, the orthophoto as texture source reaches its limits with large-scale visualizations, especially for urban landscapes. Here the orthophotos from aerial surveys contain no photo texture for the vertical façades, etc. For consistent three-dimensional visualizations the orthophoto must be developed further into a photo model. This is obtained by transferring the texture in a photograph onto the surface of a three-dimensional object model, often a CAD vector model (Section 7.5).

Chapter 7 is mainly concerned with the production of digital orthophotos and their accuracy. It will be seen that professionally created, digital orthophotos are a very accurate source of data. Before describing the production process, the distortions in the metric image (central projection) compared with a map (orthogonal projection) are discussed.

7.1 Perspective distortion in a metric image

Distortions in mappings from one surface to another are described in differential geometry by means of the Tissot indicatrix[1]. This also includes perspective images. This theory can be applied, however, only if the surface of the object can be described by smooth analytical functions. The surfaces of the ground and other objects can generally not be so described. More often the ground surface is represented in photogrammetry by digitizing a large set of individual points on the surface. These are then approximated by small (flat) surface elements which at their joins have no discontinuities in height but discontinuities in slope. (In geometry such surfaces are characterized by the C^0-continuity.) A different method, based on the XYZ-coordinates of closely spaced, discrete points, is therefore required to describe the distortions in the metric photograph.

Figure 7.1-1 shows an object surface defined by the Z-coordinates of a square XY grid. In the sense of a map (orthogonal projection) the square XY grid is the distortion-free image of the ground surface. The distortions in the photograph are derived by computing the points in the photograph corresponding to every corner of the grid on the ground by means of the collinearity equations (2.1-19). The result is sketched in Figure 7.1-2.

The deformation of the grid gives a very clear impression of the deformations in an aerial photograph (Figure 7.1-3).

Numerical results for the deformation at individual points (e.g. point 1 of the square $1, 2, 3, 4$ in Figure 7.1-2) can be derived from the image coordinates ξ and η and the ground coordinates X and Y of the point, as follows:

(a) length distortions in X and Y directions:

$$\lambda_x = \frac{\delta_{12}}{\Delta X}, \quad \lambda_y = \frac{\delta_{13}}{\Delta Y} \tag{7.1-1}$$

in which

$$\delta_{ik} = \sqrt{(\xi_i - \xi_k)^2 + (\eta_i - \eta_k)^2}$$

[1]Snyder, J.P.: Flattening the Earth: Two Thousand Years of Map Projections. Chicago University Press, 1993.

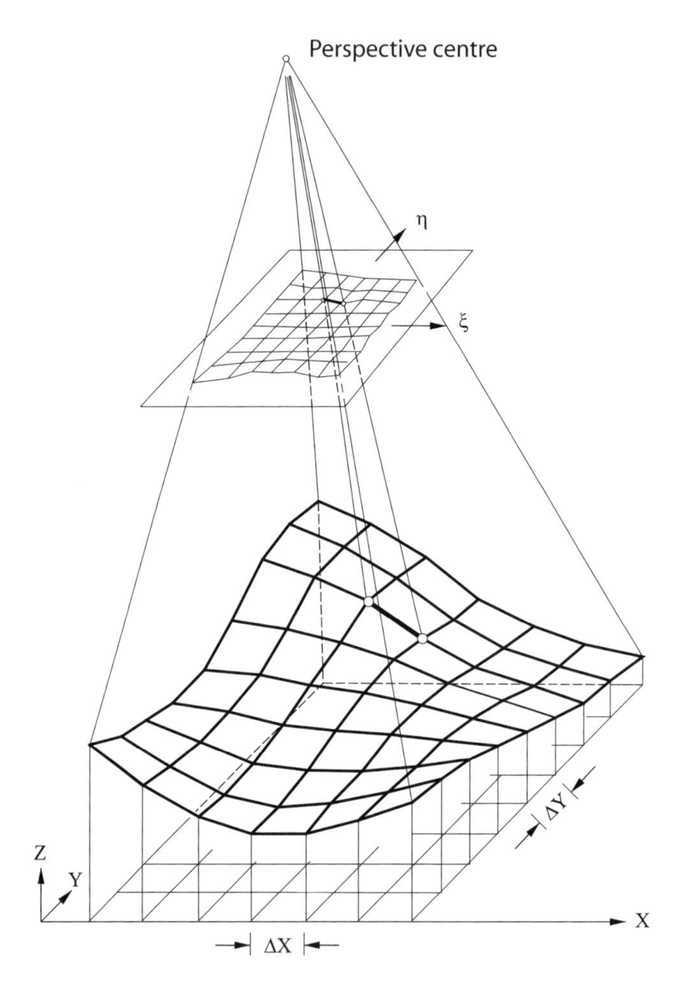

Figure 7.1-1: Relation between the square grid $\Delta X = \Delta Y$ in the XY plane and the corresponding deformed grid in the image plane

(b) distortion ω_{90} of the right angle at point 1:

$$\omega_{90} = 90° - \angle 2'1'3' \tag{7.1-2}$$

(c) extreme length distortions λ_1 and λ_2 in the directions of maximum deformation, from the two equations (see the theory of projection of two surfaces in any appropriate textbook):

$$\lambda_1^2 + \lambda_2^2 = \lambda_x^2 + \lambda_y^2 \tag{7.1-3}$$

$$\lambda_1 \lambda_2 = \lambda_x \lambda_y \cos \omega_{90} \tag{7.1-4}$$

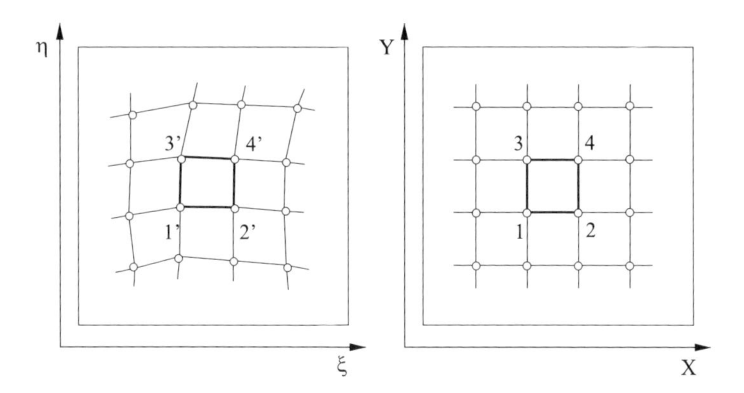

Figure 7.1-2: Grid in the photograph (central projection) and in the map (orthogonal projection)

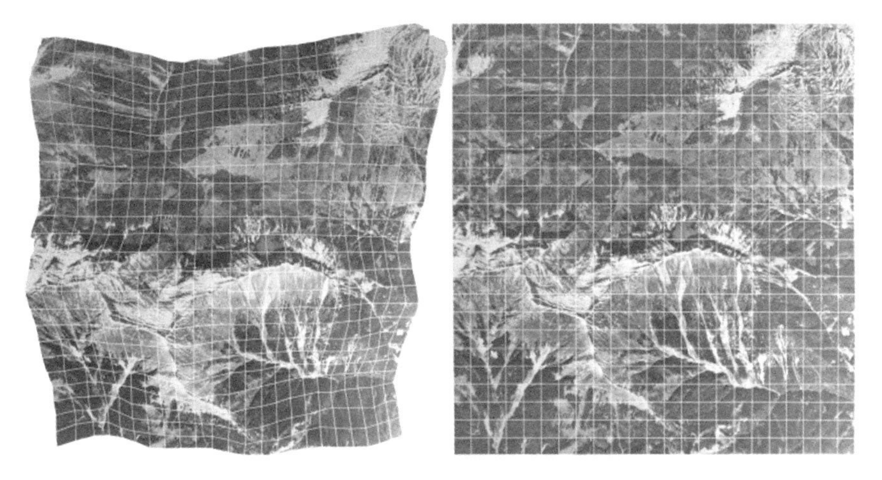

Figure 7.1-3: Deformed grid in the aerial photograph and the square grid in the orthophoto (courtesy of the Mapping Division of the Austrian Federal Office of Metrology and Surveying (BEV))

(d) maximum angular deformation 2ω:

$$\sin\omega = \frac{\lambda_1 - \lambda_2}{\lambda_1 + \lambda_2} \tag{7.1-5}$$

(e) deformation of area:

$$\lambda_1\lambda_2 = \lambda_x\lambda_y\cos\omega_{90} = \frac{\Delta 1'2'3'}{\Delta 123} \tag{7.1-6}$$

Note: since every grid point is the corner not only of one grid square, but of four squares (in general), the deformations at each grid point according to Equations (7.1-1) to (7.1-6) can be computed four times. The four values will not be exactly the same because, on the one hand, the ground surface is, in general, discontinuous at each grid point and, on the other hand, the initial Equations (7.1-1) are derived from finite differences and not from infinitely small differentials. The four values, which have only minor differences, should be averaged at each grid point.

Exercise 7.1-1. Assume the elements of interior and exterior orientation of the example at the end of Section 2.1.3 and compute from three neighbouring points P_1, P_2 and P_3 the various deformations from Equations (7.1-1) to (7.1-6), referred to a photo map at a scale of 1 : 10000. Solution:

	X [m]	Y [m]	Z [m]
1	363400	61500	575.12
2	363410	61500	576.87
3	363400	61510	578.04

Photo coordinates:

	ξ [mm]	η [mm]
1'	−33.694	92.706
2'	−33.773	93.893
3'	−34.835	92.859

Length distortions:	$\lambda_x = 1.1896$	$\lambda_y = 1.1512$
Right angle distortion:	$\omega_{90} = 90° - \angle(2'1'3') = 11°27'$	
Extreme length distortions:	$\lambda_1 = 1.2828$	$\lambda_2 = 1.0463$
Maximum angular distortion:	$2\omega = 11°40'$	
Area distortion:	$\lambda_1\lambda_2 = 1.3$	

Exercise 7.1-2. Repeat the computations under the assumption of an exactly vertical camera axis. (Solution: $\lambda_1 = 1.2760, \lambda_2 = 1.0431, 2\omega = 11°32'$.)

If the object surface is a horizontal plane, the deformations of the grid in the photograph can be seen very clearly in comparison with the XY grid in the object (Figure 7.1-4). In this diagram, the photograph is tilted about an axis running from its lower left to its upper right corner. The line of maximum slope of the photograph runs from the upper left to the lower right corner.

The relation between the square grid and the deformed grid is here defined by the collinearity equations (2.1-19) in which Z is taken as the height of the object plane (parallel to the XY plane). Instead of the collinearity equations, the projective transformation (2.1-24) can be used.

Numerical results for the deformations can also be derived by applying Equations (7.1-1) to (7.1-6) to plane object surfaces. A mathematical formulation by the Tissot indicatrix is also possible in this case. Rather than formulating such a relatively complex solution, we give below a practical approximation from which the distortions in photographs of plane objects with small angles of tilt can be derived. The

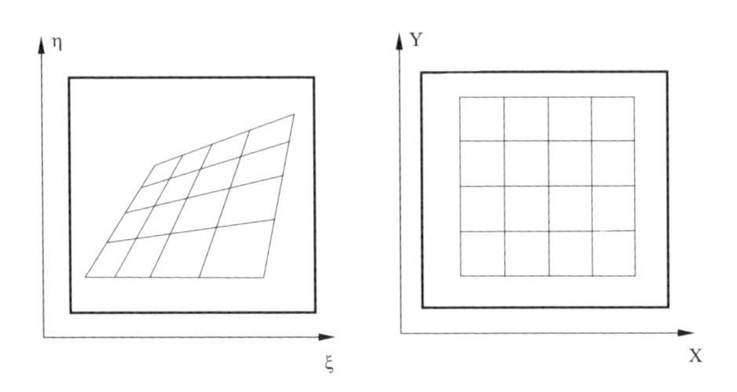

Figure 7.1-4: Grid in the photograph and in the object plane (map)

development begins with Equations (4.3-3). First, the model is transformed into the photograph by means of Equations (4.3-24). We then set $d\kappa = 0$ ($d\kappa$ produces no deformation) and $d\omega = d\varphi = v/\sqrt{2}$, i.e. v is a tilt about a diagonal of the photograph. With $\xi = \eta = \nu/\sqrt{2}$ we obtain for the image displacement $\Delta\rho = \sqrt{\Delta\xi^2 + \Delta\eta^2}$ along the deformed image diagonal:

$$\Delta\rho = \frac{\rho^2}{c}\nu \tag{7.1-7}$$

$\Delta\rho$ lies in the image diagonal perpendicular to the tilt axis, whereby $\Delta\rho$ is a displacement towards the tilt axis in the "higher" corner and away from it in the "lower" corner.

The magnitude of $\Delta\rho$ shows the extent by which a semi-diagonal $\bar{\rho}$, i.e. the distance between a point $\overline{N}'$ (image centre) and a point $\overline{P}'$ (image corner), is increased or reduced to ρ by the image tilt ν. Figure 7.1-4 shows the upper half of a tilted photograph in which $\bar{\rho}$ is reduced to ρ. Equation (7.1-7) can also be derived from the diagram of Figure 7.1-5 by simple trigonometric relations and series expansion.

Table 7.1-1 shows the image displacements $\Delta\rho$ as a function of tilt ν for three types of cameras.

	ν			
	1 gon (54′)	2 gon (1°48′)	5 gon (4°30′)	10 gon (9°)
Normal angle ($c = 30$ cm)	1.4	2.7	6.9	13.8 mm
Wide angle ($c = 15$ cm)	2.8	5.5	13.8	27.5 mm
Super wide angle ($c = 9$ cm)	4.6	9.2	22.9	45.9 mm

Table 7.1-1: Image displacements $\Delta\rho$ (in [mm]) in the corners ($\rho = 115 \times \sqrt{2}$) as a function of tilt ν and type of camera

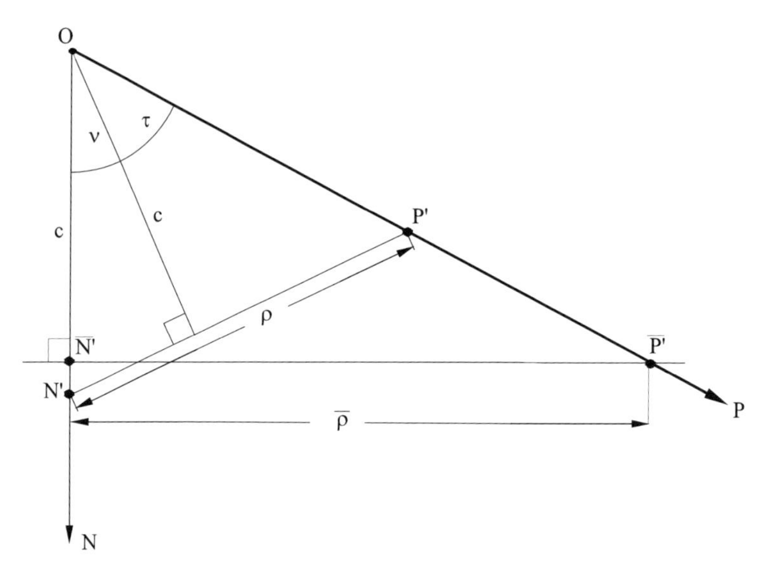

Figure 7.1-5: Reduction of the image length in an exactly vertical photograph to ρ in a photograph tilted by ν (N is the nadir point and ν the nadir distance.)

The increase or reduction of a small length $d\rho$ follows from differentiation of Equation (7.1-7):

$$d(\Delta\rho) = 2\frac{\rho}{c}\nu\, d\rho \tag{7.1-8}$$

Numerical Example. A line 10 mm in length along the diagonal in the "upper" corner of a wide-angle photograph ($c = 15$ cm) tilted by 5 gon has been reduced by:

$$d(\Delta\rho) = 2(115\sqrt{2}/150)(5\pi/200)10 = 1.7\,\text{mm}$$

In an exactly vertical photograph the line would therefore be 11.7 mm long.

Exercise 7.1-3. How long would this line be in an exactly vertical photograph if it measured 10 mm in the "lower" corner of the tilted photograph? Take the remaining data from the example above. (Solution: 8.3 mm)

Exercise 7.1-4. Calculate the reduction (increase) of length of the 10 mm length if it lies parallel to the ν axis in the "upper" ("lower") corner of the photograph. Take the remaining data from the example above. (Solution: 10.8 mm (9.2 mm) in an exactly vertical photograph.)

Exercise 7.1-5. Derive Equation (7.1-7) from the diagram of Figure 7.1-5 and show also the neglected second-order terms. (Solution: $\rho\nu^2(\tan^2\tau - 1)$, in which τ is the angle shown in Figure 7.1-5.)

Exercise 7.1-6. Use Equations (7.1-1) to (7.1-6) to compute the image deformations of the example of Exercise 7.1-1 under the assumption that the three points 1, 2 and 3 are at a height of 500 m, i.e. the object plane lies at $Z = 500$ m.

Solution: Object and image coordinates:

Point	X [m]	Y [m]	Z [m]	ξ [mm]	η [mm]
1	363400	61500	500	−32.014	88.104
2	363410	61500	500	−32.050	89.123
3	363400	61510	500	−33.030	88.070

$$\begin{aligned} &\text{Extreme length distortions:} & \lambda_1 &= 1.0200 \ \lambda_2 = 1.0162 \\ &\text{Maximum angle distortion:} & s_\omega &= 0.22\,\text{gon}\,(12') \\ &\text{Area distortion:} & \lambda_1\lambda_2 &= 1.04 \end{aligned}$$

Exercise 7.1-7. Repeat the computation under the assumption that $\Delta X_{12} = \Delta Y_{13} = 5\,\text{m}$ and compare these results with those for $\Delta X_{12} = \Delta Y_{13} = 50\,\text{m}$. Solution with $\Delta X_{12} = \Delta Y_{13} = 5\,\text{m}$:

$$\lambda_1 = 1.0208 \quad \lambda_2 = 1.0164 \quad 2\omega = 0.28\,\text{gon}\,(15')$$

Solution with $\Delta X_{12} = \Delta Y_{13} = 50\,\text{m}$:

$$\lambda_1 = 1.0202 \quad \lambda_2 = 1.0169 \quad 2\omega = 0.21\,\text{gon}\,(11')$$

What is the source of the differences? (Answer: In the initial Equations (7.1-1) finite differences are used instead of differentials.)

7.2 Orthophotos of plane objects

Orthophoto production is significantly dependent on whether or not the camera axis is perpendicular to the object plane.

7.2.1 With vertical camera axis

As already explained in Section 2.1.4, the photograph is the same as an orthophoto if the image plane and the object plane are parallel at exposure (see Figure 2.1-11 for an example). The square grid in the object plane is then imaged as a square grid in the plane of the photograph.

If an (unavoidably) small tilt error ν occurs when pointing the camera axis perpendicularly to the object plane or when arranging the image and object plane to be parallel, then Equations (7.1-7) and (7.1-8) can be used to estimate the effect of this error on the orthophoto.

Numerical Example. In producing the orthophoto shown by Figure 2.1-11, assume that the orthogonal alignment of the camera axis to the plane of the façade was possible only to an accuracy of $\nu = 0.5\,\text{gon}(27')$. The orthophoto, which has a scale of 1 : 320, then has the following maximum error:

Error along the 100 mm long semi-diagonal (7.1-7):

$$\Delta\rho = \frac{100^2}{157.65}\frac{0.5}{63.66} = 0.5\,\mathrm{mm}$$

Error in a 5 mm wide window in the corner of the orthophoto (7.1-8):

$$d(\Delta\rho) = 2\,\frac{100}{157.65}\frac{0.5}{63.66}\,5 = 0.05\,\mathrm{mm}$$

This simple procedure for creating photographs free of deformation is used mainly in architectural photogrammetry. Users must be aware, however, that the images of points outside the plane of the object will be displaced in the direction of the principal point H. The magnitude of this radial displacement $\Delta\rho$ of such objects from their bases (e.g. houses in aerial photographs) can be derived from Figure 7.2-1:

$$\Delta\rho = \Delta R \frac{c}{Z_0} = \Delta Z \frac{\rho}{c\, m_B} \tag{7.2-1}$$

Table 7.2-1a shows the maximum image displacements for typical scales in terrestrial photogrammetry for the Wild P31 camera (Table 3.8-2), while Table 7.2-1b shows those for typical scales and cameras in aerial photogrammetry.

Exercise 7.2-1. Repeat the computation of the Exercise 2.1-7 with the aid of Equation (7.2-1).

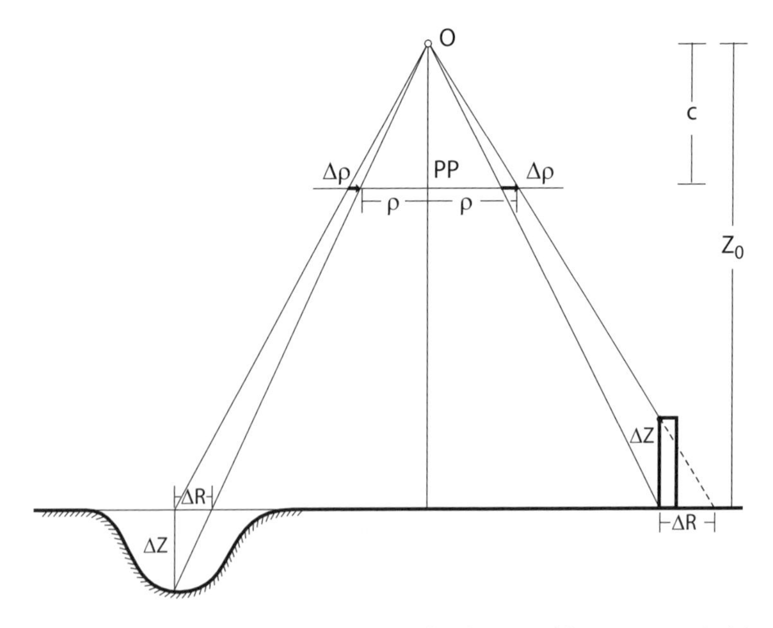

Figure 7.2-1: Radial image displacement $\Delta\rho$ of points outside an assumed object plane (exactly vertical photograph)

m_B	ΔZ 1 cm	5 cm	10 cm	25 cm	100 cm
Normal angle ($\rho = 83$ mm, $c = 202$ mm)					
25	0.16	0.8	1.6	4.1	16.5 mm
100	0.04	0.2	0.4	1.0	4.1 mm
250	0.02	0.1	0.2	0.4	1.6 mm
500	0.01	0.04	0.1	0.2	0.8 mm
Wide angle ($\rho = 82$ mm, $c = 100$ mm)					
25	0.3	1.6	3.3	8.2	32.5 mm
100	0.1	0.4	0.8	2.1	8.2 mm
250	0.03	0.2	0.3	0.8	3.3 mm
500	0.012	0.1	0.2	0.4	1.6 mm
Super-wide angle ($\rho = 75$ mm, $c = 45$ mm)					
25	0.7	3.3	6.7	16.6	66.5 mm
100	0.2	0.9	1.7	4.1	16.6 mm
250	0.1	0.4	0.7	1.6	6.6 mm
500	0.03	0.2	0.3	0.8	3.3 mm

(a) Typical scales for terrestrial photogrammetry (Wild P31)

m_B	ΔZ 1 m	5 m	10 m	25 m	100 m	250 m
Normal angle (23 cm $\times$ 23 cm, $c = 30$ cm)						
5000	0.1	0.5	1.1	2.7	10.8	27.1 mm
10000	0.05	0.3	0.5	1.4	5.4	13.6 mm
25000	0.02	0.1	0.2	0.5	2.2	5.4 mm
50000	0.01	0.1	0.1	0.3	1.1	2.7 mm
Wide angle (23 cm $\times$ 23 cm, $c = 15$ cm)						
5000	0.2	1.1	2.2	5.4	21.7	54.2 mm
10000	0.1	0.5	1.1	2.7	10.8	27.1 mm
25000	0.04	0.2	0.4	1.1	4.4	10.8 mm
50000	0.02	0.1	0.2	0.5	2.2	5.4 mm
Super-wide angle (23 cm $\times$ 23 cm, $c = 9$ cm)						
5000	0.4	1.8	3.6	9.0	36.1	90.4 mm
10000	0.2	0.9	1.8	4.5	18.1	45.2 mm
25000	0.1	0.4	0.7	1.8	7.2	18.1 mm
50000	0.03	0.2	0.3	0.9	3.6	9.0 mm

(b) Typical scales for aerial photogrammetry ($\rho = 115\sqrt{2}$ mm)

Table 7.2-1: Radial image displacements $\Delta\rho$ [mm] in the corners as a function of the image scale $1 : m_B$, the distance ΔZ of the points from the assumed object plane and the type of camera

7.2.2 With tilted camera axis

The production of orthophotos from tilted photographs, where the camera axis is directed obliquely onto the object plane, is widely practised. This orthophoto production is also known as rectification. In close range work the photographs are taken on site without any special precautions and in a more or less arbitrary alignment, perhaps also handheld. Metric cameras are now less often used. Increasingly, digital semi-metric cameras and digital amateur cameras are employed (Sections 3.8.4 and 3.8.7).

Control points are necessary to rectify tilted images. Where calibrated cameras are in use, only three control points are strictly necessary for the spatial resection of the metric image (Section 4.2.1); to increase reliability and avoid extrapolation, however, it is advisable to have at least four control points positioned in the corners of the image. Where uncalibrated cameras are used, at least four control points are required for every image in order to determine the eight parameters of the projective transformation (2.1-24). For large blocks of images, the numerous control points required for either the central perspective or projective solution can be generated by means of a bundle block adjustment with known or unknown interior orientation (Section 5.3)[2].

The rectification of the distorted images is done either on the basis of a central perspective transformation (for metric images) or a projective transformation (for non-metric images) (Section 2.1.4). The technique of digital projective rectification[3] has already been presented in Section 2.2.3. Here the indirect rectification (transformation of object plane to image plane) is preferred to direct rectification (transformation of image plane to object plane). This section also explains grey level interpolation as part of the required resampling. Figure 2.2-8 shows a practical example.

When rectifying tilted images of flat objects, the dominant source of error is the departure of the actual object surface from the assumed reference plane. Figure 7.2-2 shows the radial displacements $\Delta\rho$ in the tilted plane of the photograph and the corresponding displacements Δr in the orthophoto resulting from the perspective rectification of the tilted photograph. The displacements Δr in the orthophoto are radial from the imaged nadir point[4] N^O, which is the pole of the zenith or the vanishing point of all verticals, in the case of aerial photographs.

Thus, if we assume a horizontal ground plane, the positions of buildings, trees etc. in a rectified orthophoto will only be correct at ground level. Roofs, tree crowns, etc. are displaced from their true positions. Similarly, in rectified photographs of building façades details in front of or behind the plane of the façade will be displaced radially from the "nadir point".

[2] As a-priori knowledge, parallel and/or orthogonal straight lines, as well as much other information about an object, can be used in place of control points to make the rectification. There are many solutions on this topic. See Hartley, R.I., Zisserman, A.: Multiple View Geometry. Cambridge University Press, 2001, and also Section B 4.7.2, Volume 2.

[3] Analogue rectification, which has been practised for many decades and which is based on re-photographing the original images, no longer plays a significant role. It is therefore not discussed in this edition; the methods and associated instrumentation are extensively presented in earlier editions.

[4] In the general case, as opposed to that of aerial photographs, the term "nadir point" is used to mean the foot of the perpendicular from the perspective centre to the reference plane.

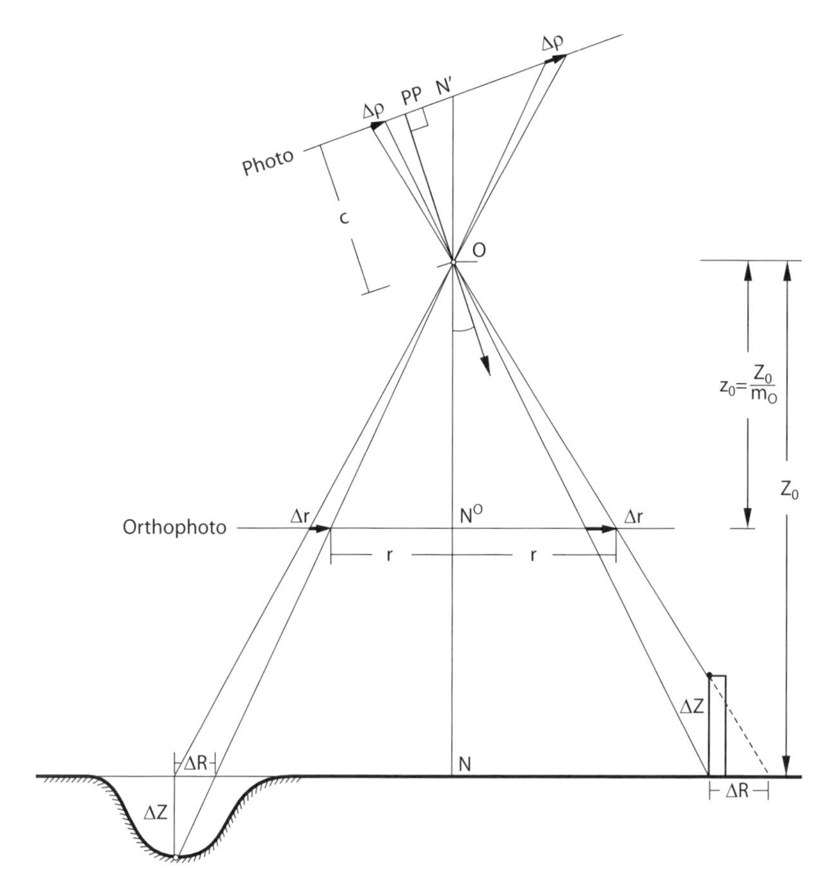

Figure 7.2-2: Radial image displacements Δr in the orthophoto of points outside the reference plane of the perspective rectification

The magnitude of the displacement, by analogy with Equation (7.2-1), is:

$$\Delta r = \Delta Z \frac{r}{Z_0} \tag{7.2-2}$$

r ... radial distance of the point from N^O, which is fixed in the orthophoto by the coordinates X_0 and Y_0 of the imaging station O

For near vertical photographs, the most common imaging configuration in single image photogrammetry, both formulas can be used in the same way, i.e. Equations (7.2-1) and (7.2-2). In the first formula, $\Delta \rho$ and ρ refer to the photo plane, in the second formula Δr and r refer to the orthophoto which is normally an enlargement. The numerical values in Table 7.2-1 are therefore also valid for orthophotos derived from near vertical images. It is necessary only to take account of the enlargement factor between image and orthophoto. This factor varies in practice between 3 and 6.

Numerical Example. An orthophoto has been generated from a normal-angle aerial photograph. This has a photo scale of 1 : 25000 and has been digitized at a resolution of 25 μm. The orthophoto's pixel equivalent in the ground coordinate system is 50 cm. Deviations from the assumed horizontal ground plane have a maximum of 10 m. Table 7.2-1 indicates a corresponding error of 0.2 mm in the original image. In the 5-times enlarged orthophoto, i.e. at a scale of 1 : 5000, the corresponding deviation is 1 mm, which is 5 m in the ground coordinate system. This error becomes particularly graphic when expressed in pixel units: viewed on a computer screen this error amounts to $5/0.5 = 10$ pixel!

This example and Table 7.2-1, in particular, show the limits of single-image photogrammetry. The upper part of the table indicates the limits for architectural surveys and similar applications of terrestrial photogrammetry. The lower part shows the limits for exposure configurations typical in aerial surveying. The table also expresses the fact that normal-angle photographs are better suited to orthophoto production than wide-angle photographs.

Exercise 7.2-2. A digital orthophoto is to be produced at a scale of 1 : 2500. The orthophoto format is 50 cm $\times$ 50 cm and an error of 1 mm can be tolerated. What is the maximum permissible offset of ground points from the horizontal reference plane used for the rectification? Assume near vertical photography at an image scale of 1 : 7500. There is a choice of cameras with principal distances 15 cm, 21 cm and 30 cm. (Solution: ΔZ is 3.2 m, 4.5 m and 6.4 m respectively)

Exercise 7.2-3. How large can the maximum offsets be if the image scale changes from 1 : 7500 to 1 : 15000? (Solution: ΔZ = 6.4 m, 9.0 m and 12.8 m respectively.)

Orthophoto production on the basis of a flat reference surface fails as soon as the ground has small hills and depressions. For objects with curved surfaces it is appropriate to apply differential rectification, as discussed in Section 7.3.

7.2.3 Combined projective and affine rectification

The problem discussed in this section is illustrated by Figure 7.2-3. An xy object plane is tilted with respect to the XY plane of the reference coordinate system and an orthophoto is required in this XY plane. The xy object plane and the XY plane of the reference coordinate system are affine with respect to one another. The general affine transformation is described by Equations (2.1-8). The mathematical relationship between the $\xi\eta$ image plane and the xy object plane can be expressed as a projective transformation defined by Equation (2.1-24). Orthophoto production can therefore be achieved in the following steps:

- projective rectification of the $\xi\eta$ image into the xy plane. Result: xy image.
- affine rectification of the xy image into the XY plane. Result: required orthophoto.

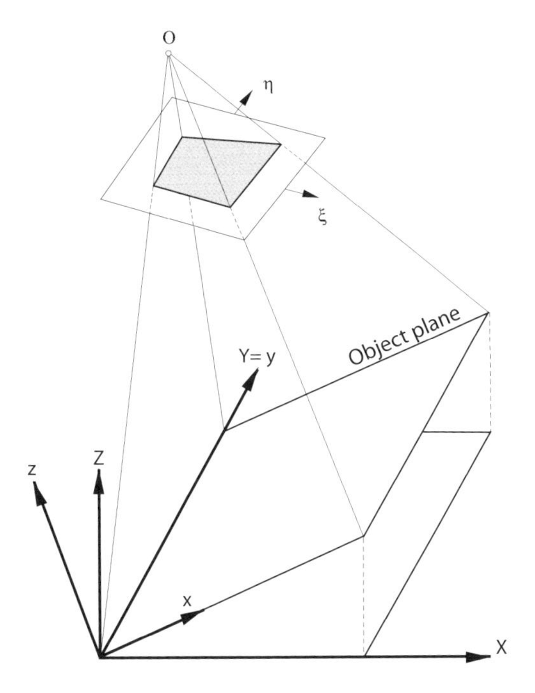

Figure 7.2-3: Orthophoto production of a tilted plane

Both steps can, however, be combined into a single step since the $\xi\eta$ image plane and the XY orthophoto plane are also related by a projective transformation[5]. The transformation parameters can be determined with the aid of at least 4 control points, i.e. $\xi\eta$ and XY coordinates are known for at least 4 points.

The combined projective and affine rectification is of considerable practical importance: in large-scale orthophotos of developed areas, the flat (tilted) roofs should be rectified in this way. (For completeness it is mentioned here that the principle of orthophoto production which takes account of buildings, as illustrated in Figure 7.3-2, also leads to orthophotos of roofscapes which are free of error.)

[5]The theory of projective transformation states that two consecutive projective transformations can be replaced by a single (third) projective transformation. The affine transformation is also a projective transformation in which the perspective centre lies at infinity (see Exercise 2.1-8 and its solution). The proof of this statement is readily given: combine Equations (2.1-8) and (2.1-24), ensuring first that the coordinate axes are identified as in Figure 7.2-3. The result is an 8-parameter projective transformation between the $\xi\eta$ image plane and the XY orthophoto plane. A more elegant derivation of this relationship is possible using homogeneous coordinates (Appendix 2.2-1).

7.3 Orthophotos of curved objects

7.3.1 Production principle

The production of orthophotos for objects with curved surfaces is based on the theory of perspective distortion in a metric image which was discussed in Section 7.1. To produce a digital orthophoto of a curved object surface, a digital model of this surface is required. For simplicity an object will be assumed which has a square grid in the XY plane ($\Delta X = \Delta Y$) and for every grid point the Z coordinate is known. (The acquisition of such models has been discussed in Sections 6.6.2 and 6.8.6.)

Orthophoto production of curved object surfaces is implemented in an indirect way, i.e. an initially empty image matrix is created in the orthophoto plane and corresponding matrix elements found in the reference image. This indirect transformation will be discussed in more detail with reference to Figure 7.1-1 and under the assumption of standard imaging conditions appropriate to aerial photogrammetry:

- first, an image matrix is defined in the XY plane, the orthophoto plane. In orthophoto production from aerial photographs, the pixel spacing of the image matrix is usually significantly finer than the grid spacing of the terrain model.
- in the second step, the Z coordinate for every pixel in the orthophoto matrix must be determined by interpolating in the ground model. The surface of a grid square is commonly assumed to be a hyperbolic paraboloid (also called an HP surface). Section 2.1.3a explains in detail how the Z coordinate can be found at any XY position within a grid square. The result of this interpolation process gives the Z coordinates for all XY orthophoto pixels.
- in the third step the image coordinates corresponding to the XYZ locations in the orthophoto grid are calculated using the collinearity equations (2.1-19) and the known elements of interior and exterior orientation.
- in the final step the grey value from the reference image corresponding to the $\xi\eta$ coordinate pair is assigned to the corresponding XY position in the orthophoto matrix. Since the $\xi\eta$ coordinate pair will not, in general, lie at the centre of a pixel in the reference image, a grey value interpolation is required, as outlined in Section 2.2.3a.

This process for producing orthophotos is relatively expensive. Approximate solutions, which have an acceptable loss of accuracy, are therefore in demand. The widely used anchor point method is a solution of this sort. The steps outlined above, which generate the $\xi\eta$ coordinate pairs in the reference image corresponding to the XY coordinate pairs in the orthophoto, are only implemented for the grid points in the ground model, as shown in Figure 7.1-1. The results for a square mesh are shown in Figure 7.3-1: the $\xi\eta$ coordinates and the XY coordinates for four points are known. (In Figure 7.3-1 the origin of the object coordinate system has been shifted to the lower left corner; the reduced object coordinates are indicated by $\overline{X}$ and $\overline{Y}$.) For the pixels in the orthophoto matrix, which are shown in the left half of Figure 7.3-1, the corresponding $\xi\eta$

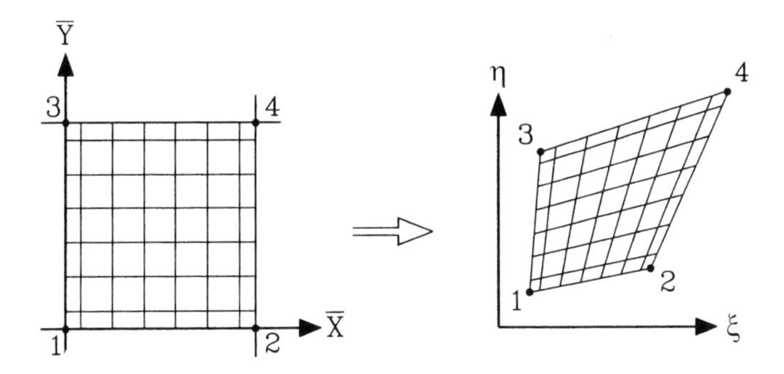

Figure 7.3-1: Bilinear transformation of a square mesh using four anchor points

coordinates can be obtained by means of a bilinear transformation:

$$\begin{aligned} \xi &= b_{01} + b_{11}\overline{X} + b_{21}\overline{Y} + b_{31}\overline{XY} \\ \eta &= b_{02} + b_{12}\overline{X} + b_{22}\overline{Y} + b_{32}\overline{XY} \end{aligned} \qquad (7.3\text{-}1)$$

The 8 coefficients b_{ik} are determined from the 4 anchor points. (For each coordinate axis, the problem corresponds to interpolation on a hyperbolic paraboloid. This is discussed in detail in Section 2.2.3a. Exercise 3.2-2 provides the solution for a bilinear transformation of fiducial marks.) The matching of corresponding points in the orthophoto and reference image by means of Equations (7.3-1) is appreciably faster than matching by means of collinearity equations, as described above. The economic advantage is even greater with larger grid spacings. However, this leads to a loss of accuracy which will be discussed in Section 7.3.2d.

Digital orthophoto production of objects with curved surfaces[6] can also be extended to ground definitions based on both grid points and break lines. This generates orthophotos which are particularly accurate on the break lines. Cadastral and land use borders etc. often run along break lines and so such orthophotos are often in demand. Orthophoto production which takes account of break lines and object edges is described in detail in Section C 1.1.2, Volume 2. The problem of Earth curvature, which must be taken into account when producing small-scale orthophotos, is also discussed in Volume 2, Section C 1.2.

Discussion of digital orthophoto production which can also handle buildings and other man-made constructions is included in Volume 2, Section C 1.1.3. Because of its

[6]In the past various devices have been constructed for analogue orthophoto production of curved surfaces. Due to their cumbersome operation, analogue orthophoto production achieved little importance. Before digital orthophoto production has become standard, analytical orthophoto production was widely applied since 1975. As in the anchor point method, the grid was transformed into the reference image and the image content inside individual quadrilaterals transferred by means of optical elements (zoom and dove prisms) onto the orthophoto plane which was a film surface wrapped on a drum. The entire transfer process was under computer control. The most widely used instrument was the Avioplan OR1 from Wild. This instrument and its principle of operation are described in earlier editions.

practical importance, this topic will also be discussed here, with the aid of Figure 7.3-2. Consider using the ground model (Figure 7.3-2, left) for orthophoto production. This model, of course, only describes the ground and not the buildings. In the resulting orthophoto, for example, a point Q on a building appears at position $\overline{Q}_0$ and not at the required position Q_0. In an accurate orthophoto ground point P is expected at position $\overline{Q}_0$. However, if a surface model is used which also contains the buildings (Figure 7.3-2, right), then an orthophoto is obtained in which point Q on the building appears at the required location Q_0 using image point Q'. If there is no visibility analysis then image point Q' appears a second time at position $P_0 = \overline{Q}_0$: there is then double imaging in the occluded areas.

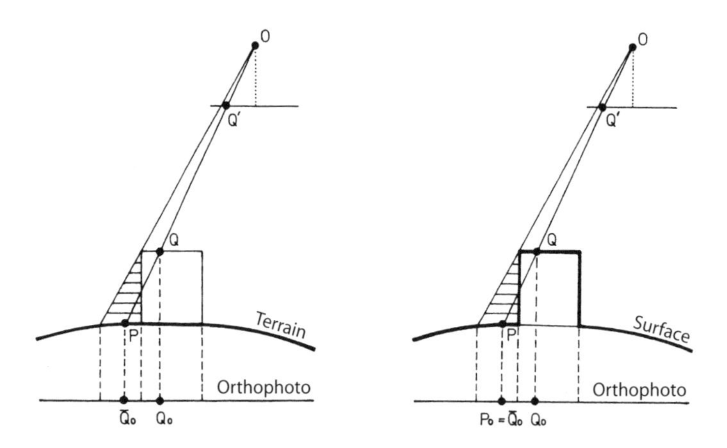

Figure 7.3-2: Digital orthophoto production without (left) and with (right) modelling of buildings

Figures 7.3-3 to 7.3-5 illustrate the various orthophoto products. Figure 7.3-3 shows a conventional orthophoto produced with a ground model. In the tall buildings in the upper half of the image it is particularly noticeable that their roofs are appreciably displaced and the vertical façades are also, incorrectly, imaged.

Figure 7.3-4 shows an orthophoto created using a surface model which also includes the man-made features. However, a visibility analysis was not undertaken and double imaging occurs. The double imaging can easily be seen in the buildings above the river; one of the two doubly imaged areas is geometrically correct for an orthophoto, the other is incorrect, as in Figure 7.3-3.

Figure 7.3-5 shows an orthophoto produced with the same surface model used to produce the orthophoto in 7.3-4. Here, however, a visibility analysis was undertaken which eliminates the false image areas in the occlusions and replaces them with dark grey values. The missing pixels can be transferred from a second photographs taken from a different location. (The publication from which Figures 7.3-3 to 7.3-5 were taken (see corresponding footnote) contains such an orthophoto).

Figure 7.3-3: Orthophoto generated using a ground model[7]

Figure 7.3-4: Orthophoto generated using a surface model which also contains buildings and man-made structures; double imaging occurs here due to the missing visibility analysis[7]

Figure 7.3-5: As 7.3-4, but including a visibility analysis[7]

[7]Taken from Amhar, F., Jansa, J., Ries, C.: IAPRS 1998.

When producing orthophotos from a block of original images there can be undesirable step changes in grey level and colour at the joins between the content of different images. These jumps are mainly the result of different viewing directions due to different imaging locations. These grey level and colour changes can be eliminated as part of the mosaicking process.

(Further information: `http://gi.leica-geosystems.com`. Kerschner, M.: ISPRS-J 56, pp. 53–64, 2001. Shiren, Y., Li, L., Peng, G.: PE&RS 55(1), pp. 49–53, 1989.)

Exercise 7.3-1. Given four points with $\xi\eta$ and XY coordinates and the XY coordinates of a fifth point:

	ξ [mm]	η [mm]	X [m]	Y [m]
1	26.19	27.14	2325.00	4525.00
2	28.87	26.21	2330.00	4525.00
3	26.08	29.32	2325.00	4530.00
4	28.96	29.01	2330.00	4530.00
5			2327.50	4526.00

Determine the eight coefficients of the bilinear transformation and the $\xi\eta$ coordinates of point 5. (Solution: $\xi_5 = 27.528$ mm, $\eta_5 = 27.175$ mm.)

Exercise 7.3-2. Consider the reversal of the transformation and transform point 5 from the $\xi\eta$ system back into the XY system. (Because this involves a quadratic equation there are two solutions; the correct solution lies inside the quadrilateral, the false solution outside.)

Additional further reading on digital orthophoto production: Baltsavias, E., Käser, Ch.: IAPRS 32(4), pp. 42–51, 1998. Schickler, W., Thorpe, A.: IAPRS 32(4), pp. 527–532, 1998. Weidner, U.: OEEPE Publication 37, pp. 307–314, 1999.

7.3.2 Orthophoto accuracy

The accuracy of differential rectification depends on the following sources of error:

(a) errors in the source material (residual errors of optical distortion, film deformation, positioning errors of the film scanner, etc.)

(b) errors in the elements of interior and exterior orientation (also control point errors if the elements of exterior orientation are determined by spatial resection (Section 4.2.1))

(c) image displacements of objects not included in the surface model (for example, roof ridges, tree crowns above the digital terrain model (DTMs) or erosion channels, embankment edges, if a DTM is used which is based on a square grid)

(d) interpolation errors in the anchor point method, due to linear interpolation between grid points rather than using the cross ratio

(e) height errors at the grid points used in the DTMs, which can arise during data acquisition and/or DTM interpolation between arbitrarily distributed points

(f) errors in approximating curved object surfaces by a grid model in which the mesh elements are defined by hyperbolic paraboloids or tilted triangles (Figure 7.1-1 shows a mesh with facets defined by hyperbolic paraboloids, Figure C 1.1-6, Volume 2, a mesh with additional triangular elements)

The dominant sources of error cause positional errors in the orthophoto which are close to zero at the principal point of the photograph (more exactly, the nadir point N^O), increase with distance from this point and are predominantly radial with respect to the principal point. (Figure 7.3-6 is a practical example from an accuracy study[8].) The largest errors therefore lie at the edges of the orthophoto, a fact which enables the entire process of production to be checked by comparing neighbouring orthophotos side by side along their common boundary.

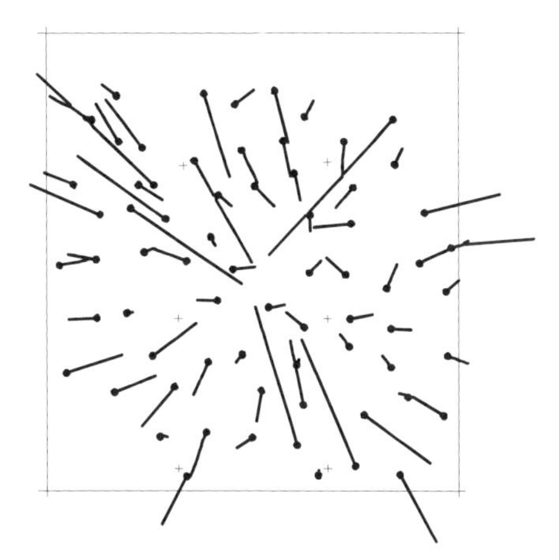

Figure 7.3-6: Planimetric errors in a differentially rectified orthophoto, derived from known coordinates of signalized check points

The following detailed comments can be made on some of the sources of error mentioned above.

With regard to c) Image displacements of objects not included in the surface model

Details not contained in the surface model used to create the orthophoto will be incorrectly imaged. When using a DTM for orthophoto production, roof ridges, tree crowns, etc. are displaced in the orthophoto. If the DTM also does not contain break lines, then

[8]Otepka, G., Duschanek, E.: Geow. Mitt., vol. 13, pp. 125–150, 1978.

embankment edges, etc. are incorrectly reproduced. Therefore the only correctly positioned points and lines in the orthophoto are the ones which are also defined in the corresponding surface model.

Figure 7.3-7 shows the situation for a tree which is not included in the DTM. The displacement of the tip of the tree in the (tilted) metric image is indicated by $\Delta\rho$. Figure 7.3-7 also shows the corresponding displacement $\Delta\rho'$ in an imaginary vertical photograph. The intersection of line P with the (sloping) ground defines a point P_0, given by the displacement ΔR in the XY object coordinate system or the displacement Δr in the orthophoto. The displacements are derived from the following ratios:

$$\frac{\Delta R}{\Delta Z - \Delta R \tan\bar{\alpha}} = \frac{\rho'}{c}$$

which can be re-arranged to give:

$$\Delta R = \frac{\Delta Z}{\dfrac{c}{\rho'} + \tan\bar{\alpha}} \tag{7.3-2}$$

If the ground slopes inwards (towards the nadir), as in Figure 7.3-7, angle $\bar{\alpha}$ is positive; if the ground slopes outwards then $\bar{\alpha}$ is negative.

The angle $\bar{\alpha}$ appearing in Equation (7.3-2) is the slope angle of the ground at point P^O in the direction of line $\overrightarrow{P^O N^O}$ (Figure 7.3-7). Since it is more convenient to use the maximum slope angle (denoted by α), the horizontal angle β between the line of maximum slope and line $\overrightarrow{P^O N^O}$ must be taken into account in Equation (7.3-2) (lower diagram, Figure 7.3-7)[9]:

$$\Delta R = \frac{\Delta Z}{\dfrac{c}{\rho'} + \tan\alpha\cos\beta} = \frac{\Delta Z}{\dfrac{h}{rm_0} + \tan\alpha\cos\beta} \tag{7.3-3}$$

It is easy to see that the cosine of angle β is required: for $\beta = 0$, i.e. the line of maximum slope is in the direction of line $\overrightarrow{P^O N^O}$, the original Equation (7.3-2) is obtained, for $\beta = 100\,\text{gon}\;(90°)$, i.e. the contour line lies in the direction of ray $\overrightarrow{P^O N^O}$, the tangent term in Equation (7.3-2) is removed by multiplication with $\cos\beta = 0$. If angle β is taken over a full circle, then the differentiation between positive and negative slope angles is unnecessary.

In the second part of the Equation (7.3-3) the ratio c/ρ' has been replaced by the ratio $H/(rm_0)$. These quantities have the following meaning:

- h ... relative flying height above point P_0. It is obtained by subtracting the ground height, available from the DTM, at location P_0^O in the orthophoto, from the Z_0 coordinate of the perspective centre of the metric image.
- r ... planimetric distance of point P_0^O from nadir point N^O, known from coordinates X_0,Y_0 of the perspective centre of the metric image.
- m_0 ... scale factor of the orthophoto

[9] Similar formulae have been given by Blachut, T. and van Wijk, M.: Ph.Eng. 36, pp. 365–374, 1970.

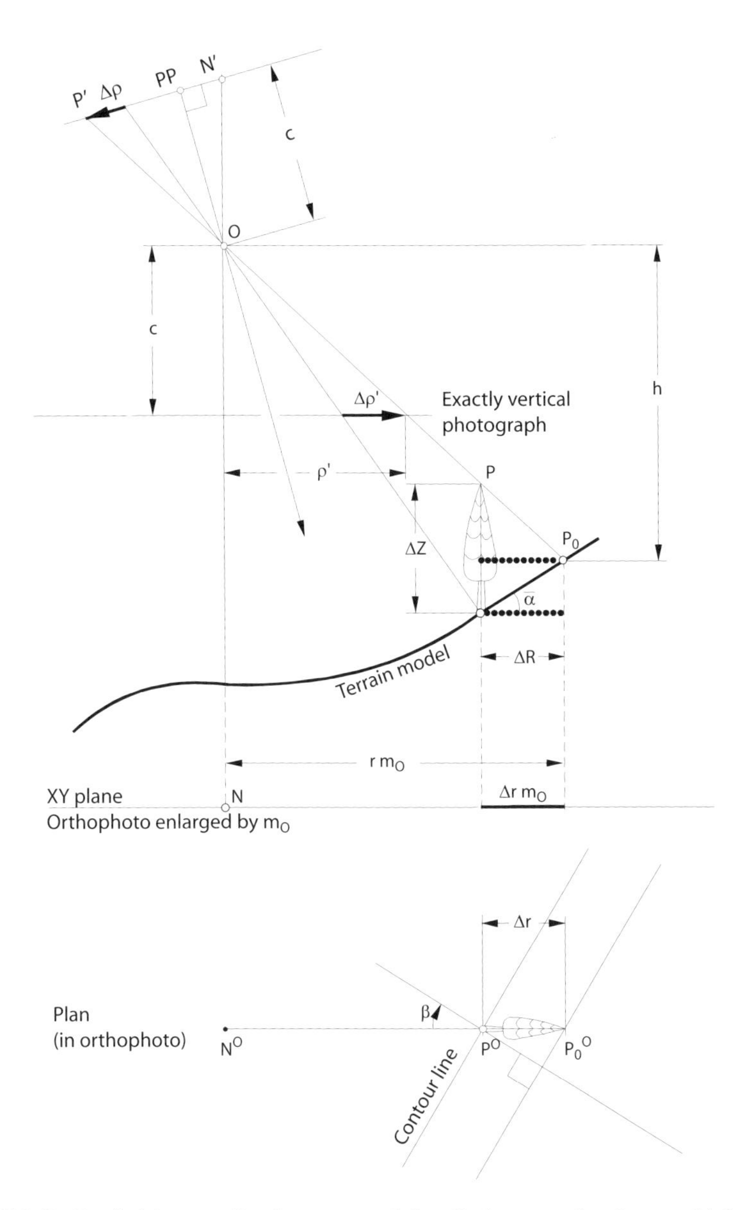

Figure 7.3-7: Radial image displacement of details in an orthophoto which were ignored in the data capture

For coarse estimations of the image displacements to be expected in the orthophoto, the first part of Equation (7.3-3) can be used, with the distance of the corresponding details (tree, etc) from the principal point of the (near vertical) metric image used as the value for ρ'. The displacement Δr in the orthophoto is derived by multiplying ΔR (Equation (7.3-3)) by the orthophoto scale $1 : m_O$.

Exercise 7.3-3. A vertical photograph ($c = 15\,\text{cm}$), taken from an absolute height of $Z_0 = 2000\,\text{m}$, has been differentially rectified to create an orthophoto at a scale of $1 : 2500$. For two 10 m high buildings, estimate the displacement of the top of the gable end in both the aerial photograph and the orthophoto. Both buildings are 10 cm away from the principal point in the aerial photograph; one building is at a height of $Z = 150\,\text{m}$ on an outward-facing slope ($\alpha = 10\,\text{gon}\,(9°), \beta = 150\,\text{gon}\,(135°)$) and the other at a height of $Z = 800\,\text{m}$ on an inward facing slope ($\alpha = 15\,\text{gon}\,(13°30'), \beta = 75\,\text{gon}\,(67°30')$). (Solution: $\Delta\rho = 0.54\,\text{mm}, \Delta r = 2.9\,\text{mm}, \Delta\rho = 0.83\,\text{mm}, \Delta r = 2.5\,\text{mm}$.)

With regard to d) Interpolation errors in the anchor point method

The anchor point method uses linear interpolation along the lines between grid points and, if there are also triangular facets in the ground model (Section C 1.1.2, Volume 2), along the sides of the triangles. Along the corresponding lines in image and object, interpolation must not be linear but must use the cross ratio (Section 2.1.5).

The interpolation error δ in the reference image, due to the anchor point method, is greatest in the middle of a line on the object, with a length identified by the variable s. It has the following value (for derivation see the second or third edition of this book):

$$\delta = \frac{s^2}{4c} \tan\alpha \tag{7.3-4}$$

$\tan\alpha$ ... slope of this line
s ... grid spacing or triangle side length of the DTM, given at the scale of the metric image

Numerical Example. Using the anchor point method, an orthophoto is to be derived from a wide-angle aerial photograph ($c = 150\,\text{mm}$), taken from a height of 3 km above the ground. The available DTM has a grid spacing of $\Delta X = \Delta Y = 50\,\text{m}$. The maximum slope of the ground is 50%.

The grid spacing in the reference image, indicated by s in Equation (7.3-4), is given by: $s = 50 \times 150/3000 = 2.5\,\text{mm}$.

$$(7.3\text{-}4) : \delta = \frac{2.5^2}{4 \times 150} 0.5 = 0.005\,\text{mm} = 5\,\mu\text{m}$$

In view of the fact that the photographs used to create the orthophotos have a pixel size between 10 and 30 μm (Section 3.4.1), this error can be ignored.

Practical note: The numerical example shows that even under extreme conditions the interpolation error arising in the anchor point method lies significantly within a pixel and is therefore of little relevance.

Exercise 7.3-4. Repeat the numerical example for a normal-angle camera (c = 300 mm). The flying height is now 6 km instead of 3 km. (Solution: 2.6 μm. Comment: The interpolation error is smaller in normal-angle images than in wide-angle images.)

With regard to e) Height errors at the grid points used in the DTMs

These errors primarily have their source in accuracy losses during data acquisition for the DTM. According to Equation (7.3-3), the ΔZ errors cause positional errors in the XY coordinate plane and orthophoto, which has been derived as indicated in Figure 7.3-7. The point P shown there can also be viewed as a grid point with height error. Three times the average error given in Section 6.7.2.3 can be used for ΔZ.

Exercise 7.3-5. Photographs for data acquisition and orthophoto production at a scale of 1 : 30000, principal distance $c = 150$ mm. At the least favourable position in the corner of the image, the ground has an inward slope of 10%; here the contour lines run parallel to the edge of the photograph. Orthophoto scale $1 : m_O = 1 : 10000$. Data acquisition takes place statically in a grid pattern.

Section 6.7.2.3: $\Delta Z = 3 \times 0.00015 \times 4500\,\text{m} = 2.0\,\text{m}$

$$\text{(7.3-3): } \Delta R = \frac{2.0}{150/140 + 0.1 \times 0.7} = 1.75\,\text{m} \;\text{ or }\; \Delta r = 1.750/10000 = 0.18\,\text{mm}$$

Exercise 7.3-6. The maximum error in the orthophoto must again be determined. The difference in this case is that normal-angle photographs (c = 300 mm) are used in place of the wide-angle photographs for the data acquisition and orthophoto production. Photo scale is again 1 : 30000. Data acquisition by static raster measurement (see Section 6.6.2) is now better by a factor of two:

Section 6.7.2.3: $\Delta Z = 3 \times 0.00015 \times 9000\,\text{m} = 4.0\,\text{m}$

Orthophoto accuracy, however, changes little:

$$\text{(7.3-3): } \Delta R = \frac{4.0}{300/140 + 0.1 \times 0.7} = 1.81\,\text{m} \;\text{ or }\; \Delta r = 1.810/10000 = 0.18\,\text{mm}$$

Result: If data acquisition and orthophoto production are done using the same image source, then orthophoto accuracy is almost independent of the camera type. The lower height accuracy provided by normal-angle images is compensated for by the fact that in normal-angle images the height error has a smaller effect on planimetric error in the orthophoto than in wide-angle images.

However, if the data acquisition is made with a wide-angle image and the orthophoto produced from a normal-angle image, then there is an increase in accuracy almost in proportion to the principal distances of the images used for data acquisition and orthophoto production, as the following example shows:

Exercise 7.3-7. Data acquisition is defined by the situation described in Exercise 7.3-5; orthophoto production, however, is done with normal-angle imagery at a scale of

1 : 30000 ($c = 300$ mm). Planimetric error in the orthophoto at a scale of 1 : 10000 is now only:

$$(7.3\text{-}3)\text{: } \Delta R = \frac{2.0}{300/140 + 0.1 \times 0.7} = 0.90\,\text{m} \text{ or } \Delta r = 900/10000 = 0.09\,\text{mm}$$

This large improvement in accuracy suggests that orthophotos could even be produced at a scale of 1 : 5000 from normal-angle photographs at 1 : 30000. It should, however, be noted that the enlargement factor from the reference image to the orthophoto would change from 1 : 3 to 1 : 6 and this would result in a loss of geometric resolution in the orthophoto. This deterioration can be avoided if the normal-angle images are taken at a very large scale (e.g. 1 : 15000 for orthophotos at 1 : 5000). The accuracy of these 1 : 5000 orthophotos is indicated in the following numerical example:

Exercise 7.3-8. Again data acquisition is done under the conditions described in Exercise 7.3-5: $m_B = 30000$, $c = 150$ mm, $\tan\alpha = 0.1$, $\beta = 50$ gon (45°), $\Delta Z = 2.0$ m. This information is used to derive orthophotos at a scale of 1 : 5000 from imagery at a scale of 1 : 15000 ($c = 300$ mm). Orthophoto accuracy is given by:

$$(7.3\text{-}3)\text{: } \Delta R = \frac{2.0}{300/140 + 0.1 \times 0.7} = 0.90\,\text{m} \text{ or } \Delta r = 900/5000 = 0.18\,\text{mm}$$

It should not be ignored, however, that the accuracy in relation to the orthophoto pixel, normally chosen to be 50 cm in Exercise 7.3-7 and 25 cm in Exercise 7.3-8, decreases from 1.8 pixels (= 0.90/0.50) to 3.6 pixels (= 0.90/0.25).

With regard to f) Errors in approximating curved object surfaces by a grid model

The maximum errors will occur where the greatest curvature occurs in the ground. In order to estimate this, we must know the detail of at least one short profile, along the line of maximum slope, in the roughest area of the ground. Figure 7.3-8 shows such a profile.

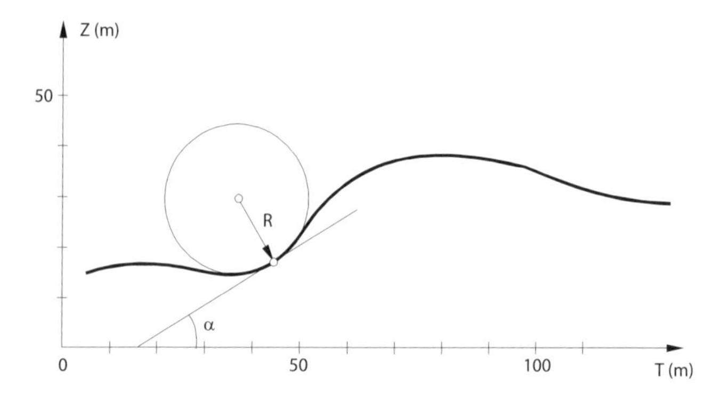

Figure 7.3-8: Ground profile with minimum radius of curvature

If this profile is digitized with a grid interval ΔT and intermediate points then interpolated linearly, the maximum height error ΔZ can be estimated from Figure 7.3-9. The the maximum separation, ΔP, between the curve and the chord is:

$$\Delta P = \frac{\Delta T^2}{8R\cos^2\alpha}$$

The following is then obtained for ΔZ:

$$\Delta Z = \frac{\Delta P}{\cos\alpha} = \frac{\Delta T^2}{8R\cos^3\alpha} \tag{7.3-5}$$

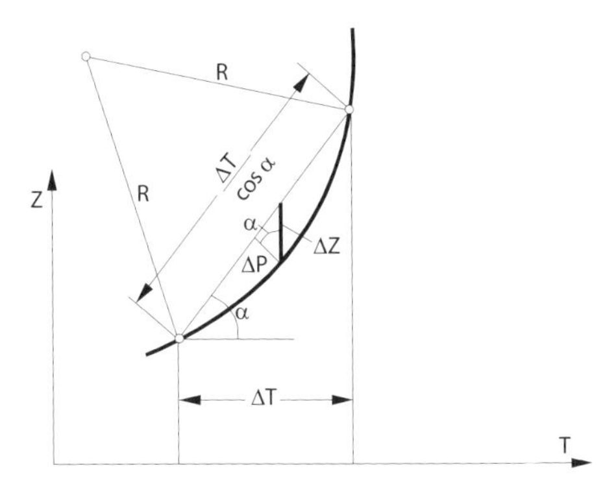

Figure 7.3-9: Height error ΔZ as a function of the radius R and grid interval ΔT

Numerical Example. In Figure 7.3-8 the radius $R = 15$ m and the angle $\alpha = 35$ gon ($31.5°$). The maximum height error ΔZ after digitizing and linear interpolation in a grid interval $\Delta T = 10$ m in the orthophoto is then, from Equation (7.3-5):

$$\Delta Z = \frac{10^2}{8 \times 15 \times 0.853^3} = 1.34\,\text{m}$$

This error in height produces an error in position ΔR which can also be calculated from Equation (7.3-3). If we assume that the image of the roughest part of the ground is 10 cm from the centre of the wide-angle photograph and that the maximum slope is directed outwards ($\beta = 200$ gon ($180°$)), then:

$$\Delta R = \frac{1.34}{300/100 - 0.613} = 0.56\,\text{m}$$

This error in position of 0.56 m can be expected at the given position if the orthophoto is produced using the anchor point method and a grid model with 10 m grid spacing. In a 1 : 5000 orthophoto this error amounts to 0.11 mm; expressed in orthophoto pixels, which are normally set to a size of 25 cm in 1 : 5000 orthophotos, the error amounts to 2.2 Pixel (= 0.56/0.25).

As can be seen from Equation (7.3-3), the error caused by approximating the ground with a grid model, primarily decreases linearly with increasing principal distance. Further, as can be seen from Equation (7.3-5), this error is strongly dependent on grid spacing ΔT. One way to reduce the error is to interpolate a very dense grid of heights from the acquired data. If a sophisticated interpolation method is employed, then the spacing of this grid can be used as the value of ΔT in estimating the error; however, if surface interpolation is done using a simple method such as triangular plane between individual measurement points, then the separation of the original measurements should be taken as the value for ΔT. If data acquisition is by digitizing contour lines, then the value of ΔT, which is here essentially independent of the interpolation method, should be the contour line separation.

The problem of the accuracy of digital height models is studied further in Tempfli, K.: ITC-Journal 1980(3), pp. 478–510, and Frederiksen, P.: IAPRS XXIII(4), pp. 284–293, 1980. and treated by spectral analysis (already briefly discussed in the context of filtering in the frequency domain, Section 3.5.2.2). The height errors predicted by this method serve then as data for Equation (7.3-3). Readers should also consult further references[10], in which typical standard deviations for complete orthophotos are derived from empirical accuracy studies. These figures naturally include all error influences so that individual influences cannot be isolated. Such information is particularly useful in practice. The root mean square error of the complete orthophoto should not exceed ± 0.2 mm for small-scale orthophotos or ± 0.4 mm for large-scale orthophotos. With respect to ground resolution, i.e. to the orthophoto pixel, the standard error should not exceed 3 pixels. (Image displacements of objects not incorporated in the ground model require separate consideration.)

Exercise 7.3-9. Draw a new test profile in which ground heights in comparison with Figure 7.3-8 are raised by a factor of 2. (In the context of a spectral analysis, amplitudes increase by a factor of 2.) Repeat the accuracy estimation in this case. (Solution: $R \approx 10\,\mathrm{m}, \alpha \approx 30\,\mathrm{gon}\,(27°), \Delta Z = 1.77\,\mathrm{m}, \Delta R = 0.74\,\mathrm{m}$.)

Exercise 7.3-10. Draw a new test profile in which ground heights in comparison with Figure 7.3-8 are rougher by a factor of 2, i.e. the T axis is compressed by a factor of 2. (In the context of a spectral analysis the wavelengths are reduced by a factor of 2.) Repeat the accuracy estimation in this case. (Solution: $R \approx 5\,\mathrm{m}, \alpha \approx 31\,\mathrm{gon}\,(28°), \Delta Z = 3.6\,\mathrm{m}, \Delta R = 1.5\,\mathrm{m}$.)

Further reading on orthophotography: Ecker, R., Kalliany, R., Otepka, G.: in Fritsch and Hobbie (Eds.): Photogrammetric Week '93, pp. 143–156, Wichmann Verlag, 1993.

[10] See the publication referenced by Figure 7.3-6 as well as Blachut, T., van Wijk, M.: IAPRS XIII (Commission II/4), Helsinki, 1976. Ducher, G.: Test on orthophoto and stereo-orthophoto accuracy. OEEPE-Publication No. 25, Frankfurt, 1991.

7.4 Analogue, analytical and digital single image analysis

Single image analysis is often done on the basis of an orthophoto and this will be discussed in the following Section 7.4.1. In the subsequent sections methods will be described which go directly from an original image to a three-dimensional reconstruction of the photographed object.

7.4.1 Analogue, analytical and digital orthophoto analysis

In many cases an orthophoto is an appropriate medium to employ at the interface between a photogrammetrist and a partner from a neighbouring discipline. Analogue analysis requires little in the way of instrumentation: a transparent film is laid on top of the orthophoto and the feature interpretations are drawn upon it; distances and areas can also be obtained from the orthophoto.

In analytical analysis the orthophoto is placed on a 2D digitizing table (2D digitizer); with the aid of a cursor the XY coordinates of identified object points are measured. If required, post-processing enables the Z coordinates to be obtained from the digital terrain model (DTM) by interpolation using the XY positions.

Digital orthophoto analysis takes place in a GIS or CAD environment. The digital orthophoto is stored on a computer. Using the cursor on the display screen, the XY coordinates of identified object points are obtained. The corresponding Z coordinates can be directly taken from the DTM which can also be accessed through the computer. In place of manual digitizing, the analysis of digital orthophotos also permits the application of the automated and semi-automated methods of digital image processing (e.g. the semi-automatic processing for plan, Section 6.8.7).

Planimetric analysis of orthophotos, whether analogue, analytical or digital, can never be more accurate than the orthophotos themselves. Their accuracy has been described in detail in Section 7.3.2. It will be mentioned again here that object elements, which do not form part of the digital surface model, can be significantly displaced in the orthophoto. Typical examples are roof ridges and tree tops (see Equation (7.3-3)).

The third dimension cannot be derived from the orthophotos alone. As already indicated, the third coordinate can be obtained from the DTM. An interesting alternative is the preparation of stereopartners for the orthophotos (details available in Section C 1.5, Volume 2). This combination of orthophoto and stereopartner, known as a stereo-orthophoto, allows the orthophoto to be viewed stereoscopically. Stereoscopic viewing of the orthophoto makes photo interpretation and analysis more comfortable, and also more exact than monoscopic viewing (Section C 1.5.3, Volume 2).

7.4.2 Analytical and digital analysis of a tilted image of a flat object

An analytical rectification starts either with measurement of image coordinates by attaching the analogue photograph to a 2D digitizer, or by locating points in an original digital image displayed on a computer screen.

Four control points are required if the photo to be rectified does not have known interior and exterior orientations. From the image and object coordinates, the eight parameters of a projective transformation can be determined by means of linear equations. All new points can then be transformed subsequently into the object coordinate system. This method has already been presented in Section 2.1.4.

If, for a given interior orientation, the exterior orientation has been determined for example by a bundle block adjustment (Section 5.3), an analytical rectification can then be implemented using the collinearity equations (2.1-20). If the flat object is parallel to the XY plane, the height of the object plane should be used as the Z coordinate. Generally, this height, and also the heights of the control points, are set to zero.

If a second metric image is available instead of either control points or exterior orientation elements, then the orientation elements necessary for the analytical rectification of an image can be determined (see Footnote 2 in Section 7.2.2).

7.4.3 Analytical and digital single image analysis of curved object surfaces

This method of analysis assumes that a digital model of the curved object surface is available. For simplicity the discussion is limited to a model consisting of Z coordinates on a square XY grid. The grid should be sufficiently dense that linear interpolation is permissible within a grid element. It will also be assumed that the interior and exterior orientation of the photos to be analysed are known. Figure 7.4-1 shows the initial configuration.

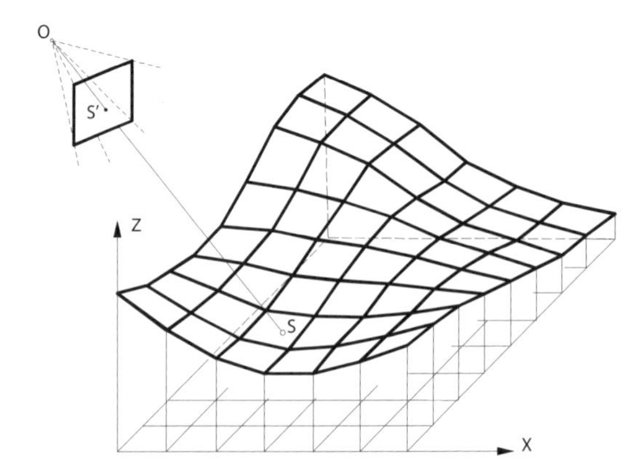

Figure 7.4-1: Oriented single image with digital surface model

The collection of identified points in the metric image defines a spatial bundle of rays (Figure 7.4-1 shows one of these points identified as S'). The intersection points S of this ray bundle with the digital surface model provides the X, Y and Z coordinates in the object coordinate system.

The XYZ coordinates of such an intersection point S can be found in the following steps:

(a) determination of the XYZ coordinates of the corresponding image point S' with the help of a spatial similarity transformation (Equation (2.1-8)) where the scale is set to unity:

$$\begin{pmatrix} X \\ Y \\ Z \end{pmatrix} = \begin{pmatrix} X_0 \\ Y_0 \\ Z_0 \end{pmatrix} + \mathbf{R} \begin{pmatrix} \xi - \xi_0 \\ \eta - \eta_0 \\ 0 - c \end{pmatrix} \tag{7.4-1}$$

$\mathbf{R} \ldots$ rotation matrix for the exterior orientation of the metric image

Tip: It helps to visualize Equation (7.4-1) if the unit matrix is (initially) used for $\mathbf{R}$

The next steps are explained with the help of Figure 7.4-2:

(b) intersection of the line $(O)(S')$ in the XY plane with the XY grid lines

(c) calculation of the Z coordinates of these intersection points using the DTM along the (vertical) profile containing the ray $\overrightarrow{OS'}$

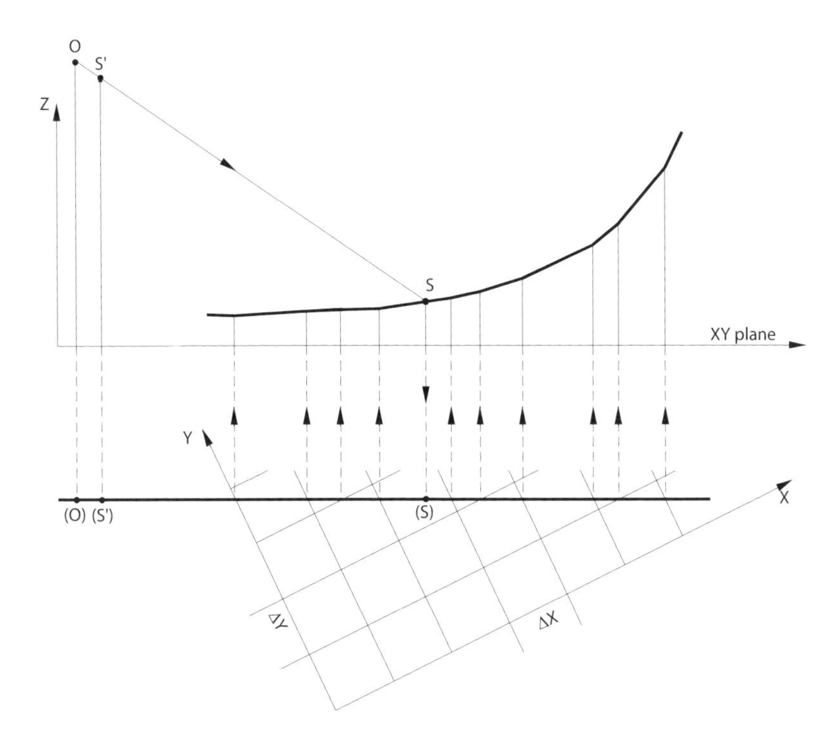

Figure 7.4-2: Vertical plane through ray $\overrightarrow{OS'}$ and the corresponding plan view

(d) intersection of the ray $\overrightarrow{OS'}$ with the vertical profile; the result is the Z coordinate of point S. Its XY coordinates are taken from the ground plan.

Exercise 7.4-1. In the numerical example of Section 2.1.3, image coordinates for both points P and Q were calculated from their object coordinates. With the aid of Equation (7.4-1), calculate the XYZ coordinates of both points in the object coordinate system. (Solution: For P: $X = 362530.713\,\text{m}, Y = 61215.863\,\text{m}, Z = 2005.589\,\text{m}$.)

Single image analytical analysis of curved object surfaces is known as monoplotting[11]. Monoplotting is used for the analysis of both aerial and terrestrial photographs. The accuracy of the analysis primarily depends on the intersection angle of the ray $\overrightarrow{OS'}$ with the surface profile in the vertical plane. When the intersecting rays are very oblique, large XY coordinate errors are generated where there are:

- minor height errors in the digital surface model
- small errors in the interior and exterior orientation of the image
- small measurement errors in the image coordinates of point S'

In general, therefore, monoplotting cannot replace stereoanalysis.

Exercise 7.4-2. From the opposite side of a valley, horizontal photographs are taken with a Wild P32 (Section 3.8.3) for the purpose of monoplotting. The side to be analysed has a slope of some 50%. How large are the XY coordinate errors if the digital ground model has an accuracy of around $\pm 25\,\text{cm}$? (Solution: 1.5 m.)

Exercise 7.4-3. How large are the XYZ coordinate errors if the measurement accuracy of image coordinates is $\pm 10\,\mu\text{m}$? The photographing range is 2 km. (Solution: $\Delta XY = 1.7\,\text{m}, \Delta Z = 0.8\,\text{m}$.)

Further reading for Section 7.4.3: Makarovic, B.: ITC-J, pp. 583–600, 1973.

7.5 Photo models

Orthophoto technology from aerial photographs reaches its limits in urban landscapes or with individual objects which are to be recorded by close range photogrammetry. The orthophoto is only an orthogonal projection onto a plane, normally the XY plane of the object coordinate system. The photographic texture on planes orthogonal to the XY plane is therefore missing in an orthophoto; where there are several overlapping object surfaces only one can be covered with a photo texture. These disadvantages can be overcome with a photo model. This is taken to mean a three-dimensional geometric model (frequently a CAD vector model) whose surface is covered by a photographic texture.

[11] The analysis of a digital orthophoto with a background DTM, as discussed in Section 7.4.1, may also be referred to as "monoplotting".

Photogrammetric recording and modelling of three-dimensional, geometric models has already been discussed in Sections 6.6.3 and 6.8.8 (see also Section C 1.7.1, Volume 2). The following discussion is limited mainly to the transfer of texture from digital images onto the surface of the three-dimensional object model. Production of a photo model will be explained using a practical example taken from a university lecture. The modelled subject is the historic Wartberg church in Lower Austria.

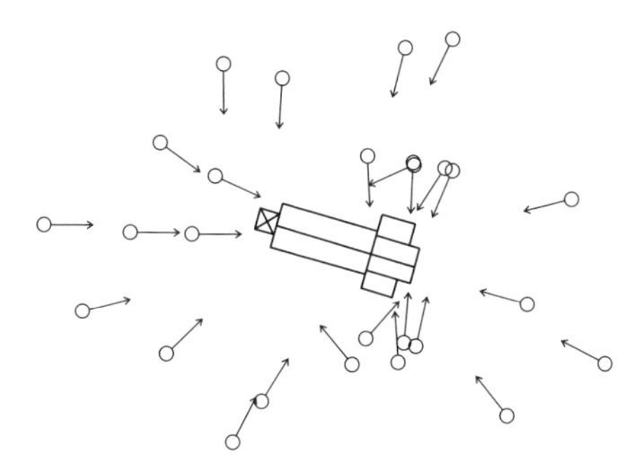

Figure 7.5-1: Camera stations and directions of camera axes

Figure 7.5-1 shows the camera stations and directions of the axes. The images were taken at different heights and zenith angles. The camera used was a calibrated Kodak DCS 460c (Section E 3.5, Volume 2) with objective lenses $f = 28$ mm and $f = 55$ mm. Pixel size is 9 μm. 510 object points were determined using a bundle adjustment (Section 5.3) to an accuracy of ± 2 cm to ± 4 cm), with every object point, on average, located in 5 metric images. Image point measurement was done manually in the ORPHEUS[12] multi-image, digital monocomparator. This software tool displays multiple images, organized in image pyramids, on a computer screen. In this system the polygonal boundaries of the flat surfaces can be defined. Figure 7.5-2 shows the rendered geometric model with artificial, oblique illumination.

To create the photo model, the polygonal borders of the flat surfaces are transformed by means of the collinearity equations into those images which are candidates for the transfer of photo texture. The texture is then taken from the image in which the associated flat surface is largest. The actual transfer of photo texture is done by indirect transformation (Section 2.2.3), i.e. an image matrix is defined in the appropriate (tilted) object plane and filled with the photo texture from the corresponding section of the original image. A bilinear grey level interpolation (Section 2.2.3a) is used for the required re-sampling. Pixel size on the object surface is chosen to be 5 cm; the quantity of data for the entire 3D model is 485 kBytes. Figure 7.5-3 shows the 3D photo model. It was exported in the VRML format (virtual reality modeling language) to enable distribution over the WWW (world wide web).

[12] `http://www.ipf.tuwien.ac.at/products/produktinfo/orient/html_hjk/orpheus_e.html`

Figure 7.5-2: Geometric model of the church at Wartberg

Figure 7.5-3: 3D photo model of the church at Wartberg

The photo texture on the roofs of such 3D photo models is frequently taken from very large-scale aerial photographs. In this example, an amateur camera was used to take images from a helicopter. The images were calibrated using a bundle block adjustment (Section 5.7). Details of the analysis can be found in the publications mentioned in the footnote[13].

Comparison of Figures 7.5-2 and 7.5-3 shows the advantages and disadvantages of a 3D photo model compared with a rendered object model using artificial surface texture:

- the main advantage of the 3D photo model is that a rich level of detail without geometric modelling can be achieved with the aid of photo texture. Particularly valued details are the windows and structure of the walls. (It should, however, be

[13]Zischinsky, T., Dorffner, L., Rottensteiner, F.: IAPRS XXXIII, pp. 959–965, Amsterdam, 2000.

noted that this level of detail is only provided for visualization and is not present in the geometric model.)

- because of the texture, which is an optical recording of the real object, 3D photo models have a very high documentation value.
- When no photo texture is available, an artificial surface texture can be added.
- the texture of a 3D photo model has all the deficiencies associated with a photograph; complex image processing operations are normally required to create uniform effects.

7.6 Static and dynamic visualizations

In static visualizations the object and viewing location and direction are fixed; in dynamic visualizations the object is fixed (the usual case here) but viewing location and/or direction vary. A dynamic visualization is put together from a dense sequence of static visualizations.

Static visualisations have already been used several times in this textbook, with a focus at infinity and a camera direction perpendicular to the XY plane. Orthophotos belong in this category. Oblique viewing directions are much more attractive, as in the photogrammetric analysis of a village in Figure 6.6-8.

Dynamic visualization is becoming increasingly important, also known as animation. Digital models of natural and man-made landscapes (e.g. terrain models, city models, orthophotos) provide the base data for the simulation of virtual flyovers. Digital models of urban landscapes and historically valuable collections of buildings (e.g. city models, 3D photo models of historic buildings, Section 7.5) provide the base data for virtual walkthroughs. Here the flight path or walking route can be pre-defined or defined interactively by the user by means of on-screen controls. A combination of this visualization (including additional linked information) with a continuous update of the user's location (for example with GPS) leads to the more comprehensive application of location based services.

Chapter 8

Laser scanning

In the past ten years, laser scanning has revolutionized both topographic mapping and close range three-dimensional object recording. Particularly in analysis, there is much in common between laser scanning and photogrammetry. In this series of books based on photogrammetric knowledge, it is therefore natural to deal also with laser scanning. The sections which follow will show the parallels between photogrammetry and laser scanning, as well as highlight the differences between both these technologies.

For purposes of terminology, and classification of laser scanning, it should be noted that in the field of remote sensing laser scanning has, for some time, been identified by the term Lidar (light detection and ranging). The term Laser Radar (radio detection and ranging) has also been used in remote sensing for the location of objects with the aid of a laser (light amplification by stimulated emission of radiation). Laser light is strongly collimated, monochromatic and coherent. Where geometric problems take precedence in remote sensing, it is common to speak less of remote sensing and more of photogrammetry (Section 1.1). From that perspective also, laser scanning can be classified as part of photogrammetry.

In the following sections we will present airborne laser scanning (Section 8.1), terrestrial laser scanning (Section 8.2) and short range laser scanning (Section 8.3).

8.1 Airborne laser scanning

8.1.1 Principle of operation

Using a laser scanner, points on the ground are sampled (Figure 8.1-1). With the aid of a narrow laser beam, a pulse of laser light from the scanner is diffusely reflected by a point on the ground surface. From the elapsed time between transmission and reception of the pulse, the distance between scanner and ground point can be determined. (The elapsed time, multiplied by the group velocity of the light pulse which is about 0.03% less than the velocity of light, gives double the distance value.)

The laser beam in the laser scanner is deflected at right angles to the direction of flight and this angle of deflection is recorded. The coordinates of the laser scanner's location and its orientation angles are required in order to convert the polar coordinates of the measured object point into XYZ coordinates. These constantly changing values are determined by means of a dynamic POS (Position and Orientation System), consisting

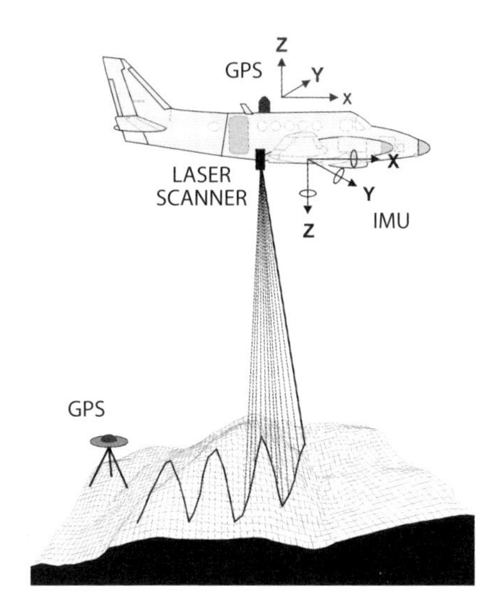

Figure 8.1-1: Scanning laser in aircraft

of GPS and an IMU. The same technology which is based on GPS and IMUs to support aerial photography is equally applicable to airborne laser scanning. This is discussed in detail in Sections 3.7.3.2 and 3.7.3.3. GPS, IMU and scanner operation are synchronized to microsecond accuracy. Of course, the relative positions, or eccentricities, of the sensors must be determined before the flight. Gyro-stabilized platforms are rarely used in laser scanning. (See Section 3.7.3.3.)

Reflections can occur at different locations along the measurement beam. The upper diagram in Figure 8.1-2 shows the arrival of the first pulse at points on the object surface. The lower diagram indicates the first echoes which return and the arrival of the pulse at lower lying object points along the beam, if these exist. This occurs if the surface scattering back the first echo is not fully extended in the beam diameter. The lower lying objects are then the source of the last echo. If the distance is measured with the first echo, the recorded laser points are on the crown of the tree (second beam from left), on the eaves of the roof (third beam from left), on a power line (fifth beam from left), etc. If the distance is measured with the last echo, then the recorded laser points lie, using the same order as before, on the tree trunk or ground, on the road, at the ground below, etc. Almost all laser scanners can measure the distance with both the first and last echoes, and record both with a small time separation. There are laser scanners which can additionally utilize any intermediate echoes for range measurement and, in this way determine a whole spectrum of distances for each laser beam. There are limits to this recording process. Depending on the performance capabilities of the particular range measuring device, there must be a small minimum separation between individual records. The separation capability depends on the pulse length and

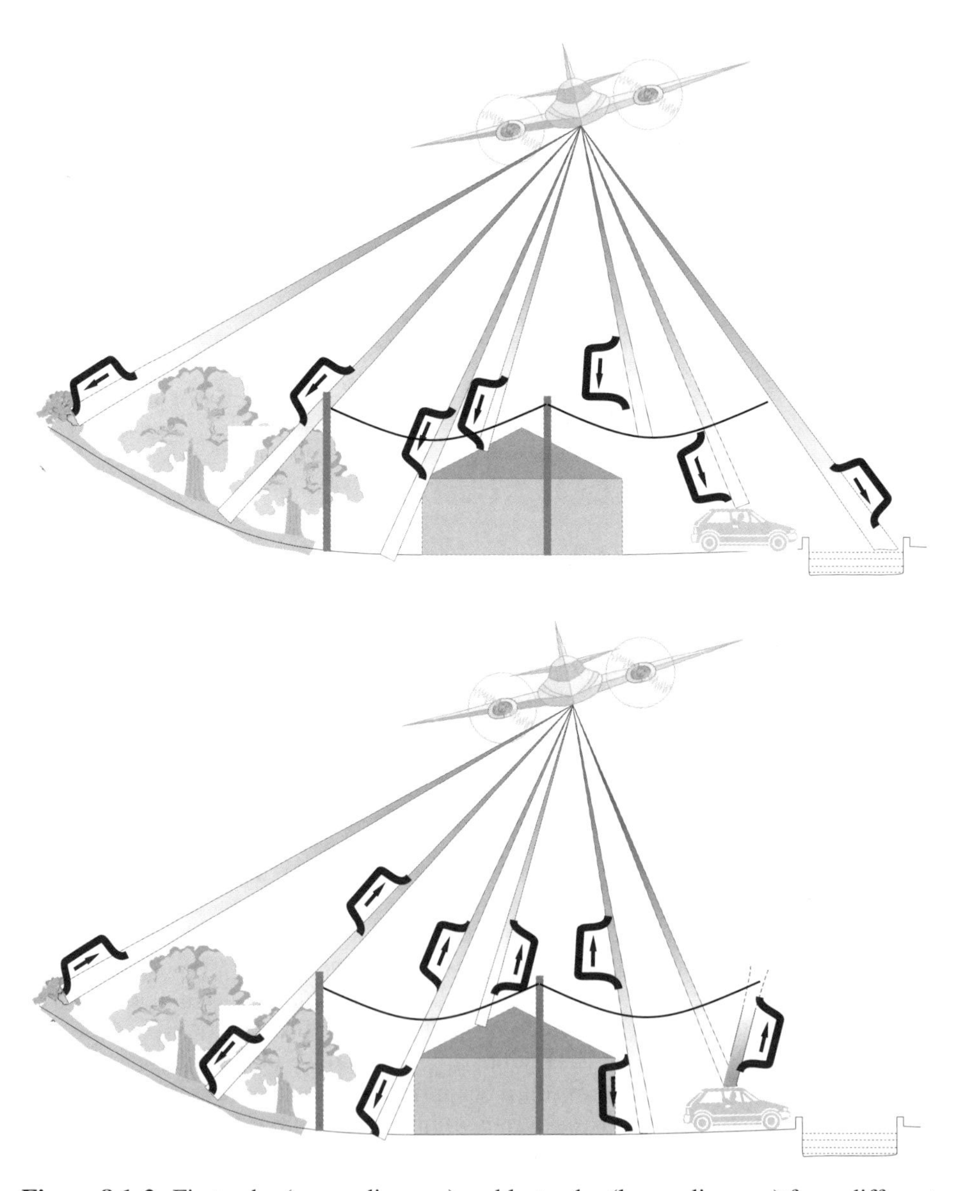

Figure 8.1-2: First echo (upper diagram) and last echo (lower diagram) from different objects, as well as reflection and high absorption

the minimum separation can, for example, amount to 1.5 m:

$$1.5\,\text{m} = \frac{300000\,\frac{\text{m}}{\text{s}} \times 10\,\text{ns}}{2} \mathrel{\hat{=}}$$

$$\text{half the measured path length} = \frac{\text{velocity of light} \times \text{shortest measurable time interval}}{2}$$

Generally, for vegetation shorter than 1.5 m high, only the top surface is recorded, even if there is an occasional echo from the ground (first beam from left in Figure 8.1-2).

It is indicated in Figure 8.1-2 that a number of the transmitted laser pulses does not return. For example there can be highly specular reflection from very smooth surfaces such as a car roof. The deflected beam can later strike another surface point which reflects diffusely and then return an echo to the laser scanner. Ranges measured in this way via indirect routes (multi path) have gross errors; they are too long (long ranges). At other surfaces, such as a clear stretch of water (right hand beam, Figure 8.1-2), absorption can be so strong and reflection so little, that insufficient energy for range measurement is returned. The absorption also depends on the angle of incidence of the laser beam: it is especially strong at water surfaces, for example, when the laser beam meets the surface perpendicularly. For this reason, laser scanner strips over water surfaces often have fewer points in the middle than at the edges of the strip.

The technical data for the ALS50-II Airborne Laser Scanner from Leica Geosystems, are summarized below.

Operating Altitude	200 m – 6000 m above ground level
Pulse rate	max. 150000 measurements per second (150 kHz)
Accuracy	(incl. 10 cm GPS error, restricted to max. 40° FOV) at 1000 m above ground: 11 cm in height, 11 cm in horizontal direction at 4000 m above ground: 15 cm in height, 44 cm in horizontal direction
Number of returns	4 (first, second, third, last)
Number of intensities	3 (first, second, last)
Intensity digitization	8 bit intensity + 8 bit automatic gain control (AGC) level
Maximum Field of View (FOV)	75° full angle
Roll stabilization	automatic adaptive, range = 75° – current FOV
Laser divergence	0.22 mrad (45″) at $1/e^2$
Recording media	300 GB Harddisk (17 hours at maximum pulse rate)
Waveform profiling	8 bits at 1 ns interval at 50 kHz (option)

Table 8.1-1: Technical data for the ALS50-II Airborne Laser Scanner

The accuracies of the laser ranging component and the GPS height determination are primarily responsible for the height accuracy indicated in Table 8.1-1 for the ALS50-II. Neither effect is strongly dependent on flying height. The dependence of height accuracy on flying height, which is clearly present, derives from the relatively long path through the atmosphere, a source of disturbance, the enlargement of the laser footprint with increasing altitude and the reduction in received energy due to spreading loss. As the footprint becomes larger, the recorded direction of the polar coordinates and the object point, to which the distance refers, fit less well together. The planimetric accuracy is influenced by the accuracy of the GPS location in plan, the accuracy of the IMU orientation and the accuracy to which the deflection angle of the laser beam is recorded. The last two mentioned error sources effect a relatively large reduction in accuracy with increase in flying height.

Further reading on the laser beam and its interaction with surfaces: Wagner, W., Ullrich, A., Ducic, V., Melzer, T., and Studnicka, N.: Gaussian decomposition and calibration of a novel small-footprint full-waveform digitising airborne laser scanner. ISPRS Journal of Photogrammetry and Remote Sensing 60(2), pp. 100–112, 2006.

8.1.2 Analysis and processing

8.1.2.1 Georeferencing

The orientation functions $X_0(t), Y_0(t), Z_0(t), \omega(t), \varphi(t), \kappa(t)$ can be determined using the GPS and IMU information and related to corresponding polar coordinates $\alpha(t)$ and $s(t)$ through the synchronization time t. The resulting XYZ coordinates of individual laser points are then given in the object coordinate system to which the reference station for the differential GPS was assigned.

$$\begin{pmatrix} X \\ Y \\ Z \end{pmatrix} = \begin{pmatrix} X_0(t) \\ Y_0(t) \\ Z_0(t) \end{pmatrix} + \mathbf{R}_{\omega(t)\varphi(t)\kappa(t)} \begin{pmatrix} 0 \\ s(t)\sin\alpha(t) \\ s(t)\cos\alpha(t) \end{pmatrix} \qquad (8.1\text{-}1)$$

The actual object coordinate system is either a global GPS coordinate system (e.g. WGS84) or the ground coordinate system. In the following text we will take the ground coordinate system as the object coordinate system. The conversion from global coordinate system to ground coordinate system, which is normally used for presentation of final results, can be achieved by using datum transformation formulae and the appropriate geoid.

In Equation (8.1-1) the zero point of the α scale has been taken as the middle of the laser scanner's field of view. This zero direction must coincide with the $\kappa(t)$ rotation axis of the rotation matrix $\mathbf{R}_{\omega(t)\varphi(t)\kappa(t)}$. A corresponding calibration, which takes into account the relative positions of GPS, IMU and laser scanner in the aircraft, will not be discussed here. However, note that companies undertaking laser scanner flights require reference heights of a number of test surfaces (often only horizontal surfaces) in order to effect a calibration for each project. These should be given in the ground coordinate system we have chosen as the object coordinate system[1].

To achieve rigorous control and improved accuracy, it is recommended to change from this direct georeferencing to an integrated sensor orientation, i.e. using control and tie points as is common when georeferencing images from line cameras (Section 5.5). In place of the type of control point used in photogrammetry, with coordinates given in the reference coordinate system, special control elements must be made available which take account of the properties of laser scanner data. As already mentioned, horizontal planes are suitable as height control elements. The equation of such a control plane should be determined by a number of points measured on the ground. It has, for example, the form:

$$Z_{\text{ref}} = a_0 + a_1 X + a_2 Y \qquad (8.1\text{-}2)$$

a_0, a_1, a_2 ... parameters of the plane determined from a least-squares adjustment of the terrestrial XYZ coordinates.

Using the GPS/IMU information and the datum transformation formulae, the direct georeferenced coordinates $X_{\text{meas}}, Y_{\text{meas}}, Z_{\text{meas}}$ are obtained from Equation (8.1-1). Using coordinates $X_{\text{meas}}, Y_{\text{meas}}$ in Equation (8.1-2), the height Z_{ref} corresponding to height Z_{meas} is obtained. The smaller the tilt of the control plane, the less is the effect of $X_{\text{meas}}, Y_{\text{meas}}$ errors on the height difference ($Z_{\text{meas}} - Z_{\text{ref}}$), required for the integrated sensor orientation.

Due to the relatively large inaccuracy in plan position of the laser points (Table 8.1-1), and their relatively large separation which, depending on quality requirements (Section 8.1.3) can be selected between 0.5 m and 3 m, control points cannot be employed for the determination of horizontal shifts. Instead, linear or planar control elements must be used. Such extended control elements could be flat roof surfaces, as shown in Figure 8.1-3. These should be steeply sloped roofs in different orientations so that horizontal stability is guaranteed in any direction. As Figure 8.1-3 suggests, three such roofs intersect in a well-defined point and can therefore be regarded as a "virtual" full control point. The equations of these control planes should be determined using terrestrial measurements.

For the integrated sensor orientation, the $X_{\text{meas}}, Y_{\text{meas}}, Z_{\text{meas}}$ coordinates of the laser points which define the respective control planes must be determined using Equation (8.1-1) and the datum transformation formulae. The perpendicular offsets of these points from the (known) control point plane are critical items of information for the integrated sensor orientation. (The determination of these offsets is explained in Section B 3.5.5.1 in Volume 2.) In the context of a least-squares adjustment, which is still to be discussed, these offsets have components in all 3 coordinate axes. The example of a height adjustment plane, discussed above with reference to Equation (8.1-2), is a special case within this 3D solution.

[1]Further reading on calibration: Kilian, J., Haala, N., Englich, M.: IAPRS 31(B3), pp. 383–388, Vienna, 1996. Morin, K., El-Sheimy, N.: Optical 3-D Measurement Techniques V (Eds.: Grün/Kahmen), pp. 88–96, Wichmann-Verlag, 2001. Burman, H.: IAPRS 34 (Part 3A, Commission III), pp. 67–72, Graz, 2002.

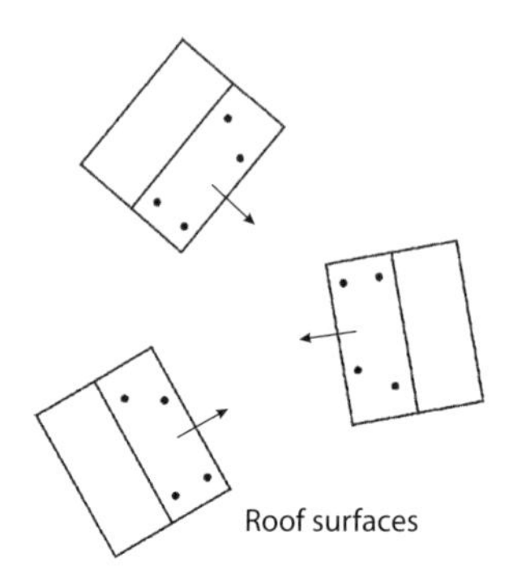

Figure 8.1-3: Three tilted planes defined by roofs, corresponding to a "control point"

The question remains open as to how the laser points belonging to each control plane can be automatically found. In general, the $X_{\text{meas}}, Y_{\text{meas}}, Z_{\text{meas}}$ point cloud is searched for planar regions which correspond to the control planes. Once the initial process of direct georeferencing has been implemented, and is close to the final result, it is possible to extract areas in the $X_{\text{meas}}, Y_{\text{meas}}, Z_{\text{meas}}$ point cloud which correspond to the points measured by ground surveys. These can be examined for planarity, for example using best-fitting planes, with removal of the outliers such as scanned points on chimneys or dormers.

This concept of control planes can also be employed when connecting overlapping laser scanner strips. The difference between a tie plane and a control plane is that the tie plane parameters are unknown, for example the values a_0, a_1, a_2 in Equation (8.1-2). Using the method of least squares, these parameters are introduced as unknowns within the solution for the integrated sensor orientation. A tie plane has its own set of parameters, regardless of whether two, three or more laser scanner strips are involved. For each scanner strip in the solution, the perpendicular offsets of the $X_{\text{meas}}, Y_{\text{meas}}, Z_{\text{meas}}$ coordinates from the corresponding plane are minimized in the usual least-squares manner.

The derivation of detailed equations for integrated sensor orientation from laser scanner data lies outside the scope of this book. Only a general outline is therefore given:

- the polar coordinate equations (8.1-1) are used in place of the collinearity equations, which are central to the integrated sensor orientation of line images (Section 5.5). These are solved with respect to the polar coordinates and must, of course, be linearized, etc. (Further details on the equations of polar coordinates can be found in Section B 3.5.3, Volume 2.)
- GPS and IMU models to handle the GPS and IMU information must be constructed according to Sections 5.4 and 5.5. Independent models for each scanner strip are recommended.
- the entire concept of georeferencing using control and tie planes, presented above,

was proposed by Kager. Details of the refinement for height adjustment can be found in a published paper[2].

Very good agreement at the edges of the laser scanner strips can be achieved with the methods presented here. The inclusion of control planes checks the parameters of the data transformation formulae in those particular interest areas. An improvement in the parameters is possible as a result. In addition, it is not necessary to take account of the geoid in the processing, unless the areas are very large. (This problem is handled in detail in Sections B 5.3 and B 5.4, Volume 2.) The IMU information can be included without modification (Section 5.6). Only the κ heading angle must be corrected by the convergence of the meridian (5.6-2). In this approach with ground coordinates defining the object coordinate system, Earth curvature should be taken into account by modifying the polar coordinates (8.1-1). A simple solution is to apply an Earth curvature correction to the Cartesian coordinates of the scanned points in a scanner plane perpendicular to the flight direction, using Equation (4.5-4). Note that A must be measured from the middle of the strip.

The guidelines of GPS/IMU-supported aerotriangulation (Section 5.4) govern the number of control planes and their arrangement. As a result of the high quality of GPS and IMU information which can now be achieved, it is sufficient to have control plane clusters at both ends of the laser scanner strips in a project. Where laser scanning tasks are designed in blocks, there should be control plane clusters, as a minimum, in the corners of the blocks. To strengthen formations of blocks, particularly if independent GPS and IMU models are employed, then laser scanner cross strips are recommended at the beginning and end of the blocks.

8.1.2.2 Derivation of terrain models

The construction of object models can take place after georeferencing. This section deals with the derivation of digital terrain models (DTMs) from the digital surface models (DSMs). Figure 6.6-10 explains the difference between these two models. Although the scanned laser points from the last echo are used to derive the DTM, there will still be many scanned points above the surface of the ground, depending on the density of vegetation and buildings. Figure 8.1-7 (left) represents a DSM. It results from a resampling (Section 2.2.3) of the irregularly distributed laser scan points on an orthogonal 2 m grid, for example using a nearest neighbour technique. The shaded relief of this surface grid model was based on oblique illumination from the top left during rendering.

The starting point for deriving DTMs from scanned points is shown in Figure 8.1-4 for a single profile. In a first step an adjusting surface to the point cloud is calculated. The example in Figure 8.1-4 shows a best-fitting curve. Two suitable methods for achieving this could be linear prediction or bivariate polynomials. With respect to these best-fitting surfaces, the residuals v_i from the least-squares adjustment are shown. In

[2]Kager, H., Kraus, K.: Optical 3-D Measurement Techniques V (Eds.: Grün/Kahmen), pp. 103–110, Wichmann Verlag, 2001.

a second least-squares adjustment, the Z coordinates of the scanned points are given special weights p_i which depend on the residuals of the first adjustment.

Such weights, calculated on the basis of a weighting function, are introduced to enable robust parameter estimation[3] as well as the reliable detection of gross errors. As a rule, a symmetrical bell curve, centred on the zero point of the residuals v, is used as the weighting function. However, such a weighting function is not suited to the robust estimation of a DTM from laser scanned points. It would, in fact, reduce in equal measure the effect on the fitted surface of points with (large) positive and (large) negative residuals. The result of such a robust adjustment would be similar to that shown in Figure 8.1-4.

For the robust estimation of DTMs from laser scanned points, a weighting function is applied in the second least-squares adjustment. This function takes account of the following properties of errors in the scanned points:

- points with positive residuals, which are probably ground points, should be given a higher weight than points with negative residuals, which are probably not ground points. (In Figure 8.1-4, two points are shown with positive and negative residuals v_i.)
- the terrain probably lies below the curve or surface fitted during the first least-squares adjustment (Figure 8.1-4), and a datum shift of the weighting function is appropriate.

A weighting function, which takes account of these requirements, has the form (Figure 8.1-5):

$$\begin{aligned} &\text{for } v_i \le g \qquad p_i = \frac{1}{1+(a|v_i - g|)^b} \qquad a, b > 0 \\ &\text{for } v_i > g \qquad p_i = 1 \end{aligned} \tag{8.1-3}$$

v_i ... residuals from the first least-squares adjustment
a, b ... parameters which control the decreasing effect of the weighting function from right to left
g ... datum shift which also reduces the effect of points with small positive residuals v_i on the second adjustment

Figure 8.1-6 shows the result of the second least-squares adjustment using the weighting function shown in Figure 8.1-5. It meets our expectations: the DTM is fitted through the low lying laser scanned points. In this combined interpolation and filtering process, points above the model have little influence. In practice, several iterations are needed to reach a final solution.

By selecting an adequate threshold value for the residuals v in the second least-squares adjustment, the original scanned points can be classified into points either on the

[3] For example, Section B 7.2.1.5, Volume 2, and Koch, K.R.: Parameter Estimation and Hypothesis Testing in Linear Models. 2nd edition, Springer 1999.

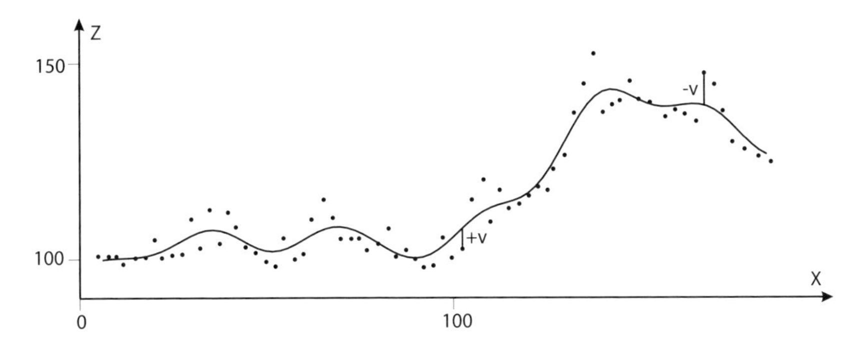

Figure 8.1-4: Best-fitting curve through the laser scanned points and residuals of the first least-squares adjustment

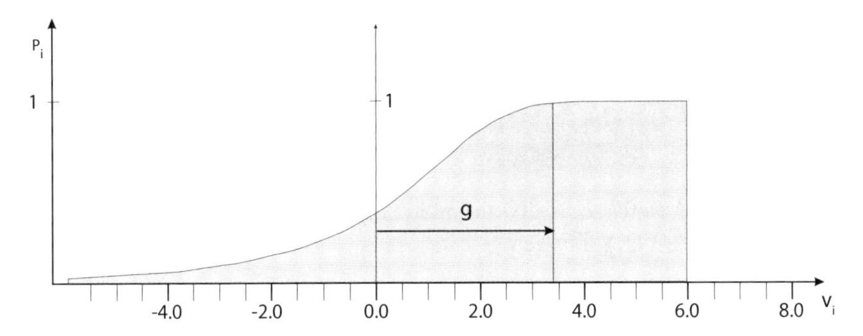

Figure 8.1-5: Weighting function for a robust least-squares adjustment which derives a DTM from laser scanned points

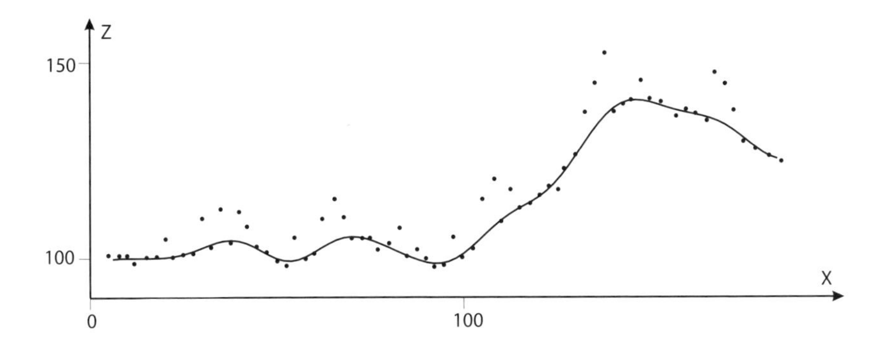

Figure 8.1-6: Results of the second least-squares adjustment using the weighting function of Figure 8.1-5 and the residuals v_i from the first least-squares adjustment in Figure 8.1-4

ground or off the ground. Strictly speaking there should be a final least-squares adjustment using only ground points without application of variable weights.

For widespread areas of dense forest and extensive building, the process of interpolation and filtering must be expanded into a hierarchical strategy, described in a number of publications[4].

Figure 8.1-7 shows the result from part of an international test[5]. Despite the relatively low point density of 0.25 points/m^2, i.e. with an average point separation of 2 m along and across the direction of flight, vegetation and buildings have been successfully eliminated, as can be seen in Figure 8.1-7. A small building on the railway embankment remains: its height is too low.

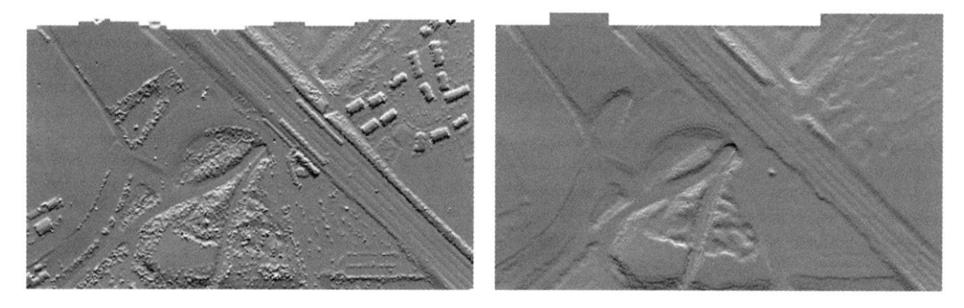

Figure 8.1-7: DSM (left) and fully automatically derived DTM (right) from a small part of the OEEPE test at Vaihingen

Figure 8.1-8 shows the DSM from a scanning flight over Vienna. It has a grid spacing of 0.5 m which is approximately the point separation in the original data. In so far as it can be defined in urban areas, the DTM can be derived from this model (Figure 8.1-9). The accuracy of this DTM, which depends on an individual area's land use, has been determined by the City Council's survey department (Magistratsabteilung 41) with the help of a large number of ground check points. Table 8.1-2 shows the results. Note that there is a distinction between standard deviation σ (see also Appendix 4.6-1) and r.m.s. (root mean square error). The r.m.s. is calculated from the actual errors ε, and these also contain systematic effects. Laser scanned DTMs, depending on surface roughness and ground cover, are generally too high.

The standard deviation of ± 1.0 cm on open streets is remarkable at first glance, particularly in respect of the fact that individual laser points have a height accuracy of only around ± 6 cm. (This accuracy corresponds to the manufacturer's specifications for the TopoSys laser scanner which was used for data acquisition.) This large improvement in accuracy can be simply explained: on surfaces with very limited curvature, the least-squares adjustment leads to an averaging of the scanned points within a limited area. It

[4]E.g. Briese, C., Pfeifer, N.: in Optical 3-D Measurement Techniques V (Eds.: Grün/Kahmen), pp. 80–87, Wichmann Verlag, 2001.

[5]Torlegard, K., Jonas, N.: OEEPE-Publication No. 40, 2001.

Figure 8.1-8: Surface model (DSM) from laser scanner data (0.5 m grid spacing)

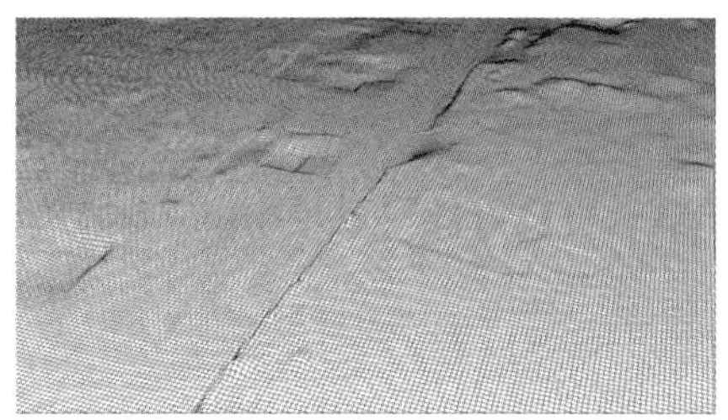

Figure 8.1-9: DTM derived using robust, hierarchical estimation

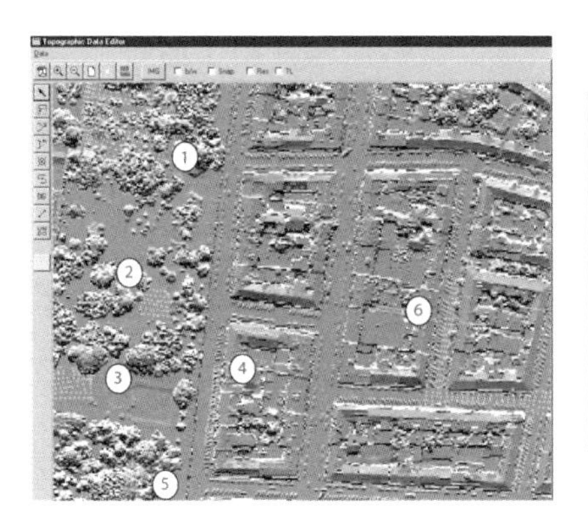

	r.m.s. [cm]	σ [cm]
Park, dense vegetation	±14.5	±11.1
Park, sparse vegetation	±11.4	±7.8
Park, open ground	±8.6	±4.5
Streets with parked cars	±9.2	±3.7
Streets without cars	±2.4	±1.0
All points	±10.5	±7.1

Table 8.1-2: Accuracies of the DTM in Figure 8.1-9 for various types of land use

is well known that the arithmetic mean has an accuracy of:

$$\sigma_M = \frac{\sigma_E}{\sqrt{n}} \tag{8.1-4}$$

σ_E ... standard deviation of a single observation
σ_M ... standard deviation of the arithmetic mean
n ... number of contributing observations

The accuracy increase from ±6 cm to ±1 cm is reached when 36 points are included in the averaging, which is readily achieved at a point separation of 0.5 m, i.e. a 3 m × 3 m area on the road.

8.1.2.3 Generation of building models

The automatic generation of building models from laser scanner data is currently a subject of intensive research. Only the results of two different methods will be presented here.

The first method starts by forming the difference between a DSM and a DTM. The difference model contains not only the buildings but also the vegetation and many other objects lying above the DTM. Building points are extracted from the difference model using special filter techniques. Within these filtered point clouds, planar areas are subsequently detected and individual planes joined together to form roofs. Figure 8.1-10 shows the results for a building block in Vienna. The data originate from the same scanning flight which provided the data for the derivation of the DTM in Figure 8.1-9. The difference model formed at the beginning of this procedure is therefore a subtraction of the data in Figure 8.1-9 from the data in Figure 8.1-8. In order not to lose the eaves of the roofs, the extraction of building models uses range measurements based on the first echo (the third ray from the left in Figure 8.1-2 illustrates this procedure).

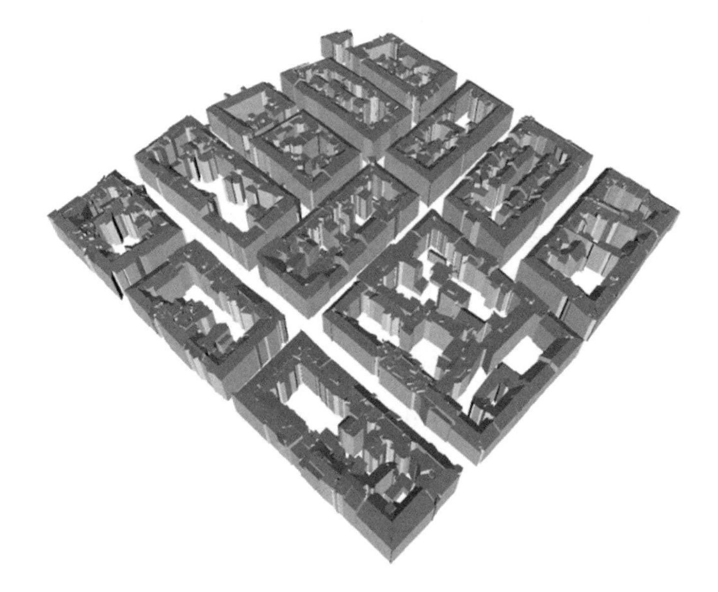

Figure 8.1-10: Building model extracted from laser scanner data

The second method was developed under contract to engineering consultancy Schmid, Vienna, by the "Advanced Computer Vision" technology centre. It has some similarities with the previous method discussed above. However, it makes use of the ground plans of buildings which are often available in two-dimensional GIS databases. Figure 8.1-11 shows the result and indicates the laser spots used. This result is a CAD model.

Further reading: Brenner, C.: DGK 530, 2000. Maas, H.-G.: IAPRS 32 (Commission 3-2 W5), pp. 193–199, Munich, 1999. Vosselman, G., Dijkman, S.: IAPRS 34 (Commission 3/W4), pp. 37–43, Annapolis, 2001. Peternell, M., Steiner, T.: Computer-Aided Design 36, pp. 333–342, 2004.

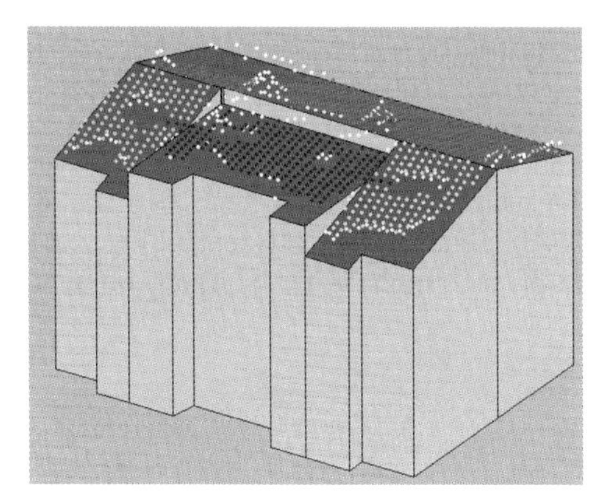

Figure 8.1-11: Building model derived from laser scanner data and a two-dimensional building ground plan

8.1.3 Comparison of two paradigms and further performance parameters of laser scanners

The term paradigm implies the pattern of approach and thought which lies at the heart of a scientific or technical field. In the following section the paradigms behind airborne laser scanning and (stereo-) photogrammetry will be compared.

The essence of (stereo-) photogrammetry is explained by the simple diagram of Figure 8.1-12. Here the geometrical reconstruction of an object in three-dimensional space, from at least two photographic images, is central to the process. One photograph defines a bundle of rays or directions. A point on the object can be reconstructed in three-dimensional space when it is intersected from at least two directions. The directions, or rays, generated by the natural light of the sun, are recorded by sensors known as "passive sensors". Within the context of a paradigm, it is irrelevant if:

- the record is made on film or an electronic imaging device.
- electronic recording is done using a two-dimensional array of detectors or a linear array in a 3-line camera.
- image points are measured stereoscopically by a human operator or digitally, using a correlation algorithm.
- a third image, or further images of the same object, or a set of images of an extended object are available.
- a calibrated metric camera or uncalibrated amateur camera are used.
- the position and orientation of the images are determined from control and tie points or elements, or from GPS and IMU measurements. (In Figure 8.1-12 "GPS/IMU" is placed in brackets to imply an option.)

The question now presents itself as to which paradigm lies at the heart of laser scanning. As in (stereo-) photogrammetry, the geometrical reconstruction of objects in three-dimensional space is foremost, although not from at least two measuring locations but from only a single location (Figure 8.1-13). In place of a bundle of rays there is a field of directions and distances, i.e. a vector field. In place of passive sensors there are active sensors. GPS/IMU information is essential; the connection of two or more scans in a spatial network can only be achieved using common surface elements and not points.

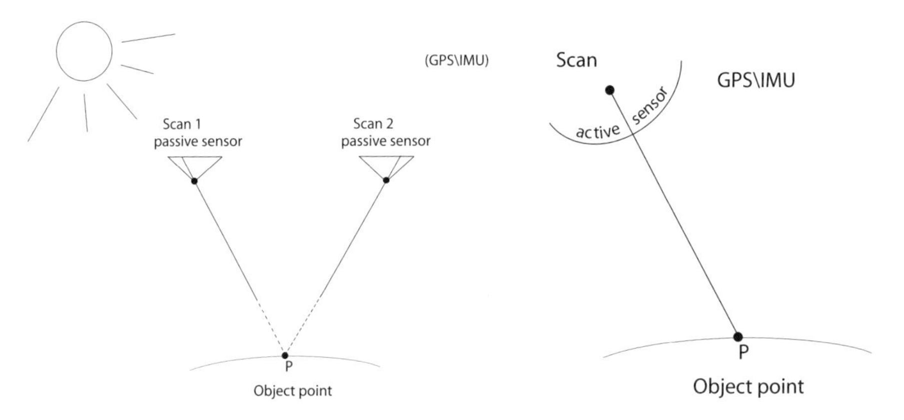

Figure 8.1-12: (Stereo-) photogrammetry paradigm

Figure 8.1-13: Airborne laser scanning paradigm

The remainder of this section will evaluate the contrasts between airborne laser scanning and (stereo-) photogrammetry from an application perspective and then present performance parameters for airborne laser scanning.

(a) in forested and built-up areas, laser scanning is superior to photogrammetry for the derivation of terrain models. In photogrammetry, a relevant ground point must be visible from at least two imaging positions but in laser scanning the view from a single recording position is sufficient.

(b) active sensing makes laser scanning independent of (natural) sunlight. Laser scanning flights can, in principle, take place at any time of day or night. Admittedly, clouds are as much of a hindrance to the near infra-red emissions used for laser scanning as they are to photogrammetric flight operations.

(c) laser scanning requires no texture on the object surface. In contrast to photogrammetry, laser scanning can be employed for areas of sand devoid of vegetation (dunes), smooth and dry cultivated fields, bright stretches of concrete, forested ground in shadow, etc. However, on very flat surfaces which give specular reflection, and surfaces with a low reflectance (clear water), there is no laser beam echo returned to the receiver.

(d) the accuracy of a DTM obtained photogrammetrically is critically dependent on the flying height and camera type (Section 6.7.2c). The accuracy of a DTM determined by laser scanning is less dependent on the flying height (Table 8.1-1). Instead, the accuracy of a laser scanned DTM is highly dependent on the point density. The point density during laser scanning is influenced by the flying height, the flying speed, the scanner's field of view and the scanner's data rate. The following rule of thumb, from empirically derived accuracies, can be used to estimate height accuracies:

$$\sigma_H\,[\mathrm{cm}] = \pm\left(\frac{6}{\sqrt{n}} + 120\tan\alpha\right) \tag{8.1-5}$$

$\tan\alpha$ ... ground slope
n ... point density per square metre

The figures of 6 cm and 120 cm in Equation (8.1-5) remain somewhat uncertain and should be further verified in the near future.

Numerical Example. To obtain country-wide DTMs, laser scans are currently flown with point separations along and across the direction of flight of around 2 m, i.e. the point density is 0.25 points m^{-2}.

$$(8.1\text{-}5):\ \sigma_H[\mathrm{cm}] = \pm\left(\frac{6}{\sqrt{0.25}} + 120\tan\alpha\right) = \pm(12 + 120\tan\alpha),$$

$$\text{i.e. for a ground slope of } 10\%\text{: } \sigma_H = \pm 24\,\mathrm{cm}$$

Roughly the same accuracy can be achieved with a wide-angle camera at a flying height of 1 km ($m_B = 6700$):

$$(6.7\text{-}5):\ \sigma_H\,[\mathrm{cm}] = \pm(15 + 100\tan\alpha),$$

$$\text{i.e. for a ground slope of } 10\%:\ \sigma_H = \pm 25\,\mathrm{cm}$$

In forested areas, a greater photogrammetric standard deviation has to be expected (compare Section 6.7.2b). Under the assumption that only 25% of scanned forest points reach the ground (i.e. the average separation of points along and across the direction of flight is 4 m), a considerable accuracy can still be achieved in a forested area:

$$(8.1\text{-}5):\ \sigma_{H,\mathrm{Forest}}\,[\mathrm{cm}] = \pm\left(\frac{6}{\sqrt{0.0625}} + 120\tan\alpha\right) = \pm(24 + 120\tan\alpha)$$

Numerical Example. Particularly for the sake of geomorphological fidelity of the DTMs, which will be examined in more detail in Section (e), laser scanning operations are flown with an average point separation of 0.5 m along and across the direction of flight. This leads to the following standard deviation σ of DTMs derived from these data:

$$(8.1\text{-}5):\ \sigma_H\,[\mathrm{cm}] = \pm\left(\frac{6}{\sqrt{4}} + 120\tan\alpha\right) = \pm(3 + 120\tan\alpha)$$

An area with ground slopes of 5%, for example, has an accuracy of $\sigma_H = \pm 9\,\mathrm{cm}$.

(e) the geomorphological quality of DTMs derived from laser scanned data is, like accuracy, crucially dependent on point density. Provided that undulating terrain is being recorded, the sampling theorem (supported illustratively by Figure 3.3-6) provides the following connection between the minimum wavelength L_{min} and the point separation Δ. (Note that in laser scanning the object is directly "digitized". In terrain scans the point separation Δ is therefore often called the Ground Sampling Distance, GSD.)

$$L_{\text{min}} \approx 3\Delta \tag{8.1-6}$$

Figure 8.1-14 shows a ground profile with minimum wavelength L_{min} and two different sets of point records with point separation Δ. It is obvious that the curve can be reconstructed from either configuration of points.

Note: with $L_{\text{min}} = 2\Delta$, an unfavourable distribution of laser points can deliver 3 identical Z values. For this reason, and in particular because of the planimetric error in the laser points, the factor 2 should be significantly increased. The factor 3 has been chosen here, 2.8 is often used.

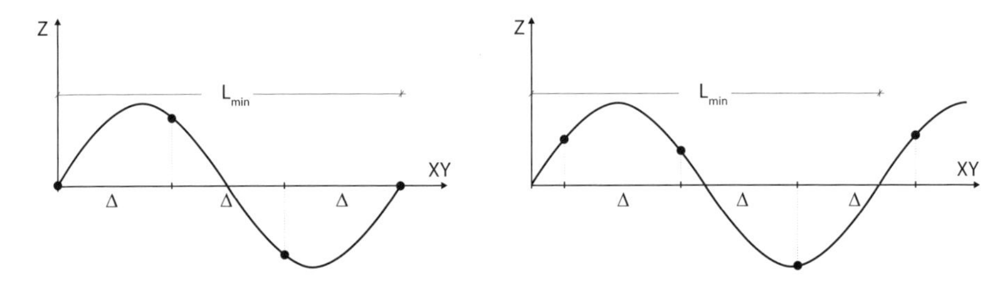

Figure 8.1-14: Minimum wavelength of a ground profile and two different arrangements of points to determine it (Δ = point separation)

Laser scanning delivers point clouds in which no structural elements (edges on the ground, etc.) are emphasized. With photogrammetric data capture, and often as an extension to automated data capture (Section 6.8.6), considerable value is placed on the structural elements. In order to ensure the prerequisites for deriving structural elements from laser point clouds, a high point density is required during data capture. The spatial relationships for the cross-section of the embankment of a dam are sketched in Figure 8.1-15. In order to determine the edges (for which there is a publication[6] describing this technique), the top and side surfaces, assumed planar, must be recorded with at least two points. The relationship between the minimum width B_{min} and the point separation Δ is:

$$B_{\text{min}} \approx 2\Delta \tag{8.1-7}$$

Figure 8.1-15 shows two different configurations of points with separation Δ when the minimum width of the crown of the embankment is B_{min}. In both cases there is a minimum of two points on the top surface.

[6]Briese, C.: IAPRS 35(B3), pp. 1097–1102, 2004.

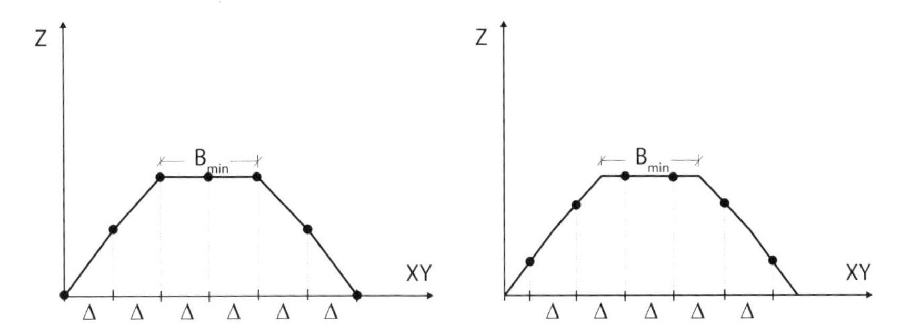

Figure 8.1-15: Minimum width of a dam profile and two different arrangements of points to determine it (Δ = point separation)

Exercise 8.1-1. An area of interest contains wavelike deformations in the ground with a minimum wavelength of 4 m and dam walls with a minimum top width of 2 m. What laser point separation should be chosen for data recording? (Solution: $\Delta_{\text{Wave}} = 1.3$ m. $\Delta_{\text{Dam}} = 1$ m, therefore use a point separation of 1 m.)

(f) compared with photogrammetry, laser scanning has two disadvantages. Firstly, planimetric resolution is appreciably worse. Laser scanning currently has a resolution in the half metre range. In contrast, photogrammetry can provide resolution in the decimetre range and occasionally better (Section 3.1.5.2). Secondly, laser scanning only provides a "range image", in contrast to photogrammetry with its multispectral information in the visible and infra-red region of the electromagnetic spectrum which facilitates delineation of land usage and object identification. It should also be mentioned that currently photogrammetry provides the better planimetric accuracy and airborne laser scanning the better height accuracy. It is therefore not surprising that, from the separate paradigms of (stereo-) photogrammetry (Figure 8.1-12) and laser scanning (Figure 8.1-13), a common paradigm can be created. This common paradigm is illustrated in Figure 8.1-16.

The (straight-line) directions of the imaging rays (Figure 8.1-12) and the (straight-line) vectors (Figure 8.1-13) have been replaced in the common paradigm (Figure 8.1-16) by symbolic (electromagnetic) waves. This serves to emphasize the physical line of sight in the common paradigm. The common paradigm can be characterized as follows:

- for every photogrammetric pixel there is also a "spectrum" of distances which consists of at least the first and last echoes.
- for every laser scanning pixel there is also a spectrum of electromagnetic radiation which, depending on sensor, is typified by natural and/or artificial radiation.
- for every pixel there is, in general, a second pixel with the identical information content but recorded from a different direction.

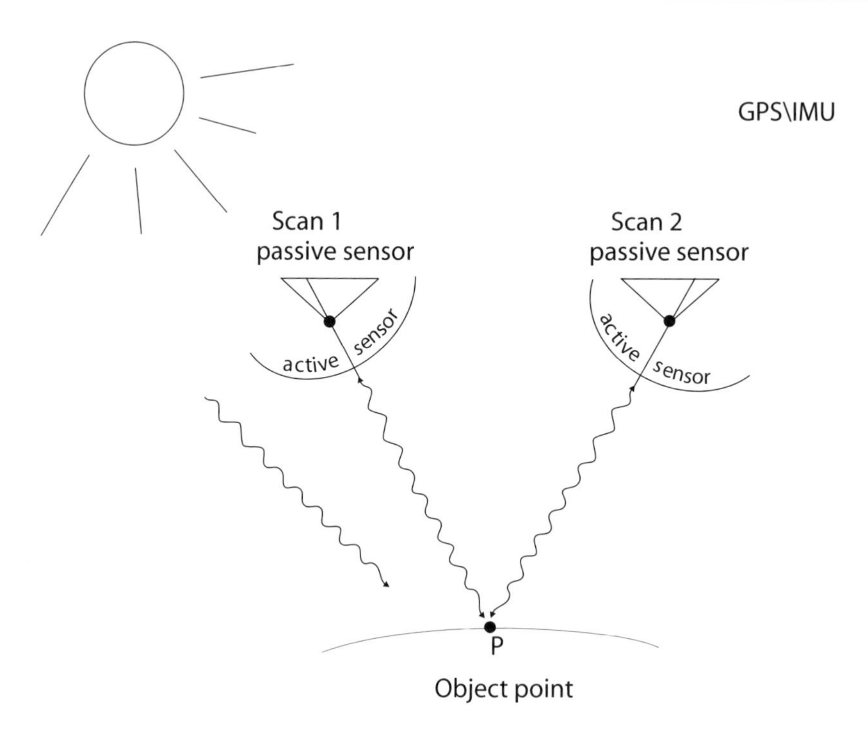

Figure 8.1-16: Common paradigm based on (stereo-) photogrammetry and laser scanning

A possible restriction to single components of the common paradigm is of considerable importance in practice (object recording using only a single measurement beam, etc).

The realization of the common paradigm can be achieved in different ways. Various alternatives, starting with simple solutions, are as follows:

- a digital orthophoto is superimposed on the DTM derived from laser data.
- the laser-based DTM is used for digital single image analysis (monoplotting, Section 7.4.3).
- the laser scanner also records the intensity of the reflected laser beam. The result is a black-and-white infra-red image (Figure 3.2-10), although one which has the poor resolution of laser scanning.
- a digital photographic camera, either a CCD array camera or a 3-line camera, is built into the aircraft, together with the laser scanner, but without synchronization between them.
- laser scanner and digital linear array camera are mounted on the same gyrostabilized platform (Section 3.7.3.3) and are synchronized with one another. With this configuration it is particularly important that the separation between the sensors is kept as small as possible.

Further reading: Baltsavias, E.: ISPRS Journal 54, pp. 138–147, 1999. Jelalian, A.: Laser radar systems. Artech House, Boston, London, 1992. Rees, W.: Physical principles of remote sensing. Cambridge University Press, 2001. Wehr, A., Lohr, U.: ISPRS Journal 54, pp. 68–82, 1999.

8.2 Terrestrial laser scanning

8.2.1 Principle of operation

Terrestrial laser scanning differs from airborne laser scanning principally in that, during the scanning process, the scanner is not moved[7]. The stationary terrestrial laser scanner therefore requires deflection mechanisms for pointing in two directions. (In airborne laser scanners the laser beam must only be deflected in a plane perpendicular to the direction of flight, Section 8.1.1.) Both deflection mechanisms are shown in Figure 8.2-1. The pulsed laser beam is sent from the range finder electronics unit (1) and meets the polygonal mirror element (3) which rotates at relatively high speed. The laser beam (2) is reflected off the mirror surfaces such that it is scanned through the vertical angle ζ. After a ζ profile has been recorded, the upper part of the instrument rotates through a small angle $\Delta\alpha$ in order to sample the neighbouring ζ profile, and so on until a full horizontal circle has been covered.

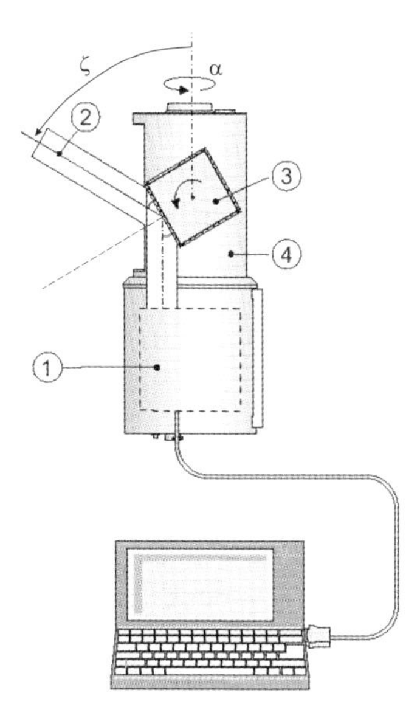

Figure 8.2-1: Principle of the Riegl terrestrial laser scanner

[7]However, mobile mapping (Figure 3.8-12), also uses laser scanners whose operation corresponds, in principle, to the airborne laser scanning concept.

In terrestrial laser scanning, the object is scanned from several measurement stations such that there is only a small overlap between the point clouds generated at each station. GPS can well serve the purpose of locating these stations and this will be discussed in Section 8.2.2. From one station one obtains the polar coordinates α, ζ and s. The data recorded from one station is, in terrestrial laser scanning, normally called a scan.

The technical data for the Riegl LMS-Z420i scanner are summarized in the following table:

Wavelength/Intensity	1550 nm (near infrared), the intensity of the echo can be recorded
Maximum frequency of vertical line scan (ζ profile)	20 Hz
Data rate	24 kHz normal scanning, 8 kHz fine scanning
detectable echos	first or last echo
maximum range	200 m for a surface reflectance $\rho > 80\%$, 60 m for $\rho > 10\%$ (see Section 3.7.5)
minimum range	2 m
FOV, field of view, for a ζ profile	80°
IFOV, instantaneous FOV	0.25 mrad(52″), i.e. the footprint has a diameter of 2.5 cm at a range of 100 m and 1.25 cm at a range of 50 m
Range accuracy	±10 mm
minimum increment in both angular directions	0.07 mrad(14″), i.e. the sampling interval for an object at 100 m range is 7 mm and at 50 m range it is 3.5 mm

Table 8.2-1: Technical data for the Riegl LMS-Z420i terrestrial laser scanner

8.2.2 Georeferencing

This section will first discuss direct georeferencing. This assumes that the laser scanner's position (X_0, Y_0, Z_0), for example with the help of GPS in static mode, and angular orientation $(\omega, \varphi, \kappa)$ have been determined. From these parameters, and the polar coordinates α, ζ, s, the XYZ coordinates in a global coordinate system can be derived:

$$\begin{pmatrix} X \\ Y \\ Z \end{pmatrix} = \begin{pmatrix} X_0 \\ Y_0 \\ Z_0 \end{pmatrix} + \mathbf{R}_{\omega\varphi\kappa} \begin{pmatrix} s \sin\zeta \cos\alpha \\ s \sin\zeta \sin\alpha \\ s \cos\zeta \end{pmatrix} \tag{8.2-1}$$

ζ … zenith angle with zero direction along the κ axis
α … horizontal angle, with zero direction defined by the zero direction for κ rotation

Comment: It is interesting to make a comparison with the similar relationship (8.1-1) for airborne laser scanning. It is particularly noticeable that no synchronization of the transformation parameters is required in terrestrial laser scanning and that the polar coordinates are derived from two deflection angles.

If the location coordinates and orientation angles are unknown (a technique known also as free stationing), then these parameters must be determined indirectly from control points (indirect sensor orientation). In this we initially assume that, in every station scan, a number of control points can be located and identified. With the Riegl LMS Z420i (Table 8.2-1) this is an automated measurement procedure provided that control points have been marked by small retro-reflecting targets. Due to the high contrast between these targets and their environment, the signals can be automatically identified in the intensity image (Table 8.2-1). After normal scanning and preliminary location of the control points, areas around the individual signals are re-scanned at a very high resolution. In the re-scanned intensity images, the $\zeta\alpha$ coordinates of the control points can be determined to sub-pixel accuracy, for example using a weighted centroid calculation (Section 6.8.5 ff.). Using the 4 nearest scanned points, the s coordinate can be determined by means of a bilinear transformation (Section 2.2.3a).

The transformation parameters X_0, Y_0, Z_0 and ω, φ, κ for one measurement station can be determined from at least 3 well distributed control points. A simple solution[8], which corresponds to the absolute orientation of the photogrammetric stereomodel (Section 4.4), starts with the conversion of polar coordinates into a local Cartesian coordinate system as follows:

$$\begin{pmatrix} x \\ y \\ z \end{pmatrix} = \begin{pmatrix} s \sin\zeta \cos\alpha \\ s \sin\zeta \sin\alpha \\ s \cos\zeta \end{pmatrix} \tag{8.2-2}$$

The continuation of this "absolute orientation" is described in detail in Section 4.1.1. The scale factor can be assigned the value of 1.

The economic advantage of indirect sensor orientation can be appreciably increased by simultaneously connecting all scans in a measurement project. For the full set of scans only a few more control points need be determined in the object coordinate system. To connect scans, retro-reflecting targets are again required which can, however, be automatically found and identified in their overlapping areas. Figure 8.2-2 illustrates three scanner stations connected through common measured points. After converting polar coordinates to Cartesian coordinates (Equation (8.2-2)), each scan has its own local Cartesian coordinate system (Figure 8.2-2 shows this only for the middle station). In addition, Figure 8.2-2 shows the global XYZ object coordinate in which a number of retro-reflecting control point targets must be measured.

The indirect sensor orientation for a complete set of scans, each with local Cartesian coordinates derived from the original polar coordinates, corresponds to the spatial block adjustment with independent models described in detail in Section 5.2. The scale factors m in the individual scans can again be assigned the value of unity. Accuracies and

[8] A more rigorous solution, which minimizes the least-squares sum of the original observations α, ζ, s, can be found in Section B 3.5.3, Volume 2.

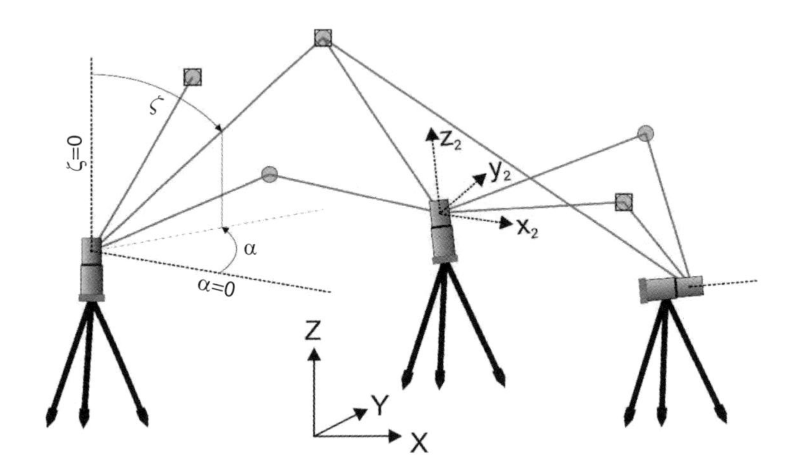

Figure 8.2-2: Connection of multiple laser scanning stations (scans)

distributions of control points can also be found in Section 5.2.3. Since terrestrial laser scanning is often used to record façades on both sides of a street, then the accuracies and control point configurations of strip triangulation are appropriate for such linear measurement tasks (Section 5.2.3.4).

In a practical example[9] with five scans, the following accuracies were achieved: ±3.5 mm for individual coordinates of a measurement station and ±7.5 mm for the individual coordinates of a tie point.

8.2.3 Connecting point clouds

Laser point clouds have no sharply defined points which could be identified in several point clouds and then used to connect them. Retro-reflecting targets, introduced in the previous section for the simultaneous connection of all scans in a measurement project, represent a relatively expensive solution.

In Section 8.2.5 we will learn a technique for locating and identifing tie points manually within the intensity images of the laser scans. To support this, an automatic fine scan can run in the background, comparable with related methods to determine tie points automatically in aerotriangulation (Section 6.8.3.5). Laser scanner images are, however, three-dimensional: the intensity image provides two dimensions, the third dimension is provided by the distances s. In this context, interest operators (Section 6.8.1.3) can also provide good service.

Single points have already been rejected for connecting the point clouds of individual airborne laser-scanned strips. Planes in corresponding regions have been used instead (Section 8.1.2.1).

[9]Pfeifer, N., Kraus, K., Schwarz, R., Ullrich, A.: Ingenieurvermessung 2000 (Eds. Schnädelbach/Schilcher), pp. 114–121, Wittwer Publishing, 2000.

The well-known ICP (iterative closest point) algorithm can connect point clouds without using individual tie points. This assumes approximately positioned and oriented point clouds. In the overlapping region between two point clouds, two subsets of points are chosen. Subsequently, for every point in one subset a corresponding point, at the shortest Euclidean distance in the other subset, is sought. In this way, many corresponding point pairs are generated which can be used to transform one point cloud onto the other, for example using a highly over-determined spatial similarity transform. The point clouds then move "closer" to one another. In a second iteration, corresponding point pairs with shortest separations are again determined and through them a second transformation is carried out. Iterations are continued until the squared sum of point pair separations is below a given threshold value. Due to the relatively large number of incorrect matches which can occur, the transformations, which are normally highly over-determined, should make use of robust parameter estimation. It is conceivable to extend the method from two point clouds to the simultaneous connection of all point clouds, comparable with block adjustment using independent models (Section 5.2).

Further reading on ICP: Besl, P., McKay, N.: IEEE Transactions on Pattern Analysis and Machine Intelligence (PAMI) 14(2), pp. 239–256, 1992. Chen, C., Hung, Y., Cheng, J.: Transactions PAMI 21(11), 1999. Pottmann, H., Leopoldseder, S., Hofer, M.: IAPRS34(3A), pp. 265–270, Graz, 2002.

8.2.4 Strategies for object modelling

The result of measurement by laser scanner is a point cloud. In the following explanation it will be assumed that the primary axis of the instrument (Figure 8.2-1) is approximately aligned to the direction of gravity. For one station or scan, the recorded point cloud is shown projected onto the $\zeta\alpha$ coordinate plane in Figure 8.2-3. The various range values s are the functional values of this "image".

By storing the measurement values as a matrix, neighbouring measurements can be quickly accessed via the matrix subscripts. The measured points have no "meaning", in contrast to a manual photogrammetric analysis or a tacheometer (total station) survey. These last named, conventional methods assign to every measured point an attribute (house corner, linking data to neighbouring point, etc.). From the laser scanner's point cloud this information must be determined after data acquisition. As an aside it is worth noting that laser scanning generates a large amount of data without attributes and, in contrast, conventional methods generate a small amount of data with manually assigned attributes.

Before we investigate further the determination of line and point information from laser scanner point clouds, we will briefly discuss the geometry implicitly contained in Figure 8.2-3. The columns contain object points which result from the intersection of a (vertical) plane with the object surface. The particular (vertical) plane contains the recording location and is defined by horizontal angle α. The line of intersection is a (vertical) profile. The distances from the measurement station to the profile points are known; the separations between profile points are defined by the $\Delta\zeta$ increments.

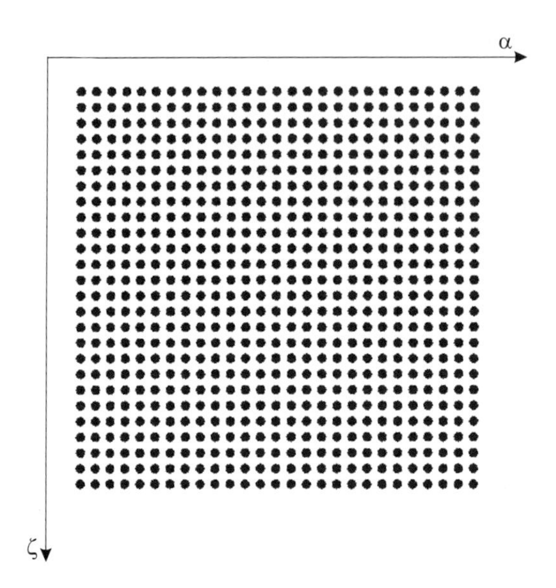

Figure 8.2-3: Laser points viewed in the $\zeta\alpha$ reference plane

A row of the matrix in Figure 8.2-3 contains object points which result from the intersection of a conical surface with the object's surface, where the cone has its apex at the recording location and opening angle 2ζ (Figure 8.2-1). The distances from the measurement station to these intersection points are known; the separations between columns of the matrix in Figure 8.2-3 is defined by the $\Delta\alpha$ increments. If the object surface is a plane then the intersection line between the current cone and the object surface is a conic section, i.e. an ellipse, parabola or hyperbola. In the $\zeta\alpha$ coordinate plane, these conic sections are drawn as straight lines along the corresponding α coordinate line. Figure 8.2-4 illustrates the laser points after transformation into the xyz coordinate system according to Equation (8.2-2). The constant interval between points has disappeared. The simple neighbourhood connections (topology) between laser points is, however, unchanged by this transformation. Transformation into the xyz system has the major advantage that planar regions can be found, for example by detecting common normal vectors to triangles created from neighbouring points. (In contrast, planar regions cannot be found in the $\zeta\alpha s$ coordinate system of Figure 8.2-3; object planes and lines are bent.) In the example here 4 planar regions can be found and their laser points are marked with different symbols in Figure 8.2-4.

After segmentation, planes bordering on one another can be intersected. This results in straight lines and points (Figure 8.2-5). The following strategy for object modelling therefore lies at the heart of laser scanning: sets of points belonging to individual, smooth surface elements (in our example these are planes) are extracted from the point cloud. By intersecting the smooth surfaces, curved edges are obtained (in our example these are straight lines) and the intersection of the edges provides the required individual object points. Conventional methods have a reversed strategy: single points are

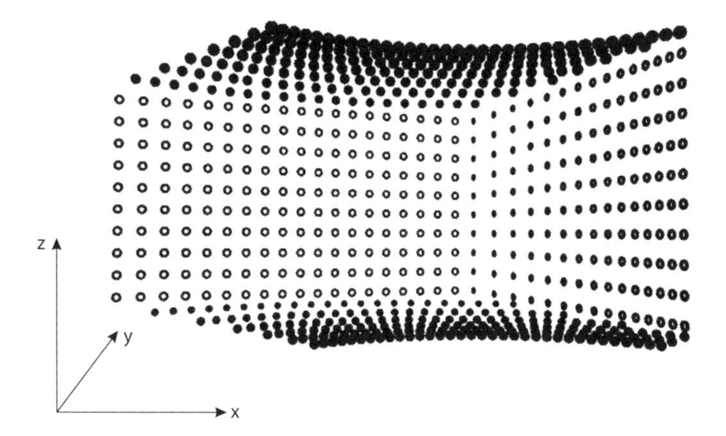

Figure 8.2-4: Laser points in Cartesian coordinate system and the result of segmentation in 4 planar regions.

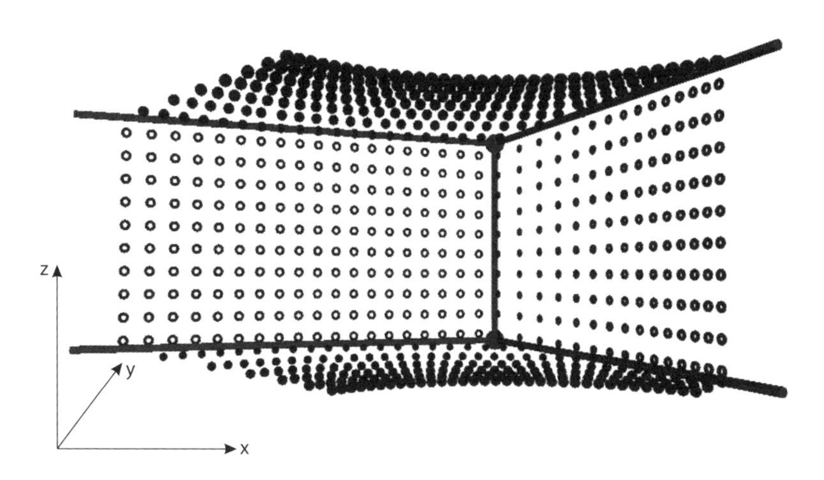

Figure 8.2-5: Intersection of surfaces and edges

measured which create lines defining the object's topology and the lines, together with individual points on smooth surfaces, define the surface elements.

The accuracy of lines and points indirectly derived from laser scans is very high because the highly redundant set of laser points leads to a significant increase in surface measurement accuracy compared with a single laser point measurement (Section 8.1.2.2). In an analogous way to robust least-squares interpolation of surfaces (Section 8.1.2.2), this reduction, or filtering, of random measurement error should be undertaken on smooth object surface elements. For the assignment of laser points to the smooth surface elements it would be advantageous if the random errors were already filtered out. In practice, therefore, segmentation and filtering are alternately used in an object reconstruction procedure.

Figure 8.2-6 shows the result of a measurement with a Riegl LMS-Z360 laser scanner. The statue of Marcus Antonius on a carriage being drawn by three lions, was scanned from several locations. Using a small number of retro-reflecting targets fixed near and around the statue, all scans were connected together and transformed into a common, global coordinate system. The surface was modelled using the software package Geomagic Studio[10]. Data can be smoothed with this software, i.e. measurement noise is filtered out, and NURBS elements formed (Non-Uniform Rational B-Splines). Details are published in: Briese, C., Pfeifer, N., Haring, A.: CIPA conference proceedings, Antalya, Turkey, 2003.

Figure 8.2-6: Results of a laser scanning measurement of the statue of Marcus Antonius

8.2.5 Integration of laser data and photographic data

The combination of laser data and photographic data is just as appropriate to terrestrial laser scanning as it is to airborne laser scanning (Section 8.1.3). For some time terrestrial laser scanners have been able to record the intensity of the reflected laser impulse, as well as its range. Figure 8.2-7 contrasts a laser scan made with a Riegl LMS-Z360 and a digital photographic image of the same scene, taken with a Kodak DCS 460c (Section E 3.5, Volume 2).

The resolution of both images is very different. The terrestrial laser scanner, like the airborne laser scanner (Section 8.1.3ff.), has a worse resolution than the camera by an order of magnitude. The laser scanner image is deceptive: it is a black-and-white

[10] www.geomagic.com/en/products/studio/

Figure 8.2-7: Laser scanner intensity image and photographic image

infra-red image (Figure 3.2-10). In this region of the spectrum there are certainly large contrasts between different types of vegetation but only small contrasts between building walls (Figure 3.7-18). In addition, the laser scanner image has no three-dimensional quality: instead of the oblique illumination by sunlight which gives the photographic image a three-dimensional look, the laser scanner image is generated by an artificial, near infra-red illumination with the same direction as the recording rays. Compared with the photographic image, however, the laser scanner image has the significant advantage that, concealed behind each pixel, is the distance to that particular pixel.

To combine laser scanner and photographic data, the same solutions are possible for terrestrial laser scanners as are given for airborne laser scanners in Section 8.1.3ff. However, the connection of different image types into the same georeferenced network should be briefly discussed. If the different images appear on the same display, as in Figure 8.2-7, then corresponding points in each can be manually (monoscopically) identified and, preferably using feature-based correlation (Section 6.8.1.4) located to sub-pixel resolution. Following this, a common, least-squares adjustment of the terrestrial laser scanning stations and the photographic image stations can be executed.

Three-dimensional photo models (Section 7.5), arising from both data sources, are particularly attractive. The geometry of the object surfaces is mainly derived from the laser data; texture is taken from the digital photographic data and transferred to the surfaces of the geometric model. A very simple type of photo model can be obtained without the need for the full effort normally required for photo models. This variant requires that a digital camera be rigidly mounted on top of the terrestrial laser scanner and its orientation, with respect to the scanner, known by calibration. The following procedure is then possible. Directly after recording, the 3D points from a laser scanning station are transformed (back projected) into the digital photographs where the corresponding colour values are extracted. The colour values are subsequently applied to small spheres and, in 3D space, placed at the positions of the scan points. This results in a 3D "sphere cloud" which gives an observer the familiar colours of a photograph. This 3D "sphere cloud" can be generated immediately after recording and then visualized in a very pleasing way.

8.3 Short range laser scanning

At very close scanning ranges, from a few decimetres up to around 2 m, a completely different scanning principle is adopted. In general terms this principle can be identified as a light-sectioning technique: a light source projects a plane of light onto an object surface, thereby generating a profile which is recorded by a digital camera with a known offset from the light source. In this technique the laser is not used for range measurement; it is simply used to create a plane of light which "draws" the above mentioned profile on the object. The various solutions for implementing the light-sectioning method are described in detail in Section C 2.3.3, Volume 2. In the following text, a very specific and efficient solution will be presented which has matured in recent years to a high level of perfection.

The principle can be very clearly explained on the basis of the horizontal "normal case" of photogrammetry. Figure 8.3-1 illustrates light sectioning with a laser scanner. This figure has close similarities with Figure 3.8-1 showing the "normal case" of terres-

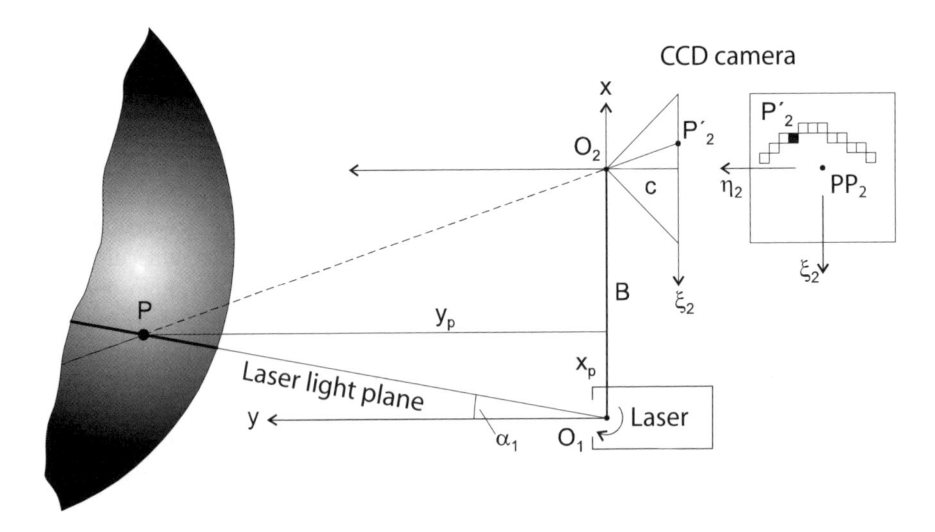

Figure 8.3-1: "Normal case" of light sectioning by a laser scanner

trial photogrammetry. It is therefore also possible to describe the method shown in Figure 8.3-1 as the "normal case" of light sectioning by laser scanning.

The metric camera at recording position O_1 in Figure 8.3-1 is here replaced by a laser scanner which generates a plane of light. This, in turn, creates a profile of light on the object. This plane of light can be rotated by an angle α_1 about the z axis of the model coordinate system (= the scanner's internal z axis). The laser scanner transmits the angle α_1 for the different light profiles to the data processing unit of the measurement system. The digital (CCD) metric camera is positioned at a distance B, which is the base length in the "normal case", along the x model coordinate axis and with its optical axis (viewing direction) parallel to the y model coordinate axis. The CCD camera makes a photographic recording of the light profile, synchronized with the transmissions of the laser light planes (at varying angles α_1). Each profile is separately recorded on a full digital image. The CCD camera must therefore be capable of a high image recording frequency.

At a particular time of recording, the digital image of the profile corresponding to the light plane transmitted at rotation angle α_1 is shown in Figure 8.3-1. The image plane, shown projecting from the negative position in Figure 2.1-6, has been rotated about the ξ_2 axis for display purposes. (Although these are digital images, the definition of image coordinate system used in Figure 3.8-1 has been used instead of the normal definition presented in Figure 2.2-3.)

The coordinates x_P, y_P, z_P, in the model coordinate system, of a point P on the imaged profile, can be determined from its image coordinates ξ_2, η_2, the principal distance c of the camera, base length B and angle α_1 of the laser light plane as follows:

$$\frac{x_P}{y_P} = \tan \alpha_1 \tag{8.3-1}$$

$$\frac{B - x_P}{y_P} = \frac{-\xi_2}{c} \tag{8.3-2}$$

The coordinate y_P is then given by Equation (8.3-3a), obtained after elimination of the x_P coordinate in Equations (8.3-1) and (8.3-2). The coordinate x_P can subsequently be derived by substituting y_P in Equation (8.3-2) to obtain Equation (8.3-3b). The third object coordinate, z_P, is calculated from the ratios in Equation (8.3-3c):

$$y_P = \frac{B}{\tan \alpha_1 - \dfrac{\xi_2}{c}} \tag{8.3-3a}$$

$$x_P = y_P \frac{\xi_2}{c} + B \tag{8.3-3b}$$

$$z_P = y_P \frac{\eta_2}{c} \tag{8.3-3c}$$

The processing of the images can be fully automated. The point sequences in images of individual profiles can be located using simple image processing algorithms. Equations (8.3-3) are then subsequently evaluated for each located point using its image coordinates ξ_2 and η_2. Although the processing strategy is not very susceptible to error it should, however, be noted that there is no check with respect to individual object points. In Equations (8.3-3) there are 3 measurement values ξ_2, η_2 and α_1 for determining the 3 coordinates x_P, y_P, z_P. In contrast, in the "normal case" of photogrammetry there are 4 measurement values ξ_1, η_1, ξ_2, η_2 in Equations (3.8-1) for the determination of the 3 coordinates x, y, z. In photogrammetry there is therefore one redundant measurement present.

The combination of geometric data with photographic data is readily possible with this construction principle, illustrated in Figure 8.3-1. After the CCD camera has taken the sequence of individually imaged profiles, a digital photograph, generally an RGB image, can then be taken. Three coordinates x_P, y_P, z_P can then be assigned to each RGB pixel of the digital image. With little effort, it is equally possible to implement the reverse process in which pixels from the digital image are assigned to the x_P, y_P, z_P coordinates to create coloured spheres (see end of Section 8.2). The best fusion results when an object surface is modelled from the laser point cloud and the complete RGB image draped over this.

A few technical details will be given below for the relatively widely used Minolta VIVID 9i scanner (`www.minolta3d.com`):

Scan range:	0.5 – 2.5 m
Laser wavelength:	690 nm
Number of 3D pixels:	640 × 480
Range accuracy:	±0.008 mm
Colour image array size (RGB):	640 × 480
Data file size for 3D information and colour values:	3.6 MByte
Recording time for 3D data scan:	2.5 s

Table 8.3-1: Technical data for the Minolta VIVID 9i close-range laser scanner

Figure 8.3-2 shows the instrument. As indicated in Figure 8.3-1, the CCD camera is mounted in the upper part of the housing. A laser light plane, which can be deflected about a horizontal axis, is located behind the lower opening.

Figure 8.3-3 shows a three-dimensional object model, which has been recorded by a Minolta VI-900. The CCD image has been transferred onto the object surface, derived from the point cloud. Since the doll's head was recorded with only a single scan, the unmeasured area in shadow at the back of the head can clearly be seen.

Figure 8.3-2: VIVID 9i 3D laser scanner from the Minolta company

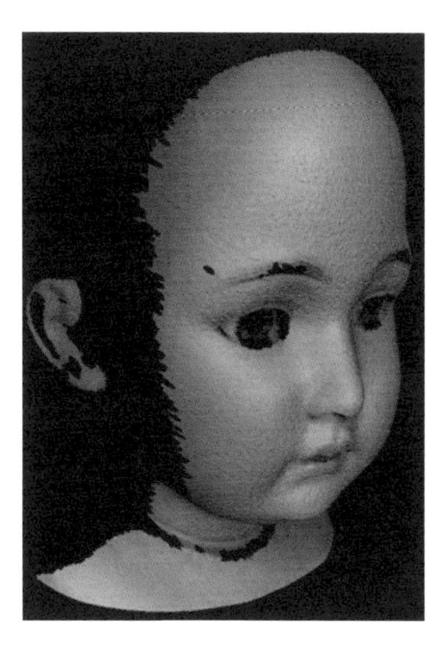

Figure 8.3-3: Doll's head measured by the Minolta VI-900 laser scanner

Appendices

Appendix to Section 2.1

2.1-1 Three-dimensional rotation matrix

For the three independent parameters of a rotation matrix in three dimensions we may use three rotation angles ω, φ and κ about the coordinate axes. Initially the xyz system is coincident with the fixed (higher order, global) coordinate system XYZ. The xyz system is then rotated in three steps. In each case an counterclockwise rotation as seen looking along the axis towards the origin is regarded as positive.

- the primary rotation ω of the xyz system about the X axis.

  Note: the primary rotation takes place about one of the coordinate axes of the fixed coordinate system.

  Question: What equation is used to transform the $x_\omega y_\omega z_\omega$ coordinates of a point into the fixed system?

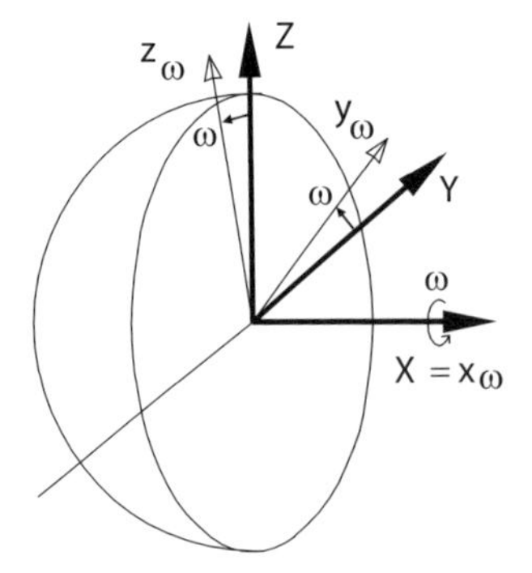

It follows from Equations (2.1-1) and (2.1-2) as well as Figure 2.1-1 that:

$$\begin{aligned} X &= x_\omega \\ Y &= y_\omega \cos\omega - z_\omega \sin\omega \\ Z &= y_\omega \sin\omega + z_\omega \cos\omega \end{aligned}$$

or

$$\mathbf{X} = \begin{pmatrix} 1 & 0 & 0 \\ 0 & \cos\omega & -\sin\omega \\ 0 & \sin\omega & \cos\omega \end{pmatrix} \begin{pmatrix} x_\omega \\ y_\omega \\ z_\omega \end{pmatrix} = \mathbf{R}_\omega \mathbf{x}_\omega \qquad (2.1\text{-}1\text{-}1)$$

- secondary rotation φ of the xyz system about the y_ω axis.

  Note: the secondary rotation takes place about the y_ω axis which has been previously rotated about the primary axis.

  Question: What equation is used to transform the $x_{\omega\varphi} y_{\omega\varphi} z_{\omega\varphi}$ coordinates of a point through the $x_\omega y_\omega z_\omega$ system into the fixed system?

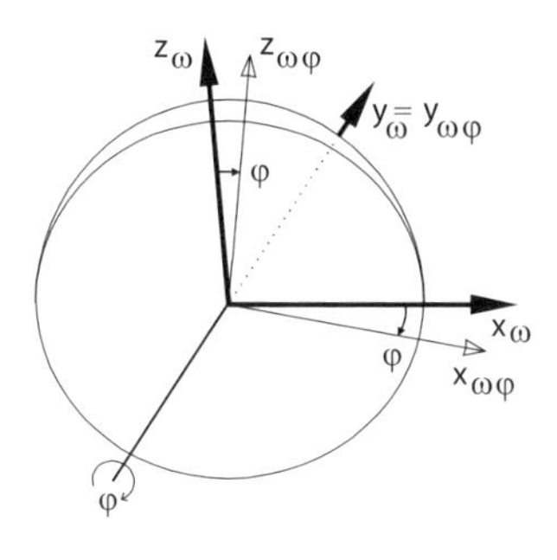

Note that in the application of Equations (2.1-1) and (2.1-2) as well as in Figure 2.1-1 a positive φ rotation in the $x_\omega z_\omega$ plane is a rotation in the direction from the z_ω axis towards the x_ω axis. Positive rotations are to be taken in cyclical order from the x axis to the y axis, from the y axis to the z and from the z axis to the x axis. This conforms with the definition that an counterclockwise rotation as seen looking along the axis towards the origin is positive.

$$
\begin{aligned}
x_\omega &= x_{\omega\varphi}\cos\varphi + z_{\omega\varphi}\sin\varphi \\
y_\omega &= y_{\omega\varphi} \\
z_\omega &= -x_{\omega\varphi}\sin\varphi + z_{\omega\varphi}\cos\varphi
\end{aligned}
$$

or

$$
\mathbf{x}_\omega = \begin{pmatrix} \cos\varphi & 0 & \sin\varphi \\ 0 & 1 & 0 \\ -\sin\varphi & 0 & \cos\varphi \end{pmatrix} \begin{pmatrix} x_{\omega\varphi} \\ y_{\omega\varphi} \\ z_{\omega\varphi} \end{pmatrix} = \mathbf{R}_\varphi \mathbf{x}_{\omega\varphi} \qquad (2.1\text{-}1\text{-}2)
$$

Substituting from (2.1-1-1) in (2.1-1-2) gives:

$$
\mathbf{X} = \mathbf{R}_\omega \mathbf{R}_\varphi \mathbf{x}_{\omega\varphi} \qquad (2.1\text{-}1\text{-}3)
$$

- tertiary rotation κ of the xyz system about the $z_{\omega\varphi}$ axis.

  Note: the tertiary rotation takes place about that axis which has previously been rotated about both the primary and the secondary axes.

  Question: What equation is used to transform the $x_{\omega\varphi\kappa}y_{\omega\varphi\kappa}z_{\omega\varphi\kappa}$ coordinates (writing $x_{\omega\varphi\kappa} = x, y_{\omega\varphi\kappa} = y, z_{\omega\varphi\kappa} = z$) of a point through the $x_{\omega\varphi}y_{\omega\varphi}z_{\omega\varphi}$ and through the $x_\omega y_\omega z_\omega$ system into the fixed system?

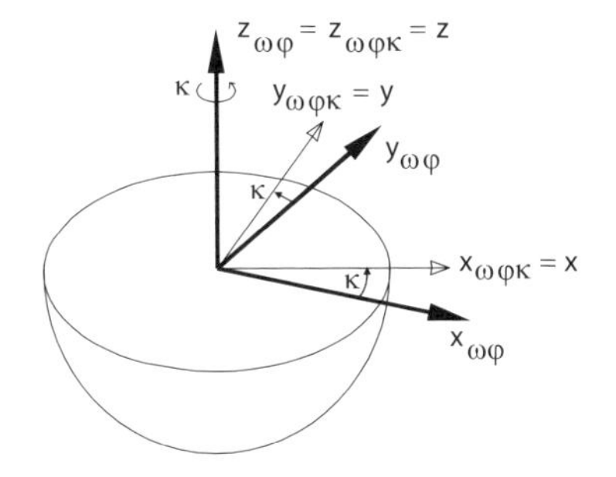

Using Equations (2.1-1) and (2.1-2) and Figure 2.1-1 it follows that:

$$
\begin{aligned}
x_{\omega\varphi} &= x_{\omega\varphi\kappa}\cos\kappa - y_{\omega\varphi\kappa}\sin\kappa \\
y_{\omega\varphi} &= x_{\omega\varphi\kappa}\sin\kappa + y_{\omega\varphi\kappa}\cos\kappa \\
z_{\omega\varphi} &= z_{\omega\varphi\kappa}
\end{aligned}
$$

or

$$
\mathbf{x}_{\omega\varphi} = \begin{pmatrix} \cos\kappa & -\sin\kappa & 0 \\ \sin\kappa & \cos\kappa & 0 \\ 0 & 0 & 1 \end{pmatrix} \begin{pmatrix} x_{\omega\varphi\kappa} \\ y_{\omega\varphi\kappa} \\ z_{\omega\varphi\kappa} \end{pmatrix} = \mathbf{R}_x \mathbf{x} \qquad (2.1\text{-}1\text{-}4)
$$

Substituting from (2.1-1-4) in (2.1-1-3) gives:

$$\mathbf{X} = \mathbf{R}_\omega \mathbf{R}_\varphi \mathbf{R}_\kappa \mathbf{x} \tag{2.1-1-5}$$

Multiplication of the three matrices $\mathbf{R}_\omega$, $\mathbf{R}_\varphi$ and $\mathbf{R}_\kappa$ to derive the single matrix $\mathbf{R}$ is performed in two steps. First $\mathbf{R}_{\varphi\kappa} = \mathbf{R}_\varphi \mathbf{R}_\kappa$ is found and finally $\mathbf{R}_{\omega\varphi\kappa} = \mathbf{R}_\omega \mathbf{R}_{\varphi\kappa}$:

$$\begin{pmatrix} c_\kappa & -s_\kappa & 0 \\ s_\kappa & c_\kappa & 0 \\ 0 & 0 & 1 \end{pmatrix}$$

$$\begin{pmatrix} c_\varphi & 0 & s_\varphi \\ 0 & 1 & 0 \\ -s_\varphi & 0 & c_\varphi \end{pmatrix} \quad \begin{pmatrix} c_\varphi c_\kappa & -c_\varphi s_\kappa & s_\varphi \\ s_\kappa & c_\kappa & 0 \\ -s_\varphi c_\kappa & s_\varphi s_\kappa & c_\varphi \end{pmatrix} = \mathbf{R}_{\varphi\kappa}$$

$$\begin{pmatrix} 1 & 0 & 0 \\ 0 & c_\omega & -s_\omega \\ 0 & s_\omega & c_\omega \end{pmatrix} \begin{pmatrix} c_\varphi c_\kappa & -c_\varphi s_\kappa & s_\varphi \\ c_\omega s_\kappa + s_\omega s_\varphi c_\kappa & c_\omega c_\kappa - s_\omega s_\varphi s_\kappa & -s_\omega c_\varphi \\ s_\omega s_\kappa - c_\omega s_\varphi c_\kappa & s_\omega c_\kappa + c_\omega s_\varphi s_\kappa & c_\omega c_\varphi \end{pmatrix} = \mathbf{R}_{\omega\varphi\kappa} \tag{2.1-1-6}$$

If the order of rotations is changed the sequence of multiplications of the matrices must be changed accordingly. For example, if φ is taken as the primary rotation, ω as the secondary rotation and κ as the tertiary rotation, we derive the following matrix:

$$\mathbf{R}_{\varphi\omega\kappa} = \begin{pmatrix} c_\varphi c_\kappa + s_\varphi s_\omega s_\kappa & -c_\varphi s_\kappa + s_\varphi s_\omega c_\kappa & s_\varphi c_\omega \\ c_\omega s_\kappa & c_\omega c_\kappa & -s_\omega \\ -s_\varphi c_\kappa + c_\varphi s_\omega s_\kappa & s_\varphi s_\kappa + c_\varphi s_\omega c_\kappa & c_\omega c_\varphi \end{pmatrix} \tag{2.1-1-7}$$

Exercise 2.1-14. Derive the matrix representing the resultant rotation in three dimensions if κ, φ and ω are the primary, secondary and tertiary rotations respectively, all in the positive sense.

We sometimes wish to derive the rotation angles from the elements r_{ik} of the rotation matrix. From the matrix $\mathbf{R}_{\omega\varphi\kappa}$ we wish to find the three angles ω, φ and κ as specified in Figure 2.1-5. It follows from Equation (2.1-1-6) that:

$$\tan\omega = \frac{-r_{23}}{r_{33}}, \quad \sin\varphi = r_{13}, \quad \tan\kappa = \frac{-r_{12}}{r_{11}} \tag{2.1-1-8}$$

Note: if $r_{13} > 0$ then φ could lie in either the first or the second quadrant while, if $r_{13} < 0$, φ could take a value in the third or fourth quadrants. There is, however, no ambiguity for the angles ω and κ as may be seen when one considers the following relationships from the matrix (2.1-1-6):[1]

$$\begin{aligned} \omega : r_{23} &= -\sin\omega\cos\varphi & r_{33} &= \cos\omega\cos\varphi \\ \kappa : r_{12} &= -\cos\varphi\sin\kappa & r_{11} &= \cos\varphi\cos\kappa \end{aligned}$$

[1]The unambiguous determination of the angles ω and κ is performed in many computers with the `ARCTAN2` function: `omega = ARCTAN2` $(-r_{23}, r_{33})$ = `[ ARCTAN2 (opposite, adjacent)]` The angle is given in radians with a value between $-\pi$ and π (excluding $-\pi$), counting from the x axis in a counterclockwise sense (positive count) or in a clockwise sense (negative count). Example: The function `ARCTAN2` $(-1, -1)$ produces (-2.35619); that is -150 gon or $-135°$.

Numerical Example. From the following rotation matrix which relates to rotations about the axes in the order specified in Figure 2.1-5

$$\begin{pmatrix} -0.340110 & 0.999407 & 0.004822 \\ -0.999419 & -0.034096 & 0.000621 \\ 0.000784 & -0.004798 & 0.999988 \end{pmatrix}$$

find the two sets of rotation angles ω, φ and κ:

$$\sin\varphi_1 = 0.004822 \Rightarrow \varphi_1 = 0.3070\,\text{gon}\,(16'35'') \quad \varphi_2 = 199.6930\,\text{gon}\,(179°43'25'')$$

$$\begin{aligned} 0.000621/\cos\varphi_1 &= -\sin\omega_1 = 0.0006210 & &\Rightarrow \omega_1 = 399.9605\,\text{gon} \\ 0.999988/\cos\varphi_1 &= \cos\omega_1 = 0.9999996 & &(359°57'52'') \end{aligned}$$

$$\begin{aligned} 0.000621/\cos\varphi_2 &= -\sin\omega_2 = -0.0006210 & &\Rightarrow \omega_2 = 199.9605\,\text{gon} \\ 0.999988/\cos\varphi_2 &= \cos\omega_2 = -0.9999996 & &(179°57'52'') \end{aligned}$$

$$\begin{aligned} 0.999407/\cos\varphi_1 &= -\sin\kappa_1 = 0.9994186 & &\Rightarrow \kappa_1 = 297.8292\,\text{gon} \\ -0.034091/\cos\varphi_1 &= \cos\kappa_1 = -0.0340914 & &(267°18'00'') \end{aligned}$$

$$\begin{aligned} 0.999407/\cos\varphi_2 &= -\sin\kappa_2 = -0.9994186 & &\Rightarrow \kappa_2 = 97.8292\,\text{gon} \\ -0.034091/\cos\varphi_2 &= \cos\kappa_2 = 0.0340914 & &(87°18'00'') \end{aligned}$$

Exercise 2.1-15. Repeat the above numerical example under the assumption that φ is the primary, ω the secondary and κ the tertiary rotation. Hint: The corresponding rotation matrix is shown above as Equation (2.1-1-7). (Solution: $\varphi = 0.3070\,\text{gon} = 16'35''$ OR $\varphi = 200.3070\,\text{gon}\,(180°16'35'')$; $\omega = 399.9605\,\text{gon}\,(359°57'52'')$ OR $\omega = 200.0395\,\text{gon}\,(180°57'52'')$; $\kappa = 297.8290\,\text{gon}\,(268°02'46'')$ OR $\kappa = 97.8290\,\text{gon}\,(88°02'46'')$.)

Exercise 2.1-16. The following rotations take place in the order $\omega\varphi\kappa$: $\omega = 2\,\text{gon}\,(1°48')$, $\varphi = 10\,\text{gon}\,(9°)$, $\kappa = 50\,\text{gon}\,(45°)$. In order to achieve the same resultant rotation by means of three sequential rotations in the order $\varphi\omega\kappa$, what would the individual rotations be? (Solution: $\omega = 1.9754\,\text{gon}\,(1°46'40'')$ OR $198.0246\,\text{gon}\,(178°13'20'')$, $\varphi = 10.0049\,\text{gon}\,(9°00'16'')$ OR $210.0049\,\text{gon}\,(189°00'16'')$, $\kappa = 50.3130\,\text{gon}\,(45°16'54'')$ OR $250.3130\,\text{gon}\,(225°16'54'')$.)

The above examples show that numerical problems can occur in the derivation of the rotation angles from the elements r_{ik} of the rotation matrix. These numerical problems occur when $\cos\varphi = 0$, that is when $\varphi = 100\,\text{gon} = 90°$ or when $\varphi = 300\,\text{gon} = 270°$. Fortunately such angles do not arise in aerial photogrammetry, although they are quite possible in close range applications. Similarly, numerical problems arise in the case of the $\mathbf{R}_{\varphi\omega\kappa}$ rotation matrix (Equation (2.1-1-7)) when $\cos\omega = 0$, that is when $\omega = 100\,\text{gon} = 90°$ or when $\omega = 300\,\text{gon} = 270°$.

The following are some sources of information concerning rotation matrices: ASPRS Manual of Photogrammetry, 5th ed., 2004. Shih, T.-Y.: PE&RS 56, pp. 1173–1179, 1990. Sections B 3.4.1 and B 3.4.2 in Volume 2.

2.1-2 Mathematical relationship between image and object coordinates (collinearity condition)

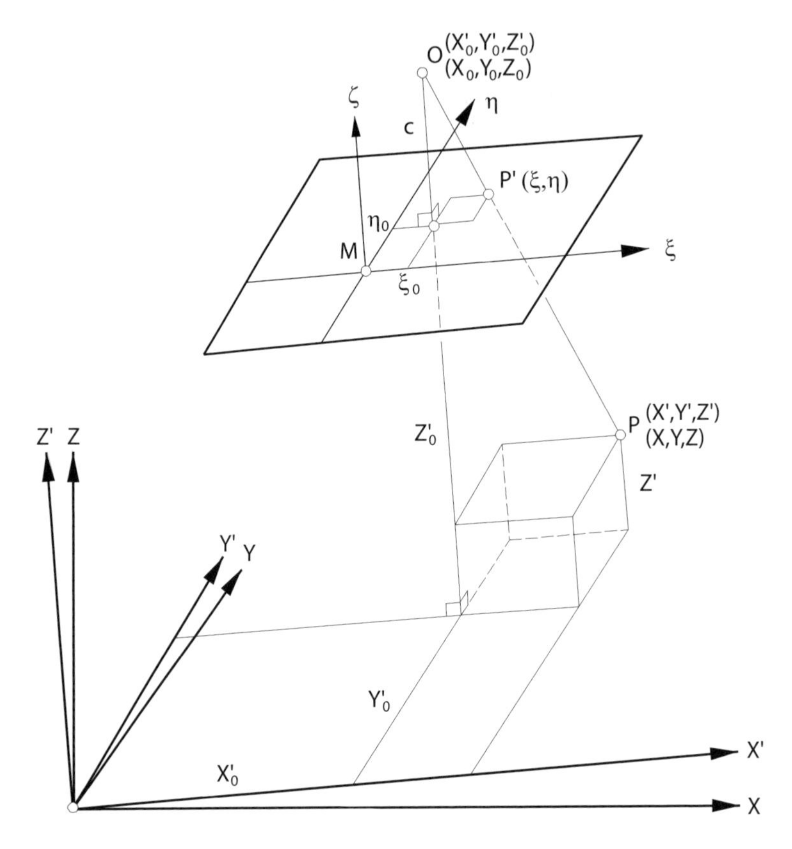

Figure 2.1-2-1: Collinearity of the camera station O, the image point P' and the object point P

When the photograph is taken, the object point P, the image point P' and the camera station O lie on a straight line. The above illustrated collinearity condition, as it is called in photogrammetry, can be expressed in the $X'Y'Z'$ coordinate system in Equations (2.1-2-1)[2]; the $X'Y'Z'$ coordinate system is parallel to the $\xi\eta\zeta$ image coordinate system, both equally tilted with respect to the XYZ object system. In the image coordinate system $\zeta = 0$ for all image points and $\zeta = c$ for the perspective centre.

$$\begin{aligned} \frac{\xi - \xi_0}{c} &= \frac{X' - X'_0}{Z'_0 - Z'} \\ \frac{\eta - \eta_0}{c} &= \frac{Y' - Y'_0}{Z'_0 - Z'} \end{aligned} \qquad (2.1\text{-}2\text{-}1)$$

[2]Zavoti (Geodetic and Geophysical Research Institute of HAS (GGKI), Sopron, Hungary) provided the idea for this derivation.

$$\xi = \xi_0 - c\frac{X' - X'_0}{Z' - Z'_0}$$
$$\eta = \eta_0 - c\frac{Y' - Y'_0}{Z' - Z'_0} \quad (2.1\text{-}2\text{-}2)$$

If the rotation matrix for the image, and therefore also for the $X'Y'Z'$ system, is $\mathbf{R}$ with respect to the XYZ object coordinate system, it follows from the collinearity of O, P' and P and from Equation (2.1-11) that:

$$\begin{pmatrix} X - X_0 \\ Y - Y_0 \\ Z - Z_0 \end{pmatrix} = \begin{pmatrix} r_{11} & r_{12} & r_{13} \\ r_{21} & r_{22} & r_{23} \\ r_{31} & r_{32} & r_{33} \end{pmatrix} \begin{pmatrix} X' - X'_0 \\ Y' - Y'_0 \\ Z' - Z'_0 \end{pmatrix} \quad (2.1\text{-}2\text{-}3)$$

Premultiplying Equation (2.1-2-3) by $\mathbf{R}^{-1} = \mathbf{R}^{\top}$ and substituting in Equation (2.1-2-2) from Equation (2.1-2-3) one obtains:

$$\begin{aligned} \xi &= \xi_0 - c\frac{r_{11}(X - X_0) + r_{21}(Y - Y_0) + r_{31}(Z - Z_0)}{r_{13}(X - X_0) + r_{23}(Y - Y_0) + r_{33}(Z - Z_0)} \\ &= \xi_0 - c\frac{Z_x}{N} \\ \eta &= \eta_0 - c\frac{r_{12}(X - X_0) + r_{22}(Y - Y_0) + r_{32}(Z - Z_0)}{r_{13}(X - X_0) + r_{23}(Y - Y_0) + r_{33}(Z - Z_0)} \\ &= \eta_0 - c\frac{Z_y}{N} \end{aligned} \quad (2.1\text{-}2\text{-}4)$$

Since the $X'Y'Z'$ system is parallel to the $\xi\eta\zeta$ image system, the elements r_{ik} are

- the cosines of the angles between the axes of the image and the mapping coordinate systems (see Equation (2.1-10))
- functions of the angles ω, φ and κ through which the camera was tilted with respect to the object system at the moment of exposure (see Equation (2.1-13) or Appendix 2.1-1)

For the sake of compact notation in the partial differential coefficients given in Appendix 2.1-3, in Equations (2.1-2-4), above, we write Z_x and Z_y for the numerators and N for the denominator.

The collinearity equations are sometimes required in another form, a similarity transformation which can be taken directly from Figure 2.1-2-1 (for details see Section B 3.5.1 in Volume 2):

$$\mathbf{X} - \mathbf{X}_0 = \begin{pmatrix} X - X_0 \\ Y - Y_0 \\ Z - Z_0 \end{pmatrix} = m\mathbf{R} \begin{pmatrix} \xi - \xi_0 \\ \eta - \eta_0 \\ -c \end{pmatrix} = m\mathbf{R}(\boldsymbol{\xi} - \boldsymbol{\xi}_0) \quad (2.1\text{-}2\text{-}5)$$

m ... a scale factor, the ratio of the length OP in object space to the length OP' in image space

2.1-3 Differential coefficients of the collinearity equations

For a solution of the non-linear equations (2.1-2-4), using Newton's (iterative) Method, almost invariably together with the method of least squares, one requires the following partial differential coefficients in which the dependence of the elements r_{ik} on the angles ω, φ and κ (shown in Equation (2.1-13)) is introduced:

$$\frac{\partial\xi}{\partial c} = -\frac{Z_x}{N} = a_1 \qquad \frac{\partial\eta}{\partial c} = -\frac{Z_y}{N} = b_1$$

$$\frac{\partial\xi}{\partial X_0} = -\frac{c}{N^2}(r_{13}Z_x - r_{11}N) = a_2 \qquad \frac{\partial\eta}{\partial X_0} = -\frac{c}{N^2}(r_{13}Z_y - r_{12}N) = b_2$$

$$\frac{\partial\xi}{\partial Y_0} = -\frac{c}{N^2}(r_{23}Z_x - r_{21}N) = a_3 \qquad \frac{\partial\eta}{\partial Y_0} = -\frac{c}{N^2}(r_{23}Z_y - r_{22}N) = b_3$$

$$\frac{\partial\xi}{\partial Z_0} = -\frac{c}{N^2}(r_{33}Z_x - r_{31}N) = a_4 \qquad \frac{\partial\eta}{\partial Z_0} = -\frac{c}{N^2}(r_{33}Z_y - r_{32}N) = b_4$$

$$\begin{aligned}\frac{\partial\xi}{\partial\omega} &= -\frac{c}{N}\Big(((Y-Y_0)r_{33} - (Z-Z_0)r_{23})\frac{Z_x}{N} \\ &\quad - (Y-Y_0)r_{31} + (Z-Z_0)r_{21}\Big) = a_5\end{aligned}$$

$$\begin{aligned}\frac{\partial\eta}{\partial\omega} &= -\frac{c}{N}\Big(((Y-Y_0)r_{33} - (Z-Z_0)r_{23})\frac{Z_y}{N} \\ &\quad - (Y-Y_0)r_{32} + (Z-Z_0)r_{22}\Big) = b_5\end{aligned}$$

$$\frac{\partial\xi}{\partial\varphi} = \frac{c}{N}\left((Z_x\cos\kappa - Z_y\sin\kappa)\frac{Z_x}{N} + N\cos\kappa\right) = a_6$$

$$\frac{\partial\eta}{\partial\varphi} = \frac{c}{N}\left((Z_x\cos\kappa - Z_y\sin\kappa)\frac{Z_y}{N} - N\sin\kappa\right) = b_6$$

$$\frac{\partial\xi}{\partial\kappa} = -\frac{c}{N}Z_y = a_7 \qquad \frac{\partial\eta}{\partial\kappa} = \frac{c}{N}Z_x = b_7$$

$$\frac{\partial\xi}{\partial X} = -\frac{c}{N^2}(Nr_{11} - Z_x r_{13}) = a_8 \qquad \frac{\partial\eta}{\partial X} = -\frac{c}{N^2}(Nr_{12} - Z_y r_{13}) = b_8$$

$$\frac{\partial\xi}{\partial Y} = -\frac{c}{N^2}(Nr_{21} - Z_x r_{23}) = a_9 \qquad \frac{\partial\eta}{\partial Y} = -\frac{c}{N^2}(Nr_{22} - Z_y r_{23}) = b_9$$

$$\frac{\partial\xi}{\partial Z} = -\frac{c}{N^2}(Nr_{31} - Z_x r_{33}) = a_{10} \qquad \frac{\partial\eta}{\partial Z} = -\frac{c}{N^2}(Nr_{32} - Z_y r_{33}) = b_{10}$$

Exercise 2.1-17. Consider in what way the derivations of the differential coefficients will change if the primary and secondary rotations are interchanged. The primary rotation will therefore become φ and the secondary rotation ω. (Solution: They change as follows: $(Y - Y_0) \Rightarrow -(X - X_0)$, $r_{23} \Rightarrow -r_{13}$, $r_{21} \Rightarrow -r_{11}$, $\sin\kappa \Rightarrow -\cos\kappa$, $\cos\kappa \Rightarrow \sin\kappa$.)

Exercise 2.1-18. Give the differential coefficients for the case of an exactly vertical, unrotated, image. (Solution (examples of the changes): $\partial\xi/\partial X = \partial\eta/\partial Y = -c/(Z - Z_0)$, $\partial\xi/\partial Y = \partial\eta/\partial X = 0$, $\partial\xi/\partial Z = c(X - X_0)/(Z - Z_0)^2$, $\partial\eta/\partial Z = c(Y - Y_0)/(Z - Z_0)^2$.)

Appendix to Section 2.2

2.2-1 Derivation of Formula (2.2-5) using homogeneous coordinates

The projective relationship (2.1-24) between the image plane and an object plane can be given as follows using homogeneous coordinates:

$$\begin{pmatrix} \overline{X} \\ \overline{Y} \\ \overline{W} \end{pmatrix} = \begin{pmatrix} a_1 & a_2 & a_3 \\ b_1 & b_2 & b_3 \\ c_1 & c_2 & c_3 \end{pmatrix} \begin{pmatrix} \xi \\ \eta \\ 1 \end{pmatrix} \qquad \overline{\mathbf{X}} = \mathbf{A}\boldsymbol{\xi} \tag{2.2-1-1}$$

One derives the original coordinates from the homogeneous coordinates as follows:

$$X = \overline{X}/\overline{W} \qquad Y = \overline{Y}/\overline{W} \tag{2.2-1-2}$$

The solution for the image coordinates from equations (2.2-1-1) is:

$$\begin{pmatrix} \bar{\xi} \\ \bar{\eta} \\ \bar{\zeta} \end{pmatrix} = \mathbf{A}^{-1} \begin{pmatrix} X \\ Y \\ 1 \end{pmatrix} \tag{2.2-1-3}$$

Hence

$$\xi = \frac{\bar{\xi}}{\bar{\zeta}} \qquad \eta = \frac{\bar{\eta}}{\bar{\zeta}} \tag{2.2-1-4}$$

The inverse of matrix $\mathbf{A}$ is:

$$\mathbf{A}^{-1} = \frac{1}{\det \mathbf{A}} \begin{pmatrix} b_2c_3 - b_3c_2 & a_3c_2 - a_2c_3 & a_2b_3 - a_3b_2 \\ b_3c_1 - b_1c_3 & a_1c_3 - a_3c_1 & a_3b_1 - a_1b_3 \\ b_1c_2 - b_2c_1 & a_2c_1 - a_1c_2 & a_1b_2 - a_2b_1 \end{pmatrix} \tag{2.2-1-5}$$

Note: since (X, Y) and (ξ, η) in (2.2-1-2) and (2.2-1-4), respectively, are derived by fractions which cancel out the scale information in $\mathbf{A}$, scale must be fixed in some way. By setting $c_3 = 1$ and taking Equations (2.2-1-3), (2.2-1-4) and (2.2-1-5) together one arrives directly at the relationship (2.2-5).

Note, however, that fixing the scale by setting $c_3 = 1$ is only appropriate as long as $c_3 \neq 0$. This will happen in the case where the line at infinity of the object plane is mapped to an image line passing through the origin of the image coordinate system.

Appendix to Section 4.1

4.1-1 Estimation by the method of least squares

We are given a set of linear equations of which the first is:

$$l_1 = a_{11}x_1 + a_{12}x_2 + \cdots + a_{1u}x_u \tag{4.1-1-1}$$

a_{ik} ... known coefficients
l_i ... known absolute terms described below as "observations"
x_k ... the unknowns, u in number, of the system of equations

If there are exactly u observations, l_i, then we have a consistent system of linear equations which may be solved directly for the u unknowns x_k. If the set of equations is expressed in matrix form as $\mathbf{l} = \mathbf{A}\mathbf{x}$ then the solution is $\mathbf{x} = \mathbf{A}^{-1}\mathbf{l}$. In order both to provide a check on errors in the observations and to increase the accuracy of the results, it is advisable that we take n observations where $n > u$. We then have a set of inconsistent equations which implies an estimation problem; this can be solved, for example, by introducing the condition that the sum of the squares of the residuals v of the observations should be a minimum. We may then write a system of observation equations:

$$\mathbf{v} = \mathbf{A}\hat{\mathbf{x}} - \mathbf{l} \tag{4.1-1-2}$$

The u unknowns $\hat{x}_k$ are then found by applying the following minimum condition:

$$\mathbf{v}^\top\mathbf{v} = \min. \overset{(4.1\text{-}1\text{-}2)}{\Rightarrow} (\mathbf{A}\hat{\mathbf{x}} - \mathbf{l})^\top(\mathbf{A}\hat{\mathbf{x}} - \mathbf{l}) = \hat{\mathbf{x}}^\top\mathbf{A}^\top\mathbf{A}\hat{\mathbf{x}} - 2\mathbf{l}^\top\mathbf{A}\hat{\mathbf{x}} + \mathbf{l}^\top\mathbf{l}$$

Consequently the solution is[3]:

$$\begin{aligned} \frac{\partial(\mathbf{v}^\top\mathbf{v})}{\partial\hat{\mathbf{x}}} &= 2\hat{\mathbf{x}}^\top\mathbf{A}^\top\mathbf{A} - 2\mathbf{l}^\top\mathbf{A} = 0 \\ \hat{\mathbf{x}} &= (\mathbf{A}^\top\mathbf{A})^{-1}\mathbf{A}^\top\mathbf{l} \end{aligned} \tag{4.1-1-3}$$

The matrix $(\mathbf{A}^\top\mathbf{A})$ is known as the matrix of the normal equations; the matrix $\mathbf{A}$ is the matrix of the observation equations or, in modern terminology, the design matrix.

The solution shown above is described in the literature as "adjustment by indirect observations". This algorithm is all that is needed for almost all of the estimation problems[4] dealt with in this book. Consequently, other approaches to adjustment are not discussed. Just two extensions to the method of "adjustment by indirect observations" are required, one to deal with the case when the initial equations (4.1-1-1) are non-linear and the other when the observations are made with differing accuracies.

[3]The observations $\mathbf{l}$ are stochastic variables (they therefore contain random errors) the calculated unknowns are simply estimates. The customary notation for an estimated result, with the "hat", $\hat{\mathbf{x}}$, is not uniformly used in this book; it is frequently written just as $\mathbf{x}$.

[4]It is also common in surveying and photogrammetry to refer to "adjustment" when errors are iteratively "adjusted" to some minimum condition in order to achieve an optimal estimation of the unknowns.

Let the first equation of a system of non-linear equations be expressed as:

$$l_1 = f(x_1, x_2, \ldots, x_u) \tag{4.1-1-4}$$

A set of non-linear equations is commonly solved by Newton's method in which improved approximations to the unknowns are found, in an iterative process, by deriving a set of linear equations as follows. This lends itself to the method of adjustment by indirect observations which deals only with linear equations. The right hand side of Equation (4.1-1-4) is expanded in a Taylor's series around initial estimates $x_1^0, x_2^0, \ldots, x_u^0$; we limit the expansion to linear terms:

$$l_1 = f(x_1^0, x_2^0, \ldots, x_u^0) + \left(\frac{\partial f}{\partial x_1}\right)^0 dx_1 + \left(\frac{\partial f}{\partial x_2}\right)^0 dx_2 + \cdots + \left(\frac{\partial f}{\partial x_u}\right)^0 dx_u \tag{4.1-1-5}$$

A comparison with Equation (4.1-1-1) leads to the following parallels:

$a_{ik} \triangleq \left(\dfrac{\partial f}{\partial x_k}\right)$ in which these coefficients are formed by partial differentiation and then evaluated numerically using the current estimates of the unknowns, $x_1^0, x_2^0, \ldots, x_u^0$. The result is simply a number which corresponds to the coefficient a_{ik} of the linear case. (Why?)

$l_i \triangleq \bar{l}_i - f(x_1^0, x_2^0, \ldots, x_u^0) = \bar{l}_i - l_1^0$ in which the value of the function $f(x_1^0, x_2^0, \ldots, x_u^0)$, computed using the current estimates of the unknowns, is denoted by l_1^0; the actual observations are denoted by $\bar{l}_i$.

$x_k \triangleq dx_k$ in which dx_k is a small correction to the current estimate x_k^0 of the unknown x_k.

Given these correspondences, it may be seen that Equation (4.1-1-5) is of the same form as the linear Equation (4.1-1-1) and may be solved in the same manner by forming the normal Equations (4.1-1-3).

In the second of the extensions to the method of "adjustment by indirect observations" we consider the case in which we have differing accuracies in the observations l_i. The accuracy is expressed in terms of the standard deviation σ_i of an observation or by its variance σ_i^2. In the surveying literature the term "root mean square error" is often preferred to standard deviation. These different accuracies are taken into account in an adjustment by indirect observations by means of weights, as they are called, p_i, which are summarized in a weight matrix $\mathbf{P}_{ll}$.

$$\mathbf{P}_{ll} = \begin{pmatrix} 1/\sigma_1^2 & & & \\ & 1/\sigma_2^2 & & \\ & & \ddots & \\ & & & 1/\sigma_n^2 \end{pmatrix} \tag{4.1-1-6}$$

The minimum condition then becomes: $\mathbf{v}^\top \mathbf{P}_{ll}\mathbf{v} = \min.$, which leads to the solution:

$$\hat{\mathbf{x}} = (\mathbf{A}^\top \mathbf{P}_{ll}\mathbf{A})^{-1}\mathbf{A}^\top \mathbf{P}_{ll}\mathbf{l} \tag{4.1-1-7}$$

The (estimated) unknowns $\hat{\mathbf{x}}$, which can be derived from either of the Equations (4.1-1-3) or (4.1-1-7), may be substituted in Equation (4.1-1-2) in order arrive at the estimated residuals $\mathbf{v}$ for all the observations $\mathbf{l}$.

From all the residuals $\mathbf{v}$ we can derive an estimate for the standard deviation of unit weight, also called root of the reference variance of an observation which has weight 1 in the weight matrix, (4.1-1-6):

$$\hat{\sigma}_0 = \sqrt{\frac{\mathbf{v}^\top \mathbf{P}_{ll}\mathbf{v}}{n-u}} \tag{4.1-1-8}$$

With this standard deviation of unit weight $\hat{\sigma_0}$ and the weight coefficients q_{kk} the standard error $\hat{\sigma}_{x_k}$ of the individual unknowns x_k may be estimated:

$$\hat{\sigma}_{x_k} = \hat{\sigma}_0\sqrt{q_{kk}} \tag{4.1-1-9}$$

The weight coefficients q_{kk} are simply the corresponding elements from the main diagonal of the inverse of the normal-equation matrix:

$$\mathbf{Q}_{xx} = (\mathbf{A}^\top \mathbf{P}_{ll}\mathbf{A})^{-1} \tag{4.1-1-10}$$

The special case in which there is no redundancy, that is when $n = u$, is also of interest. In this case the result is the solution of the following linear equation system:

$$\mathbf{A}\mathbf{x} = \mathbf{l} \qquad \mathbf{x} = \mathbf{A}^{-1}\mathbf{l} \tag{4.1-1-11}$$

The rule of propagation of errors may be applied to the last equation:

$$\mathbf{Q}_{xx} = \mathbf{A}^{-1}(\mathbf{A}^{-1})^\top = \mathbf{A}^{-1}(\mathbf{A}^\top)^{-1} = (\mathbf{A}^\top \mathbf{A})^{-1} \tag{4.1-1-12}$$

Comparison of this equation with Equation (4.1-1-10) means that one gets the accuracy of the unknowns from a system of linear equations in a roundabout way from a least squares adjustment. It is necessary to adopt a predetermined a priori accuracy for $\hat{\sigma}_0$.

Literature: Mikhail, E.: Observations and Least Squares. IEP-A Dun-Donnelly Publisher, New York, 1976. Koch, K.R.: Parameter Estimation and Hypothesis Testing in Linear Models. 2nd ed., Springer 2006.

Appendix to Section 4.2

4.2-1 Direct Linear Transformation (DLT) with homogeneous coordinates

The relationship (4.2-3) between an image plane and a three-dimensional object can be written as follows using homogeneous coordinates:

$$\begin{pmatrix} \bar{\xi} \\ \bar{\eta} \\ \bar{\zeta} \end{pmatrix} = \begin{pmatrix} a_1 & a_2 & a_3 & a_4 \\ b_1 & b_2 & b_3 & b_4 \\ c_1 & c_2 & c_2 & c_4 \end{pmatrix} \begin{pmatrix} X \\ Y \\ Z \\ 1 \end{pmatrix} \qquad \bar{\boldsymbol{\xi}} = \underset{(3,4)}{\mathbf{A}}\,\mathbf{X} \qquad (4.2\text{-}1\text{-}1)$$

Relationship (4.2-3) is then obtained from the homogeneous coordinates $\bar{\xi}, \bar{\eta}$ and $\bar{\zeta}$ as follows:

$$\xi = \bar{\xi}/\bar{\zeta} \qquad \eta = \bar{\eta}/\bar{\zeta} \qquad (4.2\text{-}1\text{-}2)$$

Note: as has already been mentioned in Section 2.1.3, since the 3×4 matrix of Equation (4.2-1-1) cannot be inverted, a single picture is insufficient for the reconstruction of a three-dimensional object with its XYZ coordinates.

Appendix to Section 4.3

4.3-1 Differential coefficients for the coplanarity equations

The coplanarity condition for the relative orientation using independent rotations results in Equation (4.3-11). One obtains the differential coefficients, which are necessary, for example, for Equation (4.3-12), through differentiation of Equation (4.3-11) with respect to the individual angles. Let us begin with $d\omega_2$:

$$\frac{\partial D}{\partial \omega_2} = p_{1,y}\frac{\partial p_{2,z}}{\partial \omega_2} - p_{1,z}\frac{\partial p_{2,y}}{\partial \omega_2} \tag{4.3-1-1}$$

As an illustration the derivative $\partial p_{2,z}/\partial\omega_2$ will be developed in full. The Equations (2.1-13) and (4.3-9) give:

$$p_{2,z} = \begin{pmatrix} \xi_2 & \eta_2 & -c \end{pmatrix} \begin{pmatrix} \sin\omega_2 \sin\kappa_2 - \cos\omega_2 \sin\varphi_2 \cos\kappa_2 \\ \sin\omega_2 \cos\kappa_2 + \cos\omega_2 \sin\varphi_2 \cos\kappa_2 \\ \cos\omega_2 \cos\varphi_2 \end{pmatrix} \tag{4.3-1-2}$$

Differentiation with respect to $\partial\omega_2$ leads to:

$$\frac{\partial p_{2,z}}{\partial \omega_2} = \begin{pmatrix} \xi_2 & \eta_2 & -c \end{pmatrix} \begin{pmatrix} \cos\omega_2 \sin\kappa_2 + \sin\omega_2 \sin\varphi_2 \cos\kappa_2 \\ \cos\omega_2 \cos\kappa_2 - \sin\omega_2 \sin\varphi_2 \sin\kappa_2 \\ -\sin\omega_2 \cos\varphi_2 \end{pmatrix} \tag{4.3-1-3}$$

Since the column vector of Equation (4.3-1-3) corresponds to the second row of the rotation matrix (2.1-13), the relationship (4.3-1-3) simplifies to the y component of the $\mathbf{p}_2$ vector (4.3-9), that is, $p_{2,y}$. The derivative $\partial p_{2,y}/\partial\omega_2$ from Equation (4.3-1-1) simplifies, after similar considerations, to the negative z component of the $\mathbf{p}_2$ vector (4.3-9), that is, $-p_{2,z}$. As a result the complete derivative from (4.3-1-1) is found:

$$\frac{\partial D}{\partial \omega_2} = p_{2,y}\, p_{1,y} + p_{2,z}\, p_{1,z}$$

The following derivative is obtained after comparable considerations:

$$\frac{\partial D}{\partial \varphi_2} = -\cos\omega_2\, p_{2,x}\, p_{1,y} - \sin\omega_2\, p_{2,x}\, p_{1,z}$$

We obtain the following derivative in a similar way:

$$\frac{\partial D}{\partial \kappa_2} = (\xi_2 r_{2,32} - \eta_2 r_{2,31})p_{1,y} - (\xi_2 r_{2,22} - \eta_2 r_{2,21})p_{1,z}$$

It should first be noted, when it comes to differentiation of D with respect to φ_1 and κ_1, that, using $\omega = 0$, the rotation matrix (2.1-13) can be simplified. The two remaining derivatives are:

$$\frac{\partial D}{\partial \varphi_1} = -p_{1,x}\, p_{2,y}$$

$$\frac{\partial D}{\partial \kappa_1} = (\xi_1 r_{1,22} - \eta_1 r_{1,21})p_{2,z} - (\xi_1 r_{1,32} - \eta_1 r_{1,31})p_{2,y}$$

Exercise 4.3-3. If both rotation matrices, $\mathbf{R}_1$ and $\mathbf{R}_2$, are replaced by the unit matrix, these derivatives are considerably simplified. Equation (4.3-12), which applies to the relative orientation of highly tilted photographs, then becomes Equation (4.3-6), the equation for near-vertical photographs. Verify this conversion.

Appendix to Section 4.6

4.6-1 The empirical determination of standard deviations and tolerances

The standard deviation σ is soundly based in statistics. Sixty-eight percent of the actual errors lie within the range $\pm\sigma$. In the related disciplines of civil engineering, mechanical engineering and so on, tolerances are preferred. The definition of tolerance is, however, relatively arbitrary. A widely used tolerance t is the 2σ tolerance:

$$t_{95} = \pm 1.97\sigma \approx \pm 2\sigma \tag{4.6-1-1}$$

As one can deduce from the Gaussian error distribution (normal distribution), about 95% of the true errors lie within this 2σ range. A somewhat more conservative tolerance is:

$$t_{99} = \pm 2.58\sigma \approx \pm 2.5\sigma \tag{4.6-1-2}$$

But there are always real errors even outside this 2.5σ tolerance. We can tell from the Gaussian error distribution that there are about 1% more. Nevertheless, the tolerance t_{99} is used in such a way that all real errors are presumed to lie within these bounds.

Not infrequently, when photogrammetric services are called for, an accuracy is requested without defining the measure of accuracy. Associated professions prefer tolerances. If tolerance is adopted as a measure of quality in calling for services it must be stated whether it is a question of t_{95} (4.6-1-1) or t_{99} (4.6-1-2) or some other tolerance.

Assessment of the extent to which the specified tolerance t_{ref} has been met is carried out with the help of check measurements of superior accuracy. To start with, an r.m.s. value (root mean square error) (4.6-1-3) can be calculated from the differences which arise:

$$\text{r.m.s.} = \sqrt{\frac{\sum \varepsilon\varepsilon}{n}} \tag{4.6-1-3}$$

n ... number of check measurements

Frequently the r.m.s. is referred to as the absolute accuracy since it reflects all error influences. If, from the disparities ε, one removes a possible systematic contribution by averaging, the differences are diminished; from the differences reduced in this way, which we denote by v, one obtains the standard deviation as follows:

$$\sigma_m = \sqrt{\frac{\sum vv}{n-1}} \tag{4.6-1-4}$$

The standard deviation σ_m is also frequently called the relative accuracy. Continuing the analysis, the tolerance t_m can be calculated with Equation (4.6-1-1) or (4.6-1-2) and compared with the specified tolerance t_{ref}. One can also check directly whether 95% of the differences v, found after removal of a possible systematic part from the differences ε, lie within the specified tolerance $t_{95,\text{ref}}$.

Check measurements of a superior accuracy are very costly. In many cases one is satisfied with a second set of measurements of the same accuracy as the first measurements. The agreed tolerance t_{ref} can be checked from the differences d which arise. In this case one first computes the standard deviation $\sigma_{d,m}$ of the differences (in the same way as explained above, one can also separate out the systematic part);

$$\sigma_{d,m} = \sqrt{\frac{\sum dd}{n}} \tag{4.6-1-5}$$

$n \ldots$ number of repeated measurements

Since two measurements of approximately equal accuracy contribute to the differences d, the standard error of a single measurement, the actual standard error, comes to:

$$\sigma_m = \sigma_{d,m}/\sqrt{2} \tag{4.6-1-6}$$

The tolerance t_m can then be found from Equation (4.6-1-1) or (4.6-1-2) and compared with the stipulated tolerance t_{ref}.

In conclusion, it should be emphasized that a great deal of attention should be paid to systematic errors when evaluating standard errors and tolerances empirically.

Numerical Example. When calling for photogrammetric point determination a tolerance of $t_{95,\text{ref}} = \pm 10\,\text{cm}$ is demanded. A value of $\sigma_{d,m} = \pm 8.5\,\text{cm}$ arises from the differences d of repeated measurements (no systematic component contributed to the differences d). Is the specified tolerance $t_{95,\text{ref}} = \pm 10\,\text{cm}$ met?

$$\begin{aligned} (4.6\text{-}1\text{-}6)\text{:}\quad \sigma_m &= 8.5\,\text{cm}/\sqrt{2} = 6.0\,\text{cm} \\ (4.6\text{-}1\text{-}1)\text{:}\quad t_{95,m} &= \pm 12\,\text{cm} \end{aligned}$$

The specified tolerance $t_{95,\text{ref}}$ is therefore not met. The photogrammetric point determination has to be repeated, possibly with a larger photo scale.

Completion of the references

ASPRS: American Society for Photogrammetry and Remote Sensing.

AVN: Allgemeine Vermessungs-Nachrichten, Wichmann Verlag, Karlsruhe, Germany.

BuL: Bildmessung und Luftbildwesen, Wichmann Verlag, Karlsruhe, Germany.

Can. Surv.: The Canadian Surveyor, The Canadian Institute of Surveying, Ottawa.

DGK: Deutsche Geodätische Kommission bei der Bayerischen Akademie der Wissenschaften, Verlag Bayerische Akademie der Wissenschaften, Munich, Germany.

Geow. Mitt. der TU Wien: Geowissenschaftliche Mitteilungen der Studienrichtung Vermessung und Geoinformation der TU Wien, Austria.

GIS: Geo-Informations-Systeme, Wichmann Verlag, Karlsruhe, Germany.

IAPRS or IAPR or IAP: International Archives of Photogrammetry (and Remote Sensing) of ISPRS (ISP).

ISPRS: International Society for Photogrammetry and Remote Sensing.

ISPRS-J: ISPRS Journal of Photogrammetry and Remote Sensing, Elsevier Science Publishers, Amsterdam, The Netherlands.

ITC-J: ITC Journal, Enschede, The Netherlands.

JEK: Jordan, Eggert, Kneissl, Handbuch der Vermessungskunde, 10th ed., J.B. Metzlersche Verlagsbuchhandlung, Stuttgart, Germany.

Mitt. der TU Graz: Mitteilungen der Geodätischen Institute der TU Graz, Austria.

Na.Ka.Verm.: Nachrichten aus dem Karten- und Vermessungswesen, Institut für Angewandte Geodäsie, Frankfurt/Main, Germany.

OEEPE Publ.: Publications of the Organisation Européenne d'Études Photogrammétriques Expérimentales (European Organisation for Experimental Photogrammetric Research), now EuroSDR (European Spatial Data Research), `http://www.eurosdr.org`.

ÖZfV or ÖZ: Österreichische Zeitschrift für Vermessungswesen und Photogrammetrie, Vienna, Austria.

PE&RS or Photogr.Eng. or Ph.Eng.: Photogrammetric Engineering and Remote Sensing, Journal of the American Society of Photogrammetry.

PFG: Photogrammetrie, Fernerkundung und Geoinformation. E. Schweizerbart'sche Verlagsbuchhandlung (Nägele und Obermiller), Stuttgart, Germany.

Ph.Rec. or The Photogrammetric Record: The Photogrammetric Society, London, United Kingdom.

Phia: Photogrammetria, Elsevier Scientific Publishing Company, Amsterdam, The Netherlands.

UNISURV: The school of surveying. The University of New South Wales, Sydney, Australia.

VPK: Vermessung, Photogrammetrie, Kulturtechnik; Schweizerischer Verein für Vermessungswesen und Kulturtechnik, Schweizerische Gesellschaft für Photogrammetrie etc., Zürich, Switzerland.

VT: Vermessungstechnik, Zeitschrift für Geodäsie, Photogrammetrie und Kartographie der (ehem.) DDR, VEB Verlag für Bauwesen, Berlin, (former) GDR.

VGI: Österreichische Zeitschrift für Vermessung und Geoinformation, Vienna, Austria.

Wiss. Arb. Geod. u. Photogr. TH Hannover: Wissenschaftliche Arbeiten der Fachrichtung Vermessungswesen der Universität Hannover, Germany.

ZfV: Zeitschrift für Vermessungswesen, Konrad Wittwer Verlag, Stuttgart, Germany.

zfv: Zeitschrift für Geodäsie, Geoinformation und Landmanagement, Wißner-Verlag, Augsburg, Germany.

ZPF: Zeitschrift für Photogrammetrie und Fernerkundung, Herbert Wichmann Verlag, Karlsruhe, Germany.

Index

B

C

G

H

N

O

P

Q

R

S

T

U

V

W

Z